Lecture Notes in Computer Scien

Commenced Publication in 1973
Founding and Former Series Editors:
Gerhard Goos, Juris Hartmanis, and Jan van Leeuwen

Jürgen Dix   Stephen J. Hegner (Eds.)

Foundations of Information and Knowledge Systems

4th International Symposium, FoIKS 2006
Budapest, Hungary, February 14-17, 2006
Proceedings

 Springer

Volume Editors

Jürgen Dix
Clausthal University of Technology
Dept. of Computer Science
Julius-Albert-Str. 4, 38678 Clausthal-Zellerfeld, Germany
E-mail: dix@tu-clausthal.de

Stephen J. Hegner
Umeå University
Department of Computing Science
90187 Umeå, Sweden
E-mail: hegner@cs.umu.se

Library of Congress Control Number: 2005939041

CR Subject Classification (1998): H.2, H.3, H.5, I.2.4, F.3.2, G.2

LNCS Sublibrary: SL 3 – Information Systems and Application, incl. Internet/Web and HCI

ISSN   0302-9743
ISBN-10   3-540-31782-1 Springer Berlin Heidelberg New York
ISBN-13   978-3-540-31782-1 Springer Berlin Heidelberg New York

Springer is a part of Springer Science+Business Media

springeronline.com

© Springer-Verlag Berlin Heidelberg 2006

Typesetting: Camera-ready by author, data conversion by Scientific Publishing Services, Chennai, India
Printed on acid-free paper   SPIN: 11663881   06/3142   5 4 3 2 1 0

Preface

This volume contains the papers presented at the 4th International Symposium on Foundations of Information and Knowledge Systems (FoIKS 2006), which was held at the Alfréd Rényi Institute of Mathematics, Hungarian Academy of Sciences, Budapest, Hungary, from February 14 to 17, 2006.

FoIKS is a biennial event with a focus on the theoretical foundations of information and knowledge systems. The goal is to bring together researchers working on the theoretical foundations of information and knowledge systems, as well as to attract researchers working in mathematical fields such as discrete mathematics, combinatorics, logics and finite model theory who are interested in applying their theories to research on database and knowledge base theory.

FoIKS took up the tradition of the conference series Mathematical Fundamentals of Database Systems (MFDBS), which enabled East-West collaboration in the field of database theory. The first FoIKS symposium was held in Burg, Spreewald (Germany) in 2000, the second FoIKS symposium was held in Salzau Castle (Germany) in 2002, and the third FoIKS symposium was held in Vienna (Austria) in 2004. Former MFDBS conferences were held in Dresden (Germany) in 1987, Visegrád (Hungary) in 1989 and in Rostock (Germany) in 1991. Proceedings of these previous MFDBS and FoIKS events were published by Springer as volumes 305, 364, 495, 1762, 2284, and 2942 of the LNCS series, respectively.

The FoIKS symposium is intended to be a forum for intensive discussions. For this reason the time slots for long and short contributions are 50 and 30 minutes, respectively, followed by 20 and 10 minutes for discussions, respectively. Each such discussion is led by the author of another paper, who is asked in advance to prepare a focused set of questions and points for further elaboration.

The FoIKS 2006 call for papers solicited contributions dealing with the foundational aspects of information and knowledge systems, including the following topics:

- Mathematical Foundations: discrete methods, Boolean functions, finite model theory
- Database Design: formal models, dependency theory, schema translations, desirable properties
- Query Languages: expressiveness, computational and descriptive complexity, query languages for advanced data models, classifications of computable queries
- Semi-structured Databases and WWW: models of Web databases, querying semi-structured databases, Web transactions and negotiations
- Security in Data and Knowledge Bases: cryptography, steganography, information hiding
- Integrity and Constraint Management: verification, validation, and enforcement of consistency, triggers

- Information Integration: heterogeneous data, views, schema dominance and equivalence
- Database and Knowledge Base Dynamics: models of transactions, models of interaction, updates, consistency preservation, concurrency control
- Intelligent Agents: multi-agent systems, autonomous agents, foundations of software agents, cooperative agents
- Logics in Databases and AI: non-classical logics, spatial and temporal logics, probabilistic logics, deontic logic, logic programming
- Knowledge Representation: planning, description logics, knowledge and belief, belief revision and update, non-monotonic formalisms, uncertainty
- Reasoning Techniques: theorem proving, abduction, induction, constraint satisfaction, common-sense reasoning, probabilistic reasoning, reasoning about actions

The Program Committee received 54 submissions. Each paper was carefully reviewed by at least two experienced referees, and most of the papers were reviewed by three referees. Fourteen papers were chosen for long presentations and three for short presentations. This volume contains polished versions of these papers with respect to the comments made in the reviews. The best papers will be selected for further extension and publishing in a special issue of the journal *Annals of Mathematics and Artificial Intelligence*.

We would like to thank everyone involved with FoIKS 2006 for their contribution to the success of the symposium.

Jürgen Dix
Stephen J. Hegner

Organization

Program Committee Co-chairs

Jürgen Dix   Clausthal University of Technology (Germany)
Stephen J. Hegner   Umeå University (Sweden)

Program Committee

Peter Baumgartner   Australian National University (Australia)
Leopoldo Bertossi   Carleton University (Canada)
Joachim Biskup   University of Dortmund (Germany)
Piero Bonatti   University of Naples (Italy)
Stefan Brass   University of Halle–Wittenberg (Germany)
Guozhu Dong   Wright State University (USA)
Thomas Eiter   Vienna University of Technology (Austria)
Burkhard Freitag   University of Passau (Germany)
Floris Geerts   University of Edinburgh (UK) and
                           University of Limburg (Belgium)
Sven Hartmann   Massey University (New Zealand)
Lauri Hella   University of Tampere (Finland)
Gyula Katona   Alfréd Rényi Institute of Mathematics (Hungary)
Gabriele Kern-Isberner   University of Dortmund (Germany)
Hans-Joachim Klein   University of Kiel (Germany)
Christoph Koch   Saarland University (Germany)
Nicola Leone   University of Calabria (Italy)
Mark Levene   Birbeck University of London (UK)
Ilkka Niemelä   Helsinki University of Technology (Finland)
Domenico Saccà   University of Calabria (Italy)
Attila Sali   Alfréd Rényi Institute of Mathematics (Hungary)
Vladimir Sazonov   University of Liverpool (UK)
Nicole Schweikardt   Humboldt University (Germany)
Dietmar Seipel   University of Würzburg (Germany)
Nicolas Spyratos   University of Paris-South (France)
Bernhard Thalheim   University of Kiel (Germany)
Rodney Topor   Griffith University (Australia)
José María Turull-Torres   Massey University (New Zealand)
Andrei Voronkov   University of Manchester (UK)
Marina de Vos   University of Bath (UK)

External Referees

Marcelo Arenas	Dan Olteanu
Matthias Beck	Martin Otto
Bernhard Beckert	Viet Phan-Luong
Alfredo Cuzzocrea	Andrea Pugliese
Alex Dekhtyar	Francesco Ricca
Esra Erdem	Flavio Rizzolo
Wolfgang Faber	Francesco Scarcello
Flavio A. Ferrarotti	Michele Sebag
Michael Fink	Stefanie Scherzinger
Alfredo Garro	Roman Schindlauer
Giuseppe De Giacomo	Peggy Schmidt
Martin Grohe	Cristina Sirangelo
Michael Guppenberger	Viorica Sofronie-Stokkermans
Giovambattista Ianni	Umberto Straccia
Mirjana Jaksic	Andrea G.B. Tettamanzi
Eiji Kawaguchi	Cesare Tinelli
Kevin Keenoy	Hans Tompits
Henning Koehler	Yannis Tzitzikas
Markus Lehmann	Domenico Ursino
Hans-Joachim Lenz	Jing Wang
Sebastian Link	Stefan Woltran
Irina Lomazova	Dell Zhang
Carlo Meghini	

Local Arrangements

Dezső Miklós   Alfréd Rényi Institute of Mathematics (Hungary)

Publicity and Website Management

Markus Kirchberg   Massey University (New Zealand)

Table of Contents

The Semijoin Algebra

Jan Van den Bussche

Universiteit Hasselt, Agoralaan D,
B-3590 Diepenbeek, Belgium
jan.vandenbussche@uhasselt.be

Abstract. When we replace, in the classical relational algebra, the join operator by the semijoin operator, we obtain what we call the semijoin algebra. We will show that, when only equi-joins are used, the semijoin algebra is equivalent with the guarded fragment of first-order logic, and thus it inherits many of the nice properties of the latter logic. When more general theta-joins are used, however, we move outside the realm of guarded logics, and we will show how the notion of guarded bisimilarity can be extended accordingly. Last but not least, we show how the semijoin algebra can be used as a tool to investigate the complexity of queries expressed in the relational algebra, where we are mainly interested in whether or not a relational algebra expression for the query needs to produce intermediate results of nonlinear size. For example, we will show that the division operation cannot be expressed by a linear relational algebra expression.

This talk is a survey of work done in collaboration with Dirk Leinders, Jerzy Tyszkiewicz, and Maarten Marx.

References

1. D. Leinders, M. Marx, J. Tyszkiewicz and J. Van den Bussche. The semijoin algebra and the guarded fragment. *Journal of Logic, Language and Information*, 14:331–343, 2005.
2. D. Leinders and J. Van den Bussche. On the complexity of division and set joins in the relational algebra. *Proceedings 24th ACM Symposium on Principles of Database Systems*, pages 76–83. ACM Press, 2005.
3. D. Leinders, J. Tyszkiewicz and J. Van den Bussche. On the expressive power of semijoin queries. *Information Processing Letters*, 91:93–98, 2004.

J. Dix and S.J. Hegner (Eds.): FoIKS 2006, LNCS 3861, p. 1, 2006.
© Springer-Verlag Berlin Heidelberg 2006

Equational Constraint Solving Via a Restricted Form of Universal Quantification

Javier Álvez* and Paqui Lucio

Dpto. de Lenguajes y Sistemas Informáticos, UPV-EHU, Paseo Manuel de
Lardizabal, 1, 20080-San Sebastián, Spain
{jibalgij, jiplucap}@si.ehu.es

Abstract. In this paper, we present a syntactic method for solving first-
order equational constraints over term algebras. The presented method
exploits a novel notion of *quasi-solved* form that we call *answer*. By allowing a restricted form of universal quantification, *answers* provide a more
compact way to represent solutions than the purely existential solved
forms found in the literature. *Answers* have been carefully designed to
make satisfiability test feasible and also to allow for boolean operations,
while maintaining expressiveness and user-friendliness. We present detailed algorithms for (1) satisfiability checking and for performing the
boolean operations of (2) negation of one *answer* and (3) conjunction
of *n* *answers*. Based on these three basic operations, our solver turns
any equational constraint into a disjunction of *answers*. We have implemented a prototype that is available on the web.

Keywords: equality, constraint satisfaction, solver, term algebra,
answer.

1   Introduction

An *equational constraint* is an arbitrary first-order formula built over a signature
Σ of function symbols and equality as unique predicate symbol. Equational constraints are interpreted over term algebras. An *equational solving method* takes as
input an equational constraint and produces the set of all its solutions or, more
precisely, some particular representation of it. Syntactic methods are *rewriting
processes* that transform the input constraint into an equivalent disjunction of
constraints, in the so-called *solved form*, which represents its solutions. In particular, those solutions serve to decide whether the input constraint is satisfiable
or not.

On one hand, equational constraint solving is an very important tool in many
areas of automated deduction. The integration of efficient equational solvers in
theorem provers has been a challenging problem, important for many practical
applications. Equational constraints can be used for restricting the set of ground

---

* Corresponding author.

J. Dix and S.J. Hegner (Eds.): FoIKS 2006, LNCS 3861, pp. 2–21, 2006.

instances in order to define more eficient resolution mechanism (cf. [6]). In automated model building, equational constraints play a crucial role for model representation (e.g. [5, 11]). On the other hand, equational constraint solving may be applied for several purposes in the areas of functional/logic programming and databases. Therein, many problems related to semantics and implementation issues can be reduce to equational constraint solving problems. Some well known examples include the problems of answering negative goals, decidindg whether a case-definition is complete, evaluating a boolean conjunctive query on a relational database, etc. Besides, equational constraint solving have been found useful in other areas such as formal verification tools, computational linguistic, machine learning, program transformation, etc.

It is well known that the *free equality theory*[1], originally introduced by Malcev in [15], is non-elementary (see [10, 21]). Besides, the inherent complexity of the satisfiability problem of equational problems (i.e. where the quantifier prefix is of the form $\forall^*\exists^*$) for finite signature is studied in [19]. The most well known algorithms for equational solving [9, 14, 15] and later extensions to richer theories (see [20]) are based on quantifier elimination with solved forms that combine equations and disequations. Negation should be allowed since, for example, the constraint $\forall v(\ x \neq f(v,v)\ )$ cannot be finitely represented without negation (disequations). As opposed to negation, universal quantification can be dropped from any equational formula by the well-known *quantifier elimination technique*. As a consequence, most solved form notions (see [8] for a survey) are boolean combinations of certain kind of existential formulas whose satisfiability test is trivial even in the case of finite signature. However, in exchange for the simplicity of the test, the solver must remove all universal quantifiers. This often requires application of the so-called[2] *Explosion Rule* ([9]) that we recall in Fig. 1, that implies substitution of a formula by a disjunction of as many formulas as there are function symbols in the finite signature Σ. A more compact

$$(\text{Exp})\quad \forall\overline{y}(\ \varphi\ )\ \longmapsto\ \bigvee_{f\in\Sigma}\exists\overline{z}\forall\overline{y}(\ \varphi\wedge w=f(\overline{z})\ )$$

if there exists an equation $x=t$ or a disequation $x\neq t$ such that

some $y_i\in\overline{y}$ occurs in t, $\overline{z}$ are fresh and Σ is finite

Fig. 1. The Explosion Rule (Exp)

representation of solutions reduces the blow up of the number of disjuncts along the quantifier elimination process, which in turn improves the method. At the same time, the basic operations for managing this more expressive notion must not be expensive. We propose a notion of *quasi-solved form*, called *answer*, that allows a restricted form of universal quantification, which is enough to avoid the

---

[1] Also called the theory of term algebra and Clark's equational theory.

[2] That is, the *Weak Domain Closure Axiom* in the nomenclature of [14].

above rule (Exp) and offers a more compact representation of solutions. *Answers* have been carefully designed to make satisfiability test feasible and also to allow for boolean operations (of negation and conjunction), while retaining expressiveness and user-friendliness. The idea of gaining efficiency via restricted forms of universal quantification has been already proposed in [18] and in [16].

A very preliminary version of this work was presented as [1]. We have implemented (in Prolog) a prototype of the general constraint solver. It is available at ~http://www.sc.ehu.es/jiwlucap/equality_constraints.html.

Outline of the paper. In the next section, we recall some useful definitions and denotational conventions. Section 3 is devoted to the details of the notion of *answer* and some examples. In Section 4, we introduce the *answer* satisfiability test with some illustrative examples. In Section 5, we show how the two other basic operations on *answers* —conjunction and negation— can be efficiently performed. Besides, we make use of these basic operations (together with the quantifier elimination technique) to provide a solving method for general equational constraints. We give a summarizing example in Section 6. Finally, we present some concluding remarks and briefly discuss some related work.

2   Definitions and Notation

Let us fix a denumerable set of variables X. Given a (finite or infinite) signature Σ, a Σ-term is a variable from X, or a constant, or a function symbol of arity n applied to n terms. A term is *ground* if it contains no variable symbols. $\mathcal{T}(\Sigma)$ stands for the algebra of all ground Σ-terms or Herbrand universe, whereas $\mathcal{T}(\Sigma, X)$ is used to denote the set of all Σ-terms. We denote by $Var(t)$ the set of all variables occurring in t and $t(\overline{v})$ denotes that $Var(t) \subseteq \overline{v}$. A term is *linear* if it contains no variable repetitions. A bar is used to denote tuples of objects. Subscripts are used to denote the components of a tuple and superscripts are used to enumerate tuples. For example, x_j denotes a component of $\overline{x}$, whereas $\overline{x}^1, \ldots, \overline{x}^j, \ldots, \overline{x}^m$ is a tuple enumeration and x_i^j is a component of the tuple $\overline{x}^j$. Concatenation of tuples is denoted by the infix operator $\cdot$, i.e. $\overline{x} \cdot \overline{y}$ represents the concatenation of $\overline{x}$ and $\overline{y}$. When convenient, we treat a tuple as the set of its components. A Σ-equation is $t_1 = t_2$, where t_1 and t_2 are Σ-terms, whereas $t_1 \neq t_2$ is a Σ-disequation (that is also written as $\neg(t_1 = t_2)$). By a *collapsing* equation (or disequation) we mean that at least one of its terms is a variable. We abbreviate collapsing Σ-equation by Σ-CoEq, and Σ-UCD stands for universally quantified collapsing Σ-disequation. We abbreviate $\bigwedge_i t_i = s_i$ by $\overline{t} = \overline{s}$ and $\bigvee_i t_i \neq s_i$ by $\overline{t} \neq \overline{s}$. To avoid confusion, we use the symbol $\equiv$ for the metalanguage equality.

A Σ-*substitution* $\sigma \equiv \{x_1 \leftarrow t_1, \ldots x_n \leftarrow t_n\}$ is a mapping from a finite set of variables $\overline{x}$, called $domain(\sigma)$, into $\mathcal{T}(\Sigma, X)$. It is assumed that σ behaves as the identity for the variables outside $domain(\sigma)$. A substitution σ is called a Σ-*assignment* if $\sigma(x_i) \in \mathcal{T}(\Sigma)$ for all $x_i \in domain(\sigma)$. We intentionally confuse the above substitution σ with the conjunction of equations $\bigwedge_i x_i = t_i$. The (possibly ground) term $\sigma(t)$ (also denoted $t\sigma$) is called an *(ground) instance* of the term t. The *most general unifier* of a set of terms $\{s_1, \ldots, s_m\}$, denoted $mgu(\overline{s})$, is an idempotent substitution σ such that $\sigma(s_i) \equiv \sigma(s_j)$ for all $1 \leq i, j \leq m$ and for any other substitution θ with the same property, $\theta \equiv \sigma' \cdot \sigma$ holds for some substitution σ'. For tuples, $mgu(\overline{s}^1, \ldots, \overline{s}^m)$ is an abbreviation of $\sigma_1 \cdot \ldots \cdot \sigma_n$ where $\sigma_i \equiv mgu(s_i^1, \ldots, s_i^m)$ for all $1 \leq i \leq n$. The *most general common instance* of two terms t and s, denoted by $mgi(t, s)$, is the term whose set of ground instances is the intersection of both sets of ground instances (for t and s), and it can be computed using a unification algorithm.

An *equational constraint* is a first-order Σ-formula built over Σ-equations (as atoms) using the classical connectives and quantifiers. Atoms include the logical constants True and False. Equational constraints are interpreted in the term algebra $\mathcal{T}(\Sigma)$. A Σ-assignment σ satisfies a Σ-equation $t_1 = t_2$ iff $t_1 \sigma \equiv t_2 \sigma$. The logical constants, connectives and quantifiers are interpreted as usual. A *solution* of an equational constraint is a Σ-assignment that satisfies the constraint. Constraint equivalence means the coincidence of the set of solutions. We make no distinction between a set of constraints $\{\varphi_1, \ldots, \varphi_k\}$ and the conjunction $\varphi_1 \wedge \ldots \wedge \varphi_k$. We abbreviate $\forall x_1 \ldots \forall x_n$ (resp. $\exists x_1 \ldots \exists x_n$) by $\forall \overline{x}$ (resp. $\exists \overline{x}$).

3   The Notion of *Answer*

In this section, we present the notion of *answer* and give some illustrative examples. The following definition also introduces some notational conventions.

Definition 1. *Let $\overline{x}$ be a k-tuple of pairwise distinct variables. A Σ-answer for $\overline{x}$ is either a logical constant (True or False) or a formula of the form $\exists \overline{w}(\ a(\overline{x}, \overline{w})\ )$, where $a(\overline{x}, \overline{w})$ is a conjunction of Σ-CoEqs and Σ-UCDs of the form:*

$$x_1 = t_1 \wedge \ldots \wedge x_k = t_k \ \wedge \ \bigwedge_{i=1}^{n} \bigwedge_{j=1}^{m_i} \forall \overline{v}(\ w_i \neq s_{ij}(\overline{w}, \overline{v})\ ) \tag{1}$$

such that the n-tuple $\overline{w} \equiv Var(t_1, \ldots, t_k)$ is disjoint from $\overline{x}$, every term $s_{ij}(\overline{w}, \overline{v})$ neither contains the variable w_i nor is a single universal variable in $\overline{v}$, and $n, m_1, \ldots, m_n \geq 0$. □

Remark 1. We abbreviate the equational part of (1) by $\overline{x} = \overline{t}(\overline{w})$. Any equation $x_j = w_i$ such that w_i does not occur in the rest of the *answer* can be left out. The scope of each universal quantifier is restricted to one single disequation, although a universal variable can occur repeatedly in a disequation. □

The following examples show that *answers* provide a compact and explanatory description of the sets of solutions that they represent.

Example 1. Let $\Sigma \equiv \{a/0, g/1, f/2\}$. Consider the following Σ-*answer*:

$$\exists w_1 \exists w_2(\ x = f(w_1, w_2) \wedge \forall v_1(\ w_1 \neq (v_1)\ ) \wedge \forall v_2(\ w_2 \neq f(w_1, v_2)\ )\ ). \tag{2}$$

By application of the rule (Exp) for eliminating v_1 and v_2, the *answer* (2) is equivalent to the disjunction of the following six existential constraints:

1. $x = f(a, a)$
2. $\exists z_1 (\, x = f(a, g(z_1)) \,)$
3. $\exists z_1 \exists z_2 (\, x = f(a, f(z_1, z_2)) \wedge z_1 \neq a \,)$
4. $\exists z_1 \exists z_2 (\, x = f(f(z_1, z_2), a) \,)$
5. $\exists z_1 \exists z_2 \exists z_3 (\, x = f(f(z_1, z_2), g(z_3)) \,)$
6. $\exists z_1 \exists z_2 \exists z_3 \exists z_4 (\, x = f(f(z_1, z_2), f(z_3, z_4)) \wedge z_3 \neq f(z_1, z_2) \,)$ □

Example 2. The following equational constraint of signature $\Sigma \equiv \{a/0, g/1, f/2\}$:

$$\exists w_1 \exists w_2 \forall y_1 \forall y_2 (\, f(f(w_1, a), f(w_2, x_2)) \neq f(f(y_1, a), f(y_2, y_2)) \wedge$$
$$f(g(y_2), x_1) \neq f(x_2, f(y_1, y_1)) \,)$$

is equivalent to the disjunction of the following two *answers*:[3]

$$\exists w_2 (\, x_2 = w_2 \wedge \forall v (\, w_2 \neq g(v) \,) \,) \tag{3}$$
$$\exists w_1 (\, x_1 = w_1 \wedge \forall v (\, w_1 \neq f(v, v) \,) \,) \tag{4}$$

It is easy to see that the equational part of any *answer* is always satisfiable. In fact, it is an idempotent substitution. Besides, with infinitely many function symbols, one can always find an assignment that satisfies a given finite conjunction of UCDs. Therefore, *answers* are always satisfiable for infinite signatures. On the contrary, if Σ is finite, the disequational part of an *answer* is not necessarily satisfiable in $\mathcal{T}(\Sigma)$. As a consequence, we say that an *answer* is a *quasi-solved form*.

4   The Satisfiability Test

In this section, we introduce an algorithm for deciding *answer* satisfiability w.r.t. a finite signature. Notice that, for finite Σ, the set $\mathcal{T}(\Sigma)$ can be finite or infinite. The test works for both. We also give some examples of satisfiability and discuss about efficiency. In the case of finite signature Σ, as explained above, an *answer* is satisfiable iff its disequational part is. Hence, we shall concentrate in the disequational part of the tested *answer*. A Σ-*answer* $\exists \overline{w} (\, a(\overline{x}, \overline{w}) \,)$ is satisfiable in $\mathcal{T}(\Sigma)$ iff there is at least one Σ-assignment with domain $\overline{w}$ that satisfies the conjunction of UCDs inside $a(\overline{x}, \overline{w})$.

Example 3. Each of the two following conjunctions of Σ-UCDs is (individually) unsatisfiable in $\mathcal{T}(\Sigma)$ (for $\Sigma \equiv \{a/0, g/1, f/2\}$):

1. $w \neq a \wedge \forall v (\, w \neq g(v) \,) \wedge \forall v_1 \forall v_2 (\, w \neq f(v_1, v_2) \,)$

2. $w \neq a \wedge w \neq g(a) \wedge \forall v (\, w \neq g(g(v)) \,) \wedge$
$$\forall v_1 \forall v_2 (\, w \neq g(f(v_1, v_2)) \,) \wedge \forall v_1 \forall v_2 (\, w \neq f(v_1, v_2) \,)$$ □

---

[3] As said in Remark 1, we have left out $x_1 = w_1$ in (3) and $x_2 = w_2$ in (4).

The satisfiability of a conjunction of UCDs on a variable w_i could be tested using the algorithm introduced in [13] for deciding whether an implicit representation of terms can be explicitely (finitely) represented. In [13], an implicit representation is an expresion $t/t_1 \vee \ldots \vee t_n$ that represents those ground instances of t that are not instances of any t_i for $1 \leq i \leq n$. Actually, we could transform any set

$$S \equiv \{\forall \overline{v}(\ w_i \neq s_{ij}(\overline{w}, \overline{v})\ ) \mid 1 \leq i \leq n,\ 1 \leq j \leq m_i\}$$

of UCDs into an equivalent[4] set of inequalities on the involved tuple $\overline{w}$, using a fresh function symbol $\underline{c}$ as tuple constructor and fresh variables $\overline{u}$, as follows:

$$\mathsf{Tup(S)} \equiv \{\ \forall \overline{v} \forall \overline{u}(\ \underline{c}(\overline{w}) \neq \underline{c}(\overline{w}\sigma_{ij})\ ) \mid 1 \leq i \leq n,\ 1 \leq j \leq m_i, \tag{5}$$
$$\sigma_{ij} \equiv \theta_{ij} \cup \{w_i \leftarrow s_{ij}\theta_{ij}\} \text{ where } \theta_{ij} \equiv \{w_k \leftarrow u_k \mid 1 \leq k \neq i \leq m_j\}\}.$$

Then, by treating the tuple constructor $\underline{c}$ as just another function symbol (except for the explosion rule), the algorithm *uncover* of [13] could be invoked, as $uncover(\underline{c}(\overline{w}), \underline{c}(\overline{w})\sigma_{11}, \ldots, \underline{c}(\overline{w})\sigma_{nm_n})$, in order to decide if there exists an explicit representation of the implicit representation $\underline{c}(\overline{w})\backslash\underline{c}(\overline{w})\sigma_{11}\vee\ldots\vee\underline{c}(\overline{w})\sigma_{nm_n}$. The algorithm *uncover* refines such implicit representations to explicit ones (if it is possible), by applying the explosion rule only to linear[5] terms. For example, being the signature $\{a/0, b/0, f/2\}$, the implicit representation

$$\underline{c}(v_1, a) \setminus \underline{c}(a, a) \vee \underline{c}(b, v_2)$$

can be explicitly represented by $\{\underline{c}(f(u_1, u_2), a)\}$, whereas $\underline{c}(v_1, v_2) \setminus \underline{c}(v_3, v_3)$ cannot be explicitly represented.

Our *answer* satisfiability test exploits the same idea but, for efficiency, it does not exhaustively obtain the set of all possible elements represented by the tuple of variables $\underline{c}(\overline{w})$, as the call $uncover(\underline{c}(\overline{w}), \underline{c}(\overline{w})\sigma_{11}, \ldots, \underline{c}(\overline{w})\sigma_{nm_n})$ does. Instead, it works by value elimination in two steps. Each step takes into account a different (syntactic) subclass of UCDs. Moreover, we use a slightly simplified version of *uncover* (see Figure 2) since term linearity is an invariant condition in our *uncover* runs. Figure 3 outlines our *answer* satisfiability test for a finite signature Σ. We denote by $Ext(w_i)$ —namely, the extension of w_i— the collection of all ground terms that w_i could take as value. As in [13], we use fresh variables for finitely representing infinite variable extensions. For instance, $\{v\}$ represents the whole $\mathcal{T}(\Sigma)$'s universe (for any Σ) and $\{f(v_1, a), f(b, v_2)\}$ an infinite subset when Σ contains $\{a, b, f\}$. The basic idea is that a UCD disallows some values in $Ext(w_i)$. The first step treats the UCDs $\forall \overline{v}(\ w_i \neq s_{ij}(\overline{v})\ )$ without existential variables (that is, $Var(s_{ij}) \cap \overline{w} \equiv \emptyset$) and without repetitions in their universal variables $\overline{v}$. Roughly speaking, this is the only subclass of UCDs that can transform an infinite variable extension —described by a collection of linear terms— into a finite one. After the first iteration in Step 1 (see Fig. 3), the extension of each w_i can either be empty, or non-empty finite, or infinite. Notice

---

[4] This well known equivalence corresponds to the rule U_2 in [9].
[5] In [13], linear terms are called unrestricted terms.

```
partition(t, θ) is    // ū stands for a fresh tuple of variables of the adequate size
    if θ is a renaming then return ∅
    else select {z ← f(s₁,...sₕ)} ∈ θ
        let σ' ≡ {z ← f(ū)} and σᵢ ≡ {uᵢ ← sᵢ} for 1 ≤ i ≤ h
        return ⋃_{g∈Σ,g≠f} t{z ← g(ū)} ∪ partition(tσ', θ \ σ ∪ {σᵢ|1 ≤ i ≤ h})

uncover(t, tθ₁,...,tθₙ) is
    let {s₁,...,sₖ} ≡ partition(t, θ₁)
    let σᵢⱼ ≡ mgi(sᵢ, tθⱼ) for each (i,j) ∈ {1,...,k} × {2,...,n}
    return ⋃ᵏ_{r=1} uncover(sᵣ, sᵣσᵣ₂,...,sᵣσᵣₙ)
```

Fig. 2. Simplified Version of *uncover* ([13])

that $Ext(w_i)$ represents an infinite set of terms if and only if it contains (at least) one non-ground term. Then, the input *answer* is unsatisfiable if some $Ext(w_i) \equiv \emptyset$. On the contrary, the *answer* is satisfiable if every $Ext(w_i)$ is infinite. This first step is very often enough to decide the satisfiability of the input *answer*.

Example 4. The above *answers* (2), (3) and (4) are decided to be satisfiable at the first step since both $Ext(w_1)$ and $Ext(w_2)$ remain infinite. For (2) and (3), the test takes into account the first UCD (the only one in (3)). For (4), no UCD satisfies the condition, hence the test decides at once.   □

Example 5. Both constraints in Example 3 are unsatisfiable. This is decided in the first step since $Ext(w)$ becomes empty.   □

At the end of the first step, if no $Ext(w_i)$ is empty and at least one $Ext(w_i)$ is finite, the test requires a second step. The second step looks whether there exists (at least) one assignment σ with domain $\overline{w}^{fin} \equiv \{w_i|Ext(w_i) \text{ is finite}\}$ that satisfies only the subclass FinUCD of the remaining UCDs such that all their existential variables are contained in $\overline{w}^{fin}$. We denote by $Ext(\overline{w}^{fin})$ the cartesian product of all $Ext(w_i)$ such that $w_i \in \overline{w}^{fin}$.

Example 6. Let $\Sigma \equiv \{a/0, g/1, f/2\}$. Suppose that S is a set of UCDs formed by $\forall v_1(w_1 \neq g(v_1))$, $\forall v_1 \forall v_2(w_1 \neq f(v_1, v_2))$ and a large finite set S_0 of UCDs of the form $\forall v(w_i \neq f(w_j, v))$ where $i, j > 1$ and $i \neq j$. At the first step, only the two first UCDs are considered to yield $Ext(w_1) \equiv \{a\}$, whereas $Ext(w_i)$ is infinite for all $i > 1$. At the second step, FinUCD becomes empty. Hence, S is satisfiable. Notice that the exhaustive computation of *uncover* over the whole tuple of variables could be much more expensive depending on the size of S_0.   □

Theorem 1. *The satisfiability test of Figure 3 terminates for any input* (S, Σ). *Besides, it returns "satisfiable" if there exists a Σ-assignment that satisfies the set* S. *Otherwise, it returns "unsatisfiable".*

Proof. It is based on the results of [13]. See the appendix.   □

// **input:** $S \equiv \{\forall \overline{v}(\ w_i \neq s_{ij}(\overline{w}, \overline{v})\ ) \mid 1 \leq i \leq n,\ 1 \leq j \leq m_i\}$
// $\Sigma \equiv \{f_1 \backslash a_1, \ldots, f_k \backslash a_k\}$ (finite signature)

step 1: **for** $i \in \{1, \ldots, n\}$ **do**
           let $\{t_1, \ldots t_{k_i}\} \equiv \{s_{ij} \mid 1 \leq j \leq m_i, Var(s_{ij}) \cap \overline{w} \equiv \emptyset$ and each
                              $v_h \in \overline{v}$ occurs at most once in $s_{ij}\}$
               if $k_i = 0$ **then** $Ext(w_i) := \{u\}$ for fresh u
                   else $Ext(w_i) := uncover(w_i, t_1, \ldots, t_{k_i})$
               when some $Ext(w_i) \equiv \emptyset$ **exit with** *unsatisfiable*
           let $\overline{w}^{fin}$ be the tuple of all $w_i \in \overline{w}$ such that $Ext(w_i)$ is finite
           if $\overline{w}^{fin} \equiv \emptyset$ **then exit with** *satisfiable*
               else go to step 2

step 2: **let** FinUCD $\equiv \{\forall \overline{v}(\ w_i \neq s_{ij}(\overline{w}, \overline{v})\ ) \in S \mid w_i \in \overline{w}^{fin}, Var(s_{ij}) \backslash \overline{v} \subseteq \overline{w}^{fin}\}$
           if FinUCD $\equiv \emptyset$ **then exit with** *satisfiable*
               else let $\{\sigma_1, \ldots, \sigma_m\}$ be the set of substitutions such that
                   Tup(FinUCD) $\equiv \{\forall \overline{v} \forall \overline{u}(\ \overline{w} \neq \overline{w}\sigma_i\ ) \mid 1 \leq i \leq m\}$ (see (6))
           $C := \bigcup_{\overline{t} \in Ext(\overline{w}^{fin})} uncover(\overline{t}, \overline{t}\sigma_1, \ldots, \overline{t}\sigma_m)$
               if $C \not\equiv \emptyset$ **then exit with** *satisfiable*
                   else exit with *unsatisfiable*

Fig. 3. *Answer* Satisfiability Test

The introduced satisfiability test has a poor worst case performance. Actually, *answer* satisfiability is an NP-complete problem (see [19]). However, the test performs efficiently in practice because of several structural reasons that can be summed up as follows. In general, *answers* having expensive computations in both steps are unlikely. If the input *answer* contains a lot of UCDs to be treated in the first step, the extension of some variable usually becomes empty and the test stops. However, where few UCDs are treated at the first step, it is usual that most extensions remain infinite and, therefore, the second step becomes unnecessary or very cheap.[6] On the contrary, the worst case occurs when every w_i has a large finite extension, but every possible assignment violates some UCD. The combination of both properties requires a lot of UCDs to be expressed.[7]

5   Operations on *Answers* and Equational Solving

In this section, we present a method for transforming any equational constraint into a disjunction of *answers*. This solving method uses, besides the satisfiability test, two boolean operations on *answers* —conjunction and negation— which will also be presented. For general equational solving, we use the classical quantifier elimination technique. However, we keep the matrix as a disjunction of

---

[6] Remember that the second step only checks the UCDs such that all their existential variables have a finite extension (according to the first step).
[7] Since the first-order language of the free equality cannot express very "deep" properties of terms.

satisfiable *answers* on the prefix variables, instead of a quantifier-free CNF (or DNF) formula. That is, at every step we have a formula of the form:

$$Q_1\overline{u}^1 \ldots Q_m\overline{u}^m (\bigvee_{j=1}^{k} \exists\overline{w}(\, a_j(\overline{u}^1 \cdot \ldots \cdot \overline{u}^m, \overline{w}) \,) \,)$$

where each $Q_i \in \{\forall, \exists\}$ and each $a_j(\overline{u}^1 \cdot \ldots \cdot \overline{u}^m, \overline{w})$ is a satisfiable *answer* on $\overline{u}^1 \cdot \ldots \cdot \overline{u}^m$. Then, if the last block Q_m is $\exists$, it is easily eliminated by erasing from each a_j all the equations on $\overline{u}^m$. Then, we also remove every UCD containing at least one w_k which does not occur in the equational part.[8] It is worthwhile to notice that the satisfiability of each $a_j(\overline{u}^1 \cdot \ldots \cdot \overline{u}^m, \overline{w})$ guarantees that both the eliminated and the remaining part of the treated *answer* are satisfiable. In fact, the *answer* is equivalent to $(\exists\overline{u}^m \varphi_1) \wedge \varphi_2$, where $(\exists\overline{u}^m \varphi_1)$ is the eliminated part and φ_2 the remaining one. If $Q_m \equiv \forall$, double negation is applied:

$$Q_1\overline{u}^1 \ldots Q_{m-1}\overline{u}^{m-1} \neg \exists\overline{u}^m \neg (\bigvee_{j=1}^{k} \underbrace{\exists\overline{w}(\, a_j(\overline{u}^1 \cdot \ldots \cdot \overline{u}^m, \overline{w}) \,)}_{\psi_j}). \tag{6}$$

Suppose a procedure P exists that transforms the negation of a disjunction of *answers* (on some variables) into a disjunction of *answers* (on the same variables). Then, using P, the inner formula $\neg\bigvee_{j=1}^{k} \psi_j$ is transformed into a disjunction of *answers*. After that, $\exists\overline{u}^m$ is easily eliminated as above. Finally, the procedure P is again applied and the original innermost block $\forall\overline{u}^m$ is already eliminated. We implement the procedure P as follows, using two boolean operations on *answers*:

$$\neg \bigvee_{j=1}^{k} \psi_j \longmapsto \bigwedge_{j=1}^{k} \neg\psi_j \longmapsto^{(1)} \bigwedge_{j=1}^{k} \bigvee_{r=1}^{m_j} \varphi_{jr} \longmapsto \bigvee_{r=1}^{m} \bigwedge_{j=1}^{k_r} \beta_{jr} \longmapsto^{(2)} \bigvee_{r=1}^{m} \bigvee_{h=1}^{n_r} \gamma_{hr}$$

(1) The negation of an *answer* ψ_j yields a disjunction $\bigvee_{r=1}^{m_j} \varphi_{jr}$ of *answers*.
(2) The conjunction $\bigwedge_{j=1}^{k_r} \beta_{jr}$ of *answers* yields a disjunction $\bigvee_{h=1}^{n_r} \gamma_{hr}$ of *answers*.

(UD) $\neg\exists\overline{v}[\overline{x} = \overline{t}(\overline{w}, \overline{v}^1) \wedge \varphi] \longmapsto \forall\overline{v}^1[\overline{x} \neq \overline{t}(\overline{w}, \overline{v}^1)] \vee \exists\overline{v}^1[\overline{x} = \overline{t}(\overline{w}, \overline{v}^1) \wedge \forall\overline{v}^2 \neg\varphi]$

where $\overline{v}^1 \equiv Var(\overline{t}) \cap \overline{v}$ and $\overline{v}^2 \equiv \overline{v}\backslash\overline{v}^1$

Fig. 4. Auxiliary Transformation Rule (UD)

Moreover, both boolean operations preserve the variables for which source and target *answers* are obtained. In the next two subsections, we give the successive steps that constitute each transformation. It is easy to see that each step preserves equivalence. For that, we use the auxiliary transformation rule (UD) of Fig. 4.

---

[8] It has just disappeared with the previously eliminated equations.

Proposition 1. *The transformation rule (UD) of Figure 4 is correct.*

Proof. See the appendix.   □

5.1 Negation of One *Answer*

Now, we show how to transform the negation of an *answer* for $\overline{x}$ into an equivalent disjunction of *answers* for $\overline{x}$. First, we apply the transformation rule (UD):

$$\neg\exists\overline{w}[\,\overline{x} = \overline{t}(\overline{w}) \wedge \bigwedge_{i=1}^{n}\bigwedge_{j=1}^{m_i} \forall\overline{v}(\,w_i \neq s_{ij}(\overline{w},\overline{v})\,)\,] \longmapsto_{(UD)}$$

$$\neg\exists\overline{w}[\,\overline{x} = \overline{t}(\overline{w})\,] \vee \exists\overline{w}[\,\overline{x} = \overline{t}(\overline{w}) \wedge \bigvee_{i=1}^{n}\bigvee_{j=1}^{m_i}\exists\overline{v}(\,w_i = s_{ij}(\overline{w},\overline{v})\,)\,].$$

By the rule (UD), the first disjunct $\neg\exists\overline{w}(\,\overline{x} = \overline{t}(\overline{w})\,)$ is transformed into:

$$\forall\overline{w}^1[\,x_1 \neq t_1(\overline{w}^1)\,] \vee \exists\overline{w}^1[\,x_1 = t_1(\overline{w}^1) \wedge \forall\overline{w}^2(\,x_2 \neq t_2(\overline{w}^1\cdot\overline{w}^2)\,)\,] \vee \ldots \vee$$
$$\exists\overline{w}^1\ldots\exists\overline{w}^{n-1}[\,x_1 = t_1(\overline{w}^1\cdot\ldots\cdot\overline{w}^{n-1}) \wedge \ldots \wedge x_{n-1} = t_{n-1}(\overline{w}^1\cdot\ldots\cdot\overline{w}^{n-1})$$
$$\wedge\,\forall\overline{w}^n(\,x_n \neq t_n(\overline{w}^1\cdot\ldots\cdot\overline{w}^n)\,)\,] \tag{7}$$

Then, we replace the variables x_i in the disequations by new variables w'_i, adding the corresponding equation $x_i = w'_i$. The result is already a disjunction of *answers* for $\overline{x}$. For the second disjunct, we first lift the internal disjunctions:

$$\bigvee_{i=1}^{n}\bigvee_{j=1}^{m_i}\exists\overline{w}[\,\overline{x} = \overline{t}(\overline{w}) \wedge \exists\overline{v}(\,w_i = s_{ij}(\overline{w},\overline{v})\,)\,]$$

and then substitute each $s_{ij}(\overline{w},\overline{v})$ for w_i in $t(\overline{w})$:

$$\bigvee_{i=1}^{n}\bigvee_{j=1}^{m_i}\exists\overline{w}\exists\overline{v}(\,\overline{x} = \overline{t}(\overline{w})\{w_i \leftarrow s_{ij}(\overline{w},\overline{v})\}\,).$$

5.2 Conjunction of k *Answers*

Using unification, we transform a conjunction of k *answers* for $\overline{x}$:

$$\bigwedge_{i=1}^{k}\exists\overline{w}^i[\,\overline{x} = \overline{t}^i(\overline{w}^i) \wedge \bigwedge_{h=1}^{n}\bigwedge_{j=1}^{m_h}\forall\overline{v}(\,w_h^i \neq s_{hj}^i(\overline{w}^i,\overline{v})\,)\,]$$

into an equivalent disjunction of *answers* for $\overline{x}$. If the $mgu(\overline{t}_1(\overline{w}^1),\ldots,\overline{t}_k(\overline{w}^k))$ does not exist, the disjunction is equivalent to False. Otherwise, we get a substitution σ that is used for joining the equational parts as follows:

$$\exists\overline{w}^1\ldots\exists\overline{w}^k[\,\overline{x} = \sigma(\overline{t}^1(\overline{w}^1)) \wedge \bigwedge_{h=1}^{n}\bigwedge_{j=1}^{m_h}\forall\overline{v}(\,\sigma(w_h^i) \neq \sigma(s_{hj}^i(\overline{w}^i,\overline{v}))\,)\,].$$

Now, letting $\overline{w} \equiv \overline{w}^1 \cdot \ldots \cdot \overline{w}^k$ we have a constraint of the form:

$$\exists \overline{w}[\ \overline{x} = t'(\overline{w}) \wedge \bigwedge_{h=1}^{n} \bigwedge_{j=1}^{m_h} \neg \exists \overline{v}(\ t_h(\overline{w}) = r_{hj}(\overline{w}, \overline{v})\ )\ ].$$

Let $\sigma_{hj} \equiv mgu(t_h(\overline{w}), r_{hj}(\overline{w}, \overline{v}))$. If σ_{hj} does not exist, $\neg \exists \overline{v}(\ t_h(\overline{w}) = r_{hj}(\overline{w}, \overline{v})\ )$ is equivalent to True. Otherwise, since σ_{hj} is an idempotent substitution, each answer $\neg \exists \overline{v}(\ \sigma_{hj}\ )$ can be transformed into a disjunction $\bigvee_{k=1}^{n_{hj}} \exists \overline{z}(\ a_k(\overline{w}, \overline{z})\ )$ of *answers* for $\overline{w}$ by (UD) as we show in Subsect. 5.1 (see (7)). Hence, the constraint has the form:

$$\exists \overline{w}[\ \overline{x} = t'(\overline{w}) \wedge \bigwedge_{h=1}^{n} \bigwedge_{j=1}^{m_h} \bigvee_{k=1}^{n_{hj}} \exists \overline{z}(\ a_k(\overline{w}, \overline{z})\ )\ ]. \tag{8}$$

Finally, we apply distribution and (recursively) conjunction of *answers* for $\overline{w}$:

$$\exists \overline{w}[\ \overline{x} = t'(\overline{w}) \wedge \bigvee_{r=1}^{m} \exists \overline{z}(\ b_r(\overline{w}, \overline{z})\ )\ ].$$

Then, we lift the internal disjunction and substitute the equational part of each *answer* $b_r(\overline{w}, \overline{z})$ in $t'(\overline{w})$.

Notice that, the only blow-up (in the number of *answers*) could be produced in the transformation of (8), by the distribution and conjunction of *answers*. However, in practice, the blow-up is often non-significant. This is due to the fact that the above $\bigvee_{k=1}^{n_{hj}} \exists \overline{z}(a_k(\overline{w}, \overline{z}))$ are obtained by using the rule (UD). Since the rule (UD) yields mutually excluding constraints, many internal conjunctions of *answers* for $\overline{w}$ are reduced to False at once.

6 A Complete Example

In this section, we demonstrate the application of our solving method to the following equational constraint on x_1, x_2, x_3 (free variables):

$$\forall y_1 \exists w_1 \forall y_2 (\ f(x_1, g(y_2)) \neq f(f(w_1, x_2), a) \wedge w_1 \neq f(y_1, y_1) \wedge$$
$$\exists w_2 \forall y_3 (\ f(x_2, a) \neq f(g(y_3), w_1)\ ) \wedge$$
$$\forall y_4 \forall y_5 (\ f(x_1, x_2) \neq f(g(x_3), f(y_4, y_5))\ )\ )$$

First, after a preliminary treatment, we obtain the following disjunction of *answers* for x_1, x_2, x_3, y_1, w_1 (as matrix) prefixed by $\forall y_1 \exists w_1$:

$$\forall y_1 \exists w_1 (\ \exists \overline{z}(\ x_1 = z_1 \wedge x_2 = z_2 \wedge x_3 = z_3 \wedge y_1 = z_4 \wedge w_1 = z_5 \wedge$$
$$z_5 \neq f(z_4, z_4) \wedge \forall y_4 \forall y_5 (\ z_2 \neq f(y_4, y_5)\ ) \wedge \forall y_2 (\ z_2 \neq g(y_2)\ )\ ) \vee$$
$$\exists \overline{z}(\ x_1 = z_1 \wedge x_2 = z_2 \wedge x_3 = z_3 \wedge y_1 = z_4 \wedge w_1 = z_5 \wedge$$
$$z_5 \neq f(z_4, z_4) \wedge z_1 \neq g(z_3) \wedge \forall y_3 (\ z_2 \neq g(y_3)\ )\ ) \vee$$
$$\exists \overline{z}(\ x_1 = z_1 \wedge x_2 = z_2 \wedge x_3 = z_3 \wedge y_1 = z_4 \wedge w_1 = z_5 \wedge$$
$$z_5 \neq f(z_4, z_4) \wedge \forall y_4 \forall y_5 (\ z_2 \neq f(y_4, y_5)\ ) \wedge z_5 \neq a\ ) \vee$$
$$\exists \overline{z}(\ x_1 = z_1 \wedge x_2 = z_2 \wedge x_3 = z_3 \wedge y_1 = z_4 \wedge w_1 = z_5 \wedge$$
$$z_5 \neq f(z_4, z_4) \wedge z_1 \neq g(z_3) \wedge z_5 \neq a\ )\ ).$$

Notice that the prefix is shorter (than the prefix of the prenex-DNF form) because *answers* allow universal quantification.

Now, quantifier elimination is successively applied until the prefix is erased. The innermost block $\exists w_1$ is easily eliminated by removing the CoEq $w_1 = z_5$ and all the UCDs involving z_5. Thus, we have a disjunction of *answers* for x_1, x_2, x_3, y_1 prefixed by $\forall y_1$ which, by double negation, is equivalent to:

$$\neg \exists y_1 (\ \neg \exists \overline{z} (\ x_1 = z_1 \wedge x_2 = z_2 \wedge x_3 = z_3 \wedge y_1 = z_4 \wedge \forall y_4 \forall y_5 (\ z_2 \neq f(y_4, y_5)\ )\ \wedge$$
$$\forall y_2 (\ z_2 \neq g(y_2)\ )\ )\ \wedge$$
$$\neg \exists \overline{z} (\ x_1 = z_1 \wedge x_2 = z_2 \wedge x_3 = z_3 \wedge y_1 = z_4 \wedge z_1 \neq g(z_3)\ \wedge$$
$$\forall y_3 (\ z_2 \neq g(y_3)\ )\ )\ \wedge$$
$$\neg \exists \overline{z} (\ x_1 = z_1 \wedge x_2 = z_2 \wedge x_3 = z_3 \wedge y_1 = z_4 \wedge \forall y_4 \forall y_5 (\ z_2 \neq f(y_4, y_5)\ )\ )\ \wedge$$
$$\neg \exists \overline{z} (\ x_1 = z_1 \wedge x_2 = z_2 \wedge x_3 = z_3 \wedge y_1 = z_4 \wedge z_1 \neq g(z_3)\ )\ ).$$

Then, each of the four negated *answers* for the variables x_1, x_2, x_3, y_1 produces a disjunction of *answers* for the same variables, as follows:

$$\neg \exists y_1 (\ [\ \exists \overline{z} (x_1 = z_1 \wedge x_2 = f(z_2, z_3) \wedge x_3 = z_4 \wedge y_1 = z_5)\ \vee$$
$$\exists \overline{z} (x_1 = z_1 \wedge x_2 = g(z_2) \wedge x_3 = z_3 \wedge y_1 = z_4)\ ]\ \wedge$$
$$[\ \exists \overline{z} (x_1 = g(z_1) \wedge x_2 = z_2 \wedge x_3 = z_1 \wedge y_1 = z_4)\ \vee$$
$$\exists \overline{z} (x_1 = z_1 \wedge x_2 = g(z_2) \wedge x_3 = z_3 \wedge y_1 = z_4)\ ]\ \wedge$$
$$[\ \exists \overline{z} (x_1 = z_1 \wedge x_2 = f(z_2, z_3) \wedge x_3 = z_4 \wedge y_1 = z_5)\ ]\ \wedge$$
$$[\ \exists \overline{z} (x_1 = g(z_1) \wedge x_2 = z_2 \wedge x_3 = z_1 \wedge y_1 = z_3)\ ]\ )$$

By distributing conjunction over disjunction, we obtain the following disjunction of conjunctions of *answers*:

$$\neg \exists y_1 (\ [\ \exists \overline{z} (x_1 = z_1 \wedge x_2 = f(z_2, z_3) \wedge x_3 = z_4 \wedge y_1 = z_5)\ \wedge$$
$$\exists \overline{z} (x_1 = g(z_1) \wedge x_2 = z_2 \wedge x_3 = z_1 \wedge y_1 = z_4)\ \wedge$$
$$\exists \overline{z} (x_1 = z_1 \wedge x_2 = f(z_2, z_3) \wedge x_3 = z_4 \wedge y_1 = z_5)\ \wedge$$
$$\exists \overline{z} (x_1 = g(z_1) \wedge x_2 = z_2 \wedge x_3 = z_1 \wedge y_1 = z_3)\ ]\ \vee$$
$$[\ \exists \overline{z} (x_1 = z_1 \wedge x_2 = f(z_2, z_3) \wedge x_3 = z_4 \wedge y_1 = z_5)\ \wedge$$
$$\exists \overline{z} (x_1 = z_1 \wedge x_2 = g(z_2) \wedge x_3 = z_3 \wedge y_1 = z_4)\ \wedge$$
$$\exists \overline{z} (x_1 = z_1 \wedge x_2 = f(z_2, z_3) \wedge x_3 = z_4 \wedge y_1 = z_5)\ \wedge$$
$$\exists \overline{z} (x_1 = g(z_1) \wedge x_2 = z_2 \wedge x_3 = z_1 \wedge y_1 = z_3)\ ]\ \vee$$
$$[\ \exists \overline{z} (x_1 = z_1 \wedge x_2 = f(z_2, z_3) \wedge x_3 = z_4 \wedge y_1 = z_5)\ \wedge$$
$$\exists \overline{z} (x_1 = z_1 \wedge x_2 = g(z_2) \wedge x_3 = z_3 \wedge y_1 = z_4)\ \wedge$$
$$\exists \overline{z} (x_1 = z_1 \wedge x_2 = f(z_2, z_3) \wedge x_3 = z_4 \wedge y_1 = z_5)\ \wedge$$
$$\exists \overline{z} (x_1 = g(z_1) \wedge x_2 = z_2 \wedge x_3 = z_1 \wedge y_1 = z_3)\ ]\ \vee$$
$$[\ \exists \overline{z} (x_1 = z_1 \wedge x_2 = f(z_2, z_3) \wedge x_3 = z_4 \wedge y_1 = z_5)\ \wedge$$
$$\exists \overline{z} (x_1 = z_1 \wedge x_2 = g(z_2) \wedge x_3 = z_3 \wedge y_1 = z_4)\ \wedge$$

$$\exists \overline{z}(x_1 = z_1 \wedge x_2 = f(z_2, z_3) \wedge x_3 = z_4 \wedge y_1 = z_5) \wedge$$
$$\exists \overline{z}(x_1 = g(z_1) \wedge x_2 = z_2 \wedge x_3 = z_1 \wedge y_1 = z_3) \,] \,).$$

It is easy to see that, in the latter three conjunctions of *answers*, the variable x_2 makes impossible the unification that is required to perform a conjunction of *answers*. Hence, the three conjunctions are transformed to False at once. Whereas the first conjunction yields the following satisfiable *answer* for $\overline{x} \cdot y_1$:

$$\neg \exists y_1 (\ \exists \overline{z}(\ x_1 = g(z_1) \wedge x_2 = f(z_2, z_3) \wedge x_3 = z_1 \wedge y_1 = z_4\ )\ ).$$

Then, the block $\exists y_1$ and the equation $y_1 = z_4$ can be eliminated. Finally, the negation of the resulting *answer* for $\overline{x}$, that is:

$$\neg \exists \overline{z}(\ x_1 = g(z_1) \wedge x_2 = f(z_2, z_3) \wedge x_3 = z_1\ )$$

yields the following disjunction of two *answers* for $\overline{x}$:

$$\exists \overline{w}(\ x_1 = w_4 \wedge x_2 = f(w_2, w_3) \wedge x_3 = w_1 \wedge w_4 \neq g(w_1)\ ) \vee$$
$$\exists \overline{w}(\ x_1 = w_2 \wedge x_2 = w_3 \wedge x_3 = w_1 \wedge \forall v_1 \forall v_2(\ w_3 \neq f(v_1, v_2)\ )\ ).$$

7   Conclusions and Related Work

The notion of *answer* provides a sufficiently compact representation of solutions while retaining user-friendliness and efficient performance of basic operations. In particular, we give detailed algorithms for *answer* satisfiability checking (for finite signature), negation of one *answer* and conjunction of several *answers*. This combination of features makes *answers* a suitable notion of (quasi-)solved form for achieving a good trade-off between time and space efficiency in theorem proving methods for logics with equality. *Answers* are particularly suitable for methods that require some equality constraint notation more expressive than substitutions. We have shown how the quantifier elimination method takes advantage of using *answers* in this sense, given a new method for general equality constraint solving. This method applies to both finite and infinite signatures. The only difference is that satisfiability checking is not needed in the latter case.

   Answer is an intermediate notion between purely existential solved forms of, for example, [14, 9] and *substitutions with exceptions* of [4]. *Answers* allow a kind of restricted universal quantification which, besides being more expressive, allows one to confine the role of the explosion rule to the satisfiability test. In this process, since universal quantifiers are not eliminated, we never blow up the tested *answer* via the explosion rule. Explosion is only implicitly used (in the satisfiability test) for computing the indispensable variable extensions. The methods presented in [14] and [9] are both based in quantifier elimination with explicit usage of the explosion rule, although they use two different notions of solved form. The method of [14], for finite signatures, is based on using the explosion rule to perform the conjunction of boolean combinations of basic formulas. For instance, $\exists w(\ x = w\ ) \wedge \neg \exists u(\ x = f(g(f(u, u)), u)\ )$ is solved by explosion of the first conjunct and then by unification. This operation produces a (signature-dependent) number of existential disjuncts. For example, if the signature also contains the constant a, it yields $x = a, \exists w(x = f(a, w)), \exists w(x = f(g(a), w)), \ldots$. Our method yields a (signature-independent) disjunction of *answers*, that is $\exists w(x = w \wedge \forall u(w \neq f(g(f(u, u)), u)))$ for the just above example. Similarly, the proposal of [9], which has been implemented by N. Peltier (see [16, 17]), uses explosion to eliminate all the universal quantifiers. Explosion increases the number of disjuncts in a ratio proportional to the signature. Besides, this blow up interacts with the "CNF-to-DNF" transformation. We cannot avoid the latter blow up, but we benefit from the smaller number of conjuncts. As a consequence, our solutions are always (except for very simple examples) shorter and computationally simpler than the ones given by the system in [17]. Actually, to improve this system, a (very different from ours) restricted form of universal quantification is proposed in [16]. This new solved form allows to replace explicit explosion with the so-called *binary explosion*, which is signature-independent and yields a binary blow up of the formula. Unfortunately, this improvement has not been yet incorporated to [17].

The closest work to our own can be found in [4], where two notions of solved form are provided. They are called *substitutions with exceptions* (SwEs for short) and *constrained substitution*. Both involve universal quantification and require a satisfiability test in the case of finite signature. However, there are significant differences, the most important one being that universal quantification is more restricted in *answers* than in both solved forms of [4]. The following discussion is applicable to both notions of solved form, even though in the sequel we will only mention SwEs. With respect to *answers*, SwEs provide a more compact representation of solutions, but the basic operations for handling them, in particular the satisfiability test, become intricate. More precisely, a SwE is an expression of the form $\sigma_0[\overline{x}, \overline{w}^0] - \{\sigma_1[\overline{x}, \overline{w}^1], \ldots, \sigma_k[\overline{x}, \overline{w}^k]\}$ where $\sigma[\overline{z}, \overline{y}]$ denotes a substitution on domain $\overline{z}$ such that $\overline{y} \equiv Var(\sigma(\overline{z}))$. A SwE of the above form is interpreted as the equality constraint $\exists \overline{w}^0(\sigma_0 \wedge \forall \overline{w}^1(\neg\sigma_1) \wedge \ldots \wedge \forall \overline{w}^k(\neg\sigma_k))$. Notice that each $\neg\sigma_i$ is a disjunction of disequations. For example, the following SwE:

$$\{x_1 \leftarrow f(a, w_1), x_2 \leftarrow g(w_1), x_3 \leftarrow f(w_2, w_1)\}$$
$$-\{ \{x_1 \leftarrow f(a, y), x_2 \leftarrow f(a, y), x_3 \leftarrow g(y)\},$$
$$\{x_1 \leftarrow f(a, g(y_1)), x_2 \leftarrow g(y_2) \leftarrow x_3, f(y_3, v)\} \}$$

corresponds to the equality constraint:

$$\exists w_1 \exists w_2(\ x_1 = f(a, w_1) \wedge x_2 = g(w_1) \wedge x_3 = f(w_2, w_1) \wedge \tag{9}$$
$$\forall y(x_1 \neq f(a, y) \vee x_2 \neq f(a, y) \vee x_3 \neq g(y)) \wedge$$
$$\forall y_1 \forall y_2 \forall y_3 \forall v(x_1 \neq f(a, g(y_1)) \vee x_2 \neq g(y_2) \vee x_3 \neq f(y_3, v)) \ )$$

In *answers*, universal quantification is restricted to affect one disequation, instead of a disjunction of disequations. Since universal quantification does not distribute over disjunction, this is not a minor difference, especially when testing satisfiability. Actually, [4] introduces a method for solving a *system of equations and disequations* with the proviso that a satisfiability test is given. There, it is shown that, for testing satisfiability, it is not enough to check that a substitution is an instance of another. Instead, it is necessary to check whether each instance of the former is an instance of the latter substitution that, in general, requires an infinite number of checks. The solving method that we introduce here provides an easy way for transforming any SwE into a (possibly empty, if unsatisfiable) disjunction of satisfiable *answers*. For example, our constraint solver transforms the above SwE, really the equality constraint (9), into the following *answer*:

$$\exists w_1 \exists w_2 (\ x_3 = f(w_2, w_1) \wedge x_2 = g(w_1) \wedge x_1 = f(a, w_1) \wedge \forall v_1 (\ w_1 \neq g(v_1)\ ))$$

Two other notions of solved form that allow for restricted forms of universal quantification were introduced in [18, 19] and [7].

The approach of [18, 19] is more interested in complexity results and in efficient checking of the satisfiability of equational problems than in computing their solutions. In [18, 19], the set of solutions of an equational problem is expressed by a restricted form of $\exists^*\forall^*$-CNF (called PFE-form), whereas our disjunction of *answers* is a $\exists^*\forall^*$-DNF formula. In order to illustrate that point, we borrow from [18, 19] the following $\exists^*\forall^*$-CNF equational constraint (where $\overline{z}$ and $\overline{y}$ stands for z_1, z_2, z_3, z_4 and y_1, y_2, y_3, y_4):

$$\exists \overline{z} \forall \overline{y}[\ (\ f(z_1, g(z_2)) = f(y_1, z_3) \vee g(y_1) = z_3 \vee f(a, y_2) = f(z_2, g(y_4)) \tag{10}$$
$$\vee f(g(y_2), z_1) \neq f(y_3, g(y_1)) \vee g(z_3) \neq z_2\ ) \wedge$$
$$(\ f(y_1, a) = f(g(z_2), a) \vee f(g(z_4), y_1) \neq f(z_2, a) \vee g(y_1) \neq g(g(y_3))\ ) \wedge$$
$$(\ f(g(z_2), z_1) = f(g(y_1), y_1) \vee g(y_2) = y_3 \vee f(a, y_4) = f(z_2, g(y_2))$$
$$\vee f(f(a, y_3), g(y_2)) \neq f(z_1, y_4) \vee g(z_2) \neq g(f(a, z_3)) \vee z_4 \neq y_1\ )]$$

In [18, 19], these equational constraints are tranformed (in polinomial time) into the so-called PFE-form. The PFE-form of the (10) is:

$$\exists \overline{z}[\ \exists u_1 \exists u_2 (\ [\ z_1 = g(u_1) \wedge z_2 = g(z_3) \wedge g(u_1) = z_3\ ] \vee$$
$$\forall y_1 \forall v[\ z_1 \neq g(y_1) \vee z_2 \neq g(v) \vee z_3 \neq v\ ]) \qquad \wedge$$
$$(\ [\ z_1 = f(a, z_3) \wedge z_2 = f(a, z_3) \wedge f(g(f(a, z_3)), f(a, u_2)) = f(g(z_4), z_4)\ ] \vee$$
$$[\ z_1 = f(a, u_2) \wedge z_2 = f(a, z_3) \wedge f(a, g(a)) = f(z_2, g(a))\ ] \vee$$
$$\forall y_3 \forall v(\ z_1 \neq f(a, y_3) \vee z_2 \neq f(a, v) \vee z_3 \neq v)\ )]$$

The satisfiability of PFE-forms can be checked by a non-deterministic algorithm ([18, 19]) in polinomial-time. Our solver proceeds in a very different way. In particular, we obtain *answers* for constraints with free variables. For the above $\exists^*\forall^*$-CNF constraint (10), our solver first tranforms the inner $\forall^*$-CNF into the disjunction of the following ten *answers* for $\overline{z}$ (for easier reading, each w_i is considered to be existentially quantified):

1. $z_1 = f(a, w_3) \wedge z_2 = w_1 \wedge z_3 = w_2 \wedge \forall v_1 (\ w_3 \neq g(v_1)\ ) \wedge w_1 \neq f(a, w_2)$
2. $z_1 = w_4 \wedge z_2 = w_5 \wedge z_3 = w_6 \wedge \forall v_2 (\ w_4 \neq g(v_2)\ ) \wedge \forall v_3 (\ w_4 \neq f(a, v_3)\ )$
3. $z_1 = f(a, g(w_9)) \wedge z_2 = w_7 \wedge z_3 = w_8 \wedge w_7 \neq f(a, w_8)$

4. $z_1 = w_{10} \wedge z_2 = w_{10} \wedge z_3 = w_{11} \wedge z_4 = w_{10} \wedge \forall v_4(w_{10} \neq g(v_4))$
5. $z_1 = g(w_{14}) \wedge z_2 = w_{12} \wedge z_3 = w_{13} \wedge \forall v_5(w_{12} \neq g(v_5))$
6. $z_1 = g(w_{15}) \wedge z_2 = g(g(w_{15})) \wedge z_3 = g(w_{15})$
7. $z_1 = g(w_{16}) \wedge z_2 = g(w_{16}) \wedge z_3 = w_{17} \wedge z_4 = g(w_{16}) \wedge \forall v_6(w_{16} \neq g(v_6))$
   $\wedge\, w_{17} \neq w_{16}$
8. $z_1 = g(w_{20}) \wedge z_2 = g(w_{18}) \wedge z_3 = w_{19} \wedge \forall v_7(w_{18} \neq g(v_7)) \wedge w_{19} \neq w_{18}$
9. $z_1 = g(g(w_{22})) \wedge z_2 = g(g(w_2 2)) \wedge z_3 = w_{21} \wedge z_4 = g(g(w_{22})) \wedge w_{21} \neq g(w_{22})$
10. $z_1 = g(w_{25}) \wedge z_2 = g(g(w_{24})) \wedge z_3 = w_{23} \wedge w_{23} \neq g(w_{24})$

Then, the existencial closure (by $\exists \bar{z} \exists \bar{w}$) of the disjunction of the above ten *answers* is easily reduced to True by checking that each answer is satisfiable.

The main goal of [7] is the efficient decidability of equational formulas with a long prefix of quantifiers, focusing on infinite signatures. Because of this focus, they do not deal with the satisfiability test. Besides, the notion of solved-form of [7] allows unrestricted nesting of negation and quantification.

We believe that *answers* could be helpful for development and improvement of resolution- and instance-based methods. On the one hand, for example, the *resolution-based method* presented in [6] can be easily adapted to deal with *answers* instead of $\exists^*\forall^*$-constraints. Moreover, all the $\exists^*\forall^*$-constraints used in the several examples of [6] are really *answers*. With such adjustment, the interesting method of [6] could benefit from compactness (in particular, avoiding explosion rule) and superior performance of basic operations. On the other hand, it seems worth investigating the usefulness of *answers* for the area of growing interest of *instance-based methods*. See [2] for a recent and good summary and for references. In conclusion, we offer some pointers to future developments and applications of the notion of *answer* to the latter area. First, in [12](Sec. 4), it is pointed out that, for redundancy elimination, "*it might be useful to employ elaborate constraint notations such as the ones proposed in [6]*" and *answers* seem to be even better suited for that goal. Second, the notion of *context* used in [3] could be represented by atoms constrained by (disjunction of) *answers*. Then, negation and conjunction of *answers* would become basic for building *context unifiers* and the satisfiability test becomes essential for guiding the procedure.

Acknowledgment. This work has been partially supported by Spanish Projects TIC2001-2476-C03 and TIN2004-079250-C03-03. We also would like to thank the anonymous referees for their valuable comments.

References

1. J. Álvez and P. Lucio. Equational constraint solving using quasi-solved forms. In *Proceedings of the 18th International Workshop on Unification (UNIF'04)*, Cork, Ireland, 5 June 2004.
2. P. Baumgartner and G. Stenz. Instance based methods. *Tutorial T3.- 2nd Int. Joint Conf. on Automated Reasoning, IJCAR 2004, Cork, Ireland*, 2004.
3. P. Baumgartner and C. Tinelli. The Model Evolution Calculus. In Franz Baader, editor, *CADE-19 – The 19th Int. Conf. on Automated Deduction*, volume 2741 of *Lecture Notes in Artificial Intelligence*, pages 350–364. Springer, 2003.

4. W. L. Buntine and H.-J. Bürckert. On solving equations and disequations. *Journal of the ACM*, 41(4):591–629, 1994.

5. Ricardo Caferra and Nicolas Zabel. Extending resolution for model construction. In *JELIA '90: Proceedings of the European workshop on Logics in AI*, pages 153–169, New York, NY, USA, 1991. Springer-Verlag New York, Inc.

6. Ricardo Caferra and Nicolas Zabel. A method for simultaneous search for refutations and models by equational constraint solving. *J. Symb. Comput.*, 13(6):613–641, 1992.

7. A. Colmerauer and T.-B.-H. Dao. Expresiveness of full first order constraints in the algebra of finite and infinite trees. In *6th Int. Conf. of Principles and Practice of Constraint Programming CP'2000*, volume 1894 of *LNCS*, pages 172–186, 2000.

8. H. Comon. Disunification: A survey. In J.L. Lassez and G. Plotkin, editors, *Essays in Honour of Alan Robinson*, 1991.

9. H. Comon and P. Lescanne. Equational problems and disunification. *Journal of Symbolic Computation*, 7:371–425, 1989.

10. Kevin J. Compton and C. Ward Henson. A uniform method for proving lower bounds on the computational complexity of logical theories. *Ann. Pure Appl. Logic*, 48(1):1–79, 1990.

11. Christian G. Fermüller and Alexander Leitsch. Hyperresolution and automated model building. *J. Log. Comput.*, 6(2):173–203, 1996.

12. H. Ganzinger and K. Korovin. New directions in instantiation-based theorem proving. In *Proc. 18th IEEE Symposium on Logic in Computer Science*, pages 55–64. IEEE Computer Society Press, 2003.

13. Jean-Louis Lassez and Kim Marriott. Explicit representation of terms defined by counter examples. *J. Autom. Reasoning*, 3(3):301–317, 1987.

14. M. J. Maher. Complete axiomatizations of the algebras of finite, rational and infinite trees. In *Proc. of the 3rd IEEE Symp. on Logic in Computer Science*, pages 348–357. Computer Society Press, 1988.

15. A. I. Malcev. Axiomatizable classes of locally free algebras. In B. F. Wells, editor, *The Metamathematics of Algebraic Systems (Collected Papers: 1936-1967)*, volume 66, chapter 23, pages 262–281. North-Holland, 1971.

16. N. Peltier. *Nouvelles Techniques pour la Construction de Modèles finis ou infinis en Déduction Automatique*. PhD thesis, Institut National Polytechnique de Grenoble, 1997.

17. Nicolas Peltier. System description: An equational constraints solver. In *CADE-15: Proceedings of the 15th International Conference on Automated Deduction*, pages 119–123, London, UK, 1998. Springer-Verlag.

18. R. Pichler. Solving equational problems efficiently. In H. Ganzinger, editor, *Proc. 16th Int. Conf. on Automated Deduction (CADE-16)*, number 1632 in Lecture Notes in Artificial Intelligence, pages 97–111. Springer Verlag, 1999.

19. R. Pichler. On the complexity of equational problems in CNF. *Journal of Symbolic Computation*, 36:235–269, 2003.

20. T. Rybina and E. Voronkov. A decision procedure for term algebras with queues. *ACM Trans. Comput. Logic*, 2(2):155–181, 2001.

21. S. Vorobyov. An improved lower bound for the elementary theories of trees. In *Automated Deduction CADE-13 LNAI 110*, pages 275–287. Springer, 1996.

Appendix

In this section, we give the proofs of Theorem 1 and Proposition 1. In the former proof, we make use of some results of [13]. In order to make this paper self-contained, let us recall them using our terminology.

> We consider a fixed term t such that $Var(t) \equiv \overline{x}$ and a fixed collection of substitutions $\theta_1, \ldots, \theta_n$ such that $domain(\theta_j) \equiv \overline{x}$ for every $1 \leq j \leq n$.
>
> PROPOSITION 4.6. ([13]) *If each $\underline{c}(\overline{x})\theta_i$ is not a linear term, then there does not exists a finite set of terms equivalent to $t/t\theta_1 \vee \ldots \vee t\theta_n$.* $\quad\square$
>
> PROPOSITION 4.8. ([13]) *If each $\underline{c}(\overline{x})\theta_i$ is a linear term, then there exists a finite set of terms equivalent to $t/t\theta_1 \vee \ldots \vee t\theta_n$.* $\quad\square$
>
> THEOREM 4.1. ([13]) *The algorithm uncover can be used to find an equivalent finite set of terms for $t/t\theta_1 \vee \ldots \vee t\theta_n$ if one exists. Otherwise, uncover will terminate with an implicit representation.* $\quad\square$

Our proof of Theorem 1 is based on the following two lemmas:

Lemma 1. *Let t be a term such that $Var(t) \equiv \overline{x}$ and $\theta_1, \ldots, \theta_n$ be a collection of substitutions such that $domain(\theta_j) \equiv \overline{x}$ and $x_i\theta_j$ is linear for every $1 \leq j \leq n$. $uncover(t, t\theta_1, \ldots, t\theta_n)$ always terminates and yields a (possibly empty) finite set of linear terms that is equivalent to $t/t\theta_1 \vee \ldots \vee t\theta_n$.*

Proof. It is a direct consequence of Proposition 4.8 and Theorem 4.1 in [13]. $\quad\square$

Lemma 2. *Let $\overline{w}^0 \cdot \overline{w}^1$ a disjoint partition of a given tuple $\overline{w}$, σ an assignment with domain $\overline{w}^0$ and $Ext(\overline{w}^1)$ an extension of $\overline{w}^1$ that is infinite for every w_j^1. Let S_0 be a set of UCDs on $\overline{w}$ of the form $\forall \overline{v}(w_i \neq s_{ij}(\overline{w}, \overline{v}))$ that satisfies the following two properties:*

(P1) $Var(s_{ij}) \cap \overline{w} \not\equiv \emptyset$ *or some v_k appears more than once in s_{ij}*
(P2) $(\{w_i\} \cup Var(s_{ij})) \cap \overline{w}^1 \not\equiv \emptyset$.

If every term in $Ext(\overline{w}^1)$ is linear, then σ can be extended to an assignment σ' with domain $\overline{w}$ that satisfies:

$$\sigma \quad \wedge \quad S_0 \quad \wedge \bigwedge_{w_j^1 \in \overline{w}^1} \bigvee_{t \in Ext(w_j^1)} (w_j^1 = t).$$

Proof. Since $\sigma(w_i^0)$ is a ground term $t_i \in \mathcal{T}(\Sigma)$ for each $w_i \in \overline{w}^0$, we can substitute t_i for w_i^0 in S_0. Then, we obtain a finite set S of universal disequations on $\overline{w}^1$ that is trivially equivalent to $\sigma \wedge S_0$. Besides, by construction, S consists universal disequations of the following three types:

(T1) $\forall \overline{v}(w_k^1 \neq s(\overline{v}))$ where at least one v_k occurs repeatedly in s
(T2) $\forall \overline{v}(w_k^1 \neq s(\overline{w}^1, \overline{v}))$
(T3) $\forall \overline{v}(t \neq s(\overline{w}^1, \overline{v}))$ where t is a ground term and $Var(s) \cap \overline{w}^1 \not\equiv \emptyset$

and σ' should extend σ to $\overline{w}$ (hence, to $\overline{w}^1$) while satisfying:

$$\mathsf{S} \ \wedge \ \bigwedge_{w_j^1 \in \overline{w}^1} \ \bigvee_{t \in Ext(w_j^1)} (w_j^1 = t). \tag{11}$$

Since each $Ext(w_i^1)$ is described by a collection of linear terms, a finite conjunction of universal disequations of type (T1) is not able to turn $Ext(w_i^1)$ up into a finite extension (this follows from Proposition 4.6 in [13]). Besides, each disequation of type (T2) and (T3) involves at least one existential variable with infinite extension. Hence, the finite number of universal disequations of these two types cannot disallow the infinite possible assignments. Therefore, there are infinitely many σ' that satisfies (11).   □

Now, we can prove our result:

Theorem 1. *The satisfiability test of Figure 3 terminates for any input* (S, Σ). *Besides, it returns "satisfiable" if there exists a Σ-assignment that satisfies the set* S. *Otherwise, it returns "unsatisfiable".*

Proof. By Lemma 1, the satisfiability test terminates and, moreover, every term in each $Ext(w_i)$ is linear along the whole process. There are four possible exit points:

1. If some $Ext(w_i)$ becomes empty, by Lemma 1, there is not possible assignment with domain w_i that satisfies the subset of S considered at the first step. Hence, there is not assignment for $\overline{w}$ that satisfies S.
2. If $\overline{w}^{fin}$ becomes empty at the end of the first step, then, by Lemma 2, the empty assignment (with empty domain) can be extended to an assignment σ' with domain $\overline{w}$ that satisfies S. Notice that the set of UCDs that satisfies (P1) and (P2) of Lemma 2 are exactly S minus the UCDs considered at the first step.
3. If the second step returns "satisfiable", on the basis of Lemma 1, C contains the finite collection of all possible assignments to $\overline{w}^{fin}$ that satisfies the subset of UCDs involved by the two steps. Therefore, by Lemma 2, each of these possible assignments σ with domain $\overline{w}^{fin}$ can be extended to $\overline{w}$ for satisfying the whole set S.
4. If the second step returns "unsatisfiable" after checking a subset S_0 of UCDs, by Lemma 1, S_0 is unsatisfiable and, hence, S is unsatisfiable too.   □

Now, we prove the following result:

Proposition 1. *The transformation rule* (UD) *of Figure 4 is correct.*

Proof. (UD) can be obtained by successive applications of the rule (ud_0):

$$\neg\exists\overline{v}[\ x = t(\overline{w}, \overline{v}^1) \wedge \varphi\ ] \longmapsto \neg\exists\overline{v}^1[\ x = t(\overline{w}, \overline{v}^1)\ ] \vee \exists\overline{v}^1[\ x = t(\overline{w}, \overline{v}^1) \wedge \neg\exists\overline{v}^2(\ \varphi\ )\ ]$$

where $\overline{v}^1 \equiv Var(t) \cap \overline{v}$ and $\overline{v}^2 \equiv \overline{v}\backslash\overline{v}^1$. It suffices to note that the constraint:

$$\neg\exists\overline{v}^{11}[\ x_1 = t_1(\overline{w}, \overline{v}^{11})\ ] \vee \exists\overline{v}^{11}[\ x_1 = t_1(\overline{w}, \overline{v}^{11}) \wedge \neg\exists\overline{v}^{12}(\ x_2 = t_2(\overline{w}, \overline{v}^{12})\ )\ ]$$
$$\vee \ldots \vee \exists\overline{v}^{11} \ldots \exists\overline{v}^{1(n-1)}[\ x_1 = t_1(\overline{w}, \overline{v}^{11}) \wedge \ldots \wedge x_{n-1} = t_{n-1}(\overline{w}, \overline{v}^{1(n-1)})$$
$$\wedge \neg\exists\overline{v}^{1n}(\ x_n = t_n(\overline{w}, \overline{v}^{1n})\ )\ ] \tag{12}$$

is equivalent to $\neg \exists \overline{v}^1 [\, \overline{x} = \overline{t}(\overline{w}, \overline{v}^1)\,]$ where $\overline{v}^1 \equiv \overline{v}^{11} \cdot \ldots \cdot \overline{v}^{1n}$. Hence, it only remains to show that the rule (ud_0) is correct. By conjunction (and distribution) of $\neg \exists \overline{v} [\, x = t(\overline{w}, \overline{v}) \wedge \varphi\,]$ with the tautology $\neg \exists \overline{u} (\, x = t(\overline{w}, \overline{u})\,) \vee \exists \overline{u} (\, x = t(\overline{w}, \overline{u})\,)$, we obtain two disjuncts. The first disjunct is trivially equivalent to the formula $\neg \exists \overline{v}^1 (\, x = t(\overline{w}, \overline{v}^1)\,)$. The second one $\exists \overline{u} (\, x = t(\overline{w}, \overline{u})\,) \wedge \neg \exists \overline{v} [\, x = t(\overline{w}, \overline{v}) \wedge \varphi\,]$ is equivalent to:

$$\exists \overline{u} \forall \overline{v}^1 [\, x = t(\overline{w}, \overline{u}) \wedge (x \neq t(\overline{w}, \overline{v}^1) \vee \neg \exists \overline{v}^2 (\, \varphi\,))\,]$$

where the quantifier $\forall \overline{v}^1$ has been lifted to the prefix. By distribution and simplification, it yields $\exists \overline{u} \forall \overline{v}^1 [\, \overline{u} \neq \overline{v}^1 \vee (x = t(\overline{w}, \overline{u}) \wedge \neg \exists (\, \overline{v}^2 \varphi\,))\,]$. Then, by the well known rule (U_2):

$$(U_2) \qquad \forall \overline{y} (\, P \wedge (y_i \neq t \vee R)\,) \longmapsto \forall \overline{y} (\, P \wedge R\{y_i \leftarrow t\}\,)$$

(see, for example, [9]) the constraint is equivalent to:

$$\exists \overline{u} [\, x = t(\overline{w}, \overline{u}) \wedge \neg \exists \overline{v}^2 (\, \varphi\{\overline{v}^1 \leftarrow \overline{u}\}\,)\,].$$

This constraint coincides with the second disjunct in the right-hand side of (ud_0) except for $\overline{v}^1$, which has been renamed $\overline{u}$.   □

Modeling the Evolution of Objects in Temporal Information Systems

A. Artale[1], C. Parent[2], and S. Spaccapietra[3]

[1] Faculty of Computer Science, Free Univ. of Bolzano, I
artale@inf.unibz.it
[2] HEC/INFORGE, Université de Lausanne, CH
christine.parent@unil.ch
[3] Database Laboratory, Ecole Polytechnique Fédérale Lausanne, CH
stefano.spaccapietra@epfl.ch

Abstract. This paper presents a semantic foundation of temporal conceptual models used to design temporal information systems. We consider a modeling language able to express both timestamping and evolution constraints. We conduct a deeper investigation on evolution constraints, eventually devising a model-theoretic semantics for a full-fledged model with both timestamping and evolution constraints. The proposed formalization is meant both to clarify the meaning of the various temporal constructors appeared in the literature and to give a rigorous definition to notions like satisfiability, subsumption and logical implication. Furthermore, we also show how to express temporal constraints using a subset of first-order temporal logic, i.e., $\mathcal{DLR}_{\mathcal{US}}$, the description logic $\mathcal{DLR}$ extended with the temporal operators *Since* and *Until*. We show how $\mathcal{DLR}_{\mathcal{US}}$ is able to capture the various modeling constraints in a succinct way and to perform automated reasoning on temporal conceptual models.

1 Introduction

This paper is a contribution to improve modeling of temporal data, building on state of the art know-how developed by the conceptual data modeling community. Analyses of many proposals for temporal models (aiming in particular at helping designing temporal databases) and a summary of results achieved in the area can be found in two good surveys [15, 21]. The main temporal modeling features that we focus on in this paper can be summarized as:

- Timestamping. The data model should obviously distinguish between temporal and atemporal modeling constructors. This is usually realized by temporal marking of classes, relationships and attributes. In the database, these markings translate into a *timestamping* mechanism, i.e., attaching lifecycle information to classes and relationship instances, and time-varying values to attributes. In this work we consider just *validity time* (rather than transaction time [20, 27]), thus lifecycle information expresses when an object or a tuple belongs to a class or a relationship, respectively. Time-varying attributes store values together with when they hold.

J. Dix and S.J. Hegner (Eds.): FoIKS 2006, LNCS 3861, pp. 22–42, 2006.

- Evolution Constraints. They apply both to classes (status and transition constraints) and relationships (generation and cross-time constraints). *Status Classes*
  constraints [11] rule the permissible evolution of an object as member of a class
  along its lifespan. For example, an object that is an active member of a class may
  become an inactive member of the same class. *Transition* constraints [16] rule *object migration*, i.e., the possibility for an object to change its class membership
  from one class to another. For example, an object in the Student class may later
  migrate to become an object of the Faculty class. Complementary aspects of evolution are modeled through *Generation Relationships* [17] which describe the fact
  that objects in a class are generated by other objects in another (possibly the same)
  class. For example, in a company database, splitting of a department translates into
  the fact that the original department generates two (or more) new departments. Objects participating to *Cross-Time Relationships* [23] may not coexist at the time the
  relationship is asserted. For example, the grandfather-of relationship can involve a
  dead grandfather with a leaving grandchild.

This paper presents a semantic foundation for temporal data models, as a possible
response to concerns stating: *"[..] it is only by considering those conceptual models
as a mathematical object with a formal definition and semantics that they can become
useful tools for the design of databases schema and applications [..]"* [12].

We present a deeper investigation on evolution constraints, eventually devising a
model-theoretic semantics for a full-fledged conceptual model with both timestamping and evolution constraints. While timestamping aspects have been extensively discussed [3, 4, 10, 14, 22, 26], a clear formalization of evolution constraints is still missing, despite the fact that in the literature such constraints have been advocated as useful
for modeling the behavior of temporal objects [4, 25, 17, 16, 23, 26, 24].

The formalization proposed here builds on previous efforts to formalize temporal
conceptual models. Namely, we rely on a previous work to define the $\mathcal{ER}_{VT}$ model [4],
a temporal Extended Entity-Relationship (EER) model—i.e., the standard ER modeling
language enriched with ISA links, disjoint and covering constraints, and full cardinality constraints—equipped with both a linear and a graphical syntax and based on a
model-theoretic semantics. $\mathcal{ER}_{VT}$ captures timestamping constructors along with transition constraints. This work extends $\mathcal{ER}_{VT}$ with status classes, generation relationships and cross-time relationships. Another closely related work is the one of Finger
and McBrien [12]. They propose a model-theoretic formalization for the ERT model,
an EER model with timestamping but just cross-time relationships (called H-marked
relationships by the authors and introduced in a previous paper by McBrien, Seltveit
and Wrangler [23]). Our proposal modifies the semantics of cross-time relationships as
presented in [12] to comply with a crucial modeling requirement, i.e snapshot reducibility [22].

The advantage of associating a set-theoretic semantics to a language is not only to
clarify the meaning of the language's constructors but also to give a semantic definition
to relevant modeling notions. In our case, given an interpretation function to assign a
set-theoretic semantics to the (temporal) modeling constructors, we are able to give a
rigorous definition of the notions of: *schema satisfiability* when a schema admits a non
empty interpretation which guarantees that the constraints expressed by the schema are

not contradictory (similarly we define the notions of class and relationships satisfiability); *subsumption* between classes (relationships) when the interpretations of a class (relationships) is a subset of the interpretation of another class (relationship) which allows to check new ISA links; *logical implication* when a (temporal) constraint is implicitly true in the current schema thus deriving new constraints. In particular, in this paper we stress both the formalization of a constructor and the set of logical implications associated to such formalization. The obtained logical implications are generally in agreement with those mentioned in the literature on temporal conceptual models. Thus, each constructor's formalization (together with its associated logical implications) can be seen as a set of precise rules on the allowed behavior of objects, in particular regarding their evolution in time. Even if we do not address specific implementation issues, these rules can be turned into explicit integrity constraints in the form of trigger rules to be added to the schema specified by the database designer, thus enabling to check the validity of user actions involving object evolution. Since the rules are the result of a formal characterization we solve what is in our opinion a serious weakness of existing modeling approaches, i.e., without a rigorous foundation there is no guarantee that the proposed model leads to a sound system.

Finally, as a byproduct of the semantic formalization, we also show how (temporal) modeling constraints can be equivalently expressed by using a subset of first-order temporal logic, i.e., the temporal description logic $\mathcal{DLR}_{\mathcal{US}}$ [5]. $\mathcal{DLR}_{\mathcal{US}}$ is a combination of the expressive and decidable description logic $\mathcal{DLR}$ (a description logic with n-ary relationships) with the linear temporal logic with temporal operators *Since* ($\mathcal{S}$) and *Until* ($\mathcal{U}$) which can be used in front of both concepts and relations. The choice of extending $\mathcal{DLR}$ is motivated by its ability to give a logical reconstruction and an extension of representational tools such as object-oriented and conceptual data models, frame-based and web ontology languages [7, 8, 9, 19]. In this paper, we use $\mathcal{DLR}_{\mathcal{US}}$ both to capture the (temporal) modeling constructors in a succinct way, and to use reasoning techniques to check satisfiability, subsumption and logical implication. We show how $\mathcal{DLR}_{\mathcal{US}}$ axioms capture the above mentioned rules associated with each constructor's formal semantics while logical implications between $\mathcal{DLR}_{\mathcal{US}}$ axioms is a way to derive new rules. Even if full $\mathcal{DLR}_{\mathcal{US}}$ is undecidable this paper addresses interesting subsets of $\mathcal{DLR}_{\mathcal{US}}$ where reasoning becomes a decidable problem.

The paper is organized as follows. Sections 2 and 4 recall the characteristics of the $\mathcal{DLR}_{\mathcal{US}}$ description logic and the $\mathcal{ER}_{VT}$ temporal data model on which we build our proposal. Section 3 shows the modeling requirements that lead us in elaborating the rigorous definition of our evolution framework. Section 5 discusses the evolution constraints we address while Section 6 provides a formal characterization for them together with a set of logical implications and the correspondent $\mathcal{DLR}_{\mathcal{US}}$ axioms. Section 7 shows that reasoning on the full-fledged temporal setting is undecidable but provides useful scenarios where reasoning becomes decidable. Section 8 concludes the paper.

2   The Temporal Description Logic

The temporal description logic $\mathcal{DLR}_{\mathcal{US}}$ [5] combines the propositional temporal logic with *Since* and *Until* and the (non-temporal) description logic $\mathcal{DLR}$ [7]. $\mathcal{DLR}_{\mathcal{US}}$ can be

$$C \rightarrow \top \mid \perp \mid CN \mid \neg C \mid C_1 \sqcap C_2 \mid \exists^{\leq k}[U_j]R \mid$$
$$\diamond^+ C \mid \diamond^- C \mid \Box^+ C \mid \Box^- C \mid \oplus C \mid \ominus C \mid C_1 \, \mathcal{U} \, C_2 \mid C_1 \, \mathcal{S} \, C_2$$
$$R \rightarrow \top_n \mid RN \mid \neg R \mid R_1 \sqcap R_2 \mid U_i/n : C \mid$$
$$\diamond^+ R \mid \diamond^- R \mid \Box^+ R \mid \Box^- R \mid \oplus R \mid \ominus R \mid R_1 \, \mathcal{U} \, R_2 \mid R_1 \, \mathcal{S} \, R_2$$

$$\top^{\mathcal{I}(t)} = \Delta^{\mathcal{I}}$$
$$\perp^{\mathcal{I}(t)} = \emptyset$$
$$CN^{\mathcal{I}(t)} \subseteq \top^{\mathcal{I}(t)}$$
$$(\neg C)^{\mathcal{I}(t)} = \top^{\mathcal{I}(t)} \setminus C^{\mathcal{I}(t)}$$
$$(C_1 \sqcap C_2)^{\mathcal{I}(t)} = C_1^{\mathcal{I}(t)} \cap C_2^{\mathcal{I}(t)}$$
$$(\exists^{\leq k}[U_j]R)^{\mathcal{I}(t)} = \{ d \in \top^{\mathcal{I}(t)} \mid \sharp\{\langle d_1, \ldots, d_n \rangle \in R^{\mathcal{I}(t)} \mid d_j = d\} \leq k \}$$
$$(C_1 \, \mathcal{U} \, C_2)^{\mathcal{I}(t)} = \{ d \in \top^{\mathcal{I}(t)} \mid \exists v > t.(d \in C_2^{\mathcal{I}(v)} \wedge \forall w \in (t, v).d \in C_1^{\mathcal{I}(w)})\}$$
$$(C_1 \, \mathcal{S} \, C_2)^{\mathcal{I}(t)} = \{ d \in \top^{\mathcal{I}(t)} \mid \exists v < t.(d \in C_2^{\mathcal{I}(v)} \wedge \forall w \in (v, t).d \in C_1^{\mathcal{I}(w)})\}$$
$$(\top_n)^{\mathcal{I}(t)} \subseteq (\Delta^{\mathcal{I}})^n$$
$$RN^{\mathcal{I}(t)} \subseteq (\top_n)^{\mathcal{I}(t)}$$
$$(\neg R)^{\mathcal{I}(t)} = (\top_n)^{\mathcal{I}(t)} \setminus R^{\mathcal{I}(t)}$$
$$(R_1 \sqcap R_2)^{\mathcal{I}(t)} = R_1^{\mathcal{I}(t)} \cap R_2^{\mathcal{I}(t)}$$
$$(U_i/n : C)^{\mathcal{I}(t)} = \{ \langle d_1, \ldots, d_n \rangle \in (\top_n)^{\mathcal{I}(t)} \mid d_i \in C^{\mathcal{I}(t)}\}$$
$$(R_1 \, \mathcal{U} \, R_2)^{\mathcal{I}(t)} = \{ \langle d_1, \ldots, d_n \rangle \in (\top_n)^{\mathcal{I}(t)} \mid$$
$$\exists v > t.(\langle d_1, \ldots, d_n \rangle \in R_2^{\mathcal{I}(v)} \wedge \forall w \in (t, v).\langle d_1, \ldots, d_n \rangle \in R_1^{\mathcal{I}(w)})\}$$
$$(R_1 \, \mathcal{S} \, R_2)^{\mathcal{I}(t)} = \{ \langle d_1, \ldots, d_n \rangle \in (\top_n)^{\mathcal{I}(t)} \mid$$
$$\exists v < t.(\langle d_1, \ldots, d_n \rangle \in R_2^{\mathcal{I}(v)} \wedge \forall w \in (v, t).\langle d_1, \ldots, d_n \rangle \in R_1^{\mathcal{I}(w)})\}$$
$$(\diamond^+ R)^{\mathcal{I}(t)} = \{\langle d_1, \ldots, d_n \rangle \in (\top_n)^{\mathcal{I}(t)} \mid \exists v > t.\langle d_1, \ldots, d_n \rangle \in R^{\mathcal{I}(v)}\}$$
$$(\oplus R)^{\mathcal{I}(t)} = \{\langle d_1, \ldots, d_n \rangle \in (\top_n)^{\mathcal{I}(t)} \mid \langle d_1, \ldots, d_n \rangle \in R^{\mathcal{I}(t+1)}\}$$
$$(\diamond^- R)^{\mathcal{I}(t)} = \{\langle d_1, \ldots, d_n \rangle \in (\top_n)^{\mathcal{I}(t)} \mid \exists v < t.\langle d_1, \ldots, d_n \rangle \in R^{\mathcal{I}(v)}\}$$
$$(\ominus R)^{\mathcal{I}(t)} = \{\langle d_1, \ldots, d_n \rangle \in (\top_n)^{\mathcal{I}(t)} \mid \langle d_1, \ldots, d_n \rangle \in R^{\mathcal{I}(t-1)}\}$$

Fig. 1. Syntax and semantics of $\mathcal{DLR}_{\mathcal{US}}$

regarded as a rather expressive fragment of the first-order temporal logic $L^{\{\text{since, until}\}}$ (cf. [10, 18]).

The basic syntactical types of $\mathcal{DLR}_{\mathcal{US}}$ are *classes* (i.e., unary predicates, also known as *concepts*) and n-ary *relations* of arity ≥ 2. Starting from a set of *atomic classes* (denoted by CN), a set of *atomic relations* (denoted by RN), and a set of *role symbols* (denoted by U) we hereinafter define inductively (complex) class and relation expressions as is shown in the upper part of Figure 1, where the binary constructors ($\sqcap, \sqcup, \mathcal{U}, \mathcal{S}$) are applied to relations of the same arity, i, j, k, n are natural numbers, $i \leq n$, and j does not exceed the arity of R.

The non-temporal fragment of $\mathcal{DLR}_{\mathcal{US}}$ coincides with $\mathcal{DLR}$. For both class and relation expressions all the Boolean constructors are available. The selection expression $U_i/n : C$ denotes an n-ary relation whose argument named U_i ($i \leq n$) is of type C; if it is clear from the context, we omit n and write ($U_i : C$). The projection expression $\exists^{\leq k}[U_j]R$ is a generalisation with cardinalities of the projection operator over the argument named U_j of the relation R; the plain classical projection is $\exists^{\geq 1}[U_j]R$. It is also possible to use the pure argument position version of the model by replacing role symbols U_i with the corresponding position numbers i. To show the expressive power of $\mathcal{DLR}_{\mathcal{US}}$ we refer to the next Sections where $\mathcal{DLR}_{\mathcal{US}}$ is used to capture various forms of temporal constraints.

The model-theoretic semantics of $\mathcal{DLR}_{\mathcal{US}}$ assumes a flow of time $\mathcal{T} = \langle \mathcal{T}_p, < \rangle$, where $\mathcal{T}_p$ is a set of time points (or chronons) and $<$ a binary precedence relation on $\mathcal{T}_p$, is assumed to be isomorphic to $\langle \mathbb{Z}, < \rangle$. The language of $\mathcal{DLR}_{\mathcal{US}}$ is interpreted in *temporal models* over $\mathcal{T}$, which are triples of the form $\mathcal{I} \doteq \langle \mathcal{T}, \Delta^{\mathcal{I}}, \cdot^{\mathcal{I}(t)} \rangle$, where $\Delta^{\mathcal{I}}$ is non-empty set of objects (the *domain* of $\mathcal{I}$) and $\cdot^{\mathcal{I}(t)}$ an *interpretation function* such that, for every $t \in \mathcal{T}$ (in the following the notation $t \in \mathcal{T}$ is used as a shortcut for $t \in \mathcal{T}_p$), every class C, and every n-ary relation R, we have $C^{\mathcal{I}(t)} \subseteq \Delta^{\mathcal{I}}$ and $R^{\mathcal{I}(t)} \subseteq (\Delta^{\mathcal{I}})^n$. The semantics of class and relation expressions is defined in the lower part of Fig. 1, where $(u, v) = \{w \in \mathcal{T} \mid u < w < v\}$. For classes, the temporal operators $\Diamond^+$ (some time in the future), $\oplus$ (at the next moment), and their past counterparts can be defined via $\mathcal{U}$ and $\mathcal{S}$: $\Diamond^+ C \equiv \top \mathcal{U} C$, $\oplus C \equiv \bot \mathcal{U} C$, etc. The operators $\Box^+$ (always in the future) and $\Box^-$ (always in the past) are the duals of $\Diamond^+$ (some time in the future) and $\Diamond^-$ (some time in the past), respectively, i.e., $\Box^+ C \equiv \neg \Diamond^+ \neg C$ and $\Box^- C \equiv \neg \Diamond^- \neg C$, for both classes and relations. The operators $\Diamond^*$ (at some moment) and its dual $\Box^*$ (at all moments) can be defined for both classes and relations as $\Diamond^* C \equiv C \sqcup \Diamond^+ C \sqcup \Diamond^- C$ and $\Box^* C \equiv C \sqcap \Box^+ C \sqcap \Box^- C$, respectively.

A *knowledge base* is a finite set Σ of $\mathcal{DLR}_{\mathcal{US}}$ axioms of the form $C_1 \sqsubseteq C_2$ and $R_1 \sqsubseteq R_2$, with R_1 and R_2 being relations of the same arity. An interpretation $\mathcal{I}$ satisfies $C_1 \sqsubseteq C_2$ ($R_1 \sqsubseteq R_2$) if and only if the interpretation of C_1 (R_1) is included in the interpretation of C_2 (R_2) at all time, i.e. $C_1^{\mathcal{I}(t)} \subseteq C_2^{\mathcal{I}(t)}$ ($R_1^{\mathcal{I}(t)} \subseteq R_2^{\mathcal{I}(t)}$), for all $t \in \mathcal{T}$. Various *reasoning services* can be defined in $\mathcal{DLR}_{\mathcal{US}}$. A knowledge base, Σ, is *satisfiable* if there is an interpretation that satisfies all the axioms in Σ (in symbols, $\mathcal{I} \models \Sigma$). A class C (or relation R) is *satisfiable* if there is $\mathcal{I}$ such that $C^{\mathcal{I}(t)} \neq \emptyset$ (respectively, $R^{\mathcal{I}(t)} \neq \emptyset$), for some time point t. A knowledge base, Σ, *logically implies* an axiom, $C_1 \sqsubseteq C_2$, and write $\Sigma \models C_1 \sqsubseteq C_2$, if we have $\mathcal{I} \models C_1 \sqsubseteq C_2$ whenever $\mathcal{I} \models \Sigma$. In this latter case, the concept C_1 is said to be *subsumed* by the concept C_2 in the knowledge base Σ. A concept C is satisfiable, given a knowledge base Σ, if there exists a model $\mathcal{I}$ of Σ such that $C^{\mathcal{I}(t)} \neq \emptyset$ for some $t \in \mathcal{T}$, i.e. $\Sigma \not\models C \sqsubseteq \bot$.

While $\mathcal{DLR}$ knowledge bases are fully able to capture atemporal EER schemas [7, 8]—i.e., given an EER schema there is an equi-satisfiable $\mathcal{DLR}$ knowledge base—in the following Sections we show how $\mathcal{DLR}_{\mathcal{US}}$ knowledge bases can capture temporal EER schemas with both timestamping and evolution constraints.

3 Modeling Requirements

This Section briefly illustrates the requirements that are frequently advocated in the literature on temporal data models.

- **Orthogonality.** Temporal constructors should be specified separately and independently for classes, relationships, and attributes. Depending on application requirements, the temporal support must be decided by the designer.

- Upward Compatibility. This term denotes the capability of preserving the nontemporal semantics of conventional (legacy) conceptual schemas when embedded into temporal schemas.
- Snapshot Reducibility. Snapshots of the database described by a temporal schema are the same as the database described by the same schema, where all temporal constructors are eliminated and the schema is interpreted atemporally. Indeed, this property specifies that we should be able to fully rebuild a temporal database by starting from the single snapshots.

Orthogonality affects mainly timestamping [25] and $\mathcal{ER}_{VT}$ already satisfies this principle by introducing temporal marks that could be used to specify the temporal behavior of classes, relationships, and attributes in an independent way.

Upward compatibility and snapshot reducibility [22] are strictly related. Considered together, they allow to preserve the meaning of atemporal constructors. In particular, the meaning of classical constructors must be preserved in such a way that a designer could either use them to model classical databases, or when used in a genuine temporal setting their meaning must be preserved at each instant of time.

These requirements are not so obvious when dealing with evolving objects. In particular, snapshot reducibility is hard to preserve when dealing with both generation and cross-time relationships where involved object may not coexist. The formalization carried out in this paper provides a data model able to respect these requirements also in presence of evolving objects.

4   The Temporal Conceptual Model $\mathcal{ER}_{VT}$

In this Section, the temporal EER model $\mathcal{ER}_{VT}$ [3, 4] is briefly introduced. $\mathcal{ER}_{VT}$ supports timestamping for classes, attributes, and relationships. $\mathcal{ER}_{VT}$ is equipped with both a linear and a graphical syntax along with a model-theoretic semantics as a temporal extension of the EER semantics [9].

An $\mathcal{ER}_{VT}$ schema is a tuple: $\Sigma = (\mathcal{L}, \text{REL}, \text{ATT}, \text{CARD}, \text{ISA}, \text{DISJ}, \text{COVER}, \text{S}, \text{T}, \text{KEY})$, such that: $\mathcal{L}$ is a finite alphabet partitioned into the sets: $\mathcal{C}$ (*class* symbols), $\mathcal{A}$ (*attribute* symbols), $\mathcal{R}$ (*relationship* symbols), $\mathcal{U}$ (*role* symbols), and $\mathcal{D}$ (*domain* symbols). ATT is a function that maps a class symbol in $\mathcal{C}$ to an $\mathcal{A}$-labeled tuple over $\mathcal{D}$, $\text{ATT}(E) = \langle A_1 : D_1, \ldots, A_h : D_h \rangle$. REL is a function that maps a relationship symbol in $\mathcal{R}$ to an $\mathcal{U}$-labeled tuple over $\mathcal{C}$, $\text{REL}(R) = \langle U_1 : C_1, \ldots, U_k : C_k \rangle$, and k is the *arity* of R. CARD is a function $\mathcal{C} \times \mathcal{R} \times \mathcal{U} \mapsto \mathbb{N} \times (\mathbb{N} \cup \{\infty\})$ denoting cardinality constraints. We denote with $\text{CMIN}(C, R, U)$ and $\text{CMAX}(C, R, U)$ the first and second component of CARD. In Figure 2, $\text{CARD}(\text{TopManager}, \text{Manages}, \text{man}) = (1, 1)$. ISA is a binary relationship $\text{ISA} \subseteq (\mathcal{C} \times \mathcal{C}) \cup (\mathcal{R} \times \mathcal{R})$. ISA between relationships is restricted to relationships with the same arity. ISA is visualized with a directed arrow, e.g. Manager ISA Employee in Figure 2. DISJ, COVER are binary relations over $2^{\mathcal{C}} \times \mathcal{C}$, describing disjointness and covering partitions, respectively. DISJ is visualized with a circled "d" and COVER with a double directed arrow, e.g. Department, InterestGroup are both disjoint and they cover OrganizationalUnit. The set $\mathcal{C}$ is partitioned into: a set $\mathcal{C}^S$ of S*napshot classes* (the S-*marked* classes in Figure 2), a set $\mathcal{C}^M$ of M*ixed classes* (the *unmarked* classes

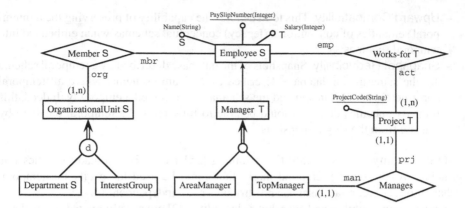

Fig. 2. The company $\mathcal{ER}_{VT}$ diagram

in Figure 2), and a set $\mathcal{C}^T$ of T*emporary classes* (the T-*marked* classes in Figure 2). A similar partition applies to the set $\mathcal{R}$. S, T are binary relations over $\mathcal{C} \times \mathcal{A}$ containing, respectively, the snapshot and temporary attributes of a class (see S, T marked attributes in Figure 2). KEY is a function that maps class symbols in $\mathcal{C}$ to their key attributes, $\text{KEY}(E) = A$. Keys are visualized as underlined attributes.

The model-theoretic semantics associated with the $\mathcal{ER}_{VT}$ modeling language adopts the *snapshot*[1] representation of abstract temporal databases and temporal conceptual models [10]. Following this paradigm, the flow of time $\mathcal{T} = \langle \mathcal{T}_p, < \rangle$, where $\mathcal{T}_p$ is a set of time points (or chronons) and $<$ is a binary precedence relation on $\mathcal{T}_p$, is assumed to be isomorphic to either $\langle \mathbb{Z}, < \rangle$ or $\langle \mathbb{N}, < \rangle$. Thus, standard relational databases can be regarded as the result of mapping a temporal database from time points in $\mathcal{T}$ to atemporal constructors, with the same interpretation of constants and the same domain.

Definition 1 ($\mathcal{ER}_{VT}$ **Semantics**). *Let Σ be an $\mathcal{ER}_{VT}$ schema. A temporal database state for the schema Σ is a tuple $\mathcal{B} = (\mathcal{T}, \Delta^{\mathcal{B}} \cup \Delta^{\mathcal{B}}_D, \cdot^{\mathcal{B}(t)})$, such that: $\Delta^{\mathcal{B}}$ is a nonempty set disjoint from $\Delta^{\mathcal{B}}_D$; $\Delta^{\mathcal{B}}_D = \bigcup_{D_i \in \mathcal{D}} \Delta^{\mathcal{B}}_{D_i}$ is the set of basic domain values used in the schema Σ; and $\cdot^{\mathcal{B}(t)}$ is a function that for each $t \in \mathcal{T}$ maps:*

- *every domain symbol D_i into a set $D_i^{\mathcal{B}(t)} = \Delta^{\mathcal{B}}_{D_i}$.*
- *Every class C to a set $C^{\mathcal{B}(t)} \subseteq \Delta^{\mathcal{B}}$.*
- *Every relationship R to a set $R^{\mathcal{B}(t)}$ of U-labeled tuples over $\Delta^{\mathcal{B}}$—i.e., let R be an n-ary relationship connecting the classes $C_1, \ldots, C_n$, $\text{REL}(R) = \langle U_1 : C_1, \ldots, U_n : C_n \rangle$, then, $r \in R^{\mathcal{B}(t)} \rightarrow (r = \langle U_1 : o_1, \ldots, U_n : o_n \rangle \wedge \forall i \in \{1, \ldots, n\}.o_i \in C_i^{\mathcal{B}(t)})$. We adopt the convention: $\langle U_1 : o_1, \ldots, U_n : o_n \rangle \equiv \langle o_1, \ldots, o_n \rangle$, when U-labels are clear from the context.*
- *Every attribute A to a set $A^{\mathcal{B}(t)} \subseteq \Delta^{\mathcal{B}} \times \Delta^{\mathcal{B}}_D$.*

$\mathcal{B}$ is said a *legal temporal database state* if it satisfies all of the constraints expressed in the schema (see [4] for full details).

---

[1] The snapshot model represents the same class of temporal databases as the *timestamp* model [21, 22] defined by adding temporal attributes to a relation [10].

Given such a set-theoretic semantics we are able to rigorously define some relevant modeling notions such as satisfiability, subsumption and derivation of new constraints by means of logical implication.

Definition 2. *Let Σ be a schema, $C \in \mathcal{C}$ a class, and $R \in \mathcal{R}$ a relationship. The following modeling notions can be defined:*

1. *C (R) is satisfiable if there exists a legal temporal database state $\mathcal{B}$ for Σ such that $C^{\mathcal{B}(t)} \neq \emptyset$ ($R^{\mathcal{B}(t)} \neq \emptyset$), for some $t \in \mathcal{T}$;*
2. *Σ is satisfiable if there exists a legal temporal database state $\mathcal{B}$ for Σ that satisfies at least one class in Σ ($\mathcal{B}$ is said a model for Σ);*
3. *C_1 (R_1) is subsumed by C_2 (R_2) in Σ if every legal temporal database state for Σ is also a legal temporal database state for C_1 ISA C_2 (R_1 ISA R_2);*
4. *A schema Σ' is logically implied by a schema Σ over the same signature if every legal temporal database state for Σ is also a legal temporal database state for Σ'.*

In the following Subsection we will show how temporal database states, $\mathcal{B}$, support defining the semantics of timestamping.

4.1 Timestamping

$\mathcal{ER}_{VT}$ is able to distinguish between *snapshot* constructors—i.e. constructors which bear no explicit specification of a given lifespan [20], which we convey by assuming a global lifespan (see Section 6.1) associated to each of their instances—*temporary* constructors—i.e. each of their instances has a limited lifespan—or *mixed* constructors—i.e. their instances can have either a global or a temporary existence. In the following, a class, relationship or attribute is called temporal if it is either temporary or mixed. The two temporal marks, S (snapshot) and T (temporary), introduced at the conceptual level, capture such temporal behavior. The semantics of timestamping can now be defined as follows (we illustrate timestamping just for classes; similar ideas are used in $\mathcal{ER}_{VT}$ to associate timestamping to both relationships and attributes):

$$o \in C^{\mathcal{B}(t)} \rightarrow \forall t' \in \mathcal{T}.o \in C^{\mathcal{B}(t')} \qquad \text{Snapshot Class}$$
$$o \in C^{\mathcal{B}(t)} \rightarrow \exists t' \neq t.o \notin C^{\mathcal{B}(t')} \qquad \text{Temporary Class}$$

The two cases are captured by the following $\mathcal{DLR}_{US}$ axioms, respectively:

$$C \sqsubseteq (\Box^+ C) \sqcap (\Box^- C) \qquad \text{Snapshot Class}$$
$$C \sqsubseteq (\Diamond^+ \neg C) \sqcup (\Diamond^- \neg C) \qquad \text{Temporary Class}$$

The distinction between snapshot, temporary and mixed constructors has been adopted in $\mathcal{ER}_{VT}$ to avoid *overloading* the meaning of un-marked constructors. Indeed, the classical distinction between temporal (using a temporal mark) and atemporal (leaving the constructor un-marked) constructors may be ambiguous in the meaning of un-marked constructors. In this classical setting, un-marking is used to model both truly atemporal constructors (i.e., snapshot classes whose instances lifespan is always equal to the whole database lifespan), as well as legacy constructors (for *upward compatibility*) where the constructor is not marked as temporal because the original data model

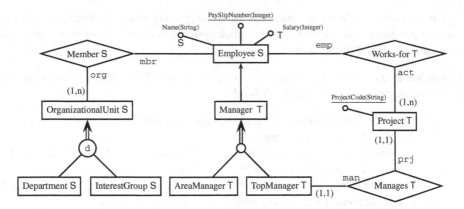

Fig. 3. The company diagram with deductions on timestamps

did not support the temporal dimension. The problem is that, due to the interaction between the various components of a temporal model, un-marked constructors can even purposely represent temporary constructors. As an example, think of an ISA involving a temporary entity (as superclass) and an un-marked entity (as a subclass). Since a designer cannot forecast all the possible interactions between the (temporal) constraints of a given conceptual schema, this ultimately means that in the classical approach *atemporality cannot be guaranteed* and this is true even for the upward compatibility.

$\mathcal{ER}_{VT}$ explicitly introduces a snapshot mark to force both atemporality and upward compatibility. As logical implication is formally defined in $\mathcal{ER}_{VT}$ (see Definition 2), missing specifications can be inferred and in particular a set of logical implications hold in the case of timestamping. For instance, in Figure 2, as `Manager` is temporary both `AreaManager` and `TopManager` are temporary, too. Because `OrganizationalUnit` is snapshot and partitioned into two sub-classes, `Department` which is snapshot and `InterestGroup`, the latter should be snapshot, too. As the temporary class `TopManager` participates in the relationships `Manages`, then the latter must be temporary, too (see [4] for an exhaustive list of deductions involving timestamps). The result of these deductions is given in Figure 3. Note that, when mapping $\mathcal{ER}_{VT}$ into a relational schema both temporary and un-marked constructors are mapped into a relation with added timestamp attributes, while snapshot constructors do not need any additional time attribute (for full details on the $\mathcal{ER}_{VT}$ relational mapping see [1]).

5   Evolution Constraints

Evolution constraints are intended to help in modeling the temporal behavior of an object. This section briefly recalls the basic concepts that have been proposed in the literature to deal with evolution, and their impact on the resulting conceptual language.

Status [25, 11] is a notion associated to temporal classes to describe the evolving status of membership of each object in the class. In a generic temporal setting, objects can be suspended and later resumed in their membership. Four different statuses can be specified, together with precise transitions between them:

- **Scheduled.** An object is scheduled if its existence within the class is known but its membership in the class will only become effective some time later. For example, a new project is approved but will not start until a later date. Each scheduled object will eventually become an active object.
- **Active.** The status of an object is active if the object is a full member of the class. For example, a currently ongoing project is an active member, at time now, of the Project class.
- **Suspended.** This status qualifies objects that exist as members of the class, but are to be seen as temporarily inactive members of the class. Being inactive means that the object cannot undergo some operations, e.g., it is not allowed to modify the values of its properties (see [11] for more details). For example, an employee taking a temporary leave of absence can be considered as a suspended employee. A suspended object was in the past an active one.
- **Disabled.** It is used to model expired objects in a class. A disabled object was in the past a member of the class. It can never again become a non-disabled member of that class (e.g., an expired project cannot be reactivated).

Transitions [16, 25] have been introduced to model the phenomenon called *object migration*. A transition records objects migrating from a *source* class to a *target* class. At the schema level, it expresses that the instances of the source class may *migrate* into the target class. Two types of transitions have been considered: *dynamic evolution*, when objects cease to be instances of the source class, and *dynamic extension*, otherwise. For example considering the company schema (Figure 3), if we want to record data about the promotion of area managers into top managers we can specify a dynamic evolution from the class AreaManager to the class TopManager. We can also record the fact that a mere employee becomes a manager by defining a dynamic extension from the class Employee to the class Manager (see Figure 5).

Generation relationships [25, 17, 24] express that (sets of) objects in a target class may be generated from (sets of) objects in a source class. While transitions involve object instances bearing the same oid, object instances linked by generation relationships necessarily bear different oids. Depending whether the source objects are preserved (as member of the source class) or disabled, we distinguish between a *production* and a *transformation*, respectively. Cardinality constraints can be added to specify the cardinality of sets involved in a generation. For example (see Figures 3,6), if we want to record the fact that (a group of) managers propose a new project a production relationship between Manager and Project can be introduced. Let us now assume that the structure of the departments of the company is dynamic, e.g., some departments may either merge or split and be replaced by others, and that it is useful to record these changes. One way would be to define a transformation relationship linking (a set of) existing departments to (a set of) new departments.

Cross-Time relationships [26, 23, 25] describe relationships between objects that do not coexist at the same time and possibly not at the time the relationship is asserted. There are many examples of these relationships (see Figure 7). Consider, for example, a relationship "biography" between an author and a famous person already dead, or the relationship "grandparent" that holds even if the grandparent passed away before the grandchild was born or the grandchild is not yet born. Still, considering the company schema (Figure 3), the relationship Works-for can be changed to cross-time whenever one wants to assign an employee to a project before its official launching, or if some employee keeps on working on a project after its official closure.

6   Formalizing Evolving Objects

The proposed formalization is based on a model-theoretic semantics and a correspondent set of axioms expressed using the temporal description logic $\mathcal{DLR}_{\mathcal{US}}$. This will give us both a formal characterization of the temporal conceptual modeling constructors, and the possibility to use the reasoning capabilities of $\mathcal{DLR}_{\mathcal{US}}$ to check satisfiability, subsumption and logical implications over temporal schemas. The model-theoretic semantics we illustrate here for the various evolution constraints is an extension of the one developed for the model $\mathcal{ER}_{VT}$, introduced in Section 4. The validity of the proposed formalization is justified by providing a set of logical implications which are in agreement with the derivations mentioned in the literature on temporal data modeling.

6.1   Status Classes

The evolution in the membership of an object to a temporal class is reflected in the changing values of the status of the object in the class. This evolution obeys some rules that give rise to a set of constraints. This Subsection formally capture these constraints.

Let C be a temporal (i.e., temporary or mixed) class. We capture status transition of membership in C by associating to C the following *status classes*: Scheduled-C, Suspended-C, Disabled-C. In particular, status classes are represented by the hierarchy of Figure 4 (where C may also be mixed) that classifies C instances according to their actual status. To preserve upward compatibility we do not explicitly introduce an active class, but assume by default that the name of the class itself denotes the set of active objects. i.e., Active-C $\equiv$ C. We can assume that the status classes are created automatically by the system each time a class is declared temporal. Thus, designers and users are not forced neither to introduce nor to manipulate status classes. They only have to be aware of the different statuses in the lifecycle of an object. Note that, since membership of objects into snapshot classes is global, i.e. objects are always active, the notion of status classes does not apply to snapshot classes.

To capture the intended meaning of status classes, we define ad-hoc constraints and then prove that such constraints capture their evolving behavior as described in the literature [25, 11]. First of all, disjointness and ISA constraints between statuses of a class C can be described as illustrated in Figure 4, where Top is supposed to be snapshot and represents the universe of discourse (i.e., Top$^{\mathcal{B}(t)} \equiv \Delta^{\mathcal{B}}$). Other than hierarchical constraints, the intended semantics of status classes induces the following rules that are related to their temporal behavior:

(EXISTS) *Existence persists until Disabled.*
    $o \in$ Exists-C$^{\mathcal{B}(t)} \rightarrow \forall t' > t.(o \in$ Exists-C$^{\mathcal{B}(t')} \vee o \in$ Disabled-C$^{\mathcal{B}(t')})$
(DISAB1) *Disabled persists.*
    $o \in$ Disabled-C$^{\mathcal{B}(t)} \rightarrow \forall t' > t.o \in$ Disabled-C$^{\mathcal{B}(t')}$
(DISAB2) *Disabled was Active in the past.*
    $o \in$ Disabled-C$^{\mathcal{B}(t)} \rightarrow \exists t' < t.o \in$ C$^{\mathcal{B}(t')}$

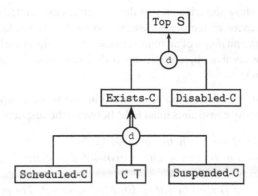

Fig. 4. Status classes

(SUSP) *Suspended was Active in the past.*
$o \in$ Suspended-$C^{\mathcal{B}(t)} \rightarrow \exists t' < t.o \in C^{\mathcal{B}(t')}$
(SCH1) *Scheduled will eventually become Active.*
$o \in$ Scheduled-$C^{\mathcal{B}(t)} \rightarrow \exists t' > t.o \in C^{\mathcal{B}(t')}$
(SCH2) *Scheduled can never follow Active.*
$o \in C^{\mathcal{B}(t)} \rightarrow \forall t' > t.o \notin$ Scheduled-$C^{\mathcal{B}(t')}$

$\mathcal{DLR}_{\mathcal{US}}$ axioms are able to fully capture the hierarchical constraints of Figure 4 (see [4] for more details). Moreover, the above semantic equations are captured by the following $\mathcal{DLR}_{\mathcal{US}}$ axioms:

(EXISTS) Exists-C $\sqsubseteq \Box^+$(Exists-C $\sqcup$ Disabled-C)
(DISAB1) Disabled-C $\sqsubseteq \Box^+$Disabled-C
(DISAB2) Disabled-C $\sqsubseteq \Diamond^-$C
(SUSP)   Suspended-C $\sqsubseteq \Diamond^-$C
(SCH1)   Scheduled-C $\sqsubseteq \Diamond^+$C
(SCH2)   C $\sqsubseteq \Box^+\neg$Scheduled-C

As a consequence of the above formalization, scheduled and disabled status classes can be true only over a single interval, while active and suspended can hold at set of intervals (i.e., an object can move many times back and forth from active to suspended status and viceversa). In particular, the following set of new rules can be derived.

Proposition 1 (Status Classes: Logical Implications). *The following logical implications hold given the above formalization of status classes:*

(SCH3) *Scheduled persists until active:* Scheduled-C $\sqsubseteq$ Scheduled-C $\mathcal{U}$ C.
    Together with axiom (SCH2), we can conclude that Scheduled-C is true just on a
    single interval.
(SCH4) *Scheduled cannot evolve directly to Disabled:* Scheduled-C $\sqsubseteq \oplus\neg$Disbled-C.
(DISAB3) *Disabled was active but it will never become active anymore:*
    Disabled-C $\sqsubseteq \Diamond^-(C \sqcap \Box^+\neg C)$.

In the following we show the adequacy of the semantics associated to status classes to describe: *a*) the behavior of temporal classes involved in ISA relationships; *b*) the notions of *lifespan, birth* and *death* of an object; *c*) the object migration between classes; *d*) the relationships that involve objects existing at different times (both generation and cross-time relationships).

Isa vs. status. When an ISA relationship is specified between two temporal classes, say *B* ISA *A*, then the following constraints must hold between the respective status classes:

(ISA1) *Objects active in B must be active in A.* $B \sqsubseteq A$

(ISA2) *Objects suspended in B must be either suspended or active in A.*
Suspended-B $\sqsubseteq$ Suspended-A $\sqcup$ A

(ISA3) *Objects disabled in B must be either disabled, suspended or active in A.*
Disabled-B $\sqsubseteq$ Disabled-A $\sqcup$ Suspended-A $\sqcup$ A

(ISA4) *Objects scheduled in B cannot be disabled in A.*
Scheduled-B $\sqsubseteq$ ¬Disabled-A

(ISA5) *Objects disabled in A, and active in B in the past, must be disabled in B.*
Disabled-A $\sqcap \Diamond^-$ B $\sqsubseteq$ Disabled-B

The formalization of status classes provided above is not sufficient to guarantee properties (ISA1-5)[2]. We need to further assume that the system behaves under the *temporal* ISA *assumption*: Each time an ISA between two temporal classes holds (B ISA A), then an ISA between the respective existence status classes (Exists-B ISA Exists-A) is automatically added by the system. Now, we are able to prove that points (ISA1-5) above are entailed by the semantics associated to status classes under the temporal ISA assumption.

Proposition 2 (Status Classes Vs. ISA: Logical Implications). *Let* A, B *be two temporal classes such that* B ISA A, *then properties (ISA1-5) are valid logical implications.*

(ISA1) Obviously true since B ISA A holds, and both A, B are considered active.

(ISA2) Let $o \in$ Suspended-B$^{\mathcal{B}(t_0)}$, since Suspended-B ISA Exists-B, and (by temporal ISA assumption) Exists-B ISA Exists-A, then, $o \in$ Exists-A$^{\mathcal{B}(t_0)}$. On the other hand, by (SUSP), $\exists t_1 < t_0.o \in B^{\mathcal{B}(t_1)}$, and then, $o \in A^{\mathcal{B}(t_1)}$. Then, by (SCH2), $o \notin$ Scheduled-A$^{\mathcal{B}(t_0)}$. Thus, due to the disjoint covering constraint between active and suspended classes, either $o \in A^{\mathcal{B}(t_0)}$ or $o \in$ Suspended-A$^{\mathcal{B}(t_0)}$.

(ISA3) Let $o \in$ Disabled-B$^{\mathcal{B}(t_0)}$, then, by (DISAB2), $\exists t_1 < t_0.o \in B^{\mathcal{B}(t_1)}$. By B ISA A and A ISA Exists-A, then, $o \in$ Exists-A$^{\mathcal{B}(t_1)}$. By (EXISTS) and the disjointness between existing and disabled classes, there are only two possibilities at point in time $t_0 > t_1$:

    1. $o \in$ Exists-A$^{\mathcal{B}(t_0)}$, and thus, by (SCH2), $o \in A^{\mathcal{B}(t_0)}$ or $o \in$ Suspended-A$^{\mathcal{B}(t_0)}$;
    or
    2. $o \in$ Disabled-A$^{\mathcal{B}(t_0)}$.

(ISA4) Let $o \in$ Scheduled-B$^{\mathcal{B}(t_0)}$, then, by (SCH1), $\exists t_1 > t_0.o \in B^{\mathcal{B}(t_1)}$, and by B ISA A, $o \in A^{\mathcal{B}(t_1)}$. Thus, by (DISAB1) and the disjointness between active and disabled states, $o \notin$ Disabled-A$^{\mathcal{B}(t_0)}$.

---

[2] We let the reader check that points 2 and 5 are not necessarily true.

(ISA5) Let $o \in$ Disabled-A$^{\mathcal{B}(t_0)}$ and $o \in B^{\mathcal{B}(t_1)}$ for some $t_1 < t_0$, then, $o \in$ Exists-B$^{\mathcal{B}(t_1)}$. By (EXISTS) and the disjointness between existing and disabled classes, there are only two possibilities at time $t_0 > t_1$: either $o \in$ Exists-B$^{\mathcal{B}(t_0)}$ or $o \in$ Disabled-B$^{\mathcal{B}(t_0)}$. By absurd, let $o \in$ Exists-B$^{\mathcal{B}(t_0)}$, then by temporal ISA assumption, $o \in$ Exists-A$^{\mathcal{B}(t_0)}$, which contradicts the assumption that $o \in$ Disabled-A$^{\mathcal{B}(t_0)}$.

Lifespan. Here we define the lifespan of objects belonging to a temporal class, together with other related notions. In particular, we define EXISTENCE$_C$, LIFESPAN$_C$, ACTIVE$_C$, BEGIN$_C$, BIRTH$_C$ and DEATH$_C$ as functions depending on the object membership to the status classes associated to a temporal class C.

The *existence time* of an object describes the temporal instants where the object is either a scheduled, active or suspended member of a given class. More formally, EXISTENCESPAN$_C : \Delta^{\mathcal{B}} \to 2^{\mathcal{T}}$, such that:

$$\text{EXISTENCESPAN}_C(o) = \{t \in \mathcal{T} \mid o \in \text{Exists-}C^{\mathcal{B}(t)}\}$$

The *lifespan* of an object describes the temporal instants where the object is an active or suspended member of a given class (thus, LIFESPAN$_C(o) \subseteq$ EXISTENCESPAN$_C(o)$). More formally, LIFESPAN$_C : \Delta^{\mathcal{B}} \to 2^{\mathcal{T}}$, such that:

$$\text{LIFESPAN}_C(o) = \{t \in \mathcal{T} \mid o \in C^{\mathcal{B}(t)} \cup \text{Suspended-}C^{\mathcal{B}(t)}\}$$

The *activespan* of an object describes the temporal instants where the object is an active member of a given class (thus, ACTIVESPAN$_C(o) \subseteq$ LIFESPAN$_C(o)$). More formally, ACTIVESPAN$_C : \Delta^{\mathcal{B}} \to 2^{\mathcal{T}}$, such that:

$$\text{ACTIVESPAN}_C(o) = \{t \in \mathcal{T} \mid o \in C^{\mathcal{B}(t)}\}$$

The functions BEGIN$_C$ and DEATH$_C$ associate to an object the first and the last appearance, respectively, of the object as a member of a given class, while BIRTH$_C$ denotes the first appearance as an active object of that class. More formally, BEGIN$_C$, BIRTH$_C$, DEATH$_C : \Delta^{\mathcal{B}} \to \mathcal{T}$, such that:

$$\text{BEGIN}_C(o) = \min(\text{EXISTENCESPAN}_C(o))$$
$$\text{BIRTH}_C(o) = \min(\text{ACTIVESPAN}_C(o)) \equiv \min(\text{LIFESPAN}_C(o))$$
$$\text{DEATH}_C(o) = \max(\text{LIFESPAN}_C(o))$$

We could still speak of existencespan, lifespan or activespan for snapshot classes, but in this case EXISTENCESPAN$_C(o) \equiv$ LIFESPAN$_C(o) \equiv$ ACTIVESPAN$_C(o) \equiv \mathcal{T}$.

6.2 Transition

Dynamic transitions between classes model the notion of object migration from a source to a target class. Two notions of dynamic transitions between classes are considered in the literature [25, 16]: *dynamic evolution*, when an object ceases to be an instance of a source class, and *dynamic extension*, when an object is still allowed to belong to the source. Concerning the graphical representation, as illustrated in Figure 5, we use a dashed arrow pointing to the target class and labeled with either DEX or DEV denoting dynamic extension and evolution, respectively.

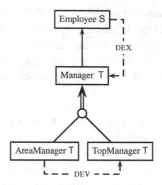

Fig. 5. Transitions employee-to-manager and area-to-top manager

In a temporal setting, objects can obviously change their membership class. Specifying a transition between two classes means that: *a.* We want to keep track of such migration; *b.* Not necessarily all the objects in the source or in the target participate in the migration; *c.* When the source class is a temporal class, migration involves only objects active or suspended. Thus, neither disabled nor scheduled objects can take part in a transition.

In the following, we present a formalization that satisfies the above requirements. Formalizing dynamic transitions as relationships would result in binary relationships linking the same object that migrates from the source to the target class. Rather than defining a relationship type with an equality constraint on the identity of the linked objects, we represent transitions by introducing a new class denoted by either DEX_{C_1,C_2} or DEV_{C_1,C_2} for dynamic extension and evolution, respectively. More formally, in case of a *dynamic extension* between classes C_1, C_2 the following semantic equation holds:

$$o \in \text{DEX}_{C_1,C_2}^{\mathcal{B}(t)} \rightarrow (o \in (\text{Suspended-C}_1{}^{\mathcal{B}(t)} \cup \text{C}_1{}^{\mathcal{B}(t)}) \wedge o \notin \text{C}_2{}^{\mathcal{B}(t)} \wedge o \in C_2^{\mathcal{B}(t+1)})$$

And the equivalent $\mathcal{DLR}_{\mathcal{US}}$ axiom is:

(DEX) $\text{DEX}_{C_1,C_2} \sqsubseteq (\text{Suspended-C}_1 \sqcup \text{C}_1) \sqcap \neg\text{C}_2 \sqcap \oplus C_2$

In case of a *dynamic evolution* between classes C_1, C_2 the source object cannot remain active in the source class. Thus, the following semantic equation holds:

$$o \in \text{DEV}_{C_1,C_2}^{\mathcal{B}(t)} \rightarrow (o \in (\text{Suspended-C}_1{}^{\mathcal{B}(t)} \cup \text{C}_1{}^{\mathcal{B}(t)}) \wedge o \notin \text{C}_2{}^{\mathcal{B}(t)} \wedge$$
$$o \in C_2^{\mathcal{B}(t+1)} \wedge o \notin C_1^{\mathcal{B}(t+1)})$$

And the equivalent $\mathcal{DLR}_{\mathcal{US}}$ axiom is:

(DEV) $\text{DEV}_{C_1,C_2} \sqsubseteq (\text{Suspended-C}_1 \sqcup \text{C}_1) \sqcap \neg\text{C}_2 \sqcap \oplus (C_2 \sqcap \neg C_1)$

Please note that, in case C_1 is a snapshot class, then, Exists-C$_1 \equiv C_1$. Finally, we formalize the case where the source (C_1) and/or the target (C_2) totally participate in a dynamic extension (at schema level we add mandatory cardinality constraints on DEX/DEV links):

$$o \in C_1^{\mathcal{B}(t)} \rightarrow \exists t' > t. o \in \mathrm{DEX}_{C_1,C_2}^{\mathcal{B}(t')} \quad \text{Source Total Transition}$$
$$o \in C_2^{\mathcal{B}(t)} \rightarrow \exists t' < t. o \in \mathrm{DEX}_{C_1,C_2}^{\mathcal{B}(t')} \quad \text{Target Total Transition}$$

The above cases are captured by the following $\mathcal{DLR}_{\mathcal{US}}$ axioms, respectively:

(STT) $C_1 \sqsubseteq \Diamond^+\mathrm{DEX}_{C_1,C_2}$ Source Total Transition
(TTT) $C_2 \sqsubseteq \Diamond^-\mathrm{DEX}_{C_1,C_2}$ Target Total Transition

In a similar way we deal with dynamic evolution constraints.

Proposition 3 (Transition: Logical Implications). *The following logical implications hold as a consequence of the transition semantics:*

1. *The classes* DEX_{C_1,C_2} *and* DEV_{C_1,C_2} *are temporary classes; actually, they hold at single time points.*
   $\mathrm{DEX}_{C_1,C_2} \sqsubseteq \oplus \neg \, \mathrm{DEX}_{C_1,C_2} \sqcap \ominus \neg\mathrm{DEX}_{C_1,C_2}$ *(similar for* DEV_{C_1,C_2}*)*
   Indeed, let $o \in \mathrm{DEX}_{C_1,C_2}^{\mathcal{B}(t)}$, *then* $o \notin C_2^{\mathcal{B}(t)}$ *and* $o \in C_2^{\mathcal{B}(t+1)}$, *thus* $o \notin \mathrm{DEX}_{C_1,C_2}^{\mathcal{B}(t+1)}$
   and $o \notin \mathrm{DEX}_{C_1,C_2}^{\mathcal{B}(t-1)}$. *Note that, the time t such that* $o \in \mathrm{DEX}_{C_1,C_2}^{\mathcal{B}(t)}$ *records when the transition event happens. Similar considerations apply for* DEV_{C_1,C_2}.
2. *Objects in the classes* DEX_{C_1,C_2} *and* DEV_{C_1,C_2} *cannot be disabled as* C_2.
   $\mathrm{DEX}_{C_1,C_2} \sqsubseteq \neg\text{Disabled-}C_2$ *(similar for* DEV_{C_1,C_2}*)*
   Indeed, since $\mathrm{DEX}_{C_1,C_2} \sqsubseteq \oplus C_2$, *i.e. objects in* DEX_{C_1,C_2} *are active in* C_2 *starting from the next point in time, then by property* (DISAB3), $\mathrm{DEX}_{C_1,C_2} \sqsubseteq \neg\text{Disabled-}C_2$. *The same holds for* DEV_{C_1,C_2}.
3. *The target class* C_2 *cannot be snapshot (it becomes temporary in case of* TTT *constraints).*
   $\mathrm{DEX}_{C_1,C_2} \sqsubseteq \Diamond^*[C_2 \sqcap (\Diamond^+\neg C_2 \sqcup \Diamond^-\neg C_2)]$
   Indeed, from (DEX), $\mathrm{DEX}_{C_1,C_2} \sqsubseteq \neg C_2 \sqcap \oplus C_2$ *(the same holds for* DEV_{C_1,C_2}*).*
4. *As a consequence of dynamic evolution, the source class,* C_1, *cannot be snapshot (and it becomes temporary in case of* STT *constraints).*
   $\mathrm{DEV}_{C_1,C_2} \sqsubseteq \Diamond^*[C_1 \sqcap (\Diamond^+\neg C_1 \sqcup \Diamond^-\neg C_1)]$
   Indeed, an object evolving from C_1 *to* C_2 *ceases to be a member of* C_1.
5. *Dynamic evolution cannot involve a class and one of its sub-classes.*
   $C_2 \sqsubseteq C_1 \models \mathrm{DEV}_{C_1,C_2} \sqsubseteq \bot$
   Indeed, from (DEV), $\mathrm{DEV}_{C_1,C_2} \sqsubseteq \oplus (C_2 \sqcap \neg C_1)$ *which contradicts* $C_2 \sqsubseteq C_1$.
6. *Dynamic extension between disjoint classes logically implies Dynamic evolution.*
   $\{\mathrm{DEX}_{C_1,C_2}, C_1 \sqsubseteq \neg C_2\} \models \mathrm{DEV}_{C_1,C_2}$

6.3 Generation Relationships

Generation relationships [25, 17] represent processes that lead to the emergence of *new instances* starting from a set of instances. Two distinct generation relationships have been introduced: *production*, when the source objects survive the generation process; *transformation*, when all the instances involved in the process are consumed. At the conceptual level we introduce two marks associated to a relationship: GP for production and GT for transformation relationships, and an arrow points to the target class (see Figure 6).

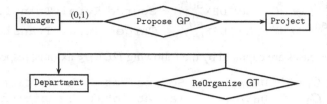

Fig. 6. Production and transformation generation relationships

We model generation as binary relationships connecting a source class to a target one: $\text{REL}(R) = \langle \texttt{source}: C_1, \texttt{target}: \texttt{Scheduled-}C_2 \rangle$. The semantics of *production relationships*, R, is described by the following equation:

$$\langle o_1, o_2 \rangle \in R^{\mathcal{B}(t)} \rightarrow (o_1 \in C_1^{\mathcal{B}(t)} \wedge o_2 \in \texttt{Scheduled-}C_2^{\mathcal{B}(t)} \wedge o_2 \in C_2^{\mathcal{B}(t+1)})$$

Thus, objects active in the source class produce objects active in the target class at the next point in time. Notice that, the use of status classes allow us to preserve snapshot reducibility. Indeed, for each pair of objects, $\langle o_1, o_2 \rangle$, belonging to a generation relationships o_1 is active in the source while o_2 is scheduled in the target. The $\mathcal{DLR}_{\mathcal{US}}$ axiom capturing the production semantics is:

(PROD) $R \sqsubseteq \texttt{source}: C_1 \sqcap \texttt{target}: (\texttt{Scheduled-}C_2 \sqcap \oplus C_2)$

The case of *transformation* is captured by the following semantic equation:

$$\langle o_1, o_2 \rangle \in R^{\mathcal{B}(t)} \rightarrow (o_1 \in C_1^{\mathcal{B}(t)} \wedge o_1 \in \texttt{Disabled-}C_1^{\mathcal{B}(t+1)} \wedge$$
$$o_2 \in \texttt{Scheduled-}C_2^{\mathcal{B}(t)} \wedge o_2 \in C_2^{\mathcal{B}(t+1)})$$

Thus, objects active in the source generate objects active in the target at the next point in time while the source objects cease to exist as member of the source. The $\mathcal{DLR}_{\mathcal{US}}$ axiom capturing the transformation semantics is:

(TRANS) $R \sqsubseteq \texttt{source}: (C_1 \sqcap \oplus \texttt{Disabled-}C_1) \sqcap \texttt{target}: (\texttt{Scheduled-}C_2 \sqcap \oplus C_2)$

Proposition 4 (Generation: Logical Implications). *The following logical implications hold as a consequence of the generation semantics:*

1. *A generation relationship, R, is temporary; actually, it is instantaneous.*
   $R \sqsubseteq \square^+ \neg R \sqcap \square^- \neg R$
   Indeed, let $\langle o_1, o_2 \rangle \in R^{\mathcal{B}(t)}$, then, since $o_2 \notin \texttt{Scheduled-}C_2^{\mathcal{B}(t+1)}$, then $\langle o_1, o_2 \rangle \notin R^{\mathcal{B}(t+1)}$. Since, $o_2 \notin C_2^{\mathcal{B}(t)}$, then $\langle o_1, o_2 \rangle \notin R^{\mathcal{B}(t-1)}$.
2. *The target class, C_2, cannot be snapshot (it becomes temporary if total participation is specified).*
   $R \sqsubseteq \texttt{target}: \lozenge^* [C_2 \sqcap (\lozenge^+ \neg C_2 \sqcup \lozenge^- \neg C_2)]$
   Indeed, let $\langle o_1, o_2 \rangle \in R^{\mathcal{B}(t)}$, then, $o_2 \notin C_2^{\mathcal{B}(t)}$ and $o_2 \in C_2^{\mathcal{B}(t+1)}$.
3. *The target class, C_2, cannot be disabled.*
   $R \sqsubseteq \texttt{target}: \neg \texttt{Disabled-}C_2$
   Indeed, let $\langle o_1, o_2 \rangle \in R^{\mathcal{B}(t)}$, then, $o_2 \in C_2^{\mathcal{B}(t+1)}$. Thus $o_2 \notin \texttt{Disabled-}C_2^{\mathcal{B}(t)}$.

4. *If R is a transformation relationship, then, C_1 cannot be snapshot (it becomes temporary if total participation is specified).*
$$R \sqsubseteq \text{source}: \Diamond^*[C_1 \sqcap (\Diamond^+ \neg C_1 \sqcup \Diamond^- \neg C_1)]$$
Indeed, C_1 will be disabled at the next point in time.

Note that, the Department class which is both the source and target of a transformation relationship (Figure 6) cannot longer be snapshot (as was in Figure 3) and it must be changed to temporary (as a consequence of this new timestamp, InterestGroup is a genuine mixed class).

6.4   Cross-Time Relationships

Cross-time relationships relate objects that are members of the participating classes at different times. The conceptual model MADS [25] allows for *synchronization* relationships to specify temporal constraints (Allen temporal relations) between the lifespan of linked objects. *Historical marks* are used in the ERT model [23] to express a relationship between objects not existing at the same time (both past and future historical marks are introduced).

This Section formalizes cross-time relationships with the aim of preserving the snapshot reducibility of the resulting model. Let us consider a concrete example. Let "biography" be a cross-time relationship linking the author of a biography with a famous person no more in existence. Snapshot reducibility says that if there is an instance (say, bio = ⟨Tulard, Napoleon⟩) of the Biography relationship at time t_0 (in particular, Tulard wrote a bio on Napoleon in 1984), then, the projection of Biography at time t_0 (1984 in our example) must contain the pair ⟨Tulard, Napoleon⟩. Now, while Tulard is a member of the class Author in 1984, we cannot say that Napoleon is member of the class Person in 1984. Our formalization of cross-time relationships proposes the use of status classes to preserve snapshot reducibility. The biography example can be solved by asserting that Napoleon is a member of the Disabled-Person class in 1984.

At the conceptual level, we mark with P,=,F (standing for Past, Now and Future, respectively) the links of cross-time relationships. Furthermore, we allow for the compound marks ⟨P,=⟩, ⟨F,=⟩ and ⟨P,=,F⟩, while just specifying = doesn't add any constraint (see Figure 7). Assuming that R is a cross-time relationship between classes C_1, C_2, then, the semantics of marking the C_1 link is:

$$\langle o_1, o_2 \rangle \in R^{\mathcal{B}(t)} \rightarrow o_1 \in \text{Disabled-C}_1{}^{\mathcal{B}(t)} \qquad \text{Strictly Past } \langle \text{P} \rangle$$
$$\langle o_1, o_2 \rangle \in R^{\mathcal{B}(t)} \rightarrow o_1 \in (C_1 \sqcup \text{Disabled-C}_1)^{\mathcal{B}(t)} \qquad \text{Past } \langle \text{P,=} \rangle$$
$$\langle o_1, o_2 \rangle \in R^{\mathcal{B}(t)} \rightarrow o_1 \in \text{Scheduled-C}_1{}^{\mathcal{B}(t)} \qquad \text{Strictly Future } \langle \text{F} \rangle$$
$$\langle o_1, o_2 \rangle \in R^{\mathcal{B}(t)} \rightarrow o_1 \in (C_1 \sqcup \text{Scheduled-C}_1)^{\mathcal{B}(t)} \qquad \text{Future } \langle \text{F,=} \rangle$$
$$\langle o_1, o_2 \rangle \in R^{\mathcal{B}(t)} \rightarrow o_1 \in (C_1 \sqcup \text{Scheduled-C}_1 \sqcup \text{Disabled-C}_1)^{\mathcal{B}(t)} \quad \text{Full-Cross } \langle \text{P,=,F} \rangle$$

The corresponding $\mathcal{DLR}_{\mathcal{US}}$ axioms are:

$$R \sqsubseteq U_1 : \text{Disabled-C}_1 \qquad \text{Strictly Past } \langle \text{P} \rangle$$
$$R \sqsubseteq U_1 : (C_1 \sqcup \text{Disabled-C}_1) \qquad \text{Past } \langle \text{P,=} \rangle$$
$$R \sqsubseteq U_1 : \text{Scheduled-C}_1 \sqcap \qquad \text{Strictly Future } \langle \text{F} \rangle$$
$$R \sqsubseteq U_1 : (C_1 \sqcup \text{Scheduled-C}_1) \qquad \text{Future } \langle \text{F,=} \rangle$$
$$R \sqsubseteq U_1 : (C_1 \sqcup \text{Scheduled-C}_1 \sqcup \text{Disabled-C}_1) \qquad \text{Full-Cross } \langle \text{P,=,F} \rangle$$

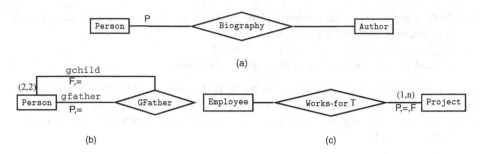

Fig. 7. Cross-Time Relationships

Proposition 5 (Cross-Time: Logical Implications). *The following logical implications hold as a consequence of the cross-time semantics (apart from point 1., we assume that C_1 (C_2) participates as either strict past or strict future):*

1. *If a relationship, R, is snapshot then historical marks reduce to the = mark (i.e., R is not a genuine cross-time relationships).*
   See next point.
2. *A cross-time relationship, R, is temporary ($R \sqsubseteq \Diamond^+ \neg R \sqcup \Diamond^- \neg R$).*
   Let assume that C_1 participates as strict past. Thus, if $\langle o_1, o_2 \rangle \in R^{\mathcal{B}(t)}$, then $o_1 \in$ Disabled-$C_1^{\mathcal{B}(t)}$ *and, by (DISAB2), $\exists t_1 < t$ s.t. $o_1 \in C_1^{\mathcal{B}(t_1)}$. Then $\langle o_1, o_2 \rangle \notin R^{\mathcal{B}(t_1)}$.*
3. *C_1 (C_2) cannot be snapshot (is temporary if total participation is specified).*
   $R \sqsubseteq U_1 : \Diamond^* [C_1 \sqcap (\Diamond^+ \neg C_1 \sqcup \Diamond^- \neg C_1)]$
   Let assume that C_1 participates as strict past. Thus, if $\langle o_1, o_2 \rangle \in R^{\mathcal{B}(t)}$, then, $o_1 \in$ Disabled-$C_1^{\mathcal{B}(t)}$. *Then, $o_1 \notin C_1^{\mathcal{B}(t)}$ while, by (DISAB2), $\exists t_1 < t$ s.t. $o_1 \in C_1^{\mathcal{B}(t_1)}$.*

7   Complexity of Reasoning on Temporal Models

As this paper shows, the temporal description logic $\mathcal{DLR}_{\mathcal{US}}$ is able to fully capture temporal schemas with both timestamping and evolution constraints. Reasoning over $\mathcal{DLR}_{\mathcal{US}}$ knowledge bases, i.e., checking satisfiability, subsumption and logical implications, turns out to be undecidable [5]. The main reason for this is the possibility to postulate that a binary relation does not vary in time. Note that, showing that temporal schemas can be mapped into $\mathcal{DLR}_{\mathcal{US}}$ axioms does not necessarily imply that reasoning over temporal schemas is an undecidable problem. Unfortunately, [2] shows that the undecidable Halting Problem can be encoded as the problem of class satisfiability w.r.t. a temporal schema with both timestamping and evolution constraints.

On the other hand, the fragment, $\mathcal{DLR}_{\mathcal{US}}^-$, of $\mathcal{DLR}_{\mathcal{US}}$ deprived of the ability to talk about temporal persistence of n-ary relations, for $n \geq 2$, is decidable. Indeed, reasoning in $\mathcal{DLR}_{\mathcal{US}}^-$ is an EXPTIME-complete problem [5]. This result gives us an useful scenario where reasoning over temporal schemas becomes decidable. In particular, if we forbid timestamping for relationships (i.e., relationships are just unmarked) reasoning on temporal models with both concept timestamping and full evolution constraints can be reduced to reasoning over $\mathcal{DLR}_{\mathcal{US}}^-$. The problem of reasoning in this setting is complete for EXPTIME since the EXPTIME-complete problem of reasoning with $\mathcal{ALC}$ knowledge bases can be captured by such schemas [6].

It is an open problem whether reasoning is still decidable by regaining timestamping for relationships (and maintaining timestamping for classes) but dropping evolution constraints. We have a strong feeling that this represents a decidable scenario since it is possible to encode temporal schemas without evolution constraints by using a combination between the description logic $\mathcal{DLR}$ and the epistemic modal logic S5. Decidability results have been proved for the sub-logic $\mathcal{ALC}_{S5}$ [13]. But, it is an open problem whether this result still holds for the more complex logic $\mathcal{DLR}_{S5}$.

8 Conclusions

In this paper we proposed a formalization of the various modeling constructors that support the design of temporal DBMS with particular attention to evolution constraints. The formalization, based on a model-theoretic semantics, has been developed with the aim to preserve three fundamental modeling requirements: Orthogonality, Upward Compatibility and Snapshot Reducibility. The introduction of status classes, which describe the evolution in the membership of an object to a temporal class, allowed us to maintain snapshot reducibility when characterizing both generations and cross-time relationships. The formal semantics clarified the meaning of the language's constructors but it also gave a rigorous definition to relevant modeling notions like: satisfiability of schemas, classes and relationships; subsumption for both classes and relationships; logical implication. Furthermore, for each constructor we presented its formalization together with the set of logical implications associated to such formalization.

Finally, we have been able to show how temporal schemas can be equivalently expressed using a subset of first-order temporal logic, i.e., $\mathcal{DLR}_{\mathcal{US}}$, the description logic $\mathcal{DLR}$ extended with the temporal operators *Since* and *Until*. Overall, we obtained a temporal conceptual model that preserves well established modeling requirements, equipped with a model-theoretic semantics where each constructor can be seen as a set of precise rules, and with the possibility to perform automated reasoning by mapping temporal schemas (without timestamping on relationships) into temporal description logic knowledge bases.

References

1. B. Ahmad. Modeling bi-temporal databases. Master's thesis, UMIST Department of Computation, UK, 2003.
2. A. Artale. Reasoning on temporal conceptual schemas with dynamic constraints. In *11th Int. Symposium on Temporal Representation and Reasoning (TIME04)*. IEEE Computer Society, 2004. Also in Proc. of the 2004 Int. Workshop on Description Logics (DL'04).
3. A. Artale and E. Franconi. Temporal ER modeling with description logics. In *Proc. of the Int. Conf. on Conceptual Modeling (ER'99)*. Springer-Verlag, November 1999.
4. A. Artale, E. Franconi, and F. Mandreoli. Description logics for modelling dynamic information. In Jan Chomicki, Ron van der Meyden, and Gunter Saake, editors, *Logics for Emerging Applications of Databases*. Lecture Notes in Computer Science, Springer-Verlag, 2003.
5. A. Artale, E. Franconi, F. Wolter, and M. Zakharyaschev. A temporal description logic for reasoning about conceptual schemas and queries. In S. Flesca, S. Greco, N. Leone, and G. Ianni, editors, *Proceedings of the 8th Joint European Conference on Logics in Artificial Intelligence (JELIA-02)*, volume 2424 of *LNAI*, pages 98–110. Springer, 2002.

6. Daniela Berardi, Diego Calvanese, and Giuseppe De Giacomo. Reasoning on UML class diagrams. *Artificial Intelligence*, 168(1–2):70–118, 2005.
7. D. Calvanese, G. De Giacomo, and M. Lenzerini. On the decidability of query containment under constraints. In *Proc. of the 17th ACM SIGACT SIGMOD SIGART Sym. on Principles of Database Systems (PODS'98)*, pages 149–158, 1998.
8. D. Calvanese, M. Lenzerini, and D. Nardi. Description logics for conceptual data modeling. In J. Chomicki and G. Saake, editors, *Logics for Databases and Information Systems*. Kluwer, 1998.
9. D. Calvanese, M. Lenzerini, and D. Nardi. Unifying class-based representation formalisms. *J. of Artificial Intelligence Research*, 11:199–240, 1999.
10. J. Chomicki and D. Toman. Temporal logic in information systems. In J. Chomicki and G. Saake, editors, *Logics for Databases and Information Systems*, chapter 1. Kluwer, 1998.
11. O. Etzion, A. Gal, and A. Segev. Extended update functionality in temporal databases. In O. Etzion, S. Jajodia, and S. Sripada, editors, *Temporal Databases - Research and Practice*, Lecture Notes in Computer Science, pages 56–95. Springer-Verlag, 1998.
12. M. Finger and P. McBrien. Temporal conceptual-level databases. In D. Gabbay, M. Reynolds, and M. Finger, editors, *Temporal Logics – Mathematical Foundations and Computational Aspects*, pages 409–435. Oxford University Press, 2000.
13. D. Gabbay, A.Kurucz, F. Wolter, and M. Zakharyaschev. *Many-dimensional modal logics: theory and applications*. Studies in Logic. Elsevier, 2003.
14. H. Gregersen and J.S. Jensen. Conceptual modeling of time-varying information. Technical Report TimeCenter TR-35, Aalborg University, Denmark, 1998.
15. H. Gregersen and J.S. Jensen. Temporal Entity-Relationship models – a survey. *IEEE Transactions on Knowledge and Data Engineering*, 11(3):464–497, 1999.
16. R. Gupta and G. Hall. Modeling transition. In *Proc. of ICDE'91*, pages 540–549, 1991.
17. R. Gupta and G. Hall. An abstraction mechanism for modeling generation. In *Proc. of ICDE'92*, pages 650–658, 1992.
18. I. Hodkinson, F. Wolter, and M. Zakharyaschev. Decidable fragments of first-order temporal logics. *Annals of Pure and Applied Logic*, 106:85–134, 2000.
19. I. Horrocks, P.F. Patel-Schneider, and F. van Harmelen. From SHIQ and RDF to OWL: The making of a web ontology language. *Journal of Web Semantics*, 1(1):7–26, 2003.
20. C. S. Jensen, J. Clifford, S. K. Gadia, P. Hayes, and S. Jajodia et al. The Consensus Glossary of Temporal Database Concepts. In O. Etzion, S. Jajodia, and S. Sripada, editors, *Temporal Databases - Research and Practice*, pages 367–405. Springer-Verlag, 1998.
21. C. S. Jensen and R. T. Snodgrass. Temporal data management. *IEEE Transactions on Knowledge and Data Engineering*, 111(1):36–44, 1999.
22. C. S. Jensen, M. Soo, and R. T. Snodgrass. Unifying temporal data models via a conceptual model. *Information Systems*, 9(7):513–547, 1994.
23. P. McBrien, A.H. Seltveit, and B. Wangler. An Entity-Relationship model extended to describe historical information. In *Proc. of CISMOD'92*, pages 244–260, Bangalore, India, 1992.
24. C. Parent, S. Spaccapietra, and E. Zimanyi. The MurMur project: Modeling and querying multi-representation spatio-temporal databases. *Information Systems*, 2005.
25. S. Spaccapietra, C. Parent, and E. Zimanyi. Modeling time from a conceptual perspective. In *Int. Conf. on Information and Knowledge Management (CIKM98)*, 1998.
26. C. Theodoulidis, P. Loucopoulos, and B. Wangler. A conceptual modelling formalism for temporal database applications. *Information Systems*, 16(3):401–416, 1991.
27. Wikipedia. Wikipedia, the free encyclopedia. Temporal Databases. see http://en.wikipedia.org/wiki/Temporal_database.

Controlled Query Evaluation with Open Queries for a Decidable Relational Submodel

Joachim Biskup[1] and Piero Bonatti[2]

[1] Fachbereich Informatik, Universität Dortmund, D-44221 Dortmund, Germany
biskup@ls6.informatik.uni-dortmund.de
[2] Sezione di Informatica, Università di Napoli "Frederico II", I-80126 Napoli, Italy
bonatti@na.infn.it

Abstract. Controlled query evaluation for logic-oriented information systems provides a model for the dynamic enforcement of confidentiality policies even if users are able to reason about a priori knowledge and the answers to previous queries. Previous foundational work simply assumes that the control mechanism can solve the arising entailment problems (no matter how complex they may be), and deals only with closed queries. In this paper, we overcome these limitations by refining the abstract model for appropriately represented relational databases. We identify a relational submodel where all instances share a fixed infinite Herbrand domain but have finite base relations, and we require finite and domain-independent query results. Then, via suitable syntactic restrictions on the policy and query languages, each entailment problem occurring in the framework can be equivalently expressed as a universal validity problem within the Bernays-Schönfinkel class, whose (known) decidability in the classical setting is extended to our framework. For both refusal and lying, we design and verify evaluation methods for open queries, exploiting controlled query evaluation of appropriate sequences of closed queries, which include answer completeness tests.

Keywords: Controlled query evaluation, Confidentiality, Refusal, Lying, Complete information system, Relational database, Open query, First-order logic, Safe query, Domain-independent query, Implication problem, Finite model theory, Bernays-Schönfinkel class, Guarded fragment, Completeness test.

1 Introduction and Survey

Controlled query evaluation for logic-oriented information systems provides a model for the dynamic enforcement of confidentiality policies even if users are able to reason about a priori knowledge and the answers to previous queries. Known foundational work [23, 10, 4, 5, 6, 7] has been based on a simple but powerful model-theoretic approach to information systems. This approach considers an instance of an information system as a interpretation in the sense of logic, a query as a sentence, and the ordinary answer as the truth value of the query w.r.t. the instance. Thus, in practical terms, only *closed queries* (with yes/no-answers) for complete information systems (which could always return the correct answer) are investigated. However, many practical information systems, like relational database systems, allow *open queries* as well. The control mechanism maintains a user log containing the sentences that represent the a priori

J. Dix and S.J. Hegner (Eds.): FoIKS 2006, LNCS 3861, pp. 43–62, 2006.

knowledge of each user (or user class) and the answers to previous queries. Whenever a new query is submitted, the control mechanism has to solve one or more *implication problems* where the sentences of the user log, some sentences related to the query, and some sentences specified in a confidentiality policy are involved. However, the implication problem is *decidable* only for appropriately restricted logics.

This paper exhibits a refinement of controlled query evaluation where open queries are handled in addition to closed queries, and where the implication problems are decidable. This refinement is based on the relational database model under the assumption that all instances share a fixed infinite Herbrand domain (an abstraction of standard data types). This assumption makes the treatment of negative information subtle, since we have to deal with implicit infinite complements of finite, explicitly represented positive information. Accordingly, we can only reconstruct a relational submodel by using a fragment of first order logic. It turns out, that the well-known Bernays-Schönfinkel class provides an appropriate fragment for all our goals. In the rest of the introduction we will further explain the problem and our proposed solution by considering some examples and explaining the basic results on controlled query evaluation.

1.1   Queries

Assume that we have a binary relation scheme *HasDisease*, say with attributes *Person* and *Disease*, and that *Pete, Tom, Lisa, . . .* and *cancer, brokenarm, cough, aids. . .* are constants. Then for instance a user can ask whether "Person *Tom* is suffering from the disease *brokenarm*" by submitting the sentence *HasDisease*(*Tom, brokenarm*) as a closed query. Subsequently he could additionally inquire whether "Person *Pete* is suffering from any of the diseases *cancer* or *aids*" by evaluating the sentence *HasDisease*(*Pete, cancer*) or *HasDisease*(*Pete, aids*). Afterwards the user might ask for more precise information submitting the sentence *HasDisease*(*Pete, cancer*). However, if the user is interested in learning "Which persons x are suffering from the disease *cancer*", then he would like to use the formula *HasDisease*(x, *cancer*) which is an open query with a free (occurence of the) variable x.

1.2   Potential Secrets, Lies, and Refusals

Within the model of controlled query evaluation, confidentiality requirements have been fully formalized, and effective methods for controlling sequences of closed queries have been developed that provably preserve confidentiality. These methods distort correct answers, if necessary, either uniformly by using only *lies* or only *refusals*, or by a *combination* thereof, following in each case a "last minute distortion" strategy.

In terms of our example, suppose that a security administrator wants to ensure that the information system never reveals that "Person *Pete* suffers from the disease *cancer*" nor that "Person *Pete* suffers from the disease *aids*", even if such a fact is actually documented to be true. In the formal approach, this requirement can be declared as a confidentiality policy instance of form *pot_sec* = {*HasDisease*(*Pete, cancer*), *HasDisease* (*Pete, aids*)}, where the sentences *HasDisease*(*Pete, cancer*) and *HasDisease*(*Pete, aids*) are called "potential secrets". This terminology indicates that if a corresponding fact is true (say, "Person *Pete* is suffering from the disease *cancer*"), then it is treated as a secret that the system has to protect; otherwise, if the fact is actually false, it is not necessary to prevent the user from believing the negation of the potential secret. Now, if the user submits the sequence of the three queries given above, then the first one, the sentence *HasDisease(Tom, brokenarm)*, can be correctly answered, while the second one, the sentence *HasDisease(Pete, cancer)* or *HasDisease(Pete, aids)*, and the third one, the sentence *HasDisease(Pete, cancer)*, might be subject to distortion. The exact behaviour of controlled query evaluation depends on the *distortion method* and on a further assumption about the user's awareness of the confidentiality policy instance. The conservative *awareness assumption* is that the user knows which sentences the system is trying to protect as potential secrets.

According to [5, 7], under this awareness assumption the distortion method using *lies* has to pretend that both *HasDisease(Pete, cancer)* or *HasDisease(Pete, aids)* and *HasDisease(Pete, cancer)* are false, even if the sentence *HasDisease(Pete, cancer)* is actually true w.r.t. to the information system instance. In that case the returned answers are "lies". Note that the first distortion arises with the second query which is weaker than any of the declared potential secrets. However, it has been shown that when using lies the disjunction of all potential secrets must always be protected too. Under the same conservative awareness assumption, the results of [5, 7] show that the distortion method using *refusals* can return the correct answer for the second query, but has to refuse the third query, independent of whether *HasDisease(Pete, cancer)* is actually true or false w.r.t. to the information system instance. Note that the refusal also happens in the latter case which per se is considered harmless. A *combined method* [6, 7], employs lies as long as possible and only then refuses, thereby enjoying the advantages of both methods. More details about controlled query evaluation of closed queries and the formal model underlying the previous and the present work are summarized in Section 3.

1.3 Open Queries

An *open query* is a formula which has at least one free occurence of a variable. Basically, the answer to such a query consists of the set of ground substitutions of the variables that make the query formula true with respect to the interpretation represented by the stored information system instance. Assuming as usual that the information system instance represents an Herbrand-like interpretation, i.e., its domain is taken from the underlying vocabulary of the logic that comprises a recursive set of constants, the ordinary evaluation of an open query of form $\Phi(x_1, ..., x_n)$ with free variables $x_1, ..., x_n$ can be simply simulated by a sequence of closed queries, each of which results from substituting the variables by constants from the vocabulary, thereby exhausting all possibilities. For example, the simple open query *HasDisease(x, cancer)* could be explored by a sequence of closed queries which starts as follows:

$\langle HasDisease(Pete, cancer), HasDisease(Tom, cancer), HasDisease(Lisa, cancer), ... \rangle.$

In doing so, we have to distinguish what is assumed about the set of constants. Each assumption might demand a different logic, in particular with respect to the class of interpretations considered for the formal definition of "implication". In the next subsections we introduce three reasonable options, the third of which will be seleted for the rest of this paper.

1.4   Fixed Finite Relevant Domain: A Special Case of Interest

The first option assumes a *fixed finite relevant domain*, used for any instance of the information system according to a public declaration, and thus known to both the control mechanism and the user. This assumption might be of some interest for special situations, mostly suitable for small and dedicated applications.

Under this option, the full relational model can be reduced to propositional logic where all pertinent decision problems are decidable. More details of the reduction and of the treatment of open queries will be presented in a full version of this paper.

1.5   Varying Finite Domains: A Discarded Case

The second option assumes a fixed *infinite* set of constants, again publicly declared in the schema, from which any instance of the information system selects a *finite subset* as its actual domain. We argue that this option is not appropriate for the task of controlled query evaluation. The main problem is the need of distinguishing whether a ground atom that is not listed in a stored relation or a query answer is false or just undefined (because it contains a constant that does not belong to the actual domain). Note that if the answer specified which atoms are false and which undefined, then the actual domain might be disclosed, which may lead to undesirable inferences.

1.6   Fixed Infinite Herbrand Domain: A Promising Generic Approach

Finally, the third option assumes a fixed *infinite* set of constants, again publicly declared in the schema, that any instance of the information system uniformly takes as its *infinite domain*. We adopt this option for the rest of this paper, since we consider it as most promising for a generic approach to make controlled query evaluation practically applicable for relational databases. First, the case of undefined truth values for ground atoms cannot occur. Second, we are not faced with combinatorial effects due to the finiteness of a domain. Third, for most applications there are no convincing reasons to restrict the usable constants to a finite subset in a specific way.

This option together with the requirement positive information is finite do neither readily fit the classical model theory of first-order logic, nor the more recently elaborated finite model theory. However, under suitable restrictions we can cast the following classical result into the favored framework: The Bernays-Schönfinkel class of formulas, which have a prenex normal form with prefix of form $\forall^* \exists^*$, has a decidable universal validity problem, both for general universal validity and for finite universal validity. The restrictions apply to the sentences the control mechanism must consider, i.e., sentences that (1) are in the user log, (2) related to queries, or (3) related to the confidentiality policy. If all pertinent sentences have a prenex normal form with prefix $\forall^*$ or $\exists^*$, then any implicational relationship $\Phi \models_{DB} \Psi$ between two of them is equivalent to the universal validity of the sentence $\neg \Phi \vee \Psi$ which is in the Bernays-Schönfinkel class. This restriction still allows to deal with queries of the positive existential relational calculus, semantic constraints like functional dependencies or join dependencies, and, as we will indicate below, completeness tests needed to evaluate open queries.

The details of the logical reconstruction of the sketched relational submodel are presented in Section 2, together with a more general insight which is considered to be important beyond the special application for controlled query evaluation:

- We prove a strong sufficient condition that for a fragment of first-order logic classical implication, finite implication, and implication w.r.t. to the class of interpretation adopted for controlled query evaluation all coincide, and are decidable.

1.7 Terminating Open Query Evaluation Under a Fixed Infinite Herbrand Domain

Given the logical setting of a fixed infinite Herbrand domain, we can refine the evaluation of an open query by means of a sequence of closed queries, ranging over all possible substitutions with constants from the domain. Clearly, we need to terminate after trying a finite number of such substitutions. Therefore, we restrict the queries to *safe* and *domain-independent* ones [1, 17]. *Safe* queries ensure finite query results. *Domain-independent* queries return the same finite result for each reasonable actual domain of the stored information system instance, including its finite *active domain*. Fortunately, all queries of the positive existential relational calculus are domain-independent.

Additionally, we have to effectively recognize when the final query result has been completely enumerated and no further substitutions have to be considered. We can achieve this requirement by a simple *completeness test*. Basically, this test just asks whether all further substitutions will make the query formula false. Using the syntactical material of the previously considered substitutions, we can easily construct a corresponding closed query, i.e., a sentence of the logic. Moreover, this sentence fits the imposed restrictions for ensuring the decidability of implications.

Completeness tests must be controlled like any other query in order to avoid harmful inferences from the communicated answers. This issue is investigated in depth in Section 4:

- We identify the simulation of an open query by a terminating finite sequence of closed queries suitable to be transferred to controlled query evaluation.
- We complement the "last minute distortion" strategy enforcing confidentiality by a "first chance distortion" strategy for terminating the simulation of open queries.
- We present controlled query evaluation methods for open queries based on lying and on refusal, respectively, and prove these methods to be secure for all sequences of queries whether they are closed or open.

2 A Relational Submodel with a Fixed Infinite Herbrand Domain

2.1 Logic Approach

Our previous work on controlled query evaluation [4, 5, 6, 7], has been based on a generic logic-oriented, model-theoretic approach to information systems, in most cases leaving open the choice of a concrete logic. In this paper we follow the previous work but adopt *first-order predicate logic* with equality, using it as a foundation for the *relational data model*, in particular for the relational calculus as query language, see for instance [1, 17].

A complete *information system* maintains a schema and an instance. The *schema DS* captures the *vocabulary* of the logic, here comprising a denumerably infinite set *dom* of *constants* and a finite set of *predicate names* (relation names). An *instance db* is a Herbrand *interpretation* which *interprets* the vocabulary of the logic, here the predicate names, assuming that the constants are denoting themselves (see e.g. [22, 21]). Concerning the domain, we take *dom* as a fixed infinite Herbrand domain. Furthermore, we only consider instances *db* that can be represented by finite stored relations, containing—as usual—the tuples that satisfy the corresponding predicate.

2.2 The Decidability Result

In this subsection, we formally define the notions introduced above, and we present the decidability result announced in the introductory Section 1. For the sake of convenience, we always identify an information system instance *db* with an interpretation I, seen as a semantic concept of formal logic. As usual, we write $I, \sigma \models \varphi$ to state that formula φ is true in I when its free variables are interpreted according to the substitution σ. We denote by $\sigma[x/v]$ the substitution that agrees with σ on all $y \neq x$ and maps x on v. When the evaluation of φ does not depend on σ (e.g., when φ is a sentence) we write $I \models \varphi$.

Let $dom = \{c_1, c_2, \ldots, c_i, \ldots\}$ be a fixed infinite set ($|dom| = \aleph_0$). In the following we consider function-free first-order languages $\mathscr{L}$ where the set of constants can be any finite subset of *dom*. Denote by $\mathrm{dom}(I)$ the domain of an interpretation I, and denote by s^I the interpretation of (predicate or constant) symbol s in I.

Definition 2.1. *An interpretation I for $\mathscr{L}$ is a* DB-interpretation *(or* DB-instance*) iff the following conditions hold:*

1. $\mathrm{dom}(I) = dom,$
2. p^I *is finite for all predicate symbols p in $\mathscr{L}$,*
3. $c_i^I = c_i$ *for all constant symbols c_i in $\mathscr{L}$.*

We study *implication* or *entailment* w.r.t. three different classes of interpretations:

$$\Gamma \models_{DB} \varphi \text{ iff for all DB-interpretations } I : I \models \Gamma \text{ implies } I \models \varphi;$$

$$\Gamma \models_{fin} \varphi \text{ iff for all finite interpretations } I : I \models \Gamma \text{ implies } I \models \varphi; \text{ and}$$

$$\Gamma \models_{gen} \varphi \text{ iff for all interpretations } I : I \models \Gamma \text{ implies } I \models \varphi.$$

Definition 2.2. *The* active domain *of an interpretation I w.r.t. a formula φ, denoted by* $\mathrm{active}_\varphi(I)$, *is the set of all $d \in \mathrm{dom}(I)$ occurring in some tuple of p^I, for some predicate p in $\mathscr{L}$, plus all c^I such that c is a constant in φ. In symbols:*

$\mathrm{active}_\varphi(I) = \{d \mid \text{ there exist } d_1, \ldots, d_n, i < n, \langle d_1, \ldots, d_i, d, d_{i+2}, \ldots, d_n \rangle \in p^I\}$
$\cup \{c^I \mid c \in \mathrm{const}(\varphi)\},$
where $\mathrm{const}(\varphi)$ *is the set of constants occurring in φ.*

Definition 2.3. *Let $Out_\varphi(x)$ be a formula stating that x does not belong to the active domain w.r.t. φ, e.g.,*
$$\forall y_1 \ldots \forall y_m. \bigwedge_p \bigwedge_{1 \leq k < \mathrm{arity}(p)} \neg p(y_1, \ldots, y_k, x, y_{k+2}, \ldots, y_{\mathrm{arity}(p)}) \wedge \bigwedge_{c \in \mathrm{const}(\varphi)} \neg x = c,$$
where m is the maximal arity of the predicates p.

Definition 2.4. *A first-order formula φ has* nominal equality *iff for all equalities $t = u$ occurring in φ, t is a variable and u is a constant. We also say that φ is a* nominal equality *formula.*

In order to prepare the main results of this section, we first state three lemmas. The proofs of these lemmas are by structural induction and suppose w.l.o.g. that only the minimal complete set of connectives $\neg, \vee$ and $\exists$ is employed.

Lemma 2.1. *Let φ be a nominal equality formula and let I be an interpretation. For all $\bar{c} \in \mathrm{dom}(I) \setminus \mathrm{active}_\varphi(I)$, all substitutions σ, and all variables v such that $\sigma(v) \notin \mathrm{active}_\varphi(I)$: $I, \sigma \models \varphi$ iff $I, \sigma[v/\bar{c}] \models \varphi$.*

Proof. Base case: φ is an atom. First assume φ is not an equality. If v does not exists or v is not free in φ, then the lemma trivially holds. Otherwise, by definition of $\mathrm{active}_\varphi(I)$, $I, \sigma \not\models \varphi$ and $I, \sigma[v/\bar{c}] \not\models \varphi$. Now assume φ is an equality. Since φ has nominal equality, φ is $x = c_j$, for some variable x and some constant c_j. As in the previous case, if v does not exists or v is not free in φ, then the lemma trivially holds. Otherwise, note that $\bar{c} \neq c_j^I$ because $c_j^I \in \mathrm{active}_\varphi(I)$ by definition. Then $I, \sigma \not\models \varphi$ and $I, \sigma[v/\bar{c}] \not\models \varphi$.

Induction step: The induction cases for $\neg, \vee$ follow immediately from the induction hypothesis. Now suppose φ is $\exists x. \psi$. If $x = v$ then the lemma holds trivially. Otherwise, $I, \sigma \models \varphi$ iff for some d, $I, \sigma[x/d] \models \psi$. By induction hypothesis, this is equivalent to

$$I, \sigma[x/d][v/\bar{c}] \models \psi. \tag{1}$$

Since $\sigma[x/d][v/\bar{c}] = \sigma[v/\bar{c}][x/d]$, we have that (1) is equivalent to $I, \sigma[v/\bar{c}] \models \varphi$.   □

Lemma 2.2. *Let φ be a nominal equality formula and let I be a DB-interpretation. For all $\bar{c} \in \mathrm{dom}(I) \setminus \mathrm{active}_\varphi(I)$ and all substitutions σ: $I, \sigma \models \varphi$ iff $J, \tau \models \varphi$, where J is the restriction of I to $\mathrm{active}_\varphi(I) \cup \{\bar{c}\}$, and*

$$\tau(x) = \begin{cases} \sigma(x) & \text{if } \sigma(x) \in \mathrm{active}_\varphi(I), \\ \bar{c} & \text{otherwise}. \end{cases}$$

Proof. Base case: φ is an atom. First assume φ is not an equality, and its free variables are $x_1, \ldots, x_n$. If all $\sigma(x_i)$ are in $\mathrm{active}_\varphi(I)$, then $\tau(x_i) = \sigma(x_i)$ $(1 \leq i \leq n)$, and the lemma trivially holds. If some $\sigma(x_i)$ is not in $\mathrm{active}_\varphi(I)$, then $\tau(x_i) = \bar{c} \notin \mathrm{active}_\varphi(J)$. Then $I, \sigma \not\models \varphi$ and $J, \tau \not\models \varphi$. Now assume φ is an equality. Since φ has nominal equality, φ is $x = c_j$, for some variable x and some constant c_j. As in the previous case, if all $\sigma(x_i)$ are in $\mathrm{active}_\varphi(I)$, then $\tau(x_i) = \sigma(x_i)$, and the lemma holds trivially. If some $\sigma(x_i)$ is not in $\mathrm{active}_\varphi(I)$, then $\tau(x_i) = \bar{c} \notin \mathrm{active}_\varphi(I)$. Note that $\bar{c} \neq c_j$ because $c_j \in \mathrm{active}_\varphi(I)$. Then $I, \sigma \not\models \varphi$ and $J, \tau \not\models \varphi$.

Induction step: The induction cases for $\neg, \vee$ follow immediately from the induction hypothesis. Now suppose φ is $\exists x. \psi$. If $I, \sigma \models \varphi$ then for some c_k, $I, \sigma[x/c_k] \models \psi$. By induction hypothesis, it follows that $J, \tau[x/c'] \models \psi$, where

$$c' = \begin{cases} c_k & \text{if } c_k \in \mathrm{active}_\varphi(I), \\ \bar{c} & \text{otherwise}. \end{cases}$$

Therefore $J, \tau \models \varphi$. Conversely, if $J, \tau \models \varphi$, then there exists $c_k \in \mathrm{active}_\varphi(I) \cup \{\bar{c}\}$ such that $J, \tau[x/c_k] \models \psi$. By induction hypothesis, $I, \sigma[x/c_k] \models \psi$, and hence $I, \sigma \models \varphi$.   □

Lemma 2.3. *Let φ be a nominal equality formula and let J be a finite interpretation such that $J \models \exists x.Out_\varphi(x)$. Let $f : \mathrm{dom}(J) \to dom$ be any total, injective function such that for all $c_i \in \mathrm{const}(\varphi)$, $f(c_i^J) = c_i$. Let I be the DB-interpretation such that for all predicates p, $p^I = \{\langle f(d_1),\ldots,f(d_n)\rangle \mid \langle d_1,\ldots,d_n\rangle \in p^J\}$. Then for all substitutions σ: $J,\sigma \models \varphi$ iff $I,(f\circ\sigma) \models \varphi$, where $(f\circ\sigma)(x) = f(\sigma(x))$.*

Proof. Base case: φ is an atom. The claim is an immediate consequence of the definitions of I and f.

Induction step: The induction cases for $\neg$ and $\vee$ are immediate from the induction hypothesis. Now assume that φ is $\exists y.\psi$. If $J,\sigma \models \exists y.\psi$, then there exists $d \in \mathrm{dom}(J)$ such that $J,\sigma[y/d] \models \psi$. Note that $f\circ(\sigma[y/d]) = (f\circ\sigma)[y/f(d)]$. Therefore, by induction hypothesis, $I,(f\circ\sigma)[y/f(d)] \models \psi$, and hence $I,(f\circ\sigma) \models \varphi$.

Conversely, suppose that $J,\sigma \not\models \exists y.\psi$. Then for all $d \in \mathrm{dom}(J)$, we have $J,\sigma[y/d] \not\models \psi$. It follows by induction hypothesis that

$$\text{for all } d \in \mathrm{dom}(J), \quad I,(f\circ\sigma)[y/f(d)] \not\models \psi. \tag{2}$$

In order to complete the proof, it suffices to show that there is no $c_k \in dom$ outside the range of f such that $I,(f\circ\sigma)[y/c_k] \models \psi$. Suppose such a c_k exists. Then, by Lemma 2.1, $I,(f\circ\sigma)[y/\bar{c}] \models \psi$, for all $\bar{c} \in dom \setminus \mathrm{active}_\varphi(I)$. One such $\bar{c}$ must belong to the range of f, because J satisfies $\exists x.Out_\varphi(x)$. This contradicts (2). □

Definition 2.5. *For all formulae φ, let UNA_φ (the unique name assumption for φ) be the following sentence:*

$$\bigwedge\{c_i \neq c_j \mid \{c_i,c_j\} \in \mathrm{const}(\varphi) \text{ and } i < j\}.$$

Theorem 2.1. *For all nominal equality formulae φ, the following are equivalent:*

1. *φ is satisfied by a DB-interpretation I;*
2. *φ is satisfied by a finite model J of $\exists x.Out_\varphi(x)$ and UNA_φ.*

Proof. Assume that statement 1 holds. Note that $\mathrm{active}_\varphi(I)$ is finite. By Lemma 2.2, there exists a finite restriction J of I satisfying φ, and such that $\mathrm{dom}(J) = \mathrm{active}_\varphi(I) \cup \{\bar{c}\}$ where $\bar{c} \notin \mathrm{active}_\varphi(I) = \mathrm{active}_\varphi(J)$. This implies that J satisfies $\exists x.Out_\varphi(x)$. Moreover, J satisfies UNA_φ because J is a restriction of I and I is a DB-interpretation. This proves that statement 1 implies statement 2.

Now assume that statement 2 holds. Let f be any function satisfying the assumptions of Lemma 2.3. Note that such an f exists because J satisfies UNA_φ. Then, by Lemma 2.3, φ is satisfied by a DB-interpretation I, i.e. statement 1 holds. □

In the following let $\neg\mathscr{L} = \{\neg\varphi \mid \varphi \in \mathscr{L}\}$.

Theorem 2.2. *Let $\mathscr{L}$ be a fragment of first-order logic with nominal equality, closed under $\vee$, enjoying the finite model property, and capable of expressing $\neg\exists x.Out_\varphi(x)$ and $\neg\mathrm{UNA}_\varphi$ (up to logical equivalence) for all $\mathscr{L}$-formulae φ. Then the restrictions to*

$\neg \mathscr{L} \times \mathscr{L}$ of the entailment relations $\models_{gen}$, $\models_{DB}$ and $\models_{fin}$ are all decidable. Moreover, for all $\psi \in \neg \mathscr{L}$ and all $\varphi \in \mathscr{L}$, the following are equivalent:

1. $\psi \models_{DB} \varphi$;
2. $\psi \wedge \exists x.Out_\varphi(x) \wedge UNA_\varphi \models_{gen} \varphi$;
3. $\psi \wedge \exists x.Out_\varphi(x) \wedge UNA_\varphi \models_{fin} \varphi$.

Proof. Since $\mathscr{L}$ enjoys the finite model property, relations $\models_{gen}$ and $\models_{fin}$ are equivalent and decidable over $\neg \mathscr{L} \times \mathscr{L}$. Morever, by Theorem 2.1, $\psi \models_{DB} \varphi$ is equivalent to $\psi \wedge \exists x.Out_\varphi(x) \wedge UNA_\varphi \models_{fin} \varphi$, provided that $\neg \psi$ and φ are in $\mathscr{L}$, and that $\neg \exists x.Out_\varphi(x)$ and $\neg UNA_\varphi$ can be expressed in $\mathscr{L}$. As a consequence, statements 1, 2, and 3 are mutually equivalent. The decidability of $\models_{DB}$ over $\neg \mathscr{L} \times \mathscr{L}$ follows from this equivalence. $\square$

The above theorem applies in particular to the nominal equality subset of a fragment derived from the Bernays-Schönfinkel class and of the guarded fragment. It applies also to many description logics, including *ALCIO* with boolean role operators. Applying the theorem for the Bernays-Schönfinkel class, we can still use the formulas for the following items:

- a positive or a negative answer to a closed query of the positive existential calculus that allows to express conjunction, disjunction, and existential quantification (equivalent to conjunctive queries with union, see [1], Thm. 5.4.10);
- a potential secret, expressed by the same kind of formulas,
- a semantic constraint declared as a full dependency, in particular a functional dependency or a join dependency;
- arbitrary conjunctions and disjunctions of the items described before.

3 Formal Model for Controlled Query Evaluation

3.1 Ordinary Query Evaluation

We treat two kinds of *queries*, closed and open ones. A *closed query* is a sentence in the restricted language. Given a DB-interpretation *db* (stored as a DB-instance with implicit domain) and a *suitable* sentence Φ (issued as a closed query), Φ is either *true* (*valid*) or *false* in *db*, or in other words, the interpretation is either a *model* of the sentence, denoted by $db \models \Phi$, or not. When a user issues a closed query Φ against the schema *DS*, the (ordinary) *query evaluation* $eval^*(\Phi)$ determines the pertinent case w.r.t. the current DB-instance *db*, or equivalently either Φ or its negation:

$$eval^*(\Phi) : DS \to \wp\{\Phi, \neg \Phi\} \text{ with}$$
$$eval^*(\Phi)(db) := \text{if } db \models \Phi \text{ then } \{\Phi\} \text{ else } \{\neg \Phi\}^1. \tag{3}$$

An *open query* is a formula in the restricted language that contains at least one (occurence of a) free variable. Given a DB-interpretation *db* (stored as a DB-instance) and formula $\Phi(x_1, ..., x_n)$ with free variables $x_1, ..., x_n$ (issued as an open query), for

---

[1] In the rest of the paper, by abuse of notations we will often identify the singleton sets with their respective single elements.

each substitution of $x_1,...,x_n$ by constants $c_1,...,c_n$ of the implicit domain, the resulting sentence $\Phi(c_1,...,c_n)$ is either *true* (*valid*) or *false* in *db*. When a user issues an open query $\Phi(x_1,...,x_n)$ against the schema *DS*, conceptually the (ordinary) *query evaluation* $eval^*(\Phi(x_1,...,x_n))$ determines for all possible substitutions the pertinent case w.r.t. the current DB-instance *db* and includes the result in the final output:

$$eval^*(\Phi(x_1,...,x_n)) : DS \to \wp\{\Psi \mid \Psi \text{ is sentence in the logic}\} \text{ with}$$
$$eval^*(\Phi(x_1,...,x_n))(db) :=$$
$$\{\ \Phi(c_1,...,c_n) \mid c_1,...,c_n \in dom \text{ and } db \models \Phi(c_1,...,c_n)\} \cup$$
$$\{\neg\Phi(c_1,...,c_n) \mid c_1,...,c_n \in dom \text{ and } db \not\models \Phi(c_1,...,c_n)\}. \tag{4}$$

3.2 Generalized Controlled Query Evaluation

Controlled query evaluation consists of two steps. First, the correct answer is judged by some *censor* and then, depending on the output of the censor, some *modificator* is applied. For a visualization, refer to Table 1. In order to assist the censor, the system maintains a *user log*, denoted by *log*, which represents the explicit part of the *user's assumed knowledge*. Formally, *log* is declared to be a set of sentences. The log is meant to contain all the sentences that the user is assumed to hold true in the DB-instance, in particular publicly known *semantic constraints*. Additionally, the log records the sentences returned as *answers to previous queries*.

The censor bases its decisions on a *confidentiality policy instance* that is declared as a set *pot_sec* of sentences *pot_sec* $= \{\Psi_1,...,\Psi_k\}$, formulated in the restricted language, where each sentence is called a *potential secret*. A potential secret Ψ specifies the following (refer to [5,7] for further motivation): A user, knowing only the apriori knowledge log_0 and the answers returned so far (i.e. the current value of *log*), should not be able to exclude that $\neg\Psi$ is true in the actual DB-instance, or speaking otherwise, this case should be possible for him. In this paper, we assume that the actual confidentiality policy instance *pot_sec* is *known* to the user.

Formally we will describe an approach to *(generalized) controlled query evaluation* by a family of (possibly) partial functions $gen\_con\_eval(Q, log_0)$, each of which has two parameters: a (possibly infinite) sequence of closed and open queries $Q = \langle\Phi_1(x_{1,1},...,x_{1,n_1}),...,\Phi_i(x_{i,1},...,x_{i,n_i}),...\rangle$, and an initial user log log_0. The inputs to any such function are "admissible" pairs (db, pot_sec) where *db* is a DB-instance, and *pot_sec* is an instance of the confidentiality policy. The admissibility of an argument pair (db, pot_sec) is determined by some formal *precondition* associated with the function. The function returns an answer sequence to the user, and updates the user log as a side effect.

The response to a query $\Phi(c_1,...,c_n)$ might state that the censor required to refuse to answer. We will denote this event by putting the string $mum_{\Phi(c_1,...,c_n)}$ into the answer. However, the system does not need to remember the refused parts of the answer (which are implicitly handled by condition (a) in the definition of confidentiality below).

Summarizing in symbols,

$$gen\_con\_eval(Q, log_0)(db, pot\_sec) = \langle(ans_1, log_1), (ans_2, log_2),...,(ans_i, log_i),...\rangle,$$

where the side effect on the user log, including discarding refusals, is described by $log_i := log_{i-1} \cup mum\_removal(ans_i)$.

Table 1. Architecture of generalized controlled query evaluation

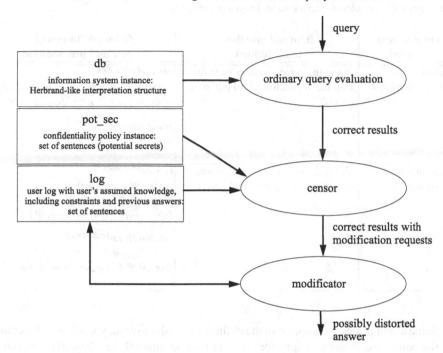

Finally, a *censor* is formally described by a function with arguments of the form (Ψ, log, db, pot_sec), where Ψ is a *sentence*, i.e a *closed query* or a *candidate sentence* to be put into the answer of an open query. The censor returns a value of the set $\{\texttt{pass}, \texttt{lie}, \texttt{refuse}\}$ for indicating the required action of the modificator.

Controlled query evaluation aims at preserving the potential secrets under all circumstances. This requirement is formally captured as follows.

Definition 3.1. *Let gen_con_eval(Q, log_0) describe a specific (generalized) controlled query evaluation with precond as associated precondition for "admissible" arguments, and pot_sec be a "known[2]" policy instance.*

gen_con_eval(Q, log_0) is defined to preserve confidentiality *with respect to pot_sec iff*
for all finite prefixes Q' of Q, for all DB-instances db_1 of the information system such that (db_1, pot_sec) satisfies precond, and for all $\Psi \in pot\_sec$,
there exists a DB-instance db_2 such that (db_2, pot_sec) also satisfies precond and such that the following properties hold:

(a) [same answers]
   $gen\_con\_eval(Q', log_0)(db_1, pot\_sec) = gen\_con\_eval(Q', log_0)(db_2, pot\_sec)$;
(b) [false potential secrets] $eval^*(\Psi)(db_2) = \neg\Psi$.

More generally, gen_con_eval(Q, log_0) is defined to preserve confidentiality *iff it preserves confidentiality with respect to all "admissible" policy instances.*

---

[2] This awareness assumption is formalized by fixing *pot_sec* in part 1 of Definition 3.1 (whereas the supposedly unknown DB-instance is varying).

Table 2. Enforcement methods of lying and refusal, censors, and preconditions of controlled query evaluation of closed queries under known potential secrets

enforcement method	informal question to be decided	formal definition of censor and precondition
uniform lying [10], [5–section 4]	adding the correct answer harmful w.r.t. the disjunction of all potential secrets?	$censor^{\text{ps,known},L}$: $log \cup eval^*(\Phi)(db) \models_{DB} pot\_sec\_disj$ with $pot\_sec\_disj := \bigvee_{\Psi \in pot\_sec} \Psi$; $precondition^{\text{ps,known},L}$: $log_0 \not\models_{DB} pot\_sec\_disj$
uniform refusal [5–section 3]	adding the correct or false answer harmful w.r.t. any potential secret?	$censor^{\text{ps,known},R}$: $(\text{exists } \Psi \in pot\_sec)$ $[log \cup eval^*(\Phi)(db) \models_{DB} \Psi$ or $log \cup \neg eval^*(\Phi)(db) \models_{DB} \Psi]$; $precondition^{\text{ps,known},R}$: $db \models log_0$ and $(\text{for all } \Psi \in pot\_sec)\,[log_0 \not\models_{DB} \Psi]$

Some assumptions are implicit in the definition: (i) the user may know the algorithm of the controlled evaluation function, (ii) the user is rational, i.e., he derives nothing besides what is implied by his knowledge and the behavior of the database.

3.3   Known Results on Closed Queries

Some known foundational work on controlled query evaluation of closed queries under known potential secrets is summarized in Table 2. For more details the reader is referred to the original presentations and the surveys given in [7]. In order to exploit the known confidentiality claims, we have to verify that the corresponding proofs remain valid for the settings we opt for in this paper. Basically, in these proofs we only use the following properties of "logical implication": transitivity, upwards monotonicity in the first argument, downwards monotonicity in the second argument, and a basic consistency lemma. This lemma says the following: If both $\Phi \cup \{\varphi\} \models_{DB} \Psi$ and $\Phi \cup \{\neg\varphi\} \models_{DB} \Psi$, then $\Phi \models_{DB} \Psi$. All these properties hold indeed for the adopted setting.

4   Open Queries Under Fixed Infinite Domain

4.1   Simulating Correct Query Evaluation

We can easily simulate the *correct* evaluation of an open query (i.e., without any confidentiality check) by a finite sequence of closed queries, as explained in the following. Let us suppose, that the query formula looks like $\Phi(x)$, where the x is a free variable[3]

---

[3] For the sake of simplicity, we present all results assuming just one variable. The more general case of n variables can be treated analogously.

which ranges over the infinite set of constants *dom* taken as fixed domain. The ordinary query evaluation basically returns all substitutions of x with an element of *dom* that make the formula $\Phi(x)$ true in the stored DB-instance, more specifically the ground formulas resulting from such substitutions. The simulation can effectively proceed roughly as follows:

- Select an enumeration sequence of *dom*, and start the enumeration.
- In round j, for the currently enumerated substitution c_j of x, submit the corresponding closed query $\Phi(c_j)$. If the answer is *no* then continue. Otherwise, output $\Phi(c_j)$ and either exit or continue according to a *completeness test*.
- The completeness test consists in asking whether all further substitutions will make the formula false. Using the syntactic material of the previous substitutions, we can easily construct such a closed query, denoted by $Complete(\Phi(x), j)$. This sentence says: "For all x: if x is different from the substitutions $c_1, \ldots, c_j$ considered so far, then $\neg\Phi(x)$".
- On exit, as a closed-world statement, output $Complete(\Phi(x), j)$ together with the ground formulas $\neg\Phi(c_k)$ $(k \leq j)$ corresponding to negative answers.

By our assumptions, for a nonempty query result[4] this simulation is guaranteed to exit immediately after having enumerated all substitutions that actually make the query formula true.

4.2   A Running Example

We will use the following items for a running example, or appropriate variants of it:

- *Vocabulary*: We have just one relation name P for a stored base relation and the set of constants $dom = \{c_1, c_2, c_3, \ldots\}$.
- *DB-instance*: The base relation contains just one tuple, representing that $P(c_1)$ is true and all other sentences $P(c_i)$ for $i \neq 1$ are false, i.e.,
  $db = \{P(c_1), \neg P(c_2), \neg P(c_3), \neg P(c_4), \ldots\}$.
- *Open query*: The user wants to know the state of the base relation, i.e., he submits the open query $P(x)$.
- *Known confidentiality policy instance*: We declare two potential secrets, namely the elements of $pot\_sec = \{P(c_1), \neg P(c_3)\}$.
- *Initial user log*: We suppose the semantic constraint that the base relation is nonempty, i.e., that $log_0 = \{(\exists x)P(x)\}$.

Example 4.1. For *correct query evaluation*, the simulation given above immediately exits in round 1, after having output $P(c_1)$ together with $Complete(P(x), 1)$ as appropriate closed world statement.

For any secure *controlled query evaluation*, we can already make the following observations. Knowing a priori the semantic constraint, the user expects at least one positive output. Moreover, being aware of the policy instance, the user expects to never see the sentence $P(c_1)$ in the positive part of the answer, as well as to never see the sentence $\neg P(c_3)$ captured by the closed world statement.

---

[4] In order to deal with an empty query result, we just have to add an initial completeness test in a preparing round 0.

4.3 Lying Method

We base our design of a lying method on the following features:

- We reuse the lying censor for closed queries under known potential secrets [5, 7]. This censor guarantees the invariant $log \not\models_{DB} pot\_sec\_disj$, where $pot\_sec\_disj :=$ $\bigvee_{\Psi \in pot\_sec} \Psi$, just by checking this condition after tentatively adding the correct answer to the log.
- We use a *fixed* enumeration of the set *dom* of constants. This enumeration is supposed to be known to the user.
- Though we are going to submit some completeness sentences $Complete(\Phi(x), j)$ as closed queries to the censor, as a fundamental design decision, we *never* explicitly return a lied completeness sentence. Any other sentence can be distorted, but if a completeness sentence is explicitly returned in the last round of controlled query evaluation of an open query, then it should always be correct! Speaking otherwise: we block this statement for lying in order to exit from the infinite enumeration of the constants in a convincing way.
- However, before stopping the enumeration, we *implicitly* suggest statements of the form $\neg Complete(\Phi(x), j)$. These suggestions can be lies.
- Simultanously blocking the final completeness sentence for lying and requiring to exit the enumeration after finitely many rounds has a price: We possibly have to distort other closed queries as some kind of "exit (enabling) lies".
- These distortions reasonably follow a "first chance distortion" strategy.

We now propose a lying method for controlled query evaluation of an open query $\Phi(x)$ applied to a DB-instance *db* and a known policy instance *pot_sec*:

1. **Phase 1:** [enumerate the domain up to round k such that c_k is the last constant that makes the open query $\Phi(x)$ true[5]]
   (a) Let $k := Min\{i \mid Complete(\Phi(x), i)$ is true in $db\}$;
   (b) `for` $j := 1..k$:
       treat $\Phi(c_j)$ as closed query, i.e.,
       `if` $log \cup eval^*(\Phi(c_j)(db)) \not\models_{DB} pot\_sec\_disj$
           `then` [correct answer] add $eval^*(\Phi(c_j))(db)$ to the answer set and to the log
           `else` [lied answer] add $\neg eval^*(\Phi(c_j))(db)$ to the answer set and to the log

2. **Phase 2:** [determine an "enumeration excess" for exiting: if needed, we further enumerate the domain beyond the last element belonging to the correct answer, in order to find suitable lies (those with the empty "literal sign", see below) for exiting with a true completeness statement.]
   `if` $(log \cup \{Complete(\Phi(x), k)\}) \not\models_{DB} pot\_sec\_disj$
       `then`
   (a) determine $l \leq k$ such that
       $\Phi(c_l), \neg\Phi(c_{l+1}), \ldots, \neg\Phi(c_k)$ has been returned;
       [$\Phi(c_l)$ is the last positive answer.]

---

[5] For an empty query result, we set $k := 0$ in (a), and we skip (b).

(b) remove $\neg\Phi(c_{l+1}),\ldots,\neg\Phi(c_k)$ from the answer set and from the log;

(c) add $\{Complete(\Phi(x),l)\}$ to the answer set and to the log;

(d) exit

   `else`

(a) determine a nonempty "enumeration excess", defined by

$k+1,k+2,\ldots,l$ with $k < l$ and

"literal signs" $\neg_{k+1},\neg_{k+2},\ldots,\neg_{l-1}$ being either "$\neg$" or empty

such that

$(log \cup \{\neg_{k+1}\Phi(c_{k+1}),\neg_{k+2}\Phi(c_{k+2}),\ldots,\Phi(c_l)\} \cup \{Complete(\Phi(x),l)\})$

$\not\models_{DB} pot\_sec\_disj;$

(b) add $\{\neg_{k+1}\Phi(c_{k+1}),\neg_{k+2}\Phi(c_{k+2}),\ldots,\Phi(c_l)\} \cup \{Complete(\Phi(x),l)\}$

to the answer set and to the log;

[Note that the last element $\Phi(c_l)$ is a lie.]

(c) exit.

Theorem 4.1. *The lying method for controlled query evaluation of open queries always terminates, i.e., in Phase 2 an "enumeration excess" is always guaranteed to exist.*

Proof. (Sketch) Let log_k denote the result of Phase 1. If $log_k \cup \{Complete(\Phi(x),k)\}$ $\not\models_{DB} pot\_sec\_disj$, then we have an empty "enumeration excess".

Otherwise, since log_k satisfies the invariant and by the basic consistency lemma,

$$log_k \cup \{\neg Complete(\Phi(x),k)\} \not\models_{DB} pot\_sec\_disj. \tag{5}$$

Hence there exists a witness DB-instance $db_{enum-ex}$ for (5), i.e.,

(1) $db_{enum-ex} \models log_k \cup \{\neg Complete(\Phi(x),k)\}$ and

(2) not $db_{enum-ex} \models pot\_sec\_disj$.

According to the first property, there exists $l \geq k+1$ with $db_{enum-ex} \models \Phi(c_l)$. Consider a maximal such l, which exists according to the definition of DB-instances. Then the "enumeration excess" can be built from

$$\{eval^*(\Phi(c_{k+1}))(db_{enum-ex}),\ldots,eval^*(\Phi(c_l))(db_{enum-ex})\}\ . \qquad \square$$

Theorem 4.2. *The lying method for controlled query evaluation of open queries, adding any enumeration excess, preserves confidentiality.*

Proof. (Sketch) By construction, the "final log" does not imply the disjunction of all potential secrets. Take any DB-instance db_2 such that

(1) $db_2 \models$ "final log" and

(2) not $db_2 \models \Psi$, for all $\Psi \in pot\_sec$.

db_2 produces the same result as the original db, thereby generating an empty "enumeration excess":

For db_2, Phase 1 runs until round l, as determined by Phase 2 for the original db. Then, (in Phase 2 for db_2) only the statement $Complete(\Phi(x),l)$ is added. $\qquad \square$

The theorems on termination and preservation of confidentiality indicate that a security administrator has some options to direct the distortions that may be necessary for securely returning a correct completeness sentence. As an example, the administrator might try to keep the enumeration excess minimal. Another example might be that the administrator prefers a specific "witness DB-instance" to construct an enumerations excess, say for blocking or pushing some closed sentences on his discretion, or for efficiency reasons.

Example 4.2. For our running example, we get $pot\_sec\_disj = P(c_1) \vee \neg P(c_3)$ and $k = 1$. In round 1, we find in Phase 1 that $log_0 \cup eval^*(P(c_1))(db) \supseteq \{P(c_1)\} \models_{DB} pot\_sec\_disj$, and thus the lie $\{\neg\{P(c_1)\}$ is added to the answer set and the log, i.e., $log_1 = \{(\exists x)P(x), \neg P(c_1)\}$. Concerning Phase 2, we can make two observations. First, $Complete(P(x), 1) \models_{DB} \neg P(c_3)$ and thus also $Complete(P(x), 1) \models_{DB} pot\_sec\_disj$. Second, $log_1 \cup \{Complete(P(x), 1)\}$ would be equivalent to $\{(\exists x)P(x), (\forall x)\neg P(x)\}$ and thus inconsistent. Each of the observations alone already show the need for determining an enumeration excess. A minimal enumeration excess extends to the next two rounds 2 and 3, yielding $\{\neg P(c_2), P(c_3), Complete(P(x), 3)\}$. Then the final answer set is $\{\neg\{P(c_1), \neg P(c_2), P(c_3), Complete(P(x), 3)\}$.

4.4 Refusal Method

We base our design of a refusal method on the following features:

- We reuse the refusal censor for closed queries under known potential secrets [5, 7]. This censor guarantees the invariant $log \not\models_{DB} \Psi$, for all $\Psi \in pot\_sec$, by checking the following condition for the query Φ:
  for all $\Psi \in pot\_sec : log \cup \{\Phi\} \not\models_{DB} \Psi$ and $log \cup \{\neg\Phi\} \not\models_{DB} \Psi$.
- We use a *fixed* enumeration of the set *dom* of constants. This enumeration is supposed to be known to the user.
- We return explicit answers up to some round k in the enumeration and, if possible, additionally a correct completeness sentence $Complete(\Phi(x), k)$.
- In each considered round j, we submit also the completeness sentence $Complete(\Phi(x), j)$ as a closed query to the censor. There might be two cases.
  The good case is that the censor allows to return the truth value of the completeness sentence, and accordingly we do so. If the completeness sentence is true, then we exit; otherwise we continue the enumeration with $j + 1$.
  The bad case is that the censor forbids to return the truth value of the completeness sentence. Then we refuse on the completeness sentence and *immediately* exit.
- The immediate exit in the bad case is justified by the observation that for one of the two possible reasons for refusal all further closed queries would be refused, too.

We now propose a refusal method for controlled query evaluation of an open query $\Phi(x)$ applied to a DB-instance db and a known policy instance pot_sec:

1. $j := 1$;
2. loop [enumerate the domain: in each round, submit first the current closed query
   and then the current completeness sentence to the censor]
   [submit the current closed query $\Phi(c_j)$ to the censor]
   if   for all $\Psi \in pot\_sec : log \cup \{\Phi(c_j)\} \not\models_{DB} \Psi$ and $log \cup \{\neg\Phi(c_j)\} \not\models_{DB} \Psi$
     then [correct answer] add $eval^*(\Phi(c_j))(db)$ to the answer set and to the log
     else [refused answer] add $\text{mum}_{\Phi(c_j)}$ to the answer set;
   [submit the current completeness sentence to the censor]
   if for all $\Psi \in pot\_sec$:
     $log \cup \{Complete(\Phi(x), j)\} \not\models_{DB} \Psi$ and $log \cup \{\neg Complete(\Phi(x), j)\} \not\models_{DB} \Psi$
    then
     if $Complete(\Phi(x), j)$ is true in db
     then
      (i) add $\{Complete(\Phi(x), j)\}$ to the answer set and to the log;
      (ii) exit
     else [if $Complete(\Phi(x), j)$ is false in db]
      (i) add $\{\neg Complete(\Phi(x), j)\}$ to the answer set and to the log ;
      (ii) $j := j + 1$ [continue the enumeration]
    else
     (i) add $\{\text{mum}_{Complete(\Phi, j)}\}$ to the answer set;
     (ii) exit
   endloop.

The method always terminates by construction.

Theorem 4.3. *The refusal method for controlled query evaluation of open queries preserves confidentiality.*

Proof. (Sketch) Consider any $\Psi \in pot\_sec$. By construction, the "final log" does not imply Ψ. Take any DB-instance db_2 such that

(1) $db_2 \models$ "final log" and
(2) $db_2 \models \neg\Psi$.

We claim that db_2 produces the same result as the original db. The proof is by induction on the enumeration. Initially, the algorithm is started for db and db_2 with the same log_0. For step j, by induction hypothesis, the current log is the same for db and db_2. Thus, the instance independent refusal censor treats both the current closed query $\Phi(c_j)$ and the current completeness statement $Complete(\Phi(x), j)$ for db and db_2 the same. So we are left with checking the situation that the censor allows to return the truth value of the completeness statement.

 Case 1: $Complete(\Phi(x), j)$ is true in db. Then this statement is added to the log of db, and thus the statement is true in db_2, according to the construction of db_2.

 Case 2: $Complete(\Phi(x), j)$ is false in db. Then $\neg Complete(\Phi(x), j)$ is added to the log of db, and thus the latter statement is true in db_2, according to the construction of db_2. Hence $Complete(\Phi(x), j)$ is false in db_2. □

Example 4.3. For our running example, in round 1 we find first that $log_0 \cup$ $eval^*(P(c_1))(db) \supseteq \{P(c_1)\} \models_{DB} P(c_1)$, and thus the refusal $\text{mum}_{P(c_1)}$ is added to the answer set. Then we find that $Complete(P(x),1) \models_{DB} \neg P(c_3)$ and thus the refusal $\text{mum}_{Complete(P(x),1)}$ is added to the answer set, and an immediate exit results.

More generally, we observe that the method always immediately exists if we have a negative potential secret $\neg P(c_j)$ with $1 < j$, since $Complete(P(x),1) \models_{DB} \neg P(c_j)$.

Example 4.4. If we had a variant of the running example with only $pot\_sec = \{P(c_1)\}$, then also $\text{mum}_{P(c_1)}$ and $\text{mum}_{Complete(P(x),1)}$ would be returned. The latter claim holds since $\{(\exists x)P(x), Complete(P(x),1)\} \models_{DB} P(c_1)$. Again, an immediate exit occurs.

Example 4.5. Consider now another variant with the base relation containing $P(c_1)$ and also $P(c_3)$, i.e., $db = \{P(c_1), \neg P(c_2), P(c_3), \neg P(c_4), ...\}$ and $pot\_sec = \{P(c_3)\}$. In the first two rounds, the correct answers $P(c_1)$ and $\neg P(c_2)$ are returned. In round 3, the refusal $\text{mum}_{P(c_3)}$ is due. Since $\{(\exists x)P(x), P(c_1), \neg Complete(P(x),1), \neg P(c_2), \neg Complete$ $(P(x),2), Complete(P(x),3)\} \models_{DB} P(c_3)$, the refusal $\text{mum}_{Complete(P(x),3)}$ is returned, and the method exits.

A first partial justification for the immediate exit on a refused completeness statement can be based on the following observation: If the refusal method exits with a refusal on the current completeness statement due to detecting that

$$log \cup \{\neg Complete(\Phi(x),j)\} \models_{DB} \Psi, \text{ for some } \Psi \in pot\_sec,$$

then all further (suppressed) considerations of closed query instances would result in refusals.

The method illustrated above follows the heuristics for refusals that proved to be necessary in the old framework to preserve confidentiality. So far, we do not have a full analysis of the variant in which the algorithm exits only when

$$log \cup \{Complete(\Phi(x),j)\} \models_{DB} \Psi, \text{ for some } \Psi \in pot\_sec.$$

5   Conclusions, Further Research, and Related Work

Assuming an *abstract* logic-oriented setting, previous conceptual work has demonstrated how sequences of *closed* queries can be evaluated in a controlled way such that the sentences of a specified confidentiality policy instance can be kept secret. This paper identifies a concrete and practically relevant *relational submodel*, where all DB-instances share a *fixed infinite* Herbrand domain, and proves that for an interesting class of fragments of first-order logic, reasoning with DB-instances is equivalent both to classical entailment and to finite model reasoning. Consequently, we obtain a fairly general extension of standard *decidability* results to the class of DB-interpretations, that in this paper's examples is applied to the Bernays-Schönfinkel fragment.

As a further contribution, the paper shows how the controlled evaluation approaches based on lies and refusals can be extended to *open* queries. For an open query, a user always expects a finite query answer together with a *completeness sentence* as a *closed world statement*. However, such a completeness sentence might violate the confidentiality policy. The solutions we proposed can be roughly summarized as follows. For uniform *lying*, we can always return a correct and confidentiality preserving completeness sentence (that may be left implicit in real implementations). For uniform *refusal*, in order to preserve confidentiality we might be forced to abort the evaluation and refuse the completeness sentence. For example, this unfortunate behavior occurs if the policy instance requires to protect negative facts. There are various topics for further research, including the following ones.

Combined method for open queries: Our attempts to extend the combined control method to open queries along the lines studied in this paper failed so far. This is the second time we find difficulties in extending the combined method, see [7].

The sequence problem: The behavior of controlled query evaluation depends on the order in which closed queries are evaluated. This feature stems from the "last minute distortion" strategy. As a consequence, the answers to an open query are not uniquely determined by the DB-instance, from the point of view of a user who does not know the actual enumeration sequence. However, for our reasoning about security, we have assumed a fixed enumeration, made public and thus known by the (adversary) user.

A *generic alternative enforcement method*: Our methods aim at considering completeness sentences as late as possible, in order to avoid early blocking of answer generation. As an alternative, at the price of cooperativeness, we might employ a simple generic method that examines an appropriately determined completeness sentence right at the beginning and only subsequently uses any of the methods for closed queries to inspect the ordinary query result.

Alternative policy models: We adopted the model of "known potential secrets". We would like to extend our results to the alternative variants, discussed and surveyed in [7], thereby investigating the "naive reduction" of secrecies to potential secrets, and the impact of weakening the awareness assumption into "unknown policy instance".

Incomplete information systems: The foundational work has been extended to deal with incomplete information systems as well, where a valid response to a query mightbe "query answer unknown" [8, 9]. The third alternative for correct answers demands a proof-theoretic approach, but provides additional flexibility for distortions. We would like to exploit the relational setting of this paper for queries to incomplete informations.

Less restricted relational submodels: The model adopted in Section 2 defines the needed restrictions of first-order logic in terms of prefixes of prenex normal forms. Our decidability result also applies to recent work on "guarded fragments" [19, 3] and thus suggests a way to achieve stronger expressiveness.

Foundation of inference control: Controlled query evaluation is part of a larger research field, known as inference control, see [18] for a recent introduction. Major contributions deal with relational databases, e.g. [11, 12, 13, 14, 15, 24, 25]. Unfortunately, there is no obvious and agreed model to systematically compare the relative achievements. We suggest to exploit the model outlined in Section 2 for this purpose.

References

1. Abiteboul, S., Hull, R, Vianu, V.: Foundations of Databases, Addison-Wesley, 1995.
2. Ackermann, W.: Solvable Cases of the Decision Problem, North-Holland, Amsterdam, 1968.
3. Andreka, H., Nemeti, I., van Benthem, J.: Modal languages and bounded fragments of predicate logic, Journal of Philosophical Logic 27,3 (1998), pp. 217-274.

4. Biskup, J.: For unknown secrecies refusal is better than lying, Data and Knowledge Engineering, 33 (2000), pp. 1-23.
5. Biskup, J., Bonatti, P.A.: Lying versus refusal for known potential secrets, Data and Knowledge Engineering, 38 (2001), pp. 199-222.
6. Biskup, J., Bonatti, P.A.: Controlled query evaluation for known policies by combining lying and refusal, Annals of Mathematics and Artificial Intelligence, 40 (2004), pp. 37-62.
7. Biskup, J., Bonatti, P.A.: Controlled query evaluation for enforcing confidentiality in complete information systems, Int. Journal of Information Security 3,1 (2004), pp. 14-27.
8. Biskup, J., Weibert, T.: Refusal in incomplete databases, Research Directions in Data and Applications Security XVII, Kluwer, Boston etc., 2004, pp. 143-157.
9. Biskup, J., Weibert, T.: Keeping secrets in incomplete databases, Workshop on Foundations of Computer Security, LICS 05, http://www.cs.chalmers.se/~andrei/FCS05/, Chicago, 2005.
10. Bonatti, P.A., Kraus, S., Subrahmanian, V.S.: Foundations of secure deductive databases, IEEE Transactions on Knowledge and Data Engineering 7,3 (1995), pp. 406-422.
11. Brodsky, A., Farkas, C., Jajodia, S.: Secure databases: constraints, inference channels, and monitoring disclosures, IEEE Transactions on Knowledge and Data Engineering 12,6 (2000), pp. 900-919.
12. Brodsky, A., Farkas, C., Wijesekera, D., Wang, X. S.: Constraints, inference channels and secure databases, Principles and Practice of Constraint Programming - CP 2000, 6th International Conference, Singapore, September 18-21, 2000. Lecture Notes in Computer Science 1894, Springer, Berlin etc., 2000, pp. 98-113.
13. Cuppens, F., Gabillon, A.: Cover story management, Data and Knowledge Engineering 37 (2001), pp. 177-201.
14. Dawson, S., De Capitani di Vimercati, S., Lincoln, P., Samarati, P.: Minimal data upgrading to prevent inference and association attacks, Proc. of the 18th ACM SIGMOD-SIGACT-SIGART Symposium on Principles of Database Systems (PODS), 1999, pp. 114-125.
15. Dawson, S., De Capitani di Vimercati, S., Samarati, P.: Specification and enforcement of classification and inference constraints, IEEE Symposium on Security and Privacy 1999, pp. 181-195.
16. Ebbinghaus, H.-D., Flum, J.: Finite Model Theory, Springer, Berlin etc., 1995.
17. Elmasri, R., Navathe, S.B.: Fundamentals of Database Systems (3rd edition), Addison-Wesley, 2000.
18. Farkas, C., Jajodia, S.: The inference problem: a survey, ACM SIGKDD Explorations Newsletter 4,2 (2002), pp. 6-11.
19. Graedel, E.: On the restraining power of guards, Journal of Symbolic Logic 64,4 (1999), pp. 1719-1742.
20. Libkin, L.: Elements of Finite Model Theory, Springer, Berlin etc., 2004.
21. Lloyd, J.W.: Foundations of Logic Programming, Springer, 1987.
22. Shoenfield, J.R.: Mathematical Logic, Addison-Wesley, Reading etc., 1967.
23. Sicherman, G.L., de Jonge, W., van de Riet, R.P.: Answering queries without revealing secrets, ACM Transactions on Database Systems 8,1 (1983), pp. 41-59.
24. Su, T.A., Ozsoyoglu, G.: Controlling FD and MVD inferences in multilevel relational database systems. IEEE Trans. on Knowledge and Data Engineering, 3(4):474-485, (1991).
25. Winslett, M, Smith, K., Qian, X.: Formal query languages for secure relational databases, ACM Transactions on Database Systems 19,4 (1994), pp. 626-662.

Iterative Modification and Incremental Evaluation of Preference Queries*

Jan Chomicki

Dept. of Computer Science and Engineering,
University at Buffalo, Buffalo, NY 14260-2000
chomicki@cse.buffalo.edu

Abstract. We present here a formal foundation for an iterative and incremental approach to constructing and evaluating preference queries. Our main focus is on *query modification*: a query transformation approach which works by revising the preference relation in the query. We provide a detailed analysis of the cases where the order-theoretic properties of the preference relation are preserved by the revision. We consider a number of different revision operators: union, prioritized and Pareto composition. We also formulate algebraic laws that enable incremental evaluation of preference queries.

1 Introduction

The notion of *preference* is common in various contexts involving decision or choice. Classical utility theory [10] views preferences as *binary relations*. This view has recently been adopted in database research [7, 8, 20, 22], where preference relations are used in formulating *preference queries*. In AI, various approaches to compact specification of preferences have been explored [6]. The semantics underlying such approaches typically relies on preference relations between worlds.

Preferences can be embedded into database query languages in several different ways. First, [7, 8, 20, 22] propose to introduce a special operator *"find all the most preferred tuples according to a given preference relation."* This operator is called *winnow* in [7, 8]. A special case of winnow is called *skyline* [5] and has been recently extensively studied [25, 3]. Second, [1, 17] assume that preference relations are defined using numeric utility functions and queries return tuples ordered by the values of a supplied utility function. It is well-known that numeric utility functions cannot represent all strict partial orders [10], not even those that occur in database applications in a natural way [8]. For example, utility functions cannot capture skylines. Also, ordered relations go beyond the classical relational model of data. The evaluation and optimization of queries over such relations requires significant changes to relational query processors and optimizers [18]. On the other hand, winnow can be seamlessly combined with any relational operators.

---

* Research supported by NSF grant IIS-0307434.

We adopt here the first approach, based on winnow, within the preference query framework of [8] (a similar model was described in [20]). In this framework, preference relations between tuples are defined by first-order logical formulas.

Example 1. Consider the relation $Car(Make, Year)$ and the following preference relation $\succ_{C_1}$ between Car tuples:

within each make, prefer a more recent car,

which can be defined as follows:

$$(m, y) \succ_{C_1} (m', y') \equiv m = m' \wedge y > y'.$$

The winnow operator ω_{C_1} returns for every make the most recent car available. Consider the instance r_1 of Car in Figure 1a. The set of tuples $\omega_{C_1}(r_1)$ is shown in Figure 1b.

	Make	Year
t_1	VW	2002
t_2	VW	1997
t_3	Kia	1997

(a)

	Make	Year
t_1	VW	2002
t_3	Kia	1997

(b)

Fig. 1. (a) The Car relation; (b) Winnow result

In this paper, we focus on preference queries of the form $\omega_\succ(R)$, consisting of a single occurrence of winnow. Here $\succ$ is a preference relation (typically defined by a formula), and R is a database relation. The relation R represents the space of possible choices. We also briefly discuss how our results can be applied to more general preference queries.

Past work on preference queries has made the assumption that preferences are *static*. However, this assumption is often not satisfied. User preferences change, sometimes as a direct consequence of evaluating a preference query. Therefore, we view preference querying as a *dynamic, iterative process*. The user submits a query and inspects the result. The result may be satisfactory, in which case the querying process terminates. Or, the result may be too large or too small, contain unexpected answers, or fail to contain expected answers. If the user is not satisfied with the query result, she has several further options:

Modify and resubmit the query. This is appropriate if the user decides to refine or change her preferences. For example, the user may have started with a partial or vague concept of her preferences [26]. We consider here query modification consisting of *revising* the preference relation $\succ$, although, of course, more general transformations may also be envisioned.

Update the database. This is appropriate if the user discovers that there are more (or fewer) possible choices than originally envisioned. For example, in comparison shopping the user may have discovered a new source of relevant data.

In this context we pursue the following research challenges:

Defining a repertoire of suitable preference relation revisions. In this work, we consider revisions obtained by *composing* the original preference relation with a new preference relation, and *transitively closing* the result (to guarantee transitivity). We study different composition operators: union, and prioritized and Pareto composition. Those operators represent several basic ways of combining preferences and have already been incorporated into preference query languages [8, 20]. The operators reflect different user attitudes towards *preference conflicts*. (A conflict is, intuitively, a situation in which two preference relations order the same pair of tuples differently.) Union ignores conflicts (and thus such conflicts need to be prevented if we want to obtain a preference relation which is a strict partial order). Prioritized composition resolves preference conflicts by consistently giving priority to one of the preference relations. Pareto composition resolves conflicts in a symmetric way. We emphasize that revision is done using composition because we want the revised preference relation to be uniquely defined in the same first-order language as the original preference relation. Clearly, the revision repertoire that we study in this paper does not exhaust all meaningful scenarios. One can also imagine approaches where axiomatic properties of preference revisions are studied, as in belief revision [13].

Identifying essential properties of preference revisions. We claim that revisions should preserve the order-theoretic properties of the original preference relations. For example, if we start with a preference relation which is a strict partial order, the revised relation should also have those properties. This motivates, among others, transitively closing preference relations to guarantee transitivity. Preserving order-theoretic properties of preference relations is particularly important in view of the iterative construction of preference queries where the output of a revision can serve as the input to another one. We study both necessary and sufficient conditions on the original and revising preference relations that yield the preservation of their order-theoretic properties. Necessary conditions are connected with the absence of preference conflicts. However, such conditions are typically not sufficient and stronger assumptions about the preference relations need to be made. Somewhat surprisingly, a special class of strict partial orders, interval orders, plays an important role in this context. The conditional preservation results we establish in this paper supplement those in [8, 20] and may be used in other contexts where preference relations are composed, for example in the implementation of preference query languages. Another desirable property of revisions is minimality in some well-defined sense. We define minimality in terms of symmetric difference of preference relations but there are clearly other possibilities.

Incremental evaluation of preference queries. At each point of the interaction with the user, the results of evaluating *previous* versions of the given preference query are available. Therefore, they can be used to make the evaluation of the *current* query more efficient. For both the preference revision and database update scenarios, we formulate algebraic laws that validate new query evaluation plans that use materialized results of past query evaluations. The laws use order-theoretic properties of preference relations in an essential way.

Example 2. Consider Example 1. Seeing the result of the query $\omega_{C_1}(r_1)$, a user may realize that the preference relation $\succ_{C_1}$ is not quite what she had in mind. The result of the query may contain some unexpected or unwanted tuples, for example t_3. Thus the preference relation needs to be modified, for example by revising it with the following preference relation $\succ_{C_2}$:

$$(m, y) \succ_{C_2} (m', y') \equiv m = "VW" \wedge m' \neq "VW" \wedge y = y'.$$

As there are no conflicts between $\succ_{C_1}$ and $\succ_{C_2}$, the user chooses union as the composition operator. However, to guarantee transitivity of the resulting preference relation, $\succ_{C_1} \cup \succ_{C_2}$ has to be transitively closed. So the revised relation is $\succ_{C*} \equiv TC(\succ_{C_1} \cup \succ_{C_2})$. (The explicit definition of $\succ_{C*}$ is given in Example 6.) The tuple t_3 is now dominated by t_2 (i.e., $t_2 \succ_{C*} t_3$) and will not be returned to the user.

The plan of the paper is as follows. In Section 2, we define the basic notions. In Section 3, we introduce preference revision. In Section 4, we discuss query modification and the preservation by revisions of order-theoretic properties of preference relations. In Section 5, we discuss incremental evaluation of preference queries in the context of query modification and database updates. In Section 6, we consider finite restrictions of preference relations. We briefly discuss related work in Section 7 and conclude in Section 8. Some proofs are outlined. The remaining results can be proved by exhaustive case analysis.

2   Basic Notions

We are working in the context of the relational model of data. Relation schemas consist of finite sets of attributes. For concreteness, we consider two infinite, countable domains: $\mathcal{D}$ (uninterpreted constants, for readability shown as strings) and $\mathcal{Q}$ (rational numbers), but our results, except where explicitly indicated, hold also for finite domains. We assume that database instances are finite sets of tuples. Additionally, we have the standard built-in predicates.

2.1   Preference Relations

We adopt here the framework of [8].

Definition 1. *Given a relation schema $R(A_1 \cdots A_k)$ such that U_i, $1 \leq i \leq k$, is the domain (either $\mathcal{D}$ or $\mathcal{Q}$) of the attribute A_i, a relation $\succ$ is a preference relation over R if it is a subset of $(U_1 \times \cdots \times U_k) \times (U_1 \times \cdots \times U_k)$.*

Although we assume that database instances are finite, in the presence of infinite domains preference relations can be infinite.

Typical properties of a preference relation $\succ$ include [10]:

- *irreflexivity*: $\forall x.\ x \not\succ x$;
- *transitivity*: $\forall x, y, z.\ (x \succ y \land y \succ z) \Rightarrow x \succ z$;
- *negative transitivity*: $\forall x, y, z.\ (x \not\succ y \land y \not\succ z) \Rightarrow x \not\succ z$;
- *connectivity*: $\forall x, y.\ x \succ y \lor y \succ x \lor x = y$;
- *strict partial order* (SPO) if $\succ$ is irreflexive and transitive;
- *interval order* (IO) [11] if $\succ$ is an SPO and satisfies the condition

$$\forall x, y, z, w.\ (x \succ y \land z \succ w) \Rightarrow (x \succ w \lor z \succ y);$$

- *weak order* (WO) if $\succ$ is a negatively transitive SPO;
- *total order* if $\succ$ is a connected SPO.

Every total order is a WO; every WO is an IO.

Definition 2. *A preference formula (pf)* $C(t_1, t_2)$ *is a first-order formula defining a preference relation* $\succ_C$ *in the standard sense, namely*

$$t_1 \succ_C t_2 \text{ iff } C(t_1, t_2).$$

An intrinsic preference formula (ipf) *is a preference formula that uses only built-in predicates.*

By using the notation $\succ_C$ for a preference relation, we assume that there is an underlying pf C. Occasionally, we will limit our attention to ipfs consisting of the following two kinds of atomic formulas (assuming we have two kinds of variables: $\mathcal{D}$-variables and $\mathcal{Q}$-variables):

- *equality constraints*: $x = y$, $x \neq y$, $x = c$, or $x \neq c$, where x and y are $\mathcal{D}$-variables, and c is an uninterpreted constant;
- *rational-order constraints*: $x \lambda y$ or $x \lambda c$, where $\lambda \in \{=, \neq, <, >, \leq, \geq\}$, x and y are $\mathcal{Q}$-variables, and c is a rational number.

An ipf all of whose atomic formulas are equality (resp. rational-order) constraints will be called an *equality* (resp. *rational-order*) ipf. If both equality and rational-order constraints are allowed in a formula, the formula will be called *equality/rational-order*. Clearly, ipfs are a special case of general constraints [23, 19], and define *fixed*, although possibly infinite, relations.

Proposition 1. *Satisfiability of quantifier-free equality/rational-order formulas is in NP.*

Proof. Satisfiability of conjunctions of atomic equality/rational-order constraints can be checked in linear time [15]. In an arbitrary quantifier-free equality/rational-order formula negation can be eliminated. Then in every disjunction one needs to nondeterministically select one disjunct, ultimately obtaining a conjunction of atomic constraints.

Proposition 1 implies that all the properties that can be polynomially reduced to validity of equality/rational-order formulas, for example all the order-theoretic properties listed above, can be decided in co-NP.

Every preference relation $\succ$ generates an indifference relation $\sim$: two tuples t_1 and t_2 are *indifferent* ($t_1 \sim t_2$) if neither is preferred to the other one, i.e., $t_1 \not\succ t_2$ and $t_2 \not\succ t_1$. We denote by $\sim_C$ the indifference relation generated by $\succ_C$.

Composite preference relations are defined from simpler ones using logical connectives. We focus on the following basic ways of composing preference relations over the same schema:

- *union*: $t_1\ (\succ_1 \cup \succ_2)\ t_2$ iff $t_1 \succ_1 t_2 \lor t_1 \succ_2 t_2$;
- *prioritized composition*: $t_1\ (\succ_1 \rhd \succ_2)\ t_2$ iff $t_1 \succ_1 t_2 \lor (t_2 \not\succ_1 t_1 \land t_1 \succ_2 t_2)$;
- *Pareto composition*:

$$t_1\ (\succ_1 \otimes \succ_2)\ t_2 \text{ iff } (t_1 \succ_1 t_2 \land t_2 \not\succ_2 t_1) \lor (t_1 \succ_2 t_2 \land t_2 \not\succ_1 t_1).$$

We will use the above composition operators to construct revisions of given preference relations. We also consider transitive closure:

Definition 3. *The* transitive closure *of a preference relation $\succ$ over a relation schema R is a preference relation $TC(\succ)$ over R defined as:*

$$(t_1, t_2) \in TC(\succ) \text{ iff } t_1 \succ^n t_2 \text{ for some } n > 0,$$

where:

$$t_1 \succ^1 t_2 \equiv t_1 \succ t_2$$
$$t_1 \succ^{n+1} t_2 \equiv \exists t_3.\ t_1 \succ t_3 \land t_3 \succ^n t_2.$$

Clearly, in general Definition 3 leads to infinite formulas. However, in the cases that we consider in this paper the preference relation $\succ_{C*}$ will in fact be defined by a finite formula.

Proposition 2. *Transitive closure of every preference relation defined by an equality/rational-order ipf is an ipf of at most exponential size, which can be computed in exponential time.*

Proof. This is because transitive closure can be expressed in Datalog and the evaluation of Datalog programs over equality and rational-order constraints terminates in exponential time (combined complexity) [19]. □

In the cases mentioned above, the transitive closure of a given preference relation is a relation definable in the signature of the preference formula. But clearly transitive closure itself, unlike union and prioritized or Pareto composition, is not a first-order definable operator.

2.2   Winnow

We define now an algebraic operator that picks from a given relation the set of the *most preferred tuples*, according to a given preference relation.

Definition 4. [8] *If R is a relation schema and $\succ$ a preference relation over R, then the* winnow *operator is written as $\omega_\succ(R)$, and for every instance r of R:*

$$\omega_\succ(r) = \{t \in r \mid \neg \exists t' \in r.\ t' \succ t\}.$$

If a preference relation is defined using a pf C, we write simply ω_C instead of $\omega_{\succ_C}$. A *preference query* is a relational algebra query containing at least one occurrence of the winnow operator.

3   Preference Revisions

The basic setting is as follows: We have an *original* preference relation $\succ$ and revise it with a *revising* preference relation $\succ_0$ to obtain a *revised* preference relation $\succ'$. We also call $\succ'$ a *revision* of $\succ$. We assume that $\succ$, $\succ_0$, and $\succ'$ are preference relations over the same schema, and that all of them satisfy at least the properties of SPOs.

In our setting, a revision is obtained by composing $\succ$ with $\succ_0$ using union, prioritized or Pareto composition, and transitively closing the result (if necessary to obtain transitivity). However, we formulate some properties, like minimality or compatibility, in more general terms.

To define minimality, we order revisions using the symmetric difference ($\triangle$).

Definition 5. *Assume* $\succ_1$ *and* $\succ_2$ *are two revisions of a preference relation* $\succ$ *with a preference relation* $\succ_0$. *We say that* $\succ_1$ *is* closer *than* $\succ_2$ *to* $\succ$ *if* $\succ_1 \triangle \succ \; \subset \; \succ_2 \triangle \succ$.

For finite domains and SPOs, the closeness order defined above concides with the order based on the partial-order distance [4] of the revision to the original relation $\succ$.

To further describe the behavior of revisions, we define several kinds of *preference conflicts*. The intuition here is to characterize those conflicts that, when eliminated by prioritized or Pareto composition, reappear if the resulting preference relation is closed by transitivity.

Definition 6. *A* 0-conflict *between a preference relation* $\succ$ *and a preference relation* $\succ_0$ *is a pair* (t_1, t_2) *such that* $t_1 \succ_0 t_2$ *and* $t_2 \succ t_1$. *A* 1-conflict *between* $\succ$ *and* $\succ_0$ *is a pair* (t_1, t_2) *such that* $t_1 \succ_0 t_2$ *and there exist* $s_1, \ldots s_k$, $k \geq 1$, *such that* $t_2 \succ s_1 \succ \cdots \succ s_k \succ t_1$ *and* $t_1 \not\succ_0 s_k \not\succ_0 \cdots \not\succ_0 s_1 \not\succ_0 t_2$. *A* 2-conflict *between* $\succ$ *and* $\succ_0$ *is a pair* (t_1, t_2) *such that there exist* $s_1, \ldots, s_k$, $k \geq 1$ *and* $w_1, \ldots, w_m$, $m \geq 1$, *such that* $t_2 \succ s_1 \succ \cdots \succ s_k \succ t_1$, $t_1 \not\succ_0 s_k \not\succ_0 \cdots \not\succ_0 s_1 \not\succ_0 t_2$, $t_1 \succ_0 w_1 \succ_0 \cdots \succ_0 w_m \succ t_2$, *and* $t_2 \not\succ w_m \not\succ \cdots \not\succ w_1 \not\succ t_1$.

A 1-conflict is a 0-conflict if $\succ$ is an SPO, but not necessarily vice versa. A 2-conflict is a 1-conflict if $\succ_0$ is an SPO. The different kinds of conflicts are pictured in Figures 2 and 3 ($\stackrel{\backsim}{\succ}$ denotes the complement of $\succ$).

Example 3. If $\succ_0 = \{(a, b)\}$ and $\succ = \{(b, a)\}$, then (a, b) is a 0-conflict which is not a 1-conflict. If we add (b, c) and (c, a) to $\succ$, then the conflict becomes a 1-conflict ($s_1 = c$). If we further add (c, b) or (a, c) to $\succ_0$, then the conflict is not a 1-conflict anymore. On the other hand, if we add (a, d) and (d, b) to $\succ_0$ instead, then we obtain a 2-conflict.

We assume here that the preference relations $\succ$ and $\succ_0$ are SPOs. If $\succ' = TC(\succ \cup \succ_0)$, then for every 0-conflict between $\succ$ and $\succ_0$, we still obviously have $t_1 \succ' t_2$ and $t_2 \succ' t_1$. Therefore, we say that the union does not resolve any conflicts. On the other hand, if $\succ' = TC(\succ_0 \rhd \succ)$, then for each 0-conflict (t_1, t_2), $t_1 \succ_0 \rhd \succ t_2$ and $\neg(t_2 \succ_0 \rhd \succ t_1)$. In the case of 1-conflicts, we get again $t_1 \succ' t_2$ and $t_2 \succ' t_1$. But in the case where a 0-conflict is not a 1-conflict, we get

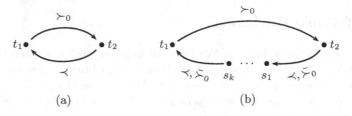

Fig. 2. (a) 0-conflict; (b) 1-conflict

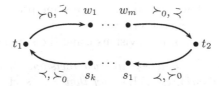

Fig. 3. 2-conflict

only $t_1 \succ' t_2$. Thus we say that prioritized composition resolves those 0-conflicts that are not 1-conflicts. Finally, if $\succ' = TC(\succ \otimes \succ_0)$, then for each 1-conflict (t_1, t_2), $\neg(t_1 \succ \otimes \succ_0 t_2)$ and $\neg(t_2 \succ \otimes \succ_0 t_1)$. We get $t_1 \succ' t_2$ and $t_2 \succ' t_1$ if the conflict is a 2-conflict, but if it is not, we obtain only $t_2 \succ' t_1$. Thus we say that Pareto composition resolves those 1-conflicts that are not 2-conflicts.

We now characterize those combinations of $\succ$ and $\succ_0$ that avoid different kinds of conflicts.

Definition 7. *A preference relation $\succ$ is i-compatible ($i = 0, 1, 2$) with a preference relation $\succ_0$ if there are no i-conflicts between $\succ$ and $\succ_0$.*

0- and 2- compatibility are symmetric. 1-compatibility is not necessarily symmetric. For SPOs, 0-compatibility implies 1-compatibility and 1-compatibility implies 2-compatibility. Examples 1 and 2 show a pair of 0-compatible relations. 0-compatibility of $\succ$ and $\succ_0$ *does not require* the acyclicity of $\succ \cup \succ_0$ or that one of the following hold: $\succ \subseteq \succ_0$, $\succ_0 \subseteq \succ$, or $\succ \cap \succ_0 = \emptyset$.

Propositions 1 and 2 imply that all the variants of compatibility defined above are decidable for equality/rational order ipfs. For example, 1-compatibility is expressed by the condition $\succ_0^{-1} \cap TC(\succ - \succ_0^{-1}) = \emptyset$ where $\succ_0^{-1}$ is the inverse of the preference relation $\succ_0$.

0-compatibility of $\succ$ and $\succ_0$ is a *necessary* condition for $TC(\succ \cup \succ_0)$ to be irreflexive, and thus an SPO. Similar considerations apply to $TC(\succ_0 \triangleright \succ)$ and 1-compatibility, and $TC(\succ \otimes \succ_0)$ and 2-compatibility. In the next section, we will see that those conditions are not *sufficient*: further restrictions on the preference relations will be introduced.

We conclude by noting that in the absence of conflicts all three notions of preference composition coincide.

Lemma 1. *For every 0-compatible preference relations $\succ$ and $\succ_0$:*

$$\succ_0 \cup \succ = \succ_0 \triangleright \succ = \succ_0 \otimes \succ$$

4   Query Modification

In this section, we study preference query modification. A given preference query $\omega_{\succ}(R)$ is transformed to the query $\omega_{\succ'}(R)$ where $\succ'$ is obtained by composing the original preference relation $\succ$ with the revising preference relation $\succ_0$, and transitively closing the result. (The last step is clearly unnecessary if the obtained preference relation is already transitive.) We want $\succ'$ to satisfy the same order-theoretic properties as $\succ$ and $\succ_0$, and to be minimally different from $\succ$. To achieve those goals, we impose additional conditions on $\succ$ and $\succ_0$.

For every $\theta \in \{\cup, \triangleright, \otimes\}$, we consider the order-theoretic properties of the preference relation $\succ' = \succ_0 \theta \succ$, or $\succ' = TC(\succ_0 \theta \succ)$ if $\succ_0 \theta \succ$ is not guaranteed to be transitive. To ensure that this preference relation is an SPO, only irreflexivity has to be guaranteed; for weak orders one has also to establish negative transitivity.

4.1   Strict Partial Orders

SPOs have several important properties from the user's point of view, and thus their preservation is desirable. For instance, all the preference relations defined in [20] and in the language Preference SQL [22] are SPOs. Moreover, if $\succ$ is an SPO, then the winnow $\omega_{\succ}(r)$ is nonempty if (a finite) r is nonempty. The fundamental algorithms for computing winnow require that the preference relation be an SPO [8]. Also, in that case incremental evaluation of preference queries becomes possible (Proposition 4 and Theorem 7).

Theorem 1. *For every 0-compatible preference relations $\succ$ and $\succ_0$ such that one is an interval order (IO) and the other an SPO, the preference relation $TC(\succ_0 \theta \succ)$, where $\theta \in \{\cup, \triangleright, \otimes\}$, is an SPO. Additionally, if the IO is a WO, then $TC(\succ_0 \theta \succ) = \succ_0 \theta \succ$.*

Proof. By Lemma 1, 0-compatibility implies that $\succ_0 \cup \succ = \succ_0 \triangleright \succ = \succ_0 \otimes \succ$. Thus, WLOG we consider only union. Assume $\succ_0$ is an IO. If $TC(\succ \cup \succ_0)$ is not irreflexive, then $\succ \cup \succ_0$ has a cycle. Consider such cycle of minimum length. It consists of edges that are alternately labeled $\succ_0$ (only) and $\succ$ (only). (Otherwise the cycle can be shortened). If there is more than one non-consecutive $\succ_0$-edge in the cycle, then $\succ_0$ being an IO implies that the cycle can be shortened. So the cycle consists of two edges: $t_1 \succ_0 t_2$ and $t_2 \succ t_1$. But this is a 0-conflict violating 0-compatibility of $\succ$ and $\succ_0$.

It is easy to see that there is no preference relation which is an SPO, contains $\succ \cup \succ_0$, and is closer to $\succ$ than $TC(\succ \cup \succ_0)$.

As can be seen from the above proof, the fact that one of the preference relations is an interval order makes it possible to eliminate those paths (and thus also cycles) in $TC(\succ \cup \succ_0)$ that interleave $\succ$ and $\succ_0$ more than once. In this way acyclicity reduces to the lack of 0-conflicts.

It seems that the interval order (IO) requirement in Theorem 1 cannot be weakened without needing to strengthen the remaining assumptions. If neither

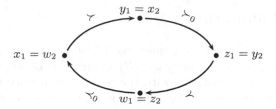

Fig. 4. A cycle for 0-compatible relations that are not IOs

of $\succ$ and $\succ_0$ is an IO, then we can find such elements $x_1, y_1, z_1, w_1, x_2, y_2, z_2, w_2$ that $x_1 \succ y_1, z_1 \succ w_1, x_1 \not\succ w_1, z_1 \not\succ y_1, x_2 \succ_0 y_2, z_2 \succ_0 w_2, x_2 \not\succ_0 w_2$, and $z_2 \not\succ_0 y_2$. If we choose $y_1 = x_2$, $z_1 = y_2$, $w_1 = z_2$, and $x_1 = w_2$, then we get a cycle in $\succ \cup \succ_0$. Note that in this case $\succ$ and $\succ_0$ are still 0-compatible. Also, there is no SPO preference relation which contains $\succ \cup \succ_0$ because each such relation has to contain $TC(\succ \cup \succ_0)$. This situation is pictured in Figure 4.

Example 4. Consider again the preference relation $\succ_{C_1}$:

$$(m, y) \succ_{C_1} (m', y') \equiv m = m' \wedge y > y'.$$

Suppose that the new preference information is captured as $\succ_{C_3}$ which is an IO but not a WO:

$$(m, y) \succ_{C_3} (m', y') \equiv m = \text{"VW"} \wedge y = 1999 \wedge m' = \text{"Kia"} \wedge y' = 1999.$$

Then $TC(\succ_{C_1} \cup \succ_{C_3})$, which properly contains $\succ_{C_1} \cup \succ_{C_3}$, is defined as the SPO $\succ_{C_4}$:

$$(m, y) \succ_{C_4} (m', y') \equiv m = m' \wedge y > y' \vee$$
$$m = \text{"VW"} \wedge y \geq 1999 \wedge m' = \text{"Kia"} \wedge y' \leq 1999.$$

For dealing with *prioritized composition*, 0-compatibility can be replaced by a less restrictive condition, 1-compatibility, because prioritized composition already provides a way of resolving some conflicts.

Theorem 2. *For every preference relations $\succ$ and $\succ_0$ such that $\succ_0$ is an IO, $\succ$ is an SPO and $\succ$ is 1-compatible with $\succ_0$, the preference relation $TC(\succ_0 \triangleright \succ)$ is an SPO.*

Proof. We assume that $TC(\succ_0 \triangleright \succ)$ is not irreflexive and consider a cycle of minimum length in $\succ_0 \triangleright \succ$. If the cycle has two non-consecutive edges labeled (not necessarily exclusively) by $\succ_0$, then it can be shortened, because $\succ_0$ is an IO. The cycle has to consist of an edge $t_1 \succ_0 t_2$ and a sequence of edges (labeled only by $\succ$): $t_2 \succ t_3, \ldots, t_{n-1} \succ t_n, t_n \succ t_1$ such that $n > 2$ and $t_1 \not\succ_0 t_n \not\succ_0 \ldots \not\succ_0 t_3 \not\succ_0 t_2$. (We cannot shorten sequences of consecutive $\succ$-edges because $\succ$ is not necessarily preserved in $\succ_0 \triangleright \succ$.) Thus (t_1, t_2) is a 1-conflict violating 1-compatibility of $\succ$ with $\succ_0$.

Clearly, there is no SPO preference relation which contains $\succ_0 \triangleright \succ$, and is closer to $\succ$ than $TC(\succ_0 \triangleright \succ)$. Violating any of the conditions of Theorem 2 may lead to a situation in which no SPO preference relation which contains $\succ_0 \triangleright \succ$ exists.

If $\succ_0$ is a WO, the requirement of 1-compatibility and the computation of transitive closure are unnecessary.

Theorem 3. *For every preference relations $\succ_0$ and $\succ$ such that $\succ_0$ is a WO and $\succ$ an SPO, the preference relation $\succ_0 \triangleright \succ$ is an SPO.*

Let's turn now to *Pareto composition*. There does not seem to be any simple way to *weaken* the assumptions in Theorem 1 using the notion of 2-compatibility. Assuming that $\succ$, $\succ_0$, or even both are IOs does not sufficiently restrict the possible interleavings of $\succ$ and $\succ_0$ in $TC(\succ_0 \otimes \succ)$ because neither of those two preference relations is guaranteed to be preserved in $TC(\succ_0 \otimes \succ)$. However, we can establish a weaker version of Theorem 3.

Theorem 4. *For every preference relations $\succ_0$ and $\succ$ such that both are WOs, the preference relation $\succ_0 \otimes \succ$ is an SPO.*

Proposition 2 implies that for all preference relations defined using equality/-rational-order ipfs, the computation of the preference relations $TC(\succ \cup \succ_0)$, $TC(\succ_0 \triangleright \succ)$ and $TC(\succ \otimes \succ_0)$ terminates. The computation of transitive closure is done in a completely database-independent way.

Example 5. Consider Examples 1 and 4. We can infer that

$$t_1 = ("\text{VW}", 2002) \succ_{C_4} ("\text{Kia}", 1997) = t_3,$$

because $("\text{VW}", 2002) \succ_{C_1} ("\text{VW}", 1999)$, $("\text{VW}", 1999) \succ_{C_3} ("\text{Kia}", 1999)$, and $("\text{Kia}", 1999) \succ_{C_1} ("\text{Kia}", 1997)$. Note that the tuples $("\text{VW}", 1999)$ and $("\text{Kia}", 1999)$ are *not* in the database.

If the conditions of Theorems 1 and 2 do not apply, Proposition 2 implies that for equality/rational order ipfs the computation of $TC(\succ \cup \succ_0)$, $TC(\succ_0 \triangleright \succ)$ and $TC(\succ \otimes \succ_0)$ yields some finite ipf $C(t_1, t_2)$. Thus the irreflexivity of the resulting preference relation reduces to the unsatisfiability of $C(t, t)$, which by Proposition 1 is a decidable problem for equality/rational order ipfs. Of course, the relation, being a transitive closure, is already transitive.

Example 6. Consider Examples 1 and 2. Neither of the preference relations $\succ_{C_1}$ and $\succ_{C_2}$ is an interval order. Therefore, the results established earlier in this section do not apply. The preference relation $\succ_{C*} = TC(\succ_{C_1} \cup \succ_{C_2})$ is defined as follows (this definition is obtained using Constraint Datalog computation):

$$(m, y) \succ_{C*} (m', y') \equiv m = m' \wedge y > y' \vee m = "\text{VW}" \wedge m' \neq "\text{VW}" \wedge y \geq y'$$

The preference relation $\succ_{C*}$ is irreflexive (this can be effectively checked). It also properly contains $\succ_{C_1} \cup \succ_{C_2}$, because $t_1 \succ_{C*} t_3$ but $t_1 \not\succ_{C_1} t_3$ and $t_1 \not\succ_{C_2} t_3$. The query $\omega_{C*}(Car)$ evaluated in the instance r_1 (Figure 1) returns only the tuple t_1.

4.2   Weak Orders

Weak orders are practically important because they capture the situation where the domain can be decomposed into layers such that the layers are totally ordered and all the elements in one layer are mutually indifferent. This is the case, for example, if the preference relation can be represented using a numeric utility function. If the preference relation is a WO, a particularly efficient (essentially single pass) algorithm for computing winnow is applicable [9].

We will see that for weak orders the transitive closure computation is unnecessary and minimal revisions are directly definable in terms of the preference relations involved.

Theorem 5. *For every 0-compatible WO preference relations $\succ$ and $\succ_0$, the preference relations $\succ \cup \succ_0$ and $\succ \otimes \succ_0$ are WO.*

For prioritized composition, we can relax the 0-compatibility assumption. This immediately follows from the fact that WOs are closed with respect to prioritized composition [8].

Proposition 3. *For every WO preference relations $\succ$ and $\succ_0$, the preference relation $\succ_0 \rhd \succ$ is a WO.*

A basic notion in utility theory is that of *representability* of preference relations using numeric utility functions:

Definition 8. *A real-valued function u over a schema R represents a preference relation $\succ$ over R iff*

$$\forall t_1, t_2 \; [t_1 \succ t_2 \text{ iff } u(t_1) > u(t_2)].$$

Such a preference relation is called utility-based.

Being a WO is a necessary condition for the existence of a numeric representation for a preference relation. However, it is not sufficient for uncountable orders [10]. It is natural to ask whether the existence of numeric representations for the preference relations $\succ$ and $\succ_0$ implies the existence of such a representation for the preference relation $\succ' = (\succ_0 \; \theta \; \succ)$ where $\theta \in \{\cup, \rhd, \otimes\}$. This is indeed the case.

Theorem 6. *Assume that $\succ$ and $\succ_0$ are WO preference relations such that*

1. *$\succ$ and $\succ_0$ are 0-compatible,*
2. *$\succ$ can be represented using a real-valued function u,*
3. *$\succ_0$ can be represented using a real-valued function u_0.*

Then $\succ' = \succ_0 \; \theta \; \succ$, where $\theta \in \{\cup, \rhd, \otimes\}$, is a WO preference relation that can be represented using any real-valued function u' such that for all x, $u'(x) = a \cdot u(x) + b \cdot u_0(x) + c$ where a and b are arbitrary positive real numbers.

Proof. By case analysis. The assumption of 0-compatibility is essential.

Surprisingly, 0-compatibility requirement cannot in general be replaced by 1-compatibility if we replace $\cup$ by $\rhd$ in Theorem 6. This follows from the fact that the lexicographic composition of one-dimensional standard orders over $\mathcal{R}$ is not representable using a utility function [10]. Thus, preservation of *representability* is possible only under 0-compatibility, in which case $\succ_0 \cup \succ \; = \; \succ_0 \rhd \succ \; = \; \succ_0 \otimes \succ$. (Lemma 1). (The results [10] indicate that for countable domains considered in this paper, the prioritized composition of WOs, being a WO, is representable using a utility function. However, that utility function is not definable in terms of the utility functions representing the given orders.)

We conclude this section by showing a general scenario in which the union of WOs occurs in a natural way. Assume that we have a numeric utility function u representing a (WO) preference relation $\succ$. The indifference relation $\sim$ generated by $\succ$ is defined as:

$$x \sim y \; \equiv \; u(x) = u(y).$$

Suppose that the user discovers that $\sim$ is too coarse and needs to be further refined. This may occur, for example, when x and y are tuples and the function u takes into account only some of their components. Another function u_0 may be defined to take into account other components of x and y (such components are called *hidden attributes* [26]). The revising preference relation $\succ_0$ is now:

$$x \succ_0 y \; \equiv \; u(x) = u(y) \wedge u_0(x) > u_0(y).$$

It is easy to see that $\succ_0$ is an SPO 0-compatible with $\succ$ but not necessarily a WO. Therefore, by Theorem 1 the preference relation $\succ \cup \succ_0$ is an SPO.

5 Incremental Evaluation

5.1 Query Modification

We show here how the already computed result of the original preference query can be reused to make the evaluation of the modified query more efficient. We will use the following result.

Proposition 4. [8] *If $\succ_1$ and $\succ_2$ are preference relations over a relation schema R and $\succ_1 \subseteq \succ_2$, then for all instances r of R:*

- $\omega_{\succ_2}(r) \subseteq \omega_{\succ_1}(r)$;
- $\omega_{\succ_2}(\omega_{\succ_1}(r)) = \omega_{\succ_2}(r)$ *if $\succ_1$ and $\succ_2$ are SPOs.*

Consider the scenario in which we iteratively modify a given preference query by revising the preference relation using only union in such a way that the revised preference relation is an SPO (for example, if the assumptions of Theorem 1 are satisfied). We obtain a sequence of preference relations $\succ_1, \ldots, \succ_n$ such that $\succ_1 \subseteq \cdots \subseteq \succ_n$.

In this scenario, the sequence of query results is:

$$r_0 = r, r_1 = \omega_{\succ_1}(r), r_2 = \omega_{\succ_2}(r), \ldots, r_n = \omega_{\succ_n}(r).$$

Proposition 4 implies that the sequence $r_0, r_1, \ldots, r_n$ is decreasing:

$$r_0 \supseteq r_1 \supseteq \cdots \supseteq r_n$$

and that it can be computed incrementally:

$$r_1 = \omega_{\succ_1}(r_0), r_2 = \omega_{\succ_2}(r_1), \ldots, r_n = \omega_{\succ_n}(r_{n-1}).$$

To compute r_i, there is no need to look at the tuples in $r - r_{i-1}$, nor to recompute winnow from scratch. The sets of tuples $r_1, \ldots, r_n$ are likely to have much smaller cardinality than $r_0 = r$.

It is easy to see that the above comments apply to all cases where the revised preference relation is a superset of the original preference relation. Unfortunately, this is not the case for revisions that use prioritized or Pareto composition. However, given a specific pair of preference relations $\succ$ and $\succ_0$, one can still effectively check whether $TC(\succ_0 \rhd \succ)$ or $TC(\succ_0 \otimes \succ)$ contains $\succ$ if the validity of preference formulas is decidable, as is the case for equality/rational-order formulas (Proposition 1).

5.2   Database Update

In the previous section we studied query modification: the query is modified, while the database remains unchanged. Here we reverse the situation: the query remains the same and the database is updated.

We consider first updates that are insertions of sets of tuples. For a database relation r, we denote by $\Delta^+ r$ the set of inserted tuples. We show how the previous result of a given preference query can be reused to make the evaluation of the same query in an updated database more efficient.

We first establish the following result.

Theorem 7. *For every preference relation $\succ$ over R which is an SPO and every instance r of R:*

$$\omega_\succ(r \cup \Delta^+ r) = \omega_\succ(\omega_\succ(r) \cup \Delta^+ r).$$

Consider now the scenario in which we have a preference relation $\succ$, which is an SPO, and a sequence of relations

$$r_0 = r, r_1 = r_0 \cup \Delta^+ r_0, r_2 = r_1 \cup \Delta^+ r_1, \ldots, r_n = r_{n-1} \cup \Delta^+ r_{n-1}.$$

Theorem 7 shows that

$$\omega_\succ(r_1) = \omega_\succ(\omega_\succ(r_0) \cup \Delta^+ r_0)$$
$$\omega_\succ(r_2) = \omega_\succ(\omega_\succ(r_1) \cup \Delta^+ r_1)$$
$$\cdots$$
$$\omega_\succ(r_n) = \omega_\succ(\omega_\succ(r_{n-1}) \cup \Delta^+ r_{n-1}).$$

Therefore, each subsequent evaluation of winnow can reuse the result of the previous one. This is advantageous because winnow returns a subset of the given relation and this subset is often much smaller than the relation itself.

Clearly, the algebraic law, stated in Theorem 7, can be used together with other, well-known laws of relational algebra and the laws specific to preference queries [8, 21] to produce a variety of rewritings of a given preference query. To see how a more complex preference query can be handled, let's consider the query consisting of winnow and selection, $\omega_\succ(\sigma_\alpha(R))$. We have

$$\omega_\succ(\sigma_\alpha(r \cup \Delta^+ r)) = \omega_\succ(\sigma_\alpha(r) \cup \sigma_\alpha(\Delta^+ r)) = \omega_\succ(\omega_\succ(\sigma_\alpha(r)) \cup \sigma_\alpha(\Delta^+ r))$$

for every instance r of R. Here again, one can use the previous result of the query, $\omega_\succ(\sigma_\alpha(r))$, to make its current evaluation more efficient. Other operators that distribute through union, for example projection and join, can be handled in the same way.

Next, we consider updates that are deletions of sets of tuples. For a database relation r, we denote by $\Delta^- r$ the set of deleted tuples.

Theorem 8. *For every preference relation $\succ$ over R and every instance r of R:*

$$\omega_\succ(r) - \Delta^- r \subseteq \omega_\succ(r - \Delta^- r).$$

Theorem 8 gives an incremental way to compute an approximation of winnow from below. It seems that in the case of deletion there cannot be an exact law along the lines of Theorem 7. This is because the deletion of some tuples from the original database may promote some originally dominated (and discarded) tuples into the result of winnow over the updated database.

Example 7. Consider the following preference relation $\succ = \{(a, b_1), \ldots, (a, b_n)\}$ and the database $r = \{a, b_1, \ldots, b_n\}$. Then $\omega_\succ(r) = \{a\}$ but $\omega_\succ(r - \{a\}) = \{b_1, \ldots, b_n\}$.

6   Finite Restrictions of Preference Relations

It is natural to consider *restrictions* of preference relations to given database instances [27]. If r is an instance of a relation schema R and $\succ$ is a preference relation over R, then $[\succ]_r = \succ \cap r \times r$ is also a preference relation over R and $\omega_{[\succ]_r}(r) = \omega_\succ(r)$.

The advantage of using $[\succ]_r$ instead of $\succ$ comes from the fact that the former depends on the database contents and can have stronger properties than the latter. For example, $[\succ]_r$ may be an SPO (or a WO), while $\succ$ is not. (Clearly, $[\succ]_r$ inherits all the order-theoretic properties of $\succ$, studied in the present paper.) Similarly, $[\succ]_r$ may be *i*-compatible with $[\succ_0]_r$, while $\succ$ is not *i*-compatible with $\succ_0$. On the other hand, $\succ$ makes more elaborate use of the preference information than $[\succ]_r$ and does not require adaptation if the input database changes.

Example 8. Let $\succ = \{(a, b)\}$, $\succ_0 = \{(b, c)\}$, $r = \{a, c\}$. Thus $\omega_\succ(r) = \omega_{[\succ]_r}(r) = \{a, c\}$. Consider revision using union, as in Theorem 1. The revised preference relation $\succ_1 = TC(\succ \cup \succ_0) = \{(a, b), (b, c), (a, c)\}$. On the other hand, $[\succ]_r = [\succ_0]_r = \emptyset$. Thus the revised preference relation $\succ_2 = TC([\succ]_r \cup [\succ_0]_r) = \emptyset$. After the revision, $\omega_{\succ_1}(r) = \{a\}$ and $\omega_{\succ_2} = \{a, c\}$. So in the latter case revision has no impact on preference. We also note that $[TC(\succ \cup \succ_0)]_r \neq TC([\succ]_r \cup [\succ_0]_r)$, and thus the correspondence between the unrestricted and the restricted preference relations no longer holds after the revision.

A related issue is that of *non-intrinsic* preference relations. Such relations are defined using formulas that refer not only to built-in predicates.

Example 9. The following preference relation is not intrinsic:

$$x \succ_{Pref} y \equiv Pref(x, y)$$

where *Pref* is a database relation. One can think of such a relation as representing *stored* preferences.

Revising non-intrinsic preference relations looks problematic. First, it is typically not possible to establish the simplest order-theoretic properties of such relations. For instance, in Example 9 it is not possible to determine the irreflexivity or transitivity of $\succ_{Pref}$ on the basis of its definition. Whether such properties are satisfied depends on the contents of the database relation *Pref*. Second, the transitive closure of a non-intrinsic preference relation may fail to be expressed as a finite formula. Again, Example 9 can be used to illustrate this point. The above problems disappear, however, if we consider $[\succ]_r$ instead of $\succ$.

7   Related Work

[16] presents a general framework for modeling change in preferences. Preferences are represented syntactically using sets of ground preference formulas, and their semantics is captured using sets of preference relations. Thanks to the syntactic representation preference revision is treated similarly, though not identically, to belief revision [13], and some axiomatic properties of preference revisions are identified. The result of a revision is supposed to be minimally different from the original preference relation (using a notion of minimality based on symmetric difference) and satisfy some additional background postulates, for example specific order axioms. [16] does not address the issue of constructing or defining revised relations, nor does it study the properties of specific classes of preference relations. On the other hand, [16] discusses also preference contraction, and domain expansion and shrinking.

In our opinion, there are several fundamental differences between belief and preference revision. In belief revision, propositional theories are revised with propositional formulas, yielding new theories. In preference revision, binary preference relations are revised with other preference relations, yielding new preference relations. Preference relations are single, finitely representable (though possibly infinite) first-order structures, satisfying order axioms. Belief revision focuses on axiomatic properties of belief revision operators and various notions of revision minimality. Preference revision focuses on axiomatic, order-theoretic properties of revised preference relations and the definability of such relations (though still taking revision minimality into account).

[28] considers revising a ranking (a WO) of a finite set of product profiles with new information, and shows that a new ranking, satisfying the AGM belief revision postulates [13], can be computed in a simple way. [26] formulates various scenarios of preference revision and does not contain any formal framework. [29] studies revision and contraction of finite WO preference relations by single pairs $t_1 \succ_0 t_2$. [12] describes minimal change revision of *rational* preference relations between propositional formulas.

Two different approaches to preference queries have been pursued in the literature: qualitative and quantitative. In the *qualitative* approach, preferences are specified using binary *preference relations* [24, 14, 7, 8, 20, 22]. In the *quantitative* utility-based approach, preferences are represented using *numeric utility functions* [1, 17], as shown in Section 4. The qualitative approach is strictly more general than the quantitative one, since one can define preference relations in terms of utility functions. However, only WO preference relations can be represented by numeric utility functions [10]. Preferences that are not WOs are common in database applications, c.f., Example 1.

Example 10. There is no utility function that captures the preference relation described in Example 1. Since there is no preference defined between t_1 and t_3 or t_2 and t_3, the score of t_3 should be equal to the scores of both t_1 and t_2. But this implies that the scores of t_1 and t_2 are equal which is not possible since t_1 is preferred over t_2.

This lack of expressiveness of the quantitative approach is well known in utility theory [10].

In the earlier work on preference queries [8, 20], one can find positive and negative results about closure of different classes of orders, including SPOs and WOs, under various composition operators. The results in the present paper are, however, new. Restricting the relations $\succ$ and $\succ_0$ (for example, assuming the interval order property and compatibility) and applying transitive closure where necessary make it possible to come up with positive counterparts of the negative results in [8]. For example, [8] shows that SPOs and WOs are in general not closed w.r.t. union, which should be contrasted with Theorems 1 and 5. In [20], Pareto and prioritized composition are defined somewhat differently from the present paper. The operators combine two preference relations, each defined over some database relation. The resulting preference relation is defined over the Cartesian product of the database relations. So such operators are not useful in the context of revision of preference relations. On the other hand, the careful design of the language guarantees that every preference relation that can be defined is an SPO.

Probably the most thoroughly studied class of qualitative preference queries is the class of *skyline* queries. A skyline query partitions all the attributes of a relation into DIFF, MAX, and MIN attributes. Only tuples with identical values of all DIFF attributes are comparable; among those, MAX attribute values are maximized and MIN values are minimized. The query in Example 1 is a very simple skyline query [5], with *Make* as a DIFF and *Year* as a MAX attribute. Without DIFF attributes, a skyline is a special case of n-ary Pareto composition.

Algorithms for evaluating qualitative preference queries are described in [8, 27], and for evaluating skyline queries, in [5, 25, 3]. [2] describes how to implement preference queries that use Pareto compositions of utility-based preference relations. In Preference SQL [22] general preference queries are implemented by a translation to SQL. [17] describes how materialized results of utility-based preference queries can be used to answer other queries of the same kind.

8   Conclusions and Future Work

We have presented a formal foundation for an iterative and incremental approach to constructing ans evaluating preference queries. Our main focus is on *query modification*, a query transformation approach which works by revising the preference relation in the query. We have provided a detailed analysis of the cases where the order-theoretic properties of the preference relation are preserved by the revision. We considered a number of different revision operators: union, prioritized and Pareto composition. We have also formulated algebraic laws that enable incremental evaluation of preference queries.

Future work includes the integration of our results with standard query optimization techniques, both rewriting- and cost-based. Semantic query optimization techniques for preference queries [9] can also be applied in this context. Another possible direction could lead to the design of a *revision language* in which richer classes of preference revisions can be specified.

One should also consider possible courses of action if the original preference relation $\succ$ and $\succ_0$ lack the property of compatibility, for example if $\succ$ and $\succ_0$ are not 0-compatible in the case of revision by union. Then the target of the revision is an SPO which is the closest to the preference relation $\succ \cup \succ_0$. Such an SPO will not be unique. Moreover, it is not clear how to obtain ipfs defining the revisions. Similarly, one could study *contraction* of preference relations. The need for contraction arises, for example, when a user realizes that the result of a preference query does not contain some expected tuples.

References

1. R. Agrawal and E. L. Wimmers. A Framework for Expressing and Combining Preferences. In *ACM SIGMOD International Conference on Management of Data*, pages 297–306, 2000.
2. W-T. Balke and U. Güntzer. Multi-objective Query Processing for Database Systems. In *International Conference on Very Large Data Bases (VLDB)*, pages 936–947, 2004.
3. W-T. Balke, U. Güntzer, and J. X. Zhang. Efficient Distributed Skylining for Web Information Systems. In *International Conference on Extending Database Technology (EDBT)*, pages 256–273, 2004.
4. K. P. Bogart. Preference Structures I: Distances Between Transitive Preference Relations. *Journal of Mathematical Sociology*, 3:49–67, 1973.

5. S. Börzsönyi, D. Kossmann, and K. Stocker. The Skyline Operator. In *IEEE International Conference on Data Engineering (ICDE)*, pages 421–430, 2001.
6. C. Boutilier, R. I. Brafman, C. Domshlak, H. H. Hoos, and D. Poole. CP-nets: A Tool for Representing and Reasoning with Conditional Ceteris Paribus Preference Statements. *Journal of Artificial Intelligence Research*, 21:135–191, 2004.
7. J. Chomicki. Querying with Intrinsic Preferences. In *International Conference on Extending Database Technology (EDBT)*, pages 34–51. Springer-Verlag, LNCS 2287, 2002.
8. J. Chomicki. Preference Formulas in Relational Queries. *ACM Transactions on Database Systems*, 28(4):427–466, December 2003.
9. J. Chomicki. Semantic Optimization of Preference Queries. In *International Symposium on Constraint Databases*, pages 133–148, Paris, France, June 2004. Springer-Verlag, LNCS 3074.
10. P. C. Fishburn. *Utility Theory for Decision Making*. Wiley & Sons, 1970.
11. P. C. Fishburn. *Interval Orders and Interval Graphs*. Wiley & Sons, 1985.
12. M. Freund. On the Revision of Preferences and Rational Inference Processes. *Artificial Intelligence*, 152:105–137, 2004.
13. P. Gärdenfors and H. Rott. Belief Revision. In D. M. Gabbay, J. Hogger, C, and J. A. Robinson, editors, *Handbook of Logic in Artificial Intelligence and Logic Programming*, volume 4, pages 35–132. Oxford University Press, 1995.
14. K. Govindarajan, B. Jayaraman, and S. Mantha. Preference Queries in Deductive Databases. *New Generation Computing*, 19(1):57–86, 2000.
15. S. Guo, W. Sun, and M.A. Weiss. Solving Satisfiability and Implication Problems in Database Systems. *ACM Transactions on Database Systems*, 21(2):270–293, 1996.
16. S. O. Hansson. Changes in Preference. *Theory and Decision*, 38:1–28, 1995.
17. V. Hristidis and Y. Papakonstantinou. Algorithms and Applications for Answering Ranked Queries using Ranked Views. *VLDB Journal*, 13(1):49–70, 2004.
18. I. F. Ilyas, R. Shah, and A. K. Elmagarmid W. G. Aref, J. S. Vitter. Rank-aware Query Optimization. In *ACM SIGMOD International Conference on Management of Data*, pages 203–214, 2004.
19. P. C. Kanellakis, G. M. Kuper, and P. Z. Revesz. Constraint Query Languages. *Journal of Computer and System Sciences*, 51(1):26–52, August 1995.
20. W. Kießling. Foundations of Preferences in Database Systems. In *International Conference on Very Large Data Bases (VLDB)*, pages 311–322, 2002.
21. W. Kießling and B. Hafenrichter. Algebraic Optimization of Relational Preference Queries. Technical Report 2003-1, Institut für Informatik, Universität Augsburg, 2003.
22. W. Kießling and G. Köstler. Preference SQL - Design, Implementation, Experience. In *International Conference on Very Large Data Bases (VLDB)*, pages 990–1001, 2002.
23. G. Kuper, L. Libkin, and J. Paredaens, editors. *Constraint Databases*. Springer-Verlag, 2000.
24. M. Lacroix and P. Lavency. Preferences: Putting More Knowledge Into Queries. In *International Conference on Very Large Data Bases (VLDB)*, pages 217–225, 1987.
25. D. Papadias, Y. Tao, G. Fu, and B. Seeger:. An Optimal and Progressive Algorithm for Skyline Queries. In *ACM SIGMOD International Conference on Management of Data*, pages 467–478, 2003.

26. P. Pu, B. Faltings, and M. Torrens. User-Involved Preference Elicitation. In *IJCAI Workshop on Configuration*, 2003.
27. R. Torlone and P. Ciaccia. Which Are My Preferred Items? In *Workshop on Recommendation and Personalization in E-Commerce*, May 2002.
28. Mary-Anne Williams. Belief Revision via Database Update. In *International Intelligent Information Systems Conference*, 1997.
29. S. T. C. Wong. Preference-Based Decision Making for Cooperative Knowledge-Based Systems. *ACM Transactions on Information Systems*, 12(4):407–435, 1994.

On the Number of Independent Functional Dependencies*

János Demetrovics[1], Gyula O.H. Katona[2], Dezső Miklós[2],
and Bernhard Thalheim[3]

[1] Computer and Automation Institute,
Hungarian Academy of Sciences, Kende u. 13-17, H-1111 Hungary
dj@ilab.sztaki.hu
[2] Alfréd Rényi Institute of Mathematics,
Hungarian Academy of Sciences, Budapest P.O.B. 127, H-1364 Hungary
ohkatona@renyi.hu, dezso@renyi.hu
[3] Institute for Computer Science and Practical Mathematics,
Christian-Albrechts-University, Olshausenstr. 40 D-24098 Kiel, Germany
thalheim@is.informatik.uni-kiel.de

Abstract. We will investigate the following question: what can be the maximum number of independent functional dependencies in a database of n attributes, that is the maximum cardinality of a system of dependencies which which do not follow from the Armstrong axioms and none of them can be derived from the remaining ones using the Armstrong axioms. An easy and for long time believed to be the best construction is the following: take the maximum possible number of subsets of the attributes such that none of them contains the other one (by the wellknown theorem of Sperner [8] their number is $\binom{n}{n/2}$) and let them all determine all the further values. However, we will show by a specific construction that it is possible to give more than $\binom{n}{n/2}$ independent dependencies (the construction will give $(1 + \frac{1}{n^2})\binom{n}{n/2}$ of them) and — on the other hand — the upper bound is $2^n - 1$, which is roughly $\sqrt{n}\binom{n}{n/2}$.

1 Introduction

Results obtained during database design and development are evaluated on two main criteria: *completeness* of and *unambiguity* of specification. Completeness requires that all constraints that must be specified are found. Unambiguity is necessary in order to provide a reasoning system. Both criteria have found their theoretical and pragmatical solution for most of the known classes of constraints. Completeness is, however, restricted by the human ability to survey large constraint sets and to understand all possible interactions among constraints.

---

* The work was supported by the Hungarian National Foundation for Scientific Research grant numbers T037846 and AT048826.

J. Dix and S.J. Hegner (Eds.): FoIKS 2006, LNCS 3861, pp. 83–91, 2006.
© Springer-Verlag Berlin Heidelberg 2006

Many database theory and application problems (e.g., data search optimization, database design) are substantially defined by the *complexity* of a database, i.e., the size of key, functional dependency, and minimal key systems. Most of the known algorithms, e.g., for normalization, use the set of all minimal keys or non-redundant sets of dependencies. Therefore, they are dependent on the cardinality of these sets. The maintenance complexity of a database depends on how many integrity constraints are under consideration. Therefore, if the cardinality of constraint sets is large, then maintenance becomes infeasible. (Two-tuple constraints such as functional dependencies require $O(m^2)$ two-tuple comparisons for relations with m elements.) Furthermore, they indicate whether algorithms are of interest for practical purposes, since the complexity of most known algorithms is measured by the input length. For instance, algorithms for constructing a minimal key are bound by the maximal number of minimal keys. The problem of deciding whether there is a minimal key with at most k attributes is NP-complete. The problem of deciding whether two sets of functional dependencies are equivalent is polynomial in the size of the two sets.

Therefore, the database design process may only be complete if all integrity constraints that cannot be derived by those that have already been specified have been specified. Such completeness is not harmful as long as constraint sets are small. The number of constraints may however be exponential in the number of attributes [2]. Therefore, specification of the *complete set of functional dependencies* may be a task that is `infeasible`. This problem is closely related to another well-known combinatoric problem presented by Janos Demetrovics during MFDBS'87 [9] and that is still only partially solved:

Problem 1. *How big the number of independent functional dependencies of an n-ary relation schema can be?*

Let R be a relational database model, and X denote the set of attributes. We say that (for two subsets of attributes A and B) $A \to B$, that is, B functionally depends on A, if in the database R the values of the attributes in A uniquely determine the values of the attributes in B. In case a some functional dependencies $\mathcal{F}$ given, the closure of $\mathcal{F}$, usually denoted by $\mathcal{F}^+$, is the set of all functional dependencies that may be logically derived from $\mathcal{F}$. E.g., $\mathcal{F}$ may be considered the obvious and important functional dependencies (like mother's name and address uniquely determine the name of a person) and then the closure, $\mathcal{F}^+$ is the set of all dependencies that can be deduced from $\mathcal{F}$.

The set of rules that are used to derive all functional dependencies implied by $\mathcal{F}$ were determined by Armstrong in 1974 and are called the *Armstrong axioms*. These rules are easily seen to be necessary and all other natural rules can be derived from them. They are the following:

- *reflexivity rule* if A is a set of attributes and B a subset of it, then $A \to B$.
- *augmentation rule* If $A \to B$ holds and C is an arbitrary set of attributes, then $A \cup C \to B \cup C$ holds as well.
- *transitivity rule* If $A \to B$ and $B \to C$ hold, then $A \to C$ holds as well.

Let us mention here, that though there are further natural rules of the dependencies, the above set is complete, that is $\mathcal{F}^+$ can always be derived from $\mathcal{F}$ using only the above three axioms. For example, union rule, that is, the natural fact that $A \to B$ and $A \to C$ imply $A \to B \cup C$ can be derived by augmenting $A \to C$ by A $(A \to A \cup C)$, augmenting $A \to B$ by C $(A \cup C \to B \cup C)$ and using transitivity for the resulting rules: $A \to A \cup C \to B \cup C$.

In this paper we will investigate the maximum possible number of independent functional dependencies of a database of n attributes. That is, the maximum of $\mathcal{F}$, where it is a system of independent, non-trivial dependencies ($A \to B$ where $B \subset A$ are not in $\mathcal{F}$) and no element of $\mathcal{F}$ can be logically derived from the other elements of $\mathcal{F}$. In this case we will call $\mathcal{F}$ independent.

Introduce the following useful notations: $[n] = \{1, 2, \ldots, n\}$. The family of all k-element subsets of $[n]$ is

$$\binom{[n]}{k}.$$

For the sake of simplicity we will denote the ith attribute by i.

It is clear that $A \to C$ and $B \to C$ can not be in $\mathcal{F}$ for a pair of subsets $A \subset B$ since then $B \to C$ would be logically obtained by another given dependency ($A \to C$), reflexivity ($A \subset B$ implies $B \to A$) and transitivity ($B \to A \to C$ implies $B \to C$). On the other hand, it is easy to see if we have a system of independent subsets of the attributes (that is, none of them containing the other one) and assume that all of them imply the whole set of attributes, this system of dependencies will be independent. This leads to the natural construction of a large set of independent dependencies by taking the maximum number of incomparable subsets of attributes, which is by Sperner's theorem [8] equal to $\binom{n}{\lfloor n/2 \rfloor}$ and let the whole set of attributes depend on all of them. This would give a set of dependencies of cardinality $\binom{n}{\lfloor n/2 \rfloor}$. On the other hand, it is easy to see that if $\mathcal{F}$ only consists of dependencies $A \to B$ with $|A| = k$ for a given constant k and for every such an $A \subset X$ there is at most one element of $\mathcal{F}$ of the form $A \to B$, then $\mathcal{F}$ is independent (a more detailed version of this argument will be given in the proof of Lemma 4). That is, the above construction will give an independent set of dependencies and a lower bound of $\binom{n}{\lfloor n/2 \rfloor}$.

However, as it will be shown by the construction of the next section, this is not the best possible bound, we can enlarge it. Still the best known lower bound is of the magnitude of $\binom{n}{\lfloor n/2 \rfloor}$ (smaller than $c\binom{n}{\lfloor n/2 \rfloor}$ for any constant $c > 1$, while the best upper bound proven in the following section is $2^n - 1$, which is roughly $\sqrt{n}\binom{n}{n/2}$. Finally, the last section of the paper will contain concluding remarks, including the answer to the following question:

Problem 2. *Is the maximum number of functional dependencies the same as the maximum number of minimal keys?*

More complexity results are discussed and proven in [5, 7, 10] or in [1, 2, 3, 6].

2   Lower Estimate: A Construction

Theorem 1. *If n is an odd prime number then one can construct*

$$\left(1 + \frac{1}{n^2} + o\left(\frac{1}{n^2}\right)\right)\binom{n}{\lfloor\frac{n}{2}\rfloor}$$

independent functional dependencies on an n-element set of attributes.

The proof will consist of a sequence of lemmas. We will also use the following proposition:

Proposition 1. (see [11]) *Assign to a functional dependency $A \to B$ the set of $2^n - 2^{|B|}$ Boolean vectors $\boldsymbol{a} = (a_1, ..., a_n)$ of the form:*

$$a_i = \begin{cases} 1, & \text{if } i \in A \\ 0 \text{ or } 1, & \text{if } i \in (B \setminus A) \text{ but not all entries } = 0 \\ 0 \text{ or } 1, & \text{otherwise.} \end{cases}$$

Then, a set of functional dependencies implies another functional dependency if and only if the Boolean vectors of the implied functional dependency are contained in the union of the sets of Boolean vectors of the given functional dependencies.

Lemma 2. *If n is an odd prime number, one can find $\frac{1}{n^2}\binom{n}{\frac{n+3}{2}}$ subsets $V_1, V_2, \ldots$ of size $\frac{n+3}{2}$ in the set $[n] = \{1, 2, \ldots n\}$ in such a way that $|V_i \cap V_j| < \frac{n-1}{2}$ holds.*

Proof. The method of the paper [4] is used. Consider the subsets $\{x_1, x_2, \ldots, x_{\frac{n+3}{2}}\}$ of integers satisfying $1 \leq x_i \neq x_j \leq n$ for $i \neq j$ and the equations

$$x_1 + x_2 + \cdots + x_{\frac{n+3}{2}} \equiv a \pmod{n}, \tag{1}$$

$$x_1 x_2 \cdots \cdots x_{\frac{n+3}{2}} \equiv b \pmod{n} \tag{2}$$

for some fixed integers a and b.

Suppose that two of them, say V_1 and V_2 have an intersection of size $\frac{n-1}{2}$. We may assume, without loss of generality, that the first two elements are different, that is $V_1 = \{x_1, x_2, \ldots, x_{\frac{n+3}{2}}\}$ and $V_2 = \{x_1', x_2', \ldots, x_{\frac{n+3}{2}}\}$. (1) and (2) imply $x_1 + x_2 \equiv x_1' + x_2'$ and $x_1 x_2 \equiv x_1' x_2' \pmod{n}$. Since the set $\{1, 2, \ldots n\}$ constitutes a field $\bmod\, n$ if n is prime, this system of equations has a unique solution, that is $x_1 = x_1', x_2 = x_2'$; the two sets are the same: $V_1 = V_2$. This contradiction proves that our sets cannot have $\frac{n-1}{2}$ common elements.

The total number of subsets of size $\frac{n+3}{2}$ is

$$\binom{n}{\frac{n+3}{2}}.$$

Each of these sets give some a and b in (1) and (2), respectively. That is, the family of all $\frac{n+3}{2}$-element sets can be divided into n^2 classes. One of these has a size at least

$$\frac{1}{n^2}\binom{n}{\frac{n+3}{2}}.$$ $\square$

We will need the notion of the *shadow* of a family $\mathcal{A} \subset \binom{[n]}{\frac{n+1}{2}}$. It will be denoted by

$$\sigma(\mathcal{A}) = \{B : |B| = \frac{n-1}{2}, \ \exists A \in \mathcal{A} : B \subset A\}.$$

A pair $\{U_1, U_2\}$ $U_i \in \binom{[n]}{\frac{n+1}{2}}$ is called *good* if $|U_1 \cap U_2| = \frac{n-1}{2}$ holds. The family $\mathcal{P} \subset \binom{[n]}{\frac{n+1}{2}}$ is a *chain* if $\mathcal{P} = \mathcal{P}_1 \cup \ldots \cup \mathcal{P}_l$ where the $\mathcal{P}_i$'s are good pairs and $\sigma(\mathcal{P}_i) \cap \sigma(\mathcal{P}_j) = \emptyset$ for $i \neq j$. The weight $w(\mathcal{P})$ of this chain is l.

Lemma 3. *There is a chain $\mathcal{P} \subset \binom{[n]}{\frac{n+1}{2}}$ chain with weight at least*

$$|\mathcal{P}| = \frac{1}{n^2}\binom{n}{\frac{n+3}{2}}.$$

Proof. Start with the family $\mathcal{V} = \{V_1, V_2, \ldots\}$ ensured by Lemma 2. In each V_i choose two different $\frac{n+1}{2}$-element subsets U_{i1} and U_{i2}. It is easy to see that this pair of subsets is good. Also, these pairs form a chain. The number of the pairs in the chain can be obtained from Lemma 2. $\square$

Lemma 4. *If $\mathcal{P} \subset \binom{[n]}{\frac{n+1}{2}}$ is a chain then the following set of functional dependencies is independent:*

$$A \to B, \quad where \quad |A| = \frac{n-1}{2}, \quad |B| = \frac{n+1}{2}, \quad A \subset B \in \mathcal{P}, \tag{3}$$

$$A \to A', \quad where \quad |A| = \frac{n-1}{2}, \quad A \notin \sigma(\mathcal{P})$$

$$and \quad A' \ is \ arbitrarily \ chosen \ so \ that \ A \subset A', \ |A'| = \frac{n+1}{2}. \tag{4}$$

(Note that in (3) we have all dependencies $A \to B$ given by the conditions, while in (4) for every remaining A we choose only one (exactly one) A' satisfying the conditions.)

Proof. We will use Proposition 1 for the proof. According to rules (3) and (4), for every $A \subset X$ with $|A| = \frac{n-1}{2}$ there are either one (B) or two (B_1 and B_2) subsets of X of size $\frac{n+1}{2}$ with $A \to B$ or $A \to B_i$, always $A \subset B$, B_i. In the first case consider one of the Boolean vector a corresponding to $A \to B$:

$$a_i = \begin{cases} 1, & \text{if } i \in A \\ 0, & \text{if } i \in (B \setminus A) \\ 0, & \text{otherwise.} \end{cases}$$

This vector has exactly $\frac{n-1}{2}$ 1 entries, and from the definition in Proposition 1 it is also clear that all Boolean vectors corresponding to all functional dependencies given by (3) and (4) have at least $\frac{n-1}{2}$ 1 entries. Therefore, the Boolean vector a may correspond to any other dependency $A' \to B'$ only with $A' = A$, which is not the case now, the dependency $A \to B$ may not be deduced from the others.

If we have both $A \to B_1$ and $A \to B_2$, consider one of the Boolean vector a corresponding to $A \to B_2$:

$$
a_i = \begin{cases}
1 \text{ , if } i \in A \\
0 \text{ , if } i \in (B_2 \setminus A) \\
1 \text{ , if } i \in (B_1 \setminus A) \\
0 \text{ , otherwise.}
\end{cases}
$$

In this case a has $\frac{n+1}{2}$ 1 entries and still all Boolean vectors corresponding to all other given functional dependencies have at least $\frac{n-1}{2}$ 0 entries. Therefore, if a Boolean vector corresponding to a given functional dependency $A' \to B'$ is equal a, the obligatory $\frac{n-1}{2}$ 1 entries must form a subset of the $\frac{n+1}{2}$ 0 entries of a, or, in other words, A' must be a subset of B_1, an $\frac{n+1}{2}$ element set. However, by the construction according to (3), for all subsets C of B_1 of size $\frac{n-1}{2}$ we have $C \to B_1$. Since B_1 is larger than C only by one element, all the corresponding Boolean vectors must have entries 1 at the positions corresponding to C and entry 0 at the only position corresponding to $B_1 \setminus C$ (this entry could be literally 0 or 1, but not all of them 1, but since it is alone, it means it should be 0). This, by Proposition 1, implies that $A \to B_2$ may not be deduced from the set of remaining given Boolean vectors.   □

Alternative proof. One can prove Lemma 4 without the use of Proposition 1, simply from the Amstrong axioms, as it follows.

Again, we will show that none of the dependencies given in the lemma can be deduced from the other ones.

Now we start with the case when the dependency $A \to B_1$ has a pair $A \to B_2(B_1 \neq B_2)$ in our system. This can happen only in case of (3): $A \to B_1$ where $\mathcal{P} = \{B_1, B_2\}, B_1 = A \cup \{b_1\}, B_2 = A \cup \{b_2\}$. We want to verify that $A \to B_1$ cannot be deduced from the other ones.

Observe that none of the other ones, $X \to Y$ satisfy both $X \subseteq B_2$ and $Y \not\subseteq B_2$. In other words

$$\text{either } X \not\subseteq B_2 \text{ or } Y \subseteq B_2 \tag{5}$$

holds for every dependency given by (3) and (4), different from $A \to B_1$. Let us see that this property is preserved by the Armstrong axioms.

It is trivial in the case of the reflexivity rule, since it gives $X \to Y$ only when $Y \subseteq X$ therefore $X \subseteq B_2$ implies $Y \subseteq B_2$.

Consider the augmentation rule. Suppose that $X \to Y$ satisfies (5) and U is an arbitrary set. If $X \not\subseteq B_2$ then the same holds for $X \cup U$, that is $X \cup U \not\subseteq B_2$. On the other hand, if $X \subseteq B_2$ then $Y \subseteq B_2$. If $U \subseteq B_2$ also holds then $Y \cup U \subseteq B_2$, if however $U \not\subseteq B_2$ then $X \cup U \not\subseteq B_2$. The dependency $X \cup U \to Y \cup U$ obtained by the augmentation rule also satisfies (5) in all of these cases.

Finally, suppose that $X \to Y$ and $Y \to Z$ both satisfy (5). We have to show the same for $X \to Z$. If $X \nsubseteq B_2$ then we are done. Suppose $X \subseteq B_2$ and $Y \subseteq B_2$. Then $Z \subseteq B_2$, must hold, as desired.

Since $A \to B_1$ does not have property (5), it cannot be deduced from the other dependencies.

Consider now the case when the distinguished dependency $A \to B_1$ has no pair $A \to B_2(B_1 \neq B_2)$ in our system. This can happen both for (3) and (4). The the proof is similar to the case above. No other dependency $X \to Y$ satisfies $X \subseteq A, Y \nsubseteq A$, that is (5) holds for them if B_2 is replaced by A. $\qquad\square$

Remark. One may think that a better construction can be made if we allow three sets B_1, B_2, B_3 of size $\frac{n+1}{2}$ with pairwise intersections $|B_1 \cap B_2| = |B_2 \cap B_3| = \frac{n-1}{2}$ where no other intersections (of these three and other sets $B_i(3 < i)$ of this size) are that big, and $A \to B_i$ holds for every subset A of B_i. Unfortunately we found a counter-example for $n > 5$.

Let $B_1 = \{1, \ldots, \frac{n+1}{2}\}, B_2 = \{2, \ldots, \frac{n+3}{2}\}, B_3 = \{3, \ldots, \frac{n+5}{2}\}, B_4 = \{1, \ldots, \frac{n-3}{2}, \frac{n+3}{2}, \frac{n+5}{2}\}$. It is easy to see that $|B_1 \cap B_2| = |B_2 \cap B_3| = \frac{n-1}{2}$, but $|B_1 \cap B_3|, |B_1 \cap B_4|, |B_2 \cap B_4|, |B_3 \cap B_4|$ are all smaller.

Introduce the notations $A_1 = \{2, \ldots, \frac{n+1}{2}\}, A_2 = \{3, \ldots, \frac{n+3}{2}\}, A_3 = \{2, 3, \ldots, \frac{n-3}{2}, \frac{n+3}{2}, \frac{n+5}{2}\}$. The the following chain of functional dependencies is obvious: $A_1 \to B_2 \to A_2 \to B_3 \to B_2 \cup B_3 \to A_3 \to B_4 \to \{1\}$. Hence we have $A_1 \to B_1$ without using it.

Proof of Theorem 1. Use Lemma 4 with the chain found in Lemma 3. It is easy to see that there is at least one $A \to C$ for every $\lfloor \frac{n}{2} \rfloor = \frac{n-1}{2}$-element subset A and the weight of the chain gives the number of those A being the left-hand side of exactly two dependencies. This gives the number of dependencies:

$$\binom{n}{\lfloor \frac{n}{2} \rfloor} + \frac{1}{n^2}\binom{n}{\frac{n+3}{2}} = \left(1 + \frac{1}{n^2} + o\left(\frac{1}{n^2}\right)\right)\binom{n}{\lfloor \frac{n}{2} \rfloor}. \qquad\square$$

3   Upper Estimate

Theorem 5. *For every n an upper bound for the maximum number of independent functional dependencies on an n-element set of attributes is $2^n - 1$.*

Proof. Let $\mathcal{F}$ be a set of independent functional dependencies on the set of attributes X. First, replace each dependency $A \to B$ in $\mathcal{F}$ by $A \to A \cup B$, obtaining $\mathcal{F}'$. We claim that $\mathcal{F}'$ is independent as well. It simply comes from the fact that by the reflexivity and augmentation axioms the two dependencies $A \to B$ and $A \to A \cup B$ are equivalent. Also, $|\mathcal{F}| = |\mathcal{F}'|$, since the images of dependencies in $\mathcal{F}$ will be different in $\mathcal{F}'$. Assume, on the contrary, that $A \to A \cup B$ is equal to $A \to A \cup C$ for $A \to B$ and $A \to C$ in $\mathcal{F}$. But then $A \cup B = A \cup C, C \subset A \cup B$ and therefore $A \to B$ implies $A \to A \cup B$ implies $A \to C$, a contradiction.

We may therefore consider only set of independent dependencies where for all $(A \to B) \in \mathcal{F}$ we have $A \subset B$. Take now the following graph G: let the vertices of the graph be all the 2^n subsets of the n attributes and for A, $B \subset X$ the edge (A, B) will be present in the G iff A to$B$ or $B \to A$ is in $\mathcal{F}$. We claim that this graph may not contain a cycle, therefore it is a forest, that is it has at most $2^n - 1$ edges, or dependencies.

Assume, on the contrary, that $A_1, A_2, A_3, \ldots, A_n = A_1$ is a cycle in G, that is for all $i = 1, \ldots n - 1$ the edge $(A_i, A_{i+1}$ is present in G, meaning that either $A_i \to A_{i+1}$ or $A_{i+1} \to A_i$ is in $\mathcal{F}$. Note that for every $i = 1, \ldots n - 1$ we have $A_{i+1} \to A_i$ since either this dependency is in $\mathcal{F}$ or in case of $(A_i \to A_{i+1}) \in \mathcal{F}$, we have that $A_i \subset A_{i+1}$, yielding $A_{i+1} \to A_i$ by reflexivity. Take now an i such that $A_i \to A_{i+1}$ (if we have no such a dependency, take the "reverse" of the cycle, $A_n, A_{n-1}, \ldots, A_1 = A_n$). This can be obtained from the other dependencies present in $\mathcal{F}$ by the transitivity chain $A_i \to A_{i-1} \to \cdots A_1 = A_n \to A_{n-1} \to \cdots \to A_{i+2} \to A_{i+1}$, contradicting the independency of the rules in $\mathcal{F}$.   □

We have an alternative proof using Proposition 1 as well.

4 Remarks, Conclusions

The main contribution of the paper is the improvement of upper and lower bounds for independent sets of functional dependencies and, thus, contributing to the solution of Problem 1.

1. The lower and upper estimates seem to be very far from each other. However, if the lower estimate is written in the form

$$c \frac{2^n}{\sqrt{n}}$$

(using the Stirling formula) then one can see that the "difference" is only a factor $\sqrt{n}$ what is negligable in comparison to 2^n. The logarithms of the lower and upper estimates are $n - \frac{1}{2} \log n$ and n.

However we strongly believe that truth is between

$$\left(1 + \frac{\alpha}{n} + o\left(\frac{1}{n}\right)\right) \binom{n}{\lfloor \frac{n}{2} \rfloor}$$

and

$$(\beta + o(1)) \binom{n}{\lfloor \frac{n}{2} \rfloor},$$

where $0 < \alpha$ and $\beta \leq 2$.

2. Theorem 1 is stated only for odd prime numbers. The assumption in Lemma 2 is only technical, we strongly believe that its statement is true for other integers, too. (Perhaps with a constant less than 1 over the n^2.) We did not really try to prove this, since the truth in Theorem 1 is more, anyway. What do we obtain from Lemma 2 if it is applied for the largest prime less than n? It is known from number theory that there is a prime p satisfying $n - n^{5/8} < p \leq n$. This will lead to an estimate where $\frac{1}{n^2}$ is replaced by

$$\frac{1}{n^2 2^{n^{5/8}}}.$$

This is much weaker than the result for primes, but it is still more than the number of functional dependencies in the trivial construction.

3. One may have the feeling that keys are the real interesting objects in a dependency system. That is, the solution of any extremal problem must be a set of keys. Our theorems show that this is not the case.

More precisely, suppose that only keys are considered in our problem, that is the maximum number of independent keys is to be determined. If this set of keys is $\{A_i \to X\}$ $(1 \le i \le m)$, then $A_i \not\subset A_j$ must be satisfied, and therefore by Sperner's theorem $m \le \binom{n}{\lfloor n/2 \rfloor}$. In this case the largest set of dependencies (keys) is provided by the keys $A \to X$, where $A \in \binom{[n]}{\lfloor \frac{n}{2} \rfloor}$.

Theorem 1 shows that the restriction to consideration of key systems during database design and development is an essential restriction. Systems of functional dependencies must be considered in parallel. Therefore, we derived a negative answer to Problem 2.

References

1. BEERI,C., DOWD, M., FAGIN, R., AND STATMAN, R., On the structure of Armstrong relations for functional dependencies. *Journal of ACM*, **31**, 1, January 1984, 30–46.

2. DEMETROVICS, J. AND KATONA, G.O.H., Combinatorial problems of database models. *Colloquia Mathematica Societatis Janos Bolyai 42*, Algebra, Cominatorics and Logic in Computer Science, Gÿor (Hungary), 1983, 331–352.

3. DEMETROVICS, J., LIBKIN, L., AND MUCHNIK, I.B., Functional dependencies and the semilattice of closed classes. *Proc. MFDBS-89*, LNCS 364, 1989, 136–147.

4. GRAHAM, R., SLOANE, N., Lower bounds for constant weight codes, *IEEE Trans. Inform. Theory* **26**(1980) 37-43.

5. KATONA, G.O.H. AND DEMETROVICS, J., A survey of some combinatorial results concerning functional dependencies in relational databases. Annals of Mathematics and Artificial Intelligence, 6, 1992.

6. MANNILA, H. AND K.-J. RÄIHÄ, K.-J., *The design of relational databases.* Addison-Wesley, Amsterdam, 1992.

7. SELEZNJEV, O. AND THALHEIM, B., On the number of minimal keys in relational databases over nonuniform domains. *Acta Cybern.*, **8**, 267–271. 1988.

8. SPERNER, E., Ein Satz über Untermengen einer endlichen Menge, *Math. Z.* **27**(1928) 544-548.

9. THALHEIM, B., Open problems in relational database theory. *Bull. EATCS*, **32**:336 – 337, 1987.

10. THALHEIM, B., Generalizing the entity-relationship model for database modeling. *JNGCS, 1990*, **3**, 3, 197 – 212.

11. THALHEIM, B., *Dependencies in Relational Databases.* Teubner-Texte zur Mathematik, B.G. Teubner Verlagsgesellschaft, Stuttgrat - Leipzig, 1991.

Arity and Alternation: A Proper Hierarchy in Higher Order Logics

Flavio A. Ferrarotti and José M. Turull Torres

Information Science Research Centre,
Department of Information Systems, Massey University,
P.O. Box 756, Wellington, New Zealand
{F.A.Ferrarotti, J.M.Turull}@massey.ac.nz

Abstract. We study the effect of simultaneously bounding the maximal arity of the higher-order variables and the alternation of quantifiers in higher-order logics, as to their expressive power on finite structures (or relational databases). Let $AA^i(r, m)$ be the class of $(i + 1)$-th order logic formulae where all quantifiers are grouped together at the beginning of the formulae, forming m alternating blocks of consecutive existential and universal quantifiers, and such that the maximal arity of the higher-order variables is bounded by r. Note that, the order of the quantifiers in the prefix may be mixed. We show that, for every $i \geq 1$, the resulting $AA^i(r, m)$ hierarchy of formulae of $(i + 1)$-th order logic is proper. From the perspective of database query languages this means that, for every $i \geq 2$, if we simultaneously increase the arity of the quantified relation variables by one and the number of alternating blocks of quantifiers by four in the fragment of higher-order relational calculus of order i, AA^{i-1}, then we can express more queries. This extends a result by J. A. Makowsky and Y. B. Pnueli who proved that the same hierarchy in second-order logic is proper. In both cases the strategy used to prove the results consists in considering formulae which, represented as finite structures, satisfy themselves. As the well known diagonalization argument applies here, this gives rise, for each order i and for each level of the $AA^i(r, m)$ hierarchy of arity and alternation, to a class of formulae which is not definable in that level, but which is definable in a higher level of the same hierarchy. We then use a similar argument to prove that the classes of $\Sigma^i_m \cup \Pi^i_m$ formulae in which the higher-order variables of all orders up to $i+1$ have maximal arity at most r, also induce a proper hierarchy in each higher-order logic of order $i \geq 3$. It is not known whether the correspondent hierarchy in second-order is proper.

1 Introduction

The study of the expressive power of different syntactically defined fragments of logics built as extensions of first-order, has received important attention throughout the development of finite model theory. A fundamental underlying question to this regard is: which kind of syntactic restrictions have impact on the expressive power of such logics over finite structures?, or equivalently: when can

J. Dix and S.J. Hegner (Eds.): FoIKS 2006, LNCS 3861, pp. 92–115, 2006.

we really express more queries over relational databases? The answers to these questions are in many cases related to important open problems in complexity theory. An example of this can be found in the well known characterization of the polynomial hierarchy in terms of prenex fragments of second-order logic.

Recall that the levels of the polynomial hierarchy are defined inductively: $\Sigma_1^p = NP$, and $\Sigma_{m+1}^p = \text{NP}^{\Sigma_m^p}$. Also recall that Σ_m^1 is the class of second-order sentences in prenex normal form in which each second-order quantifier precedes all first-order quantifiers and the second order quantifiers are arranged into at most m alternating blocks with the first block being existential. A well known result of Stokmeyer [Sto76] established that, for each $k \geq 1$, Σ_m^1 captures Σ_m^p. Thus, asking whether increasing the number of alternating blocks of quantifiers in the Σ_m^1 hierarchy allows us to express more queries, is equivalent to asking whether the polynomial hierarchy is proper. By proper we mean that for every layer in Σ_m^1 there is another layer which properly includes the former.

In second-order logic, a considerable amount of effort was devoted to the study of hierarchies defined in terms of *alternations* of quantifiers. In this line of work important results have been obtained for monadic second-order logic. Thomas [Tho82] showed that over word models monadic second-order collapses to monadic Σ_1^1. Otto [Ott95] showed that the number of existential quantifiers in monadic Σ_1^1 induces a strict hierarchy as to expressive power. That is, each layer in the hierarchy is properly included in the next one. Considering colored grids as underlying models, a compression of existential monadic second-order prefixes to a single existential quantifier was obtained in [Mat98]. In [MST02] it was shown that over finite graphs, the monadic Σ_m^1-formulae induce a strict hierarchy.

In many extensions of first-order logic, the maximum *arity* of the relation variables occurring in a formula was shown to be of a great relevance. The fragments allowing only formulae of a bounded arity in its relation variables form a natural hierarchy inside such logics, and the obvious question to be asked is whether this hierarchy is strict. An affirmative answer to this question for various extensions of first-order logic by fixed-point operators and transitive closure operators has been given by Grohe in [Gro93, Gro96]. In [Hel89, Hel92], Hella studied the notion of arity on first-order logic extended with Lindström quantifiers. In [GH96], a double arity hierarchy theorem for transitive closure logic was proven. However, in monadic second order logic it is still open whether the arity hierarchy in Σ_1^1 is strict over vocabularies of a fixed arity. The most important result obtained here is Ajtai's theorem [Ajt83] which implies that the arity hierarchy in Σ_1^1 is strict over vocabularies of arbitrary arity.

In the study of the full hierarchy Σ_m^1, Makowsky and Pnueli followed a different approach in [MP96]. They investigated the expressive power of second-order logic over finite relational structures when limitations in the arity of the second-order variables and in the number of alternations of both first-order and second-order quantifiers, are simultaneously imposed, and they proved the existence of a proper hierarchy of arity and alternation in second-order logic. Roughly speaking, the method used to prove such result consisted in considering the set

$AUTOSAT(F)$ of formulae of a given logic F which, encoded as finite structures, satisfy themselves. As the well known diagonalization argument applies here, when F is a level of the hierarchy of arity and alternation, it follows that $AUTOSAT(F)$ is not definable in F, but is definable in a higher level of the same hierarchy.

In the present article, aiming to gain a better understanding on the kind of syntactic restrictions which are relevant as to the expressive power of different fragments of higher-order logics over finite structures, we study the effect of simultaneously bounding the maximal arity of the higher-order variables and the alternation of quantifiers in formulae of higher-order logics. Let $AA^i(r, m)$ be the class of $(i + 1)$-th order logic formulae where all quantifiers of whichever order are grouped together at the beginning of the formulae, forming up to m alternating blocks of consecutive existential and universal quantifiers, and such that the maximal arity of the higher-order order variables is bounded by r. Note that, the order of the quantifiers in the prefix may be mixed.

We show that, for every $i \geq 1$, the resulting $AA^i(r, m)$ hierarchy is proper. We get our results by roughly adapting the strategy of Makowsky and Pnueli to each higher-order logic of order $i \geq 2$. From the perspective of database query languages this means that, for every $i \geq 2$, if we simultaneously increase the arity of the quantified relation variables by one and the number of alternating blocks of quantifiers by four in the fragment of higher-order relational calculus of order i, AA^{i-1}, then we can express more queries.

It is note worthy that in [HT03] it is shown that the correspondence between the polynomial hierarchy and the classes of prenex second-order formulae Σ_m^1, can be extended to higher orders. Let Σ_m^i be the class of $(i + 1)$-th order logic formulae in prenex normal form in which the quantifiers of order $i+1$ are arranged into m alternating blocks, starting with an existential block (Π_m^i is defined dually). In that article the exact correspondence between the non deterministic exponential time hierarchy and the different fragments of higher-order logics Σ_m^i was proven. As with the hierarchy Σ_m^1, it is open whether the hierarchies Σ_m^i are proper.

We also study a variation of the AA^i hierarchies, the HAA^i hierarchies, where the alternations are counted as in Σ_m^i and Π_m^i. Let $HAA^i(r, m)$ be the class of $\Sigma_m^i \cup \Pi_m^i$ formulae in which the higher-order variables of all orders up to $i + 1$ have maximal arity at most r. We prove that, for each $i \geq 2$, the $HAA^i(r, m)$ hierarchy is proper. Note that the corresponding version of this hierarchy for second-order logic, $HAA^1(r, m)$, was *not* studied in [MP96] regarding the properness of the hierarchy. It was used there (denoted as $SAA(r, m)$) to prove that, for every $r, m \geq 1$, $AUTOSAT(SAA(r, m))$ is PSpace-complete. So, it is not known whether the $HAA^1(r, m)$ hierarchy is proper. The difference between the HAA^i and the AA^i hierarchies is that in the HAA^i hierarchies the quantifiers of order $i + 1$ precede all other quantifiers in the formulae and only the alternations of quantifiers of order $i + 1$ are considered, while in the AA^i hierarchies the quantifiers of order $i + 1$ do not necessarily precede all other quantifiers and all alternations of quantifiers of whichever order are considered.

The article is organized as follows. In Section 2 we fix the notation and briefly comment on finite models, databases and logics as query languages. In Section 3 we define the syntax and semantics of the higher-order logics considered in this work. We emphasize the definition of higher-order logics as formal languages over a *finite* alphabet. This plays an important role in the encoding of formulae as finite structures which we use in Section 4, where we prove our main result on the properness of the AA^i hierarchies. In Subsection 4.1 we fix an encoding for the formulae of finite-order logic as relational structures, and we define the sets $WFF(F)$, $AUTOSAT(F)$ and $DIAG(F)$ of structures encoding well formed formulae in F, self-satisfying formulae in F, and well formed formulae in the complement of $AUTOSAT(F)$ with respect to $WFF(F)$, respectively, for a given logic F. In Subsection 4.2 we study the descriptive complexity of $WFF(F)$ for different fragments of higher-order logics. Then, we move to Subsection 4.3 where we use a diagonalization argument to give a lower bound for the definability of $AUTOSAT(F)$ for different fragments of higher-order logics. In Subsection 4.4 we give a tight upper bound for the definability of the classes $AUTOSAT(F)$, with F being the different levels of the AA^i hierarchies. Finally, in Subsection 4.5 we present our main result which is a direct consequence of the lower and upper bounds proved in Subsections 4.3 and 4.4. In Section 5 we prove the properness of the HAA^i hierarchies for $i \geq 2$, i.e., we prove that the hierarchies of $\Sigma_m^i \cup \Pi_m^i$ formulae in which the higher-order variables of all orders up to $i + 1$ have maximal arity at most r are proper for $i \geq 2$. We conclude the article with Section 6 where we examine the concept of finite model truth definitions introduced by M. Mostowski [Mos01, Mos03], and comment on its relation to our work.

2   Preliminaries

As usual ([EF99, AHV94]), we regard a *relational database schema*, as a relational vocabulary, and a *database instance* or simply *database* as a finite structure of the corresponding vocabulary. If $\mathbf{I}$ is a database or finite structure of some schema σ, we denote its domain as I. If R is a relation symbol in σ of arity r, for some $r \geq 1$, we denote as $R^{\mathbf{I}}$ the (second-order) relation of arity r which interprets the relation symbol R in $\mathbf{I}$, with the usual notion of interpretation. We denote as $\mathcal{B}_\sigma$ the class of *finite σ-structures*, or databases of schema σ.

In this paper, we consider *total* queries only. Let σ be a schema, let $r \geq 1$, and let R be a relation symbol of arity r. A *computable query of arity r and schema σ* ([CH80]), is a total recursive function $q : \mathcal{B}_\sigma \to \mathcal{B}_{\langle R \rangle}$ which preserves isomorphisms such that for every database $\mathbf{I}$ of schema σ, $dom(q(\mathbf{I})) \subseteq I$. A Boolean query is a 0-ary query. We use the notion of a *logic* in a general sense. A formal definition would only complicate the presentation and is unnecessary for our work. As usual in finite model theory, we regard a logic as a language, that is, as a set of formulas (see [EF99]). We only consider vocabularies which are purely *relational*, and for simplicity we do not allow constant symbols. If $\mathcal{L}$ is a logic and σ a relational vocabulary, we denote by $\mathcal{L}[\sigma]$ the set of $\mathcal{L}$-formulae over σ. We consider *finite* structures only. Consequently, the notion of *satisfaction*, denoted as $\models$, is related to only finite structures. By $\varphi(x_1, \ldots, x_r)$ we denote a formula of some logic whose free variables are *exactly* $\{x_1, \ldots, x_r\}$. If $\varphi(x_1, \ldots, x_r) \in \mathcal{L}[\sigma]$, $\mathbf{I} \in \mathcal{B}_\sigma$, $\bar{a}_r = (a_1, \ldots, a_r)$ is a r-tuple over $\mathbf{I}$, let $\mathbf{I} \models \varphi(x_1, \ldots, x_r)[a_1, \ldots, a_r]$ denote that φ is TRUE, when interpreted by $\mathbf{I}$, under a valuation v where for $1 \leq i \leq r \; v(x_i) = a_i$. Then we consider the set of all such valuations as follows:

$$\varphi^{\mathbf{I}} = \{(a_1, \ldots, a_r) : a_1, \ldots, a_r \in I \wedge \mathbf{I} \models \varphi(x_1, \ldots, x_r)[a_1, \ldots, a_r]\}$$

That is, $\varphi^{\mathbf{I}}$ is the relation defined by φ in the structure $\mathbf{I}$, and its arity is given by the number of free variables in φ. Formally, we say that a formula $\varphi(x_1, \ldots, x_r)$ of signature σ, *expresses* a query q of schema σ, if for every database $\mathbf{I}$ of schema σ, is $q(\mathbf{I}) = \varphi^{\mathbf{I}}$. Similarly, a sentence φ expresses a Boolean query q if for every database $\mathbf{I}$ of schema σ, is $q(\mathbf{I}) = 1$ iff $\mathbf{I} \models \varphi$. For $\varphi \in \mathcal{L}[\sigma]$ we denote by $Mod(\varphi)$ the class of finite σ-structures $\mathbf{I}$ such that $\mathbf{I} \models \varphi$. A class of finite σ-structures $\mathcal{C}$ is *definable* by a $\mathcal{L}$-sentence if $\mathcal{C} = Mod(\varphi)$ for some $\varphi \in \mathcal{L}[\sigma]$. If $\mathcal{L}$ and $\mathcal{L}'$ are logics, then $\mathcal{L} \subseteq \mathcal{L}'$ denotes that $\mathcal{L}'$ is at least as expressive as $\mathcal{L}$, i.e., all classes of models definable by an $\mathcal{L}$-sentence are also definable by an $\mathcal{L}'$-sentence. $\mathcal{L} = \mathcal{L}'$ holds if $\mathcal{L} \subseteq \mathcal{L}'$ and $\mathcal{L}' \subseteq \mathcal{L}$. $\mathcal{L} \subset \mathcal{L}'$ holds if $\mathcal{L} \subseteq \mathcal{L}'$ and $\mathcal{L} \neq \mathcal{L}'$.

3   Finite-Order Logic

Finite-order logic is an extension of first-order logic which allows to quantify over higher-order relations. We define here its syntax and semantics following the account in [Lei94]. We emphasize the fact that the set of formulae of finite-order logic can be viewed as a set of strings over a *finite* alphabet, i.e., as a formal language. This plays an important role in the encoding of formulae as finite structures which we define in Section 4.1.

Definition 1. *We define the set of* types, *as the set* **typ** *of strings over the alphabet* $\{\iota, (,), , \}$ *inductively generated by:*

- $\iota \in$ **typ***;*
- *if* $\tau_1, \ldots, \tau_r \in$ **typ** *(r $\geq$ 1), then* $(\tau_1, \ldots, \tau_r) \in$ **typ***;*
- *nothing else belongs to* **typ***.*

If $\tau_1 = \cdots = \tau_r = \iota$, *then* $(\tau_1, \ldots, \tau_r)$ *is denoted by* ι^r. *The set of types can be naturally stratified into* orders *which are inductively defined, as follows:*

- $order(\iota) = 1$
- $order((\tau_1, \ldots, \tau_r)) = 1 + max(\{order(\tau_1), \ldots, order(\tau_r)\})$

For $\tau = (\tau_1, \ldots, \tau_r)$, r *is the* arity *of the type* τ. *We associate a non-negative integer, the* maximal-arity *(ma), with each type, as follows:*

- $ma(\iota) = 0$
- $ma((\tau_1, \ldots, \tau_r)) = max(\{r, ma(\tau_1), \ldots, ma(\tau_r)\})$

Clearly, if $order(\tau) = 2$, then the maximal-arity of τ coincides with its arity. We denote $typ(i, r)$ the subset of types of order $\leq i$ and maximal-arity $\leq r$. Note that, each subset $typ(i, r)$ contains a finite amount of different types.

The intended interpretation is that objects of type ι are individuals, i.e., elements of the universe of a given model, whereas objects of type $(\tau_1, \ldots, \tau_r)$ are r-ary relations, i.e., sets of r-tuples of objects of types $\tau_1, \ldots, \tau_r$, respectively.

Definition 2. *Given a set U, the set U_τ of objects of type τ over U is defined by*

$$U_\iota = U; \qquad U_{(\tau_1, \ldots, \tau_r)} = \mathcal{P}(U_{\tau_1} \times \cdots \times U_{\tau_r})$$

Over a relational vocabulary σ, each formula of finite-order logic is a string of symbols taken from the *alphabet*

$$A = \{\neg, \vee, \wedge, \exists, \forall, (,), =, x, X, |, \iota, , \} \cup \sigma \tag{1}$$

The words that belong to the language $\{x|^n : n > 0\}$ are called *individual variables*, while the words that belong to the language $\{X\tau|^n : \tau \in \mathbf{typ} \backslash \{\iota\}$ and $n > 0\}$ are called *higher-order variables*. We call the higher-order variables of the form $X\tau|^n$, for $i = order(\tau)$ and $r = ma(\tau)$, *i-th order variables* of *maximal-arity* r. To simplify the notation we denote strings of the form $|^n$, $n > 0$, as subscripts, e.g., writing x_3 for $x|||$. In addition, we write the types of the higher-order variables as superscripts, e.g., writing $X_2^{(\iota)}$ for $X(\iota)||$. Sometimes, we omit the superscript when we denote second-order variables (i.e., variables of type ι^r, for some $r \geq 1$) if their arity is clear from the context. We use V^τ to denote any variable of type τ. So, if $\tau = \iota$ then V^τ stands for an individual variable, otherwise V^τ stands for a higher-order variable of type τ.

Definition 3. *We define the set of* well formed formulae *(wff) of finite-order logic over a relational vocabulary σ (here we do not allow constant symbols), as follows:*

   i. *If v_1 and v_2 are individual variables, then $v_1 = v_2$ is a wff.*
   ii. *If R is a relation symbol in σ of arity $r \geq 1$, and $v_1, \ldots, v_r$ are individual variables, then $R(v_1, \ldots, v_r)$ is a wff.*
   iii. *If V^τ is a higher-order variable with $\tau = (\tau_1, \ldots, \tau_r)$, then $V^\tau(V_1^{\tau_1}, \ldots, V_r^{\tau_r})$ is a wff.*
   iv. *If φ is a wff, then $(\neg\varphi)$ is a wff.*
   v. *If φ and ψ are wff, then $(\varphi \vee \psi)$ and $(\varphi \wedge \psi)$ are wff.*
   vi. *If φ is a wff and v is an individual variable, then $\exists v(\varphi)$ and $\forall v(\varphi)$ are wff.*
   vii. *If φ is a wff and V^τ is a higher-order variable, then $\exists V^\tau(\varphi)$ and $\forall V^\tau(\varphi)$ are wff.*
 viii. *Nothing else is a wff.*

The *atomic formulae* are the ones introduced by clauses (i) to (iii). The free occurrence of a variable (either an individual variable or a higher-order variable) in a formula of finite-order logic is defined in the obvious way. Thus, the set

$free(\varphi)$ of *free variables* of a formula φ is the set of both individual and higher-order variables whose occurrence in φ is not under the scope of a quantifier which binds them.

The *semantics* of formulae of finite-order logic is similar to the semantics of formulae of second-order logic, except that a valuation over a structure with universe U maps higher-order variables of type τ to objects in U_τ.

Definition 4. *Let σ be a relational vocabulary. A valuation val on a σ-structure $\mathbf{I}$ with domain I, is a function which assigns to each individual variable an element in I and to each higher-order variable V^τ, for some type $\tau \neq \iota$, an object in I_τ. Let val_0, val_1 be two valuations on a σ-structure $\mathbf{I}$, we say that val_0 and val_1 are V^τ-equivalent if they coincide in every variable of whichever type, with the possible exception of variable V^τ. We also use the notion of equivalence w.r.t. sets of variables.*

Let $\mathbf{I}$ be a σ-structure, and let val be a valuation on $\mathbf{I}$. The notion of satisfaction *in finite-order logic extends the notion of satisfaction in first-order with the following rules:*

   i. $\mathbf{I}, val \models V^{(\tau_1,\ldots,\tau_r)}(V_1^{\tau_1}, \ldots, V_r^{\tau_r})$ *iff* $(val(V_1^{\tau_1}), \ldots, val(V_r^{\tau_r})) \in val(V^\tau)$.

   ii. $\mathbf{I}, val \models \exists V^\tau(\varphi)$ *where V^τ is a higher-order variable and φ is a well formed formula, iff there is a valuation val', which is V^τ-equivalent to val, such that $\mathbf{I}, val' \models \varphi$.*

   iii. $\mathbf{I}, val \models \forall V^\tau(\varphi)$ *where V^τ is a higher-order variable and φ is a well formed formula, iff for every valuation val', which is V^τ-equivalent to val, $\mathbf{I}, val' \models \varphi$.*

The restriction of finite-order logic to formulae whose variables are all of order $\leq i$, for some $i \geq 1$, is called *i-th order logic* and is denoted by HO^i. Note that, for $i = 1$ this is first-order logic (FO), and for $i = 2$ this is second-order logic (SO). The logics of order $i \geq 2$ are known as *higher-order logics* (HO).

As in many other extensions of first-order logic, in second-order logic we can naturally associate a non-negative integer, the *arity*, with each formula. Usually, the arity of a formula of second-order logic is defined as the biggest arity of a second-order variable occurring in that formula. Taking a similar approach, we define the *maximal-arity* of a HO^i formula, $i \geq 2$, as the biggest maximal-arity of any higher-order variable occurring in that formula. For $r \geq 1$, the restriction of HO^i to formulae of maximal-arity $\leq r$ forms the fragment $HO^{i,r}$ of HO^i. Clearly, for second-order logic the maximal-arity of a formula coincides with its arity. Note that, if a variable V^τ occurs in some $HO^{i,r}$ formulae, then $\tau \in typ(i,r)$.

An easy induction using renaming of variables and equivalences such as $\neg \exists V^\tau(\varphi) \equiv \forall V^\tau(\neg\varphi)$ and $(\phi \vee \forall V^\tau(\psi)) \equiv \forall V^\tau(\phi \vee \psi)$ if V^τ is not free in ϕ, shows that each HO^i formula is logically equivalent to an HO^i formula in *prenex normal form*, i.e., to a formula of the form $Q_1 V_1 \ldots Q_n V_n(\varphi)$, where $Q_1, \ldots, Q_n \in \{\forall, \exists\}$, and where $V_1, \ldots, V_n$ are variables of order $\leq i$ and φ is a quantifier-free HO^i formula. Moreover, for every $i \geq 2$, each HO^i formula is logically equivalent to one in prenex normal form in which the quantifiers of order i precede all the remaining quantifiers in the prefix (see [HT03] for a detailed proof of this fact). Such normal form is known as *generalized Skolem normal form*, or $GSNF$. The formulae of finite-order logic which are in $GSNF$ comprise well known hierarchies whose levels are denoted Σ_m^i and Π_m^i. The class Σ_m^i consists of those HO^{i+1} formulae in $GSNF$ in which the quantifiers of order $i+1$ are arranged into at most m alternating blocks, starting with an existential block. Π_m^i is defined dually. Clearly, every HO^{i+1} formula is equivalent to a Σ_m^i formula for some m, and also to a Π_m^i formula.

4 Arity-Alternation Hierarchies in Higher-Order Logics

We define next for each order $i \geq 2$ a hierarchy in HO^i defined in terms of both, alternation of quantification and maximal-arity of the higher-order variables.

Definition 5. *(AA^i hierarchies).* *For $i, r, m \geq 1$,*

   i. $AA\Sigma^i(r, m)$ *is the restriction of $HO^{i+1,r}$ to prenex formulae with at most m alternating blocks of quantifiers, starting with an existential block. That is, $AA\Sigma^i(r, m)$ is the class of formulae $\varphi \in HO^{i+1,r}$ of the form*

$$\exists \, \overline{V_1} \, \forall \, \overline{V_2} \, \ldots \, Q_m \, \overline{V_m} \, (\psi)$$

where ψ is a quantifier-free $HO^{i+1,r}$ formula, Q_m is either $\exists$ if m is odd or $\forall$ if m is even, and for $1 \leq j \leq m$, each variable in the vector $\overline{V_j}$ is a variable of order $\leq i+1$ and maximal-arity $\leq r$.

  ii. $AA\Pi^i(r, m)$ *is defined in the same way as $AA\Sigma^i(r, m)$, but now we require that the first block consists of universal quantifiers.*

 iii. $AA^i(r, m) = AA\Sigma^i(r, m) \cup AA\Pi^i(r, m)$.

    Note that, the formulae in the AA^i hierarchies are in prenex normal form, but not necessarily in $GSNF$. Thus, the quantifiers applied to the variables of the highest order do not necessarily precede all the remaining quantifiers in the prefix, as it is the case in the Σ_m^i hierarchies. Furthermore, in the AA^i hierarchies we count every alternation of quantifiers in the prefix, independently of the order of the variables to which the quantifiers are applied to, while in the Σ_m^i hierarchies the only alternation counted are those corresponding to the quantifiers applied to the variables of the highest order.

    Makowsky and Pnueli showed that the AA^1 hierarchy imposes a proper hierarchy in second-order logic.

Theorem 1 ([MP96]). *For every $r, m \geq 1$ there are Boolean queries not expressible in $AA^1(r, m)$ but expressible in $AA^1(r + c(r), m + 4)$, where $c(r) = 1$ for $r > 1$ and $c(r) = 2$ for $r = 1$.*

They proved this result by introducing the set $AUTOSAT(F)$ of formulae in a given logic F which satisfy themselves, in certain encoding of the formulae as structures. Following a similar strategy, we show that, for every $i \geq 2$, the analogous AA^i hierarchy imposes a proper hierarchy in HO^{i+1}.

4.1   Encoding Well Formed Formulae into Relational Structures

Using *word models* (see [EF99]), it is easy to see that every formula of finite-order logic over a given relational vocabulary σ can be viewed as a finite relational structure of the following vocabulary.

$$\pi(\sigma) = \{<, R_\neg, R_\vee, R_\wedge, R_\exists, R_\forall, R_(, R_), R_=, R_x, R_X, R_|, R_\iota, R_, \} \cup \{R_a : a \in \sigma\}$$

Example 1. If $\sigma = \{E\}$ is the vocabulary of graphs and φ is the sentence $\exists X_1^{(\iota)}(\exists x_2(X_1^{(\iota)}(x_2) \vee E(x_2, x_2)))$, which using our notation for the variables corresponds to $\exists X(\iota)|(\exists x||(X(\iota)|(x||) \vee E(x||, x||)))$, then the following $\pi(\sigma)$-structure **I** encodes φ.

$$\mathbf{I} = \langle I, <^{\mathbf{I}}, R_\neg^{\mathbf{I}}, R_\vee^{\mathbf{I}}, R_\wedge^{\mathbf{I}}, R_\exists^{\mathbf{I}}, R_\forall^{\mathbf{I}}, R_(^{\mathbf{I}}, R_)^{\mathbf{I}}, R_=^{\mathbf{I}}, R_x^{\mathbf{I}}, R_X^{\mathbf{I}}, R_|^{\mathbf{I}}, R_\iota^{\mathbf{I}}, R_,^{\mathbf{I}}, R_E^{\mathbf{I}} \rangle$$

where $<^{\mathbf{I}}$ is a linear order on I, $|I| = length(\varphi)$, and for each $R_a \in \pi(\sigma)$, $R_a^{\mathbf{I}}$ contains the positions in φ carrying an a,

$$R_a^{\mathbf{I}} = \{b \in I : \text{for some } j \ (1 \le j \le |I|), a \text{ is the } j\text{-th symbol in } \varphi, \text{ and}$$

$$b \text{ is the } j\text{-th element in the order } <^{\mathbf{I}}\}$$

Moreover, instead of having a different vocabulary $\pi(\sigma)$ depending on the vocabulary σ of the formulae which we want to encode as relational structures, we can have a vocabulary ρ rich enough to describe formulae of finite-order logic for any arbitrary vocabulary σ. That is, we can fix a vocabulary ρ such that every formula of finite-order logic of whichever vocabulary σ can be viewed as a finite ρ-structure. This can be done as follows. Let ρ be the vocabulary

$$\{<, R_\neg, R_\vee, R_\wedge, R_\exists, R_\forall, R_(, R_), R_x, R_X, R_|, R_\iota, R_, R_P\}$$

We first identify every formula φ of finite-order logic over an arbitrary vocabulary σ, with a formula φ' over the vocabulary $\sigma' = \{P|^i : 1 \le i \le |\sigma| + 1\}$, where, for a predefined bijective function f from $\sigma \cup \{=\}$ to σ', φ' is the formula obtained by replacing in φ each occurrence of a relation symbol $R \in \sigma \cup \{=\}$ by the word $f(R) \in \sigma'$. We then identify every formula φ with the ρ-structure $\mathbf{I}_{\varphi'}$ corresponding to the word model for φ'.

Note that, following the previous schema, even the formulae of finite-order logic over ρ can be viewed as finite ρ-structures. This motivates the following important definition.

Definition 6. *Let F be a set of formulae of finite-order logic of vocabulary ρ, let $\rho' = \{P|^i : 1 \le i \le 15\}$ and let f be the following bijective function form $\rho \cup \{=\}$ to ρ'.*

$$\{< \mapsto P_1, = \mapsto P_2, R_\neg \mapsto P_3, R_\vee \mapsto P_4, \dots, R_\iota \mapsto P_{13}, R_, \mapsto P_{14}, R_P \mapsto P_{15}\}$$

where, for $1 \le i \le 15$, P_i denotes a string of the form $P|^i$. For every formula φ of finite-order logic over ρ, let φ' be the formula obtained by replacing in φ

each occurrence of a relation symbol $R \in \rho \cup \{=\}$ by the word $f(R) \in \rho'$. We identify every formula φ with the ρ-structure $\mathbf{I}_{\varphi'}$, where the cardinality of I is the length of φ', $<^{\mathbf{I}_{\varphi'}}$ is a linear order on I, and for each $R_a \in \rho$, $R_a^{\mathbf{I}_{\varphi'}}$ corresponds to the positions in φ' carrying an a, i.e., we identify φ with the word model for φ'.

- *We denote by $WFF(F)$ the set of finite ρ-structures $\mathbf{I}_{\varphi'}$ such that $\varphi \in F$.*
- *We denote by $AUTOSAT(F)$ the set of finite ρ-structures $\mathbf{I}_{\varphi'}$ such that $\varphi \in F$ and $\mathbf{I}_{\varphi'} \models \varphi$.*
- *We define $DIAG(F) = WFF(F) \setminus AUTOSAT(F)$.*

Let's see some concrete examples of finite ρ-structures in $WFF(F)$, $AUTOSAT(F)$ and $DIAG(F)$.

Example 2. Let $\mathbf{I}_{\varphi'}$ be the ρ-structure corresponding to the word model for $\varphi' \equiv \exists x | (\exists x || (P|(x|, x||)))$ which encodes the formula $\varphi \equiv \exists x_1 (\exists x_2 (< (x_1, x_2))$. It follows that $\mathbf{I}_{\varphi'}$ belongs to $WFF(FO[\rho])$, as φ is a wff in $FO[\rho]$. It also follows that $\mathbf{I}_{\varphi'}$ belongs to $AUTOSAT(FO[\rho])$, as $\mathbf{I}_{\varphi'} \models \varphi$. On the other hand, the ρ-structure $\mathbf{I}_{\psi'}$ corresponding to the word model for $\psi' \equiv (\neg \varphi')$, also belongs to $WFF(FO[\rho])$, as the formula $\psi \equiv (\neg \varphi)$ encoded by $\mathbf{I}_{\psi'}$ is a wff in $FO[\rho]$, but clearly $\mathbf{I}_{\psi'}$ is not in $AUTOSAT(FO[\rho])$. This means that $\mathbf{I}_{\psi'}$ is in $DIAG(FO[\rho])$. Finally, the $\mathbf{I}_{\alpha'}$ structure corresponding to the word model for $\alpha' \equiv x P |||$ is not in $WFF(FO[\rho])$, and therefore neither is in $AUTOSAT(FO[\rho])$ nor in $DIAG(FO[\rho])$, as the formula $\alpha \equiv x R_\neg$ encoded by $\mathbf{I}_{\alpha'}$ is not a well formed formulae in $FO[\rho]$.

Note that, for $i, r, m \geq 1$ and $F = AA^i(r, m)$, the sets $WFF(F), DIAG(F)$ and $AUTOSAT(F)$ are not empty.

4.2 Recognizing Well Formed Formulae

We consider now the complexity of recognizing well formed formulae of finite-order logic.

An old result by Büchi [Buc60] states that a language is regular iff it can be defined in monadic second-order logic by a Σ_1^1 sentence. In Büchi's characterization there is a block of existential quantifiers with unary second-order variables and first-order variables, followed by a block of first-order universal quantifiers, followed in turn by a first-order formula which is quantifier free. So, we have the following fact.

Fact 1. *Every regular language $\mathcal{L}$ is definable in $AA\Sigma^1(1, 2)$.*

Furthermore, the following result implies that for each context-free grammar G with set of terminal symbols T there is a $AA\Sigma^1(3, 3)$ formula φ_G of vocabulary $\pi(T) = \{<\} \cup \{R_a : a \in T\}$ such that

$$Mod(\varphi_G) = \{\mathbf{I} \in \mathcal{B}_{\pi(T)} : \mathbf{I} \text{ is a word model for some } v \in \mathcal{L}(G)\}$$

Proposition 1 ([MP96]). *Every context-free language $\mathcal{L}$ is definable in AA-$\Sigma^1(3,3)$.*

It is not difficult to see that, for $i, r, m \geq 1$ and a relational vocabulary σ, the set of wff of $AA^i(r,m)$ over σ is a context-free language. The following example illustrates this fact by showing a context-free grammar for one of these sets of formulae.

Example 3. Let $G = (N, T, P, S)$ be a context-free grammar where $N = \{S, F, B_\Sigma, B_\Pi, V, A, I\}$ is the set of nonterminal symbols, $T = \{\neg, \vee, \wedge, \exists, \forall, (,), =, x, X, |, \iota, R, , \}$ is the set of terminal symbols, S is the start symbol, and P is the following set of productions:

$S \rightarrow F$ "quantifier-free formula F"

$S \rightarrow B_\Sigma F \cup B_\Sigma B_\Pi F \cup B_\Sigma B_\Pi B_\Sigma F$ "zero, one or two alternations of quantifiers, starting with an existential block"

$S \rightarrow B_\Pi F \cup B_\Pi B_\Sigma F \cup B_\Pi B_\Sigma B_\Pi F$ "zero, one or two alternations of quantifiers, starting with an universal block"

$B_\Sigma \rightarrow \exists V \cup \exists V B_\Sigma$ "block of existential quantifiers"

$B_\Pi \rightarrow \forall V \cup \forall V B_\Pi$ "block of universal quantifiers"

$F \rightarrow A \cup (F \vee F) \cup (F \wedge F) \cup (\neg F)$ "atomic formula or compound formula"

$V \rightarrow xI \cup X(\iota)I \cup X((\iota))I$ "a variable is a first-order variable x or a higher-order variable $X\tau$, where $\tau \in typ(3,1) \setminus \{\iota\}$, plus an index"

$A \rightarrow xI = xI$ "atomic formula of the form $v_1 = v_2$"

$A \rightarrow R(xI)$ "atomic formula of the form $R(v)$"

$A \rightarrow X(\iota)I(xI) \cup X((\iota))I(X(\iota)I)$ "atomic formulae of the form $V^\tau(V_1^{\tau_1}, \ldots, V_r^{\tau_r})$ where $\tau = (\tau_1, \ldots, \tau_r) \in typ(3,1) \setminus \{\iota\}$"

$I \rightarrow | \cup |I$ "index"

The language $\mathcal{L}(G)$ generated by G is precisely the set of wff of $AA^2(1,3)$ over $\sigma = \{R\}$, where R is a unary relation symbol.

The following proposition is a consequence of the previous observations.

Proposition 2. *For $i, r, m \geq 1$, the notion of wff of $AA^i(r,m)$, is not definable in $AA\Sigma^1(1,1)$ but is definable in $AA\Sigma^1(3,3)$.*

Proof. (Sketch). Given that, for $i, r, m \geq 1$ and a relational vocabulary σ, the set of wff of $AA^i(r,m)$ over σ is a context-free language, and that, by Proposition 1, every context free language is definable in $AA\Sigma^1(3,3)$, we can conclude that the notion of wff of $AA^i(r,m)$ is definable in $AA\Sigma^1(3,3)$.

To prove that the notion of wff of $AA^i(r,m)$ is not definable in $AA\Sigma^1(1,1)$, let's assume that, for some relational vocabulary σ and some $i, r, m \geq 1$, the set of wff of $AA^i(r,m)$ over σ is definable in $AA\Sigma^1(1,1)$. Since monadic Σ_1^1 includes

$AA\Sigma^1(1,1)$, it follows that, by Büchi's characterization of regular languages, the set of wff over σ for such fragment $AA^i(r,m)$ is regular. But it is straightforward to show, by using the pumping lemma for regular languages (see [HU79]), that for every $i, r, m \geq 1$, $AA^i(r,m)[\sigma]$ is non-regular. This contradicts the assumption that the set of wff of some fragment $AA^i(r,m)$ is definable in $AA\Sigma^1(1,1)$, completing the proof. □

The next fact follows from the previous proposition by observing that, for $i, r, m \geq 1$, each set of formulae $\{\varphi' : \varphi \in AA^i(r,m)[\rho]\}$ is context-free.

Fact 2. *For $i, r, m \geq 1$, the set $WFF(AA^i(r,m)[\rho])$, i.e., the set of finite ρ-structures $\mathbf{I}_{\varphi'}$ such that $\varphi \in AA^i(r,m)[\rho]$, is not definable in $AA\Sigma^1(1,1)[\rho]$, but is definable in $AA\Sigma^1(3,3)[\rho]$.*

4.3 A Lower Bound for the Definability of AUTOSAT

We show in this section that, for every $i \geq 1$, $r \geq 3$ and $m \geq 4$, there are Boolean queries not expressible in $AA^i(r,m)$. In particular, we show that, for every $i \geq 1$, $r \geq 3$ and $m \geq 4$, $AUTOSAT(AA^i(r,m)[\rho])$ is not definable in $AA^i(r,m)[\rho]$. But we have to prove first that its complement $DIAG(AA^i(r,m)[\rho])$ is not definable in $AA^i(r,m)[\rho]$.

Proposition 3. *For $i, r, m \geq 1$ and $F = AA^i(r,m)[\rho]$, $DIAG(F)$ is not definable in F.*

Proof. Towards a contradiction, let's assume that $DIAG(F)$ is definable in F. Then there is a sentence $\psi_D \in F$ such that $Mod(\psi_D) = DIAG(F)$. From the definition of $DIAG(F)$ and from our assumption, it follows that for an arbitrary ρ-structure $\mathbf{I}_{\varphi'}$ which encodes a formula $\varphi \in F$, $\mathbf{I}_{\varphi'} \not\models \varphi$ iff $\mathbf{I}_{\varphi'} \models \psi_D$. On the other hand, there is a finite ρ-structure $\mathbf{I}_{\psi'_D}$ which encodes the sentence ψ_D. But then $\mathbf{I}_{\psi'_D} \not\models \psi_D$ iff $\mathbf{I}_{\psi'_D} \models \psi_D$, which is a contradiction. □

Proposition 4. *For $i \geq 1$, $r \geq 3$, $m \geq 4$ and $F = AA^i(r,m)[\rho]$, $AUTOSAT(F)$ is not definable in F.*

Proof. Let's assume that $AUTOSAT(F)$ is definable in F, i.e., that there is a sentence $\psi_A \in F$ such that $Mod(\psi_A) = AUTOSAT(F)$. We know by Fact 2 that there is a sentence ψ_W in $AA\Sigma^1(3,3)[\rho]$, and therefore in F, such that $Mod(\psi_W) = WFF(F)$. But then, there is a sentence φ which is in prenex normal form and which is logically equivalent to $\psi_W \wedge \neg\psi_A$. Furthermore, as ψ_W has at most three alternating blocks of quantifiers and ψ_A has at most m alternating blocks of quantifiers, there is a sentence ψ_D with at most m alternations which is equivalent to φ. It follows that $\psi_D \in F$ and $Mod(\psi_D) = DIAG(F)$ which contradicts Proposition 3 and the assumption that $AUTOSAT(F)$ is definable in F. □

Remark 1. In [MP96] it is suggested that the previous proposition is true for every level of the AA^1 hierarchy of second-order logic, i.e., for every $r, m \geq 1$, though no proof is given there. Unfortunately, it does not seem possible to prove that, using the diagonalization argument which we employed here and which was outlined in the article of Makowsky and Pnueli. The major problem is that by Fact 2 we cannot define $WFF(F)$ in $AA\Sigma^1(1,1)$. Furthermore, as we know that $WFF(F)$ is definable in $AA^1(3,3)$, but we do not know whether $WFF(F)$ is definable in $AA^1(2,3)$, we need $r \geq 3$. Finally, as the set of formulae in each level of the $AA^1(r,m)$ hierarchy is closed under negation, but not under conjunction, we need $m \geq 4$. Note that the fact that each level of the AA^i hierarchies is closed under negation seems to be not enough, since $WFF(F) \subset \mathcal{B}_\rho$. That is, there are structures of signature ρ which *do not* encode a formula from F.

We can summarize the classes of formulae for which $AUTOSAT(F)$ is *not* definable in F, as follows.

Proposition 5. *Let F be a class of formulae of finite-order logic whose syntax is context free. If $AA^1(3,4) \subseteq F$ and F is closed under negation, then $AUTOSAT(F[\rho])$ is not definable in $F[\rho]$.*

4.4   An Upper Bound for the Definability of AUTOSAT

We give in this subsection a tight upper bound for the definability of $AUTO$-$SAT(AA^i(r,m)[\rho])$ for $i,r,m \geq 1$. In order to do that, we break down the task into several subtasks. For each subtask we define a series of predicates and prove some lemmas regarding the complexity of defining such predicates. Then, using these lemmas we prove the required result.

Given that for each $HO^{i,r}$ fragment, and thus for each layer of the AA^i hierarchies, both the order and the maximal-arity of the variables are bounded, we assume, for the sake of simplicity, a vocabulary ρ with a different relation symbol for each different type of higher-order variable, i.e.,

$$\rho = \{<, R_\neg, R_\vee, R_\wedge, R_\exists, R_\forall, R_(, R_), R_x, R_|, R_,, R_P\} \cup \{R_{X^\tau} : \tau \in typ(i,r) \setminus \{\iota\}\}$$

Note that strictly speaking, ρ should be denoted as $\rho_{i,r}$, since it depends of i and r. However, to simplify the notation, we use simply ρ.

We identify every formula $\varphi \in HO^{i,r}[\rho]$ with a formula φ' over the vocabulary $\rho' = \{P|^i : 1 \leq i \leq 12 + |typ(i,r)|\}$, where φ' is the formula obtained by replacing in φ each occurrence of a relation symbol $R \in \rho \cup \{=\}$ by the word $f(R) \in \rho'$, for the following bijective function f from ρ to ρ',

$$\{< \mapsto P_1, = \mapsto P_2, R_\neg \mapsto P_3, \dots, R_P \mapsto P_{13}\} \cup \{R_{X^{\tau_j}} \mapsto P_{13+j} : \tau_j \text{ is}$$

the j-th type in the lexicographical order of $typ(i,r) \setminus \{\iota\}$.

As before, we encode each formula φ using the ρ-structure $\mathbf{I}_{\varphi'}$ which corresponds to the word model for φ'.

From now on, we call a word model of vocabulary ρ, simply a word, and if $R_a(x)$ for some unary relation symbol $R_a \in \rho$, we say that the symbol in position x of the word is a.

Definition 7. *For each $\tau \in typ(i, r)$,*

   i. $VAR^{\tau}(x)$ is a predicate indicating that the symbol in position x of a word is a symbol of a variable of type τ;

  ii. $INDEX^{\tau}(x, y_1, y_2)$ is a predicate indicating that the symbol in position x is a symbol of a variable of type τ, and the symbols in positions y_1 to y_2 encode its index;

 iii. $SAME^{\tau}(x_1, x_2)$ is a predicate indicating that the symbols in positions x_1 and x_2 are symbols of variables of type τ, and they have the same index (that is, they refer to the same variable).

Lemma 1. *The predicates $VAR^{\tau}(x)$, $INDEX^{\tau}(x, y_1, y_2)$, and $SAME^{\tau}(x_1, x_2)$ can be expressed by formulae in Σ_0^0, Σ_2^0, and $AA\Sigma^1(2,3)$, respectively.*

Proof. $VAR^{\tau}(x)$ is simply $R_{X^{\tau}}(x)$ where $R_{X^{\tau}}$ is the relation symbol in ρ used to denote variables of type τ.

For $INDEX^{\tau}(x, y_1, y_2)$, we write

$VAR^{\tau}(x) \wedge \forall z\big((y_1 < z \wedge z < y_2) \rightarrow R_|(z)\big)$
"x is a symbol of a variable of type τ and
every symbol z between y_1 and y_2 is "|"" $\wedge$

$R_|(y_1) \wedge R_|(y_2)$
"the symbols in positions y_1 and y_2 are both "|"" $\wedge$

$\forall z \neg(x < z \wedge z < y_1)$
"y_1 is the successor of x" $\wedge$

$\exists z\big(y_2 < z \wedge \forall z_2 \neg(y_2 < z_2 \wedge z_2 < z) \wedge \neg R_|(z)\big)$
"z is the successor of y_2 and the symbol in position z is not "|""

Clearly, there is a formula in Σ_2^0 which is equivalent to the previous one.

For $SAME^{\tau}(x_1, x_2)$, we write

$\exists y_1 y_2 y_3 y_4 F\Big(INDEX^{\tau}(x_1, y_1, y_2) \wedge INDEX^{\tau}(x_2, y_3, y_4) \wedge$
"F is a bijective function from the
range $y_1 - y_2$ to the range $y_3 - y_4$" $\Big)$

It is not difficult to see that there is formula in $AA\Sigma^1(2,3)$ which is equivalent to the previous one.　　　　□

We want now to encode valuations for the different variables in a formula. Given a ρ-structure $\mathbf{I}_{\varphi'}$ which encodes a formula $\varphi \in AA^i(r, m)[\rho]$, and given a valuation v on $\mathbf{I}_{\varphi'}$, for each type $\tau \in typ(i+1, r)$ the values assigned by v to all variables X^{τ} of type $\tau = (\tau_1, \ldots, \tau_k)$ which appear in φ are encoded by an object (relation) $R^{\tau'} \in I_{\tau'}$ of type $\tau' = (\iota, \tau_1, \ldots, \tau_k)$. First-order variables, i.e., variables of type ι, are a special case since the values assigned by v to the first-order variables which appear in φ are encoded by a binary relation $R \in I_{\iota^2}$ of type ι^2, with the additional restriction that each element corresponding to a first-order variable is related to exactly one element. We use $\overline{V}$ to denote a vector which contains a different variable $V^{(\iota,\tau_1,\ldots,\tau_k)}$ for each type $(\tau_1,\ldots,\tau_k) \in typ(i+1,r) \setminus \{\iota\}$, plus a second-order variable F of arity 2. Such vector $\overline{V}$ has exactly $|typ(i+1,r)|$ different variables which we use to encode valuations for the fragment $AA^i(r,m)$.

Definition 8.

i. $VAL^\tau(V^{\tau'})$ are predicates indicating that the object in $I_{\tau'}$ assigned to the variable $V^{\tau'}$ of type τ', encodes a valuation for all variables of type τ in the formula.

ii. $VAL(\overline{V})$ is a predicate indicating that the objects assigned to the variables which appear in the vector $\overline{V}$, encode a valuation for all variables occurring in the formula. Note that, VAL as well as the content of the vector $\overline{V}$ depend of the order i and the maximal arity r that we are considering.

iii. $ASSIGNS^\tau(V^{\tau'},Y^\tau,x)$ are predicates indicating that the object in I_τ assigned to the variable Y^τ is the object assigned by the valuation encoded by $V^{\tau'}$ to the variable of type τ in position x of the formula.

Lemma 2. For $\tau = (\tau_1,\ldots,\tau_k)$, $\tau \neq \iota$, $VAL^\tau(V^{\tau'})$ can be expressed by a formula in $AA\Pi^{j-1}(k+1,3)$, where $j = order(\tau)$. $VAL^\iota(F)$ can be expressed by a formula in $AA\Pi^1(2,3)$. For a given order i and maximal arity r, $VAL(\overline{V})$ can be expressed by a formula in $AA\Pi^i(r+1,3)$.

For $\tau = (\tau_1,\ldots,\tau_k)$, $\tau \neq \iota$, $ASSIGNS^\tau(V^{\tau'},Y^\tau,x)$ can be expressed by a formula in $AA\Pi^{j-1}(k+1,1)$, where $j = order(\tau)$. $ASSIGNS^\iota(F,y,x)$ can be expressed by a Σ_0^1 formula of arity 2.

Proof. For $VAL^\iota(F)$, we say that F encodes a valuation for all first-order variables in the formula.

$\forall x_1 x_2 y \big(\neg SAME^\iota(x_1,x_2) \vee (F(x_1,y) \leftrightarrow F(x_2,y)) \big)$
"If a first-order variable occurs more than once, then F assigns the same set of values to each occurrence" $\wedge$

$\forall x_1 \exists x_2 \big(VAR^\iota(x_1) \rightarrow F(x_1,x_2) \big)$
"F assigns at least one value to each first-order variable" $\wedge$

$\forall x_1 x_2 x_3 \big(F(x_1,x_2) \wedge F(x_1,x_3) \rightarrow x_2 = x_3 \big)$
"F assigns at most one value to each first-order variable"

For $VAL^\tau(V^{\tau'})$, where $\tau = (\tau_1,\ldots\tau_k)$, we say that $V^{\tau'}$, $\tau' = (\iota,\tau_1,\ldots,\tau_k)$, encodes a valuation for all higher-order variables of type τ in the formula.

$\forall x_1 x_2 X_1^{\tau_1} \ldots X_k^{\tau_k} \big(\neg SAME^\tau(x_1,x_2) \vee$
$$\big(V^{\tau'}(x_1,X_1^{\tau_1},\ldots,X_k^{\tau_k}) \leftrightarrow V^{\tau'}(x_2,X_1^{\tau_1},\ldots,X_k^{\tau_k}) \big) \big)$$
"$V^{\tau'}$ assigns the same object in I_τ to each occurrence of a same higher-order variable of type τ in the formula"

For a given order i and maximal arity r, $VAL(\overline{V})$ is written as the conjunction of the $|typ(i+1,r)|$ formulae for VAL^τ, i.e,

$$\bigwedge_{(\tau_1,\dots,\tau_k)\in typ(i+1,r)\setminus\{\iota\}} \left(VAL^{(\tau_1,\dots,\tau_k)}(V^{(\iota,\tau_1,\dots,\tau_k)})\right) \wedge VAL^\iota(F)$$

For $ASSIGNS^\iota(F,y,x)$, we say that F assigns y to the first-order variable in position x if $F(x,y)$.

For $ASSIGNS^\tau(V^{\tau'},Y^\tau,x)$, where $\tau = (\tau_1,\dots,\tau_k)$ and $\tau' = (\iota,\tau_1,\dots,\tau_k)$, the corresponding formula is as follows.

$$\forall Z_1^{\tau_1}\dots Z_k^{\tau_k}\left(V^{(\iota,\tau_1,\dots,\tau_k)}(x,Z_1^{\tau_1}\dots Z_k^{\tau_k}) \leftrightarrow Y^\tau(Z_1^{\tau_1}\dots Z_k^{\tau_k})\right)$$

"$V^{(\iota,\tau_1,\dots,\tau_k)}$ assigns Y^τ to x"   □

Definition 9.

i. $WFF(x_1,x_2)$ is a predicate indicating that the symbols in positions x_1 to x_2 form a well formed sub-formula.

ii. For $3 \leq j \leq 12 + |typ(i,r)|$, $ATOM_{P_j}(x_1,x_2)$ is a predicate indicating that the symbols in positions x_1 to x_2 form an atomic sub-formula of the form $P|^j(x)$. Note that, $P|^j$ is a word in ρ' which corresponds to a unary relation symbol in ρ.

iii. For $1 \leq j \leq 2$, $ATOM_{P_j}(x_1,x_2)$ is a predicate indicating that the symbols in positions x_1 to x_2 form an atomic sub-formula of the form $P|^j(x,y)$. Recall that $P|$ and $P||$ are the words in ρ' which correspond to the binary relation symbols $<$ and $=$, respectively.

iv. For every $\tau \in typ(i,r) \setminus \{\iota\}$, $ATOM_{X^\tau}(x_1,x_2)$ is a predicate indicating that the symbols in positions x_1 to x_2 form an atomic sub-formula of the form $V^\tau(V_1^{\tau_1},\dots,V_k^{\tau_k})$, where $\tau = (\tau_1,\dots,\tau_k)$.

v. For every $\tau \in typ(i,r) \setminus \{\iota\}$, $POS_j^\tau(x_1,x_2)$ is a predicate indicating that the symbol in position x_1 is a variable of type $\tau = (\tau_1,\dots,\tau_k)$, $k \geq j$, and the symbol in position x_2 is the variable V^{τ_j} in the j-th position of it.

vi. $NOT(x_1,x_2)$ is a predicate indicating that the symbols in positions x_1 to x_2 form a sub-formula of the form $(\neg\varphi)$.

vii. $OR(x_1,x_2,x_3)$ is a predicate indicating that the symbols in positions x_1 to x_3 form a sub-formula of the form $(\varphi_1 \vee \varphi_2)$ with x_2 the position of the $\vee$ symbol.

viii. $AND(x_1,x_2,x_3)$ is a predicate indicating that the symbols in positions x_1 to x_3 form a sub-formula of the form $(\varphi_1 \wedge \varphi_2)$ with x_2 the position of the $\wedge$ symbol.

Lemma 3. *The predicates in Definition 9 can be expressed by formulae in AA-$\Sigma^1(3,3)$.*

Proof. It follows from Proposition 1 and the fact that each predicate in Definition 9 defines in each model a language which is context-free.   □

Definition 10.

   i. $ATOMSAT(\overline{V}, x_1, x_2)$ is a predicate indicating that the symbols in positions x_1 to x_2 form an atomic sub-formula, which is satisfied by the valuation encoded by $\overline{V}$ for a structure which is the whole formula.

   ii. $QFREESAT(\overline{V}, x_1, x_2)$ is a predicate indicating that the symbols in positions x_1 to x_2 form a quantifier-free well formed sub-formula, which is satisfied by the valuation encoded by $\overline{V}$ for a structure which is the whole formula.

Lemma 4. Let $c(r) = 1$ for $r > 1$ and $c(r) = 2$ for $r = 1$. It follows that $ATOMSAT(\overline{V}, x_1, x_2)$ can be expressed by a formula in $AA\Sigma^i(r+c(r), 4)$ while $QFREESAT(\overline{V}, x_1, x_2)$ can be expressed by a formula in $AA\Sigma^i(r + c(r), 4)$.

Proof. For $ATOMSAT(\overline{V}, x_1, x_2)$ we have different cases depending on the kind of atom we are considering. If the atom is a formula of the form $P|^j(x)$, where $f(R_a) = P|^j$ for R_a a unary relation symbol in ρ, we write

$$ATOM_{P_j}(x_1, x_2) \wedge VAL(\overline{V}) \wedge$$

$$\exists yp\big(R_x(p) \wedge x_1 < p \wedge p < x_2 \wedge R_a(y) \wedge ASSIGNS^\iota(F, y, p)\big)$$

If the atom is a formula of the form $P|(x, y)$, we write

$$ATOM_{P_1}(x_1, x_2) \wedge VAL(\overline{V}) \wedge$$

$$\exists y_1 y_2 p_1 p_2 \big(R_x(p_1) \wedge R_x(p_2) \wedge x_1 < p_1 \wedge p_1 < p_2 \wedge p_2 < x_2 \wedge$$

$$y_1 < y_2 \wedge ASSIGNS^\iota(F, y_1, p_1) \wedge ASSIGNS^\iota(F, y_2, p_2)\big)$$

If the atom is a formula of the form $P||(x, y)$, we just replace $y_1 < y_2$ by $y_1 = y_2$ and $ATOM_{P_1}(x_1, x_2)$ by $ATOM_{P_2}(x_1, x_2)$ in the previous formula.

And, if the atom is a formula of the form $V^\tau(V_1^{\tau_1}, \ldots, V_k^{\tau_k})$, $\tau = (\tau_1, \ldots, \tau_k)$, we write

$$ATOM_{X^\tau}(x_1, x_2) \wedge VAL(\overline{V}) \wedge$$

$$\exists Y_1^{\tau_1} \ldots Y_k^{\tau_k} p_1 \ldots p_k \Big(V^{(\iota, \tau_1, \ldots, \tau_k)}(x_1, Y_1^{\tau_1}, \ldots, Y_k^{\tau_k}) \wedge$$

$$POS_1^\tau(x_1, p_1) \wedge \ldots \wedge POS_k^\tau(x_1, p_k) \wedge$$

$$ASSIGNS^{\tau_1}(V^{\tau_1'}, Y_1^{\tau_1}, p_1) \wedge \ldots \wedge ASSIGNS^{\tau_k}(V^{\tau_k'}, Y_k^{\tau_k}, p_k)\Big)$$

For $QFREESAT(\overline{V}, x_1, x_2)$ we use a second-order variable X and say that, if a tuple (a, b) is in X, then the symbols in positions a to b form a quantifier free sub-formula which is satisfied by the valuation encoded by $\overline{V}$ for a structure which is the whole formula.

$$\exists X\Big(X(x_1, x_2) \wedge \forall y_1 y_1\big(X(y_1, y_2) \to \big(ATOMSAT(\overline{V}, y_1, y_2) \vee$$

$$(NOT(y_1, y_2) \wedge \neg X(y_1 + 2, y_2 - 1)) \vee$$

$$(\exists y(OR(y_1, y, y_2) \wedge (X(y_1 + 1, y - 1) \vee X(y + 1, y_2 - 1)))) \vee$$

$$(\exists y(AND(y_1, y, y_2) \wedge (X(y_1 + 1, y - 1) \wedge X(y + 1, y_2 - 1))))\big)\big)\Big)\Big) \quad \square$$

Definition 11.

   i. $\exists BLOCK(x_1, x_2, x_3)$ *is a predicate indicating that the symbols in positions* x_1 *to* x_2 *form an existential quantification of some variables and the symbols in positions* $x_2 + 1$ *to* x_3 *form a well formed sub-formula.*

   ii. $\forall BLOCK(x_1, x_2, x_3)$ *is a predicate indicating that the symbols in positions* x_1 *to* x_2 *form an universal quantification of some variables and the symbols in positions* $x_2 + 1$ *to* x_3 *form a well formed sub-formula.*

   iii. $VEQUIV(\overline{V}_1, \overline{V}_2, x_1, x_2)$ *is a predicate indicating that* $VAL(\overline{V}_1)$, $VAL(\overline{V}_2)$ *and that* $\overline{V}_1$ *is S-equivalent to* $\overline{V}_2$, *with S being the set of variables appearing in the positions* x_1 *to* x_2 *of the formula.*

Lemma 5. $\exists BLOCK(x_1, x_2, x_3)$ *and* $\forall BLOCK(x_1, x_2, x_3)$ *can be expressed by formulae in* $AA\Sigma^1(3,3)$ *while* $VEQUIV(\overline{V}_1, \overline{V}_2, x_1, x_2)$ *can be expressed by a formula in* $AA^i(r+1, 4)$.

Proof. $\exists BLOCK(x_1, x_2, x_3)$ and $\forall BLOCK(x_1, x_2, x_3)$ are context free.

For $VEQUIV(\overline{V}_1, \overline{V}_2, x_1, x_2)$, we write

$$VAL(\overline{V}_1) \wedge VAL(\overline{V}_2) \wedge$$

$$\bigwedge\nolimits_{\tau \in typ(i+1,r)} \left(\forall y \Big(VAR^\tau(y) \wedge \neg \exists z (x_1 < z \wedge z < x_2 \wedge SAME^\tau(y,z)) \right.$$
$$\text{``if the variable in position } y \text{ is a variable of type } \tau$$
$$\text{which does not appear in the range } x_1 - x_2\text{''} \rightarrow$$

$$\forall Z^\tau \big(ASSIGNS^\tau(V_1^{\tau'}, Z^\tau, y) \leftrightarrow ASSIGNS^\tau(V_2^{\tau'}, Z^\tau, y) \big) \Big) \Big)$$
$$\text{``the valuations encoded by } \overline{V}_1 \text{ and } \overline{V}_2 \text{ assign the}$$
$$\text{same object from } I_\tau \text{ to the variable in position } y\text{''} \qquad \square$$

Definition 12.

   i. $\exists BLOCKSAT_m(\overline{V}, x_1, x_2)$ *is a predicate indicating that* $\overline{V}$ *encodes a valuation and the symbols in positions* x_1 *to* x_2 *form a formula with no more than m alternating blocks of quantifiers, starting with an existential quantifier, which is satisfied by* $\overline{V}$.

   ii. $\forall BLOCKSAT_m(\overline{V}, x_1, x_2)$ *is a predicate indicating that* $\overline{V}$ *encodes a valuation and the symbols in positions* x_1 *to* x_2 *form a formula with no more than m alternating blocks of quantifiers, starting with a universal quantifier, which is satisfied by* $\overline{V}$.

Lemma 6. *Let* $c(r) = 1$ *for* $r > 1$ *and* $c(r) = 2$ *for* $r = 1$. *It follows that* $\exists BLOCKSAT_m(\overline{V}, x_1, x_2)$ *and* $\forall BLOCKSAT_m(\overline{V}, x_1, x_2)$ *can be expressed by formulae in* $AA^i(r + c(r), m + 4)$.

Proof. We use induction on the number of blocks of quantifiers m. For $m = 0$, $\exists BLOCKSAT_0(\overline{V}, x_1, x_2)$ and $\forall BLOCKSAT_0(\overline{V}, x_1, x_2)$ are simply $QFREE$-$SAT(\overline{V}, x_1, x_2)$. For $m > 0$, we say that $\exists BLOCKSAT_m(\overline{V}, x_1, x_2)$, if

$$\exists \overline{V}_1 x \big(\exists BLOCK(x_1, x, x_2) \wedge$$

$$VEQUIV(\overline{V}_1, \overline{V}, x_1, x) \wedge \forall BLOCKSAT_{m-1}(\overline{V}_1, x+1, x_2))$$

And we say that $\forall BLOCKSAT_m(\overline{V}, x_1, x_2)$, if

$$\exists x \forall \overline{V}_1 (\forall BLOCK(x_1, x, x_2) \wedge$$

$$VEQUIV(\overline{V}_1, \overline{V}, x_1, x) \wedge \exists BLOCKSAT_{m-1}(\overline{V}_1, x+1, x_2)) \qquad \Box$$

Finally, the upper bound for the definability of $AUTOSAT(AA^i(r,m)[\rho])$ is as follows.

Proposition 6. *For $i, r, m \geq 1$, the class $AUTOSAT(AA^i(r,m)[\rho])$ is definable in $AA^i(r+c(r), m+4)[\rho]$ where $c(r) = 1$ for $r > 1$ and $c(r) = 2$ for $r = 1$.*

Proof. We have to show that, for every $i, r, m \geq 1$, there is a sentence $\psi_A \in AA^i(r+c(r), m+4)[\rho]$ where $c(r) = 1$ for $r > 1$ and $c(r) = 2$ for $r = 1$, such that $Mod(\psi_A) = AUTOSAT(AA^i(r,m)[\rho])$. The required formulae are simply disjunctions of the $BLOCKSAT$ formulae above.

$$\exists \overline{V} x_1 x_2 \Big(\forall z(\neg(z < x_1)) \wedge \forall z(\neg(x_2 < z))$$

"x_1 and x_2 are the first and the last element in the ordering $<$, respectively" $\wedge$

$$(\exists BLOCKSAT_0(\overline{V}, x_1, x_2) \vee \ldots \vee \exists BLOCKSAT_m(\overline{V}, x_1, x_2) \vee$$

$$\forall BLOCKSAT_0(\overline{V}, x_1, x_2) \vee \ldots \vee \forall BLOCKSAT_m(\overline{V}, x_1, x_2)) \Big) \qquad \Box$$

4.5   Main Result

Our main result stating the properness of the AA^i hierarchies follows from Propositions 4 and 6.

Theorem 2. *For every $i \geq 1$, $r \geq 3$ and $m \geq 4$, there are Boolean queries not expressible in $AA^i(r,m)$ but expressible in $AA^i(r+1, m+4)$.*

Due to the fact that we ask $r \geq 3$ and $m \geq 4$, it could seem at first sight that for the AA^1 hierarchy of second-order logic our result is less satisfactory than the result of Makowsky and Pnueli. However, this does not seem to be the case as we already explained in Remark 1.

5   The HAA^i Hierarchies

We define next a new kind of alternation-arity hierarchies in which the only alternations bounded are those corresponding to the quantifiers of the highest order.

Definition 13. *(HAA^i hierarchies) Let $i, r, m \geq 1$,*

i. $HAA\Sigma^i(r,m)$ is the class of Σ^i_m formulae in which the higher-order variables of all orders up to $i+1$ have maximal-arity at most r.

ii. $HAA\Pi^i(r,m)$ *is the class of* Π^i_m *formulae in which the higher-order variables of all orders up to* $i+1$ *have maximal-arity at most* r.

iii. $HAA^i(r,m) = HAA\Sigma^i(r,m) \cup HAA\Pi^i(r,m)$.

Note that because of (iii), for every $i \geq 1$, the $HAA^i(r,m)$ hierarchy is also closed under negation.

Regarding the relation between the AA^i hierarchies and the HAA^i hierarchies, it can be easily observed from the proof of Lemma 1 in [HT03] that, for every $i, r, m \geq 1$, $HAA^i(r,m) \supseteq AA^i(r,m)$. It is a trivial task to adapt that proof to show that, for every formula $\varphi \in AA^i(r,m)$, there is a formula φ' in GSNF which is equivalent to φ, has the same maximal-arity as φ and does not have more than m alternating blocks of quantifiers of order $i+1$.

We show next that, for $i \geq 2$, each $HAA^i(r,m)$ hierarchy is proper.

Theorem 3. *For every* $i \geq 2$, $r \geq 3$ *and* $m \geq 1$, *there are Boolean queries not expressible in* $HAA^i(r,m)$ *but expressible in* $HAA^i(max(r+1,i+2),m+1)$.

Proof. (Sketch). Clearly, for every $i \geq 2$, $r \geq 3$ and $m \geq 1$, $HAA^i(r,m) \supseteq AA^1(3,4)$. Furthermore, $HAA^i(r,m)$ is a context free language. For instance, it is not difficult to modify the context free grammar for $AA^2(1,3)$ in Example 3 to obtain a context free grammar for $HAA^2(1,3)$. We just need to add the productions $V_1 \rightarrow X((\iota))I$, $V_2 \rightarrow xI \cup X(\iota)I$, and $F \rightarrow \exists V_2(F) \cup \forall V_2(F)$, and replace the productions with nonterminals B_Σ and B_Π in the left hand side by the productions $B_\Sigma \rightarrow \exists V_1 \cup \exists V_1 B_\Sigma$ and $B_\Pi \rightarrow \forall V_1 \cup \forall V_1 B_\Pi$, respectively. Thus, given that the set of formulae in $HAA^i(r,m)$ is also closed under negation, it follows from Proposition 5 that, for $i \geq 2$, $r \geq 3$ and $m \geq 1$, $AUTOSAT(HAA^i(r,m)[\rho])$ is not definable in $HAA^i(r,m)[\rho]$.

It remains to prove that, for every $i \geq 2$, $r \geq 3$ and $m \geq 1$, $AUTOSAT(HAA^i(r,m)[\rho])$ is definable in $HAA^i(max(r+1,i+2),m+1)[\rho]$. We show next how the proof of Proposition 6 can be adapted to obtain, for every $i \geq 2$, $r, m \geq 1$, a sentence $\psi_a \in HAA^i(max(r+1,i+2),m+1)[\rho]$ such that $Mod(\psi_a) = AUTOSAT(HAA^i(r,m)[\rho])$.

The predicates used in Subsection 4.4 which are related to the identification of variables (Definition 7), the encoding of valuations (Definition 8), the identification of well formed sub-formulae (Definition 9) and the notion of satisfaction of quantifier free sub-formulae (Definition 10), remain the same.

Given the different way in which the quantifiers are arranged in the HAA^i hierarchies, it follows that the predicates that need to be adapted are only those which have to deal with quantified formulae.

First, we need to modify Definition 11. The predicates $\exists BLOCK(x_1, x_2, x_3)$ and $\forall BLOCK(x_1, x_2, x_3)$ apply now only to blocks of variables of order at most i, i.e., they hold only if the quantified variables in positions x_1 to x_2 are all of order $\leq i$. We also need to define $\exists BLOCK^i(x_1, x_2, x_3)$ and $\forall BLOCK^i(x_1, x_2, x_3)$ in order to identify blocks of quantifiers of order $i+1$. All these predicates can be expressed by formulae in $HAA\Sigma^1(3,1)$, since they all define in each model a language which is context free, and in the proof of Propositions 1 given in [MP96], the corresponding formula has a block of existentially quantified second-order variables of arity 3, followed by a first-order part.

The most delicate part of the proof is the redefinition of the predicates $\exists BLOCKSAT_m(\overline{V}, x_1, x_2)$ and $\forall BLOCKSAT_m(\overline{V}, x_1, x_2)$. The main difference with the previous case is that now these predicates hold only if the formula in positions x_1 to x_2 is a formula with no more than m alternating blocks of quantifiers of order $i + 1$, but with an arbitrary number of alternating blocks of quantifiers of order $\leq i$, which is satisfied by the valuation encoded by $\overline{V}$. Recall that $\overline{V}$ is a vector which contains a different variable $V^{(\iota, \tau_1, \ldots, \tau_k)}$ for each type $(\tau_1, \ldots, \tau_k) \in typ(i + 1, r) \setminus \{\iota\}$, plus a second-order variable F of arity 2. Recall also that $\overline{V}$ has exactly $|typ(i + 1, r)|$ different variables. We use induction on the number m of alternating blocks of quantifiers of order $i + 1$. For $m = 0$, $\exists BLOCKSAT_0(\overline{V}, x_1, x_2)$ and $\forall BLOCKSAT_0(\overline{V}, x_1, x_2)$ say that $\overline{V} = \overline{V}_1 \cup \overline{V}_2$ encodes a valuation, the variables which appear in $\overline{V}_1$ encode the sub-valuation for the variables of order $i + 1$, the variables which appear in $\overline{V}_2$ encode the sub-valuation for the variables of order $\leq i$, and the symbols in positions x_1 to x_2 form a $HO^{i,r}$ prenex formula which is satisfied by $\overline{V}$. Note that, we use here a variable $\mathcal{X}$ of order $i + 1$ and arity $|typ(i, r)| + 2$ and state that $\mathcal{X}(\overline{V}'_2, y_1, y_2)$ if the symbols in positions y_1 to y_2 form a $HO^{i,r}$ formula (possibly with free variables of order $i + 1$) which is satisfied by $\overline{V}_1 \cup \overline{V}'_2$.

$$\exists \mathcal{X} \Big(\mathcal{X}(\overline{V}_2, x_1, x_2) \wedge \forall \overline{V}'_2 y_1 y_2 \Big(\mathcal{X}(\overline{V}'_2, y_1, y_2) \to \Big(QFREESAT(\overline{V}_1 \cup \overline{V}'_2, y_1, y_2) \vee$$

$$\exists z \overline{V}''_2 \big(\exists BLOCK(y_1, z, y_2) \wedge VEQUIV(\overline{V}''_2, \overline{V}'_2, y_1, z) \wedge \mathcal{X}(\overline{V}''_2, z + 1, y_2) \big) \vee$$

$$\exists z \forall \overline{V}''_2 \big(\forall BLOCK(y_1, z, y_2) \wedge VEQUIV(\overline{V}''_2, \overline{V}'_2, y_1, z) \wedge \mathcal{X}(\overline{V}''_2, z + 1, y_2) \big) \Big) \Big) \Big)$$

For $m > 0$, $\exists BLOCKSAT_m(\overline{V}, x_1, x_2)$ and $\forall BLOCKSAT_m(\overline{V}, x_1, x_2)$ are written exactly as in the proof of Lemma 6, but replacing $\exists BLOCK(x_1, x_2, x_3)$ and $\forall BLOCK(x_1, x_2, x_3)$ with $\exists BLOCK^i(x_1, x_2, x_3)$ and $\forall BLOCK^i(x_1, x_2, x_3)$, respectively.

Since we only consider here the alternating blocks of quantifiers of order $i + 1$, the reader can easily check that the predicates $\exists BLOCKSAT_m(\overline{V}, x_1, x_2)$ and $\forall BLOCKSAT_m(\overline{V}, x_1, x_2)$ can be expressed by formulae in $HAA^i(|typ(i, r)| + 2, m + 1)$.

We can get a tighter upper bound than $|typ(i, r)| + 2$, since the arity of $\mathcal{X}$ can be significantly decreased if we encode the valuations for the variables of order $\leq i$ differently. Note that, we can encode the valuations for all variables of order 2 and arity between 1 and r by using only one variable $V^{\iota^{r+1}}$ rather than a different variable for each different arity as we did in the proof of Lemma 2. For the case of a second-order variable X of arity $k < r$, we simply use the sub-tuples formed by the elements in positions 1 to $k + 1$ of the $r + 1$-tuples in $V^{\iota^{r+1}}$ which correspond to the relation assigned to X and disregard the elements in positions $k + 2$ to $r + 1$. For variables of order ≥ 3 we can use a similar encoding. Thus we can encode a valuation for all variables of order $\leq i$ by using i variables of arity $r + 1$

and in consequence a variable $\mathcal{X}$ of arity $i+2$ instead of one of arity $|typ(i,r)|+2$. It follows that $\exists BLOCKSAT_m(\overline{V}, x_1, x_2)$ and $\forall BLOCKSAT_m(\overline{V}, x_1, x_2)$ can be expressed by formulae in $HAA^i(i+2, m+1)$ if $r+1 < i+2$ and by formulae in $HAA^i(r+1, m+1)$ otherwise.

Finally, the required formulae are simply disjunctions of the $2m$ $BLOCK$-SAT formulae above. $\square$

It seems that this result does not apply to the HAA^1 hierarchy of second-order. The main obstacle is that the valuation for first-order variables is encoded into a second-order variable. Thus the variable $\mathcal{X}$ in the proof must be of order at least 3.

6   The Concept of Finite Model Truth Definitions

A method which has been frequently used in classical model theory to compare the expressive power of logics, is Tarski's method of truth definitions [Tar33]. In particular, this method has been successfully used in higher-order logics (see [Lei94]). In finite model theory, Tarski's method has first been explored by M. Mostowski in [Mos01, Mos03]. He introduced the notion of *finite model truth definitions* (*FM-truth*, from now on) and proved a finite version of Tarski's theorem on undefinability of truth. Roughly, he proved that no *natural class of formulae* can define FM-truth for itself over *sufficiently large* finite models which have a suitable amount of arithmetical structure so that Gödelization can be carried out. It is our understanding that the concept of natural classes of formulae used by M. Mostowski, can be defined as follows.

We say that $\mathcal{L}$ is a *natural class of formulae* if, for any vocabulary σ, (i) the set of $\mathcal{L}[\sigma]$ formulae as well as the notion of satisfaction on finite σ-structures are decidable, and (ii) for each $\varphi \in \mathcal{L}[\sigma]$, if ψ is obtained from a first-order formula $\alpha \in FO[\sigma]$ by single substitution of φ in the place of a predicate in α, then ψ is equivalent to a formula in $\mathcal{L}[\sigma]$.

FM-truth definitions can be used, as in the classical case, to compare the expressive power of logics. The strategy consists in showing that a class of formulae $\mathcal{L}'$ which is known to be at least as expressive as a given natural class of formulae $\mathcal{L}$, additionally defines FM-truth for $\mathcal{L}$. If that is the case, then we have shown – using the "method of FM-truth definitions" – that $\mathcal{L}'$ is strictly more expressive than $\mathcal{L}$.

Clearly, the method of FM-truth definitions is somehow related to our work, since it can be used to establish the existence of proper hierarchies of formulae in higher-order logics by using a diagonalization argument. We discuss this relation in more detail in [FT05]. However, it is important to mention here that, the AA^i hierarchies *cannot* be proven using FM-truth definitions. This is so, because the formulae in the levels of the AA^i hierarchies are *not* natural classes of formulae, since they are *not* closed under first-order quantification.

Quite recently, in [Kol04] the notion of FM-truth definition was further discussed and compared with Vardi's concept of *combined complexity* [Var82], noting an important difference: the possibility of defining FM-truth for a logic $\mathcal{L}$

does not depend on the syntax of $\mathcal{L}$, as long as it is decidable. Furthermore, for each $i, m \geq 1$, a characterization of the logics for which the Σ_m^i class defines FM-truth, was given. In [Kol04b], FM-truth definitions were used to give a sufficient condition for the Σ_1^1 arity hierarchy to be proper over finite structures.

In contrast with the AA^i hierarchies, the layers of the HAA^i hierarchies are natural classes of formulae. Thus, for every $i, r, m \geq 1$, it follows that $HAA^i(r, m)$ does not define FM-truth for itself. We are currently investigating whether the characterization of the logics for which the Σ_m^1 class defines FM-truth which appeared in [Kol04], can help us to study whether the $HAA^1(r, m)$ hierarchy is a proper hierarchy in second-order logic.

References

[AHV94] Abiteboul, S., Hull, R. and Vianu, V., *Foundations of Databases*, Addison-Wesley, 1994.

[Ajt83] Ajtai, M., Σ_1^1-*formulae on Finite Structures*, Annals of Pure and Applied Logic, 24, pp. 1–48, 1983.

[Buc60] Büchi, J., R., *Weak second-order arithmetic and finite automata*, Z. Math. Logik und Grund. Math, 6, pp. 66–92, 1960.

[CH80] Chandra, A. K. and Harel, D., *Computable Queries for Relational Data Bases*, Journal of Computer and System Sciences, 21(2), 1980. 156–178

[EF99] Ebbinghaus, H. and Flum, J., *Finite Model Theory*, 2nd ed., Springer, 1999.

[FT05] Ferrarotti, F. A. and Turull Torres, J. M., *Arity and Alternation of Quantifiers in Higher Order Logics*, to appear as a Technical Report of the Information Systems Department of Massey University, 2005. (http://infosys.massey.ac.nz/ research/rs_techreports.html).

[Gro93] Grohe, M., *Bounded Arity Hierarchies in Fixed-Point Logics*, Proceedings of the 7th Workshop on Computer Science Lofic, volume 832 of Lecture Notes in Computer Science, pp.150–164, Springer-Verlap, 1993.

[Gro96] Grohe, M., *Arity Hierarchies*, Annals of Pure and Applied Logic, 82, pp. 103–163, 1996.

[GH96] Grohe, M., Hella, L., *A Double Arity Hierarchy Theorem for Transitive Closure Logic*, Archive for Mathematical Logic 35 157–171, 1996.

[Hel89] Hella, L., *Definability Hierarchies of Generalized Quantifiers*, Annals of Pure and Applied Logic, 43, pp. 235–271, 1989.

[Hel92] Hella, L., *Logic Hierarchies in PTIME*, In Poceedings of the 7th IEEE Symposium in Logic and Computer Science, pp. 360–368, 1992.

[HT03] Hella, L., Turull-Torres, J. M., *Expressibility of Higher Order Logics*, Electronic Notes in Theoretical Computer Science, Vol. 84, 2003.

[HU79] Hopcroft, J., Ullman, J., *Introduction to Automata Theory, Languages and Computation*, Addison-Wesley, 1979.

[Kol04] Kolodziejczyk, L. A., *Truth Definitions in Finite Models*, The Journal of Symbolic Logic, vol. 69(1), pp. 183–200.

[Kol04b] Kolodziejczyk, L. A., *A Finite Model-Theoretical Proof of a Property of Bounded Query Classes within PH*, The Journal of Symbolic Logic, vol. 69(4), pp. 1105–1116.

[Lei94] Leivant, D., *Higher order logic, Handbook of logic in artificial intelligence and logic programming*, (D. M. Gabbay et al., editors), vol. 2, Oxford University Press, pp. 228–321, 1994.

[MP96]   Makowsky, J. A., Pnueli, Y. B., *Arity and alternation in second-order logic*, Annals of Pure and Applied Logics, 78:189–202, 1996.

[Mat98]   Matz, O., *One Existential Quantifier will do in Existential Monadic Second-Order Logic Over Pictures*, in "Mathematical Foundations of Computer Science" (L. Brim, J. Gruska, and J. Zlatuška, Eds.), Lecture Notes in Computer Science, Vol.1450, pp. 751–759, Springer-Verlap, Berling, 1998.

[MST02]   Matz, O., Schweikardt N., Thomas, W., *The Monadic Quantifier Alternation Hierarchy over Grids and Graphs*, Information and Computation, 179, pp. 356–383, 2002.

[Mos01]   Mostowski, M., *On Representing Concepts in Finite Models*, Mathematical Logic Quarterly, vol.47, pp. 513–523, 2001.

[Mos03]   Mostowski, M., *On Representing Semantics in Finite Models*, Philosophical Dimensions of Logic and Science: Selected Contributed Papers from the 11th International Congress of Logic, Methodology, and Philosophy of Science, Krakow, 1999 (A. Rojszczak, J. Cachro, and G. Kurczewski, eds.), Kluwer, pp. 15–28, 2003.

[Ott95]   Otto, M., *Note on the Number of Monadic Quantifiers in Monadic Σ_1^1*, Inform. Process. Lett., 53, pp. 337–339, 1995.

[Sto76]   Stockmeyer, L., *The Polynomial Time Hierarchy*, Theoretical Computer Science, 3(1), pp. 1–22, 1976.

[Tar33]   Tarski A., *Pojęcie prawdy w językach nauk dedukcyjnych*, Warszawa, Nakładem Towarzystwa Naukowego Warszawskiego, 1933. English translation of the German version: The Concept of Truth in Formalized Languages. In: A. Tarsky, Logic, Semantics, Metamathematics, Clarendon Press, Oxford 1956, pp. 152–278.

[Tho82]   Thomas, W., *Classifying Regular Events in Symbolic Logic*, J. Comput, System Sci., 25, pp. 360–376, 1982.

[Var82]   Vardi M., *The Complexity of Relational Query Languages*, Proceedings of the 14th ACM Symposium on Theory of Computing, pp. 137–146, 1982.

Solving Abduction by Computing Joint Explanations: Logic Programming Formalization, Applications to *P2P* Data Integration, and Complexity Results

Gianluigi Greco

Dip. di Matematica, Università della Calabria, 87030 Rende, Italy
ggreco@mat.unical.it

Abstract. An extension of abduction is investigated where explanations for bunches of observations may be jointly computed by sets of interacting agents. At one hand, agents are allowed to partially contribute to the reasoning task, so that *joint explanations* can be singled out even if each agent does not have enough knowledge for carrying out abduction on its own. At the other hand, agents maintain their autonomy in choosing explanations, each one being equipped with a weighting function reflecting its perception about the reliability of set of hypotheses. Given that different agents may have different and possibly contrasting preferences, some reasonable notions of agents' agreement are introduced, and their computational properties are thoroughly studied. As an example application of the framework discussed in the paper, it is shown how to handle data management issues in Peer-to-Peer systems and, specifically, how to provide a repair-based semantics to inconsistent ones.

1 Introduction

Abduction is a well-known form of reasoning aiming at identifying the causes which some perceived observations can be explained according to. Formally, a logical theory T, a set $\mathcal{O}$ of *observations*, and a set H of *hypotheses* are given, and a subset S of H is looked for that merged with T entails $\mathcal{O}$ (short: $T \cup S \models \mathcal{O}$). The set S is said to be an *explanation* for $\mathcal{O}$, and it is usually required to satisfy some additional conditions, such as consistency with T and minimality.

Abduction was first studied by Pierce [1]. Since then, it was recognized as an important principle for common-sense reasoning, as a powerful mechanism for hypothetical reasoning in presence of incomplete information, and as a solid framework for modelling practical applications. Abduction has been deeply investigated in logic programming too (cf. [2]). In this context, the most influential definition is due to Kakas and Mancarella [3], but several other approaches have also been proposed both from proof and model-theoretic perspectives (e.g., [4, 5, 6, 7, 8]).

In this paper, we consider a generalization of the basic abductive framework to the setting where several *autonomous* agents interact with each other and *jointly* find explanations for a bunch of observations at hand. In particular, we investigate some foundational issues, by studying how an abduction problem can be solved through this kind of

J. Dix and S.J. Hegner (Eds.): FoIKS 2006, LNCS 3861, pp. 116–136, 2006.

"distributed" intelligence and how agents' reasoning mechanisms are to be modified in this novel scenario. Two key ingredients are at the basis of the proposed formalization, which are discussed below.

Firstly, agents are assumed to act in cooperative manner. This means that each agent is allowed to partially contribute to the whole explanation, i.e., to explain a proper subset of all the observations only, so that the solution of the abductive problem is the result of the merging of the individual explanations. This is particularly useful whenever each agent does not have enough knowledge for explaining all the observations by itself and coordination with other agents is required. Moreover, cooperation also means that while finding an individual explanation, each agent focuses on those hypotheses which have also been selected by some other agents. This leads to justifications which are more likely to be trustable, since the matching of partial individual explanations may be taken in favor of believing in their correctness (in the spirit, e.g., of [9]). These kinds of explanation are called *joint explanations* in the following.

Secondly, agents are assumed to have their own preferences about each possible individual explanation. Therefore, the problem of finding joint explanations is not trivialized to a standard abduction problem with a theory encompassing all the agents' individual theories. Rather, each agent is also equipped with a *weighting function* assigning a quantitative preference (i.e., an integer) to each set S of hypotheses, in order to reflect its perception about the reliability of S. Possibly, different agents have different contrasting preferences on the way the explanation should be carried out. Hence, those explanations on which all the agents *agree* are eventually selected as solutions to the abductive problem.

It is important to point out that, in the light of the observation above, our framework may be abstractly seen as a generalization of *abduction with penalization* [10, 11, 12] to multiagent systems, where a kind of multi-objective optimization has to be done. And, in fact, our choice is to pursue a game-theoretic approach to formally define the agreement semantics and suitable notions of optimality in explanations finding. In this respect, our investigation precisely faces one line of research which has been left open in [13], asking for suitable frameworks applying ideas from multi-criteria decision theory to solve conflicts among abductive agents.

In a nutshell, the main aim of the paper is to define and to subsequently investigate the intrinsic difficulty of suitable agreement semantics for the problem of computing joint explanations. In more details:

▶ In Section 3, we introduce a logic-programming framework for modelling joint explanations. Actually, we first define a basic setting, and we then enrich it by allowing agents to express preferences on the hypotheses. Given some possibly contrasting preferences, we introduce several notions of agreement among agents whose inspiration is mainly game-theoretic. For instance, we define conditions for a joint explanation to be preferred, pareto, stable, or safe.

▶ In Section 4, as an example application for the notion of joint explanations, we show how to handle data management issues in Peer-to-Peer (*P2P*) systems. In particular, we evidence that important aspects such as peers' autonomy and selfish interests, which have been only marginally modelled in previous approaches, may be elegantly taken into account while resolving conflicts in the data (violations to integrity constraints which make the system inconsistent). The proposed abductive repair approach to provide semantics for inconsistent *P2P* system is of particular interest in Data Integration contexts, and may be viewed as a generalization of [14], which has firstly proposed the use of abduction to repair inconsistent databases.

▶ The computational complexity of some important reasoning problems related to joint explanations is thoroughly investigated in Section 5. In particular, we consider two different settings, namely *agent* and *combined* complexity. The former is meant to capture the difficulty lying in the agents' interactions, since agents' theories are assumed to be fixed, whereas the latter accounts for the complexity of both agents' interactions and theories.

▶ Finally, Section 6 accounts for the discussion of relevant related works and of some concluding remarks about possible directions for further research.

2 Preliminaries

In this section, abductive agents are formalized in a very simple framework by means of stratified logic programs where classical negation is not allowed. Indeed, having agents equipped with such limited reasoning capabilities makes it easy to identify the salient features of our distributed scenario and to precisely characterize how the cost of solving an abduction problem is affected by agents' interactions. Studying joint explanations for more powerful agents is left as subject for further research.

Logic Programs. We assume the existence of alphabets of constants, variables and predicate symbols. A *term* is a constant or a variable. An *atom* is of the form $p(t_1, ..., t_k)$ where p is a k-ary predicate symbol and $t_1, ..., t_k$ are terms; the atom is *ground* if all its terms are constants; the atom is propositional if $k = 0$ (then, parenthesis are omitted in the notation). A *rule* r is a clause of the form: $a \leftarrow b_1, \cdots, b_h, \text{not } c_1, \cdots, \text{not } c_n.$, where $h, n \geq 0$, and $a, b_1, \cdots, b_h, c_1, \cdots, c_n$ are atoms. The atom a is the *head* of r, while the conjunction $b_1, ..., b_h, \text{not } c_1, \cdots, \text{not } c_n$ is the *body* of r. A rule with $k = n = 0$ is a *fact* (then, the arrow is omitted in the notation).

A *logic program* P is a finite set of rules. P is *propositional* (resp., *ground*) if all the atoms in it are propositional (resp., ground). P is *stratified*, if there is an assignment $s(\cdot)$ of integers to the predicate symbols in P, such that for each clause r in P the following holds: if p is the atom in the head of r and q (resp. *not* q) occurs in r, then $s(p) \geq s(q)$ (resp. $s(p) > s(q)$).

Let P be a stratified program. The *model-theoretic semantics* assigns to program P its unique *stable* model. It is well known that this model can be computed in polynomial time, in the case P is ground/propositional. Let W be a set of facts. Then, program P *entails* W, denoted by $P \models W$, if W is contained in the stable model of P.

Abduction in Logic Programming. Let O be a finite set of facts called *observations*. An *abductive agent* A is a pair $\langle T(A), H(A) \rangle$, where $T(A)$ is a stratified logic program encoding A's view of the world, and $H(A)$ is a finite set of facts denoting the *hypotheses* that A may use to explain facts in O. An *A-explanation* for O is a set $S \subseteq H(A)$ such that $T(A) \cup S \models O$. The following example, which will be exploited several times throughout the paper, introduces an abductive agent and shows an example explanation.

Example 1. Suppose that *John* (short: *J*) notices a friend of him, *Frank*, that is standing alone just outside a restaurant, and that is elegantly attired. John thinks that Frank is alone because he is waiting either for his girlfriend or for his brother; however, the fact that he is elegant makes the first hypothesis more reliable, in his opinion.

John's observations can be modelled by means of the facts $\mathcal{O} = \{\texttt{alone}, \texttt{elegant}\}$, whereas his mental state can be encoded by means of the pair $\langle T(J), H(J) \rangle$, where $H(J) = \{\texttt{wait\_for\_girlfriend}, \texttt{wait\_for\_brother}\}$ and $T(J)$ is the program:

> alone ← wait_for_girlfriend.
> alone ← wait_for_brother.
> elegant ← wait_for_girlfriend.

Sets $\{\texttt{wait\_for\_girlfriend}\}$ and $\{\texttt{wait\_for\_girlfriend}, \texttt{wait\_for\_brother}\}$ are the only J-explanations for $\mathcal{O}$. In particular, the former seems to be a good candidate for explaining $\mathcal{O}$, because of its minimality. Notice also that the fact wait_for_girlfriend *necessarily* occurs in any explanation.   ◁

3   Modelling Joint Explanations

In this section, we introduce the problem of finding joint explanations to observations perceived by sets of agents. Actually, we first formalize the basic framework which is eventually enriched by considering agents' preferences in finding explanations.

3.1   Basic Framework

An *abductive scenario* $\mathcal{S}$ is a pair $\langle \mathcal{A}, \mathcal{O} \rangle$, where $\mathcal{A} = \langle A_1, ..., A_n \rangle$ is a collection of abductive agents and $\mathcal{O}$ is a set of observations. Each agent A_i perceives the observations in $\mathcal{O}$ and tries to exploit its own theory $T(A_i)$ and its own hypotheses $H(A_i)$ to explain them. Moreover, A_i may also collaborate with all other agents by benefiting from their explanations. Next, we make it clear what we mean for "collaborating" in finding an explanation, by focusing on two different aspects:

- We assume that agents are allowed to partially contribute to the whole explanation, i.e., each of them may possibly find an explanation for a proper subset of $\mathcal{O}$, provided that each fact in O is explained at least by one agent.
- We assume that agents wish to focus on those hypotheses on which they can find some agreement with the other agents. In particular, if two agents, say A and B, have overlapping hypotheses, i.e., $H(A) \cap H(B) \neq \emptyset$, then their explanations, say S_A and S_B, must be such that: a fact in $H(A) \cap H(B)$ is in S_A if and only if it is in S_B as well. This requirement leads to justifications which are more likely to be trustable, since the matching of partial individual explanations may be taken in favor of believing in their correctness.

The following definition formalizes the concepts introduced so far.

Definition 1 (Joint explanation). Let $\langle \mathcal{A}, \mathcal{O} \rangle$ be an abductive scenario, with $\mathcal{A} = \langle A_1, ..., A_n \rangle$. A tuple $\langle S_1, ..., S_n \rangle$ is an $\mathcal{A}$-*joint explanation* (short: joint explanation) for $\mathcal{O}$ if the following conditions hold:

(1) $((S_i \setminus S_j) \cup (S_j \setminus S_i)) \cap (H(A_i) \cap H(A_j)) = \emptyset$, for each $i, j \in \{1, ..., n\}$;
(2) $S_i \subseteq H(A_i)$, for each $i \in \{1, ..., n\}$;
(3) there are sets $O_1, ..., O_n$ such that:
   - $T(A_i) \cup S_i \models O_i$, for each $i \in \{1..n\}$, and $\bigcup_{i=1}^{n} O_i \supseteq \mathcal{O}$.   □

Example 1 (contd.) Assume that *Mark* (short: M) and *Bob* (short: B), which are friends of Frank too, join John. Actually, Mark knows that Frank has participated to an important working-meeting this morning, so that his elegance may be well justified. Moreover, he guesses that John has still to meet some other colleagues now. Then, Mark's mental state is formalized by the pair $\langle T(M), H(M) \rangle$, where $H(M) = \{\texttt{meeting}, \texttt{wait\_for\_collegues}\}$ and $T(H)$ is the program

> alone ← wait_for_collegues.   elegant ← meeting.

However, Bob knows that Frank is going to meet his brother and, therefore, noticing him alone is not a surprise. Moreover, he has no justification for his elegance. Then, Bob's mental state is formalized by the pair $\langle T(B), H(B) \rangle$, where $H(B) = \{\texttt{wait\_for\_brother}\}$ and $T(B)$ is the program alone ← wait_for_brother.

According to our framework, the three friends constitute an abductive scenario, say $\langle \langle J, M, B \rangle, \{\texttt{alone}, \texttt{elegant}\} \rangle$, having the following joint explanations:

$S_1 = \langle \{\texttt{wait\_for\_brother}\}, \{\texttt{meeting}\}, \{\texttt{wait\_for\_brother}\} \rangle,$
$S_2 = \langle \{\texttt{wait\_for\_brother}\}, \{\texttt{meeting}, \texttt{wait\_for\_collegues}\}, \{\texttt{wait\_for\_brother}\} \rangle,$
$S_3 = \langle \{\texttt{wait\_for\_brother}, \texttt{wait\_for\_girlfriend}\}, \{\texttt{meeting}\}, \{\texttt{wait\_for\_brother}\} \rangle,$
$S_4 = \langle \{\texttt{wait\_for\_brother}, \texttt{wait\_for\_girlfriend}\}, \{\texttt{wait\_for\_collegues}\}, \{\texttt{wait\_for\_brother}\} \rangle,$
$S_5 = \langle \{\texttt{wait\_for\_brother}, \texttt{wait\_for\_girlfriend}\}, \{\texttt{meeting}, \texttt{wait\_for\_collegues}\}, \{\texttt{wait\_for\_brother}\} \rangle.$

Notice that John is now forced to exploit the fact wait_for_brother, in order to find an agreement with Bob; however, he is no longer required to make use of the hypothesis wait_for_girlfriend, since Frank's elegance may be explained by Mark (by the fact meeting).   ◁

Before leaving this section, it is relevant to note that in Definition 1, to keep things simple, we do not considered scenarios where agents do not manifest (i.e., communicate to the other agents) their own hypotheses. However, this has been done without loosing in generality. Indeed, such scenarios can be coped with by a simple syntactic expedient. As an example, assume that agent A_i does not manifest to the other agents whether an hypothesis, say $\texttt{h} \in H(A_i)$, is part of its explanation. Then, to model this scenario while keeping Definition 1 fixed, it is sufficient to replace (in both $H(A_i)$ and $T(A_i)$) the hypothesis h by a fresh hypothesis, say $\texttt{h}_{\texttt{hidden}}$, which does not occur in any set of the form $H(A_j)$, with $j \neq i$. Then, all the other agents cannot access this hypothesis, which therefore does not influence their decisions, while the semantics of the pair $\langle T(A_i), H(A_i) \rangle$ is preserved.

3.2 Individual Preferences and Agreements in Joint Explanations

In several practical situations, agents have preferences in finding explanations. We now take care of these preferences by equipping each agent A with a polynomially-computable *weighting function* γ_A assigning an integer to each possible subset of $H(A)$: The lower the number associated to a set of hypotheses S, the less trustable S.

Actually, having such weighting functions specified for each possible set of hypotheses may be often unrealistic. In practice, one may assume that functions are partially specified over a certain set $\bar{S}$ of hypotheses, so that a default weight is associated to hypotheses not in $\bar{S}$. Alternatively, one may define a function γ_A^i assigning weights to individual hypotheses, so that $\gamma_A(S) = \sum_{s \in S} \gamma_A^i(S)$.

We point out that while several approaches have already considered the notion of preference in multi-agent systems applications (e.g., [15, 16, 17]), few efforts have been paid in dealing with contrasting agents' preferences and in providing reasonable agreement semantics; this is precisely the aim of this section.

Example 1 (contd.) In our running example, assume that John's preferences are formalized through the weighting function γ_J such that: $\gamma_J(\{\texttt{wait\_for\_girlfriend}\}) = 3$, $\gamma_J(\{\texttt{wait\_for\_girlfriend}, \texttt{wait\_for\_brother}\}) = 1$, $\gamma_J(\{\texttt{wait\_for\_brother}\}) = 2$. Intuitively, John prefers minimum sized explanations and, subordinately, he is inclined to assume that Frank is waiting for his girlfriend.

Moreover, the function associated to Mark is such that: $\gamma_M(\{\texttt{meeting}\}) = 3$, $\gamma_M(\{\texttt{meeting}, \texttt{wait\_for\_collegues}\}) = 2$ and $\gamma_M(\{\texttt{wait\_for\_collegues}\}) = 1$. The function reflects that he knows that Frank had a meeting in the morning.

Finally, Bob has one hypothesis and, we let $\gamma_B(\{\texttt{wait\_for\_brother}\}) = 1$.   ◁

Given the preferences expressed by the agents, we may be interested in singling out those explanations on which agents are likely to find an agreement. Clearly enough, if all the agents assign the maximum value of their weighting functions to the same explanation, say h, then h will be the most desirable joint explanation for the observations at hand. However, agents have contrasting preferences in general and, therefore, defining a criteria for selecting some kinds of preferred explanation poses some semantics problems. Our solution approach is to exploit some standard game-theoretic arguments, which are aimed at characterizing preferred explanations as the most "rational" outcomes for agents' interactions. The basic observation is that agents are rational entities for they seek to select an explanation which gives the outcome they most prefer, given what they expect the other agents do. As an example, assume that agents $A_2, A_3, ..., A_n$ have already find an agreement on the explanation $\langle S_2, S_3, ..., S_n \rangle$; then, agent A_1 will rationally selects in the sets of all its individual explanations the one getting the maximum value for its weighting function, while still agreeing on the shared hypotheses with the other agents (cf. condition (1) in Definition 1). Since all the agents apply this form of reasoning, the following definition may be singled out.

Definition 2. Let $\langle \mathcal{A}, \mathcal{O} \rangle$ be an abductive scenario, with $\mathcal{A} = \langle A_1, ..., A_n \rangle$. A joint explanation $\langle S_1, ..., S_n \rangle$ for $\mathcal{O}$ is *preferred* if there exists no other joint explanation $\langle S_1', ..., S_n' \rangle$ and no agent A_i such that: (1) $\gamma_{A_i}(S_i') > \gamma_{A_i}(S_i)$, and (2) $S_j' = S_j$, for each $j \neq i$ with $i, j \in \{1, ..., n\}$.   □

Thus, the idea is to consider explanations which agents cannot individually "deviate" from, by selecting a more trustable set of hypotheses. Close to this approach is the framework of [18], where the semantics of multi-agent systems is provided in terms of a stable set of actions for different agents, where a set is stable if assuming an oracle that could feed each of the agents with all the other actions in the outcome, then agents' decision would be what the set envisages.

It is relevant to note that the assumption of rational behavior for the agents exploited in the definition above is also traditionally used in game theory, by leading to the definition of the solution concept of *Nash equilibrium* in non-cooperative games[1].

In particular, in Definition 2, one may view the set of hypotheses (individual explanations) as the possible actions for each agent and the weights assigned to them as the payoffs. Yet, there are some important conceptual differences here. Indeed, in classical game theory players are allowed to arbitrary chose a strategy (provided it is rational) in a fixed set of possible actions to be performed; however, in Definition 2, preferred explanations (i.e., the outcomes of the abductive problem) must satisfy the additional requirement of being joint explanations, which imposes a kind of coordination among agents and constraints the actions that can be eventually chosen. Therefore, the exploitation of some game-theoretic arguments is on a more technical level, for defining the way in which agents' individual preferences are to be maximized.

Example 1 (contd.) Consider the joint explanations S_3 and S_5. They are not preferred since John gets an incentive in selecting S_1 and S_2, respectively, without affecting the choices of the other two friends. Similarly, S_4 and S_2 are not preferred, since Mark prefers S_3 and S_1 to them, respectively. Therefore, it is easy to see that S_1 is the only preferred explanations.   ◁

We next consider a more sophisticated notion of agreement which is based upon a suitable stability criterion. Indeed, according to Definition 2, it is sufficient that any agent does not have an incentive for deviating from a given joint explanation for this explanation being considered preferred. However, in some situations, agents may coordinate which each other and exchange their beliefs and intensions, thereby forming coalitions which may influence the process of finding preferred joint explanations. To take care of these situations, in the following definition we look for joint explanations for which there is no set of agents that can "deviate" from, by jointly selecting some more trustable set of hypotheses. This is a quite strong property that makes the agents' agreement stable and the explanations very reliable.

Definition 3. Let $\langle \mathcal{A}, \mathcal{O} \rangle$ be an abductive scenario, with $\mathcal{A} = \langle A_1, ..., A_n \rangle$. A joint explanation $\langle S_1, ..., S_n \rangle$ for $\mathcal{O}$ is *stable* if there exists no other joint explanation $\langle S'_1, ..., S'_n \rangle$ and set of agents $\mathcal{K} \subseteq \{A_1, ...A_n\}$ such that: (1) $\gamma_{A_i}(S'_i) > \gamma_{A_i}(S_i)$, for each $A_i \in \mathcal{K}$, and (2) $S'_j = S_j$, for each $A_j \notin \mathcal{K}$.

Moreover, $\langle S_1, ..., S_n \rangle$ is *pareto* if there exists no other joint explanation $\langle S'_1, ..., S'_n \rangle$ such that (1) holds for $\mathcal{K} = \{A_1, ...A_n\}$.   □

Intuitively, pareto explanations are immune against deviations by the whole set of agents only, whereas stable explanation are immune against deviations by any subset of the agents. Notice, for instance, that in our running example S_1 is both a pareto and a stable explanation.

Example 2. Consider a slight modification of our running example, in which Bob has also wait_for_girlfriend as hypothesis explaining why Frank is elegant. Assume,

---

[1] And, in fact, one of the complexity results in Section 5 establishes a formal correspondence between the two settings.

also, that $\gamma_B(\{\texttt{wait\_for\_girlfriend}\}) = 2$, $\gamma_B(\{\texttt{wait\_for\_brother}\}) = 1$, and $\gamma_B(\{\texttt{wait\_for\_girlfriend}, \texttt{wait\_for\_brother}\}) = 1$. Then, the modified abductive scenario has some more joint explanations; among them, consider the following:

$$S_6 = \langle\{\texttt{wait\_for\_girlfriend}\}, \{\texttt{wait\_for\_collegues}\}, \{\texttt{wait\_for\_girlfriend}\}\rangle,$$
$$S_7 = \langle\{\texttt{wait\_for\_girlfriend}\}, \{\texttt{meeting}\}, \{\texttt{wait\_for\_girlfriend}\}\rangle,$$
$$S_8 = \langle\{\texttt{wait\_for\_girlfriend}\}, \{\texttt{meeting}, \texttt{wait\_for\_collegues}\}, \{\texttt{wait\_for\_girlfriend}\}\rangle,$$

It is easy to see that S_7 is a preferred explanation, while S_6 and S_8 are not, since Mark prefers the explanation just containing the fact $\texttt{meeting}$. Therefore, the pareto explanations are S_1 and S_7. However, S_7 is the only stable explanations. Indeed, John and Bob may jointly deviate from S_1 to S_7, because they trust more in the fact $\texttt{wait\_for\_girlfriend}$. $\triangleleft$

Competitive Agents. We conclude the description of the framework for finding joint explanations, by considering the case of competitive agents which are interested in selecting the explanations they prefer, no matter of the possibility of finding an agreement with the other agents. These agents may be viewed as "isolated", for they do not observe any explanations of the other agents. Accordingly, the explanations which are possible to be singled out are said to be "safe", precisely because they are not influenced by agents' interactions.

Definition 4. Let $\langle \mathcal{A}, \mathcal{O}\rangle$ be an abductive scenario, with $\mathcal{A} = \langle A_1, ..., A_n\rangle$, and let $\langle S_1, ..., S_n\rangle$ be a joint explanation for $\mathcal{O}$. Then, $\langle S_1, ..., S_n\rangle$ is *safe* if (1) $T(A_i) \cup S_i \models \mathcal{O}$, for each $i \in \{1, ..., n\}$; (2) $\nexists A_i \in \{A_1, ..., A_n\}$, $S_i' \subseteq H(A_i)$ s.t. $T(A_i) \cup S_i' \models \mathcal{O}$, and $\gamma_{A_i}(S_i') > \gamma_{A_i}(S_i)$. $\square$

Example 1 (contd.) It is easy to see that there are no safe explanations in our running example. Indeed, Bob cannot explain alone the whole set $\{\texttt{alone}, \texttt{elegant}\}$, thereby violating condition (1) in Definition 4. Actually, condition (2) in the same definition is also violated in both the five joint explanations. Indeed, in a competitive setting, John finds convenient to assume the hypothesis $\{\texttt{wait\_for\_girlfriend}\}$, no matter of the fact that he will not find an agreement with Bob. $\triangleleft$

3.3   Relationship Among the Notions

After that different notions of agreements have been introduced, it is natural to investigate how they are related with each other. Clearly enough, preferred, pareto, stable, and safe explanations are by definition joint explanations. Also, it is easy to see that the notion of stable explanation entails both the notions of pareto and preferred explanations. As far as existence of explanations is concerned, the existence of joint explanations does not guarantee the existence of any safe explanation. For instance, an abductive scenario where each agent does not have enough information to explain the whole set of observations does not admit, by definition, any safe explanation but may well admit some other kinds of joint explanation. Similarly, there are scenarios where joint explanations exist none of which is stable. This is not the case for preferred and pareto explanations, which are guaranteed to exist as soon as a joint explanation exists.

Proposition 1. *Let* $\langle A, \mathcal{O} \rangle$ *be an abductive scenario, with* $A = \langle A_1, ..., A_n \rangle$. *The following statements are equivalent: (a) There exists a joint explanation for* $\mathcal{O}$; *(b) There exists a preferred explanation for* $\mathcal{O}$; *(c) There exists a pareto explanation for* $\mathcal{O}$.

Proof. The implications $(c) \Rightarrow (a)$ and $(b) \Rightarrow (a)$ trivially follow from definitions of explanations. Then, we have to show that $(a) \Rightarrow (b)$ and $(a) \Rightarrow (c)$ hold as well.

Let $\langle S_1, ..., S_n \rangle$ be a joint explanation. We define a succession of joint explanations originating in $\langle S_1, ..., S_n \rangle$ and converging to a preferred one. To this aim, let $\mathbf{T}$ be a map from joint explanations to joint explanations such that:

- $\mathbf{T}(\langle S_1, ..., S_n \rangle) = \langle S_1, ..., S_n \rangle$, if $\langle S_1, ..., S_n \rangle$ is a preferred explanation, and
- $\mathbf{T}(\langle S_1, ..., S_n \rangle) = \langle S_1', ..., S_n' \rangle$, if $\langle S_1, ..., S_n \rangle$ is not preferred and $\langle S_1', ..., S_n' \rangle$ is the joint explanation satisfying conditions (1) and (2) in Definition 2.

Notice that $\mathbf{T}(\langle S_1, ..., S_n \rangle)$ is either a preferred explanation or a joint explanation in which an agent may strictly increase the value of its weighting function w.r.t. the one assumed in $\langle S_1, ..., S_n \rangle$. Therefore, the succession $\langle S_1, ..., S_n \rangle$, $\mathbf{T}(\langle S_1, ..., S_n \rangle)$, $\mathbf{T}(\mathbf{T}(\langle S_1, ..., S_n \rangle))$,... must converge to a joint explanation in which no agent can satisfy condition (1) in Definition 2. This proves that $(a) \Rightarrow (b)$ holds.

Finally, we show that $(a) \Rightarrow (c)$ holds. Indeed, let $\langle S_1, ..., S_n \rangle$ be a joint explanation such that there is an agent, say A_i, for which the value $\gamma_{A_i}(S_i)$ is the maximum over all the possible joint explanations. Then, by Definition 3, $\langle S_1, ..., S_n \rangle$ is pareto. □

4   Data Integration in *P2P* Systems

In this section, we exploit the framework for computing joint explanations to provide semantics for Peer-to-Peer (*P2P*) data management systems. We recall here that *P2P* systems are networks of *autonomous* peers representing the natural evolution of classical data integration systems and mediators (e.g., [19-22, 14]). They have, in fact, recently emerged as an effective architecture for decentralized data sharing, integration, and querying (e.g., [23-27]).

A key issue in *P2P* systems is that data stored in different autonomous peers may be mutually inconsistent, i.e., it may violate some integrity constraints issued over peer schemas to enhance their expressiveness. To remedy this problem, one can handle the inconsistency by suitably "repairing" the retrieved data. However, in order to be effective, the repair approach should consider the peculiarities of *P2P* systems and, particularly, the following two aspects:

- Peers are distributed and autonomous entities, that may interact in complicated ways before agreeing with other peers on the way the repair should be carried out. Therefore, the integration should be modelled by means of a 'decentralized' framework, where decisions are shared among peers.
- In practical applications, peers have often an a-priori knowledge about the reliability of the sources that, in turn, determines their criteria for computing repairs. E.g., peers will rarely delete tuples coming from highly reliable sources, and will try to solve conflicts by updating the less reliable sources only.

Despite the wide interest in this field, few approaches in the literature considered the issue of modelling the autonomy of the peers in choosing a repair for the system. Similarly, very few efforts have been paid for enriching *P2P* systems with peers' preferences (e.g., [26] considering trust relationships among peers), even though this is widely recognized to be a crucial aspect of *P2P* systems (cf. [27]).

We next show that the framework for computing joint explanations can be used to face these problems, by providing a surprisingly natural repair semantics for *P2P* systems. This is an important contribution of its own, witnessing some practical relevance of the notions discussed in the paper. We believe that the proposed formalization may stimulate other logic-based approaches to model *P2P* systems.

4.1 *P2P* Systems in a Nutshell

P2P systems offer transparent access to the data stored at each peer p, by means of the global schema equipped with p for modelling its domain of interest; moreover, pair of peers with the same domain of interest are interconnected by means of schema mapping rules. Formally, a *P2P* system $\mathcal{P}$ is a tuple $\langle P, \mathcal{G}, \mathcal{N}, \mathcal{M} \rangle$, where P is a non-empty set of distinct peers and $\mathcal{G}$, $\mathcal{N}$ and $\mathcal{M}$ are functions. In particular, for each peer $p \in P$,

- $\mathcal{G}(p) = \langle \Psi_p, \Sigma_p \rangle$ is the *peer schema*, where Ψ_p is a set of predicate symbols distinguished for each peer, i.e., $\Psi_p \cap \Psi_q = \emptyset$ for each $p \neq q$, and Σ_p is a stratified logic program defining the set of *integrity constraints*. Each constraint is a rule of the form $\mathrm{bad}_p \leftarrow \mathrm{Body}$, where Body is any conjunction of literals defined over the predicates in Ψ_p, and bad_p is a distinguished predicate whose derivation entails the *violation* of some constraints.[2]
- $\mathcal{N}$ is the *neighborhood* function providing a set of peers $Neigh(p) \subseteq P - \{p\}$ containing the peers (called neighbors) who are potentially interested in information stored by p;
- $\mathcal{M}(p)$ is a stratified logic program defining the set of *peer mapping assertions* of p. Actually, rules in $\mathcal{M}(p)$ are partitioned into sets $\mathcal{M}_q(p)$, for each $q \in Neigh(p)$, so that each assertion in $\mathcal{M}_q(p)$ is a logic rule whose head predicate is taken from the schema $\mathcal{G}(q)$ and whose body predicates are taken from the schema $\mathcal{G}(p)$. The intended meaning of mapping assertions is to provide a means for exchanging data among peers, and in fact, evaluating the assertions in $\mathcal{M}(p)$ is the way p exploits to publish its private information.

Example 3. Let $\mathcal{P}^o = \langle P^o, \mathcal{G}^o, \mathcal{N}^o, \mathcal{M}^o \rangle$ be a *P2P* system, where P^o consists of three peers p_1, p_2 and p_3, such that $\mathcal{N}^o(p_1) = \{p_3\}$, $\mathcal{N}^o(p_2) = \{p_3\}$ and $\mathcal{N}^o(p_3) = \emptyset$.

Figure 1 summarizes the structure of the system $\mathcal{P}^o$, and reports for each peer, its global schema, and its mapping assertions. Notice the integrity constraints on p_3 stating that a professor cannot be a student.   ◁

A *global database* for $\mathcal{P}$ is a function $\mathcal{B}$ associating to each peer p a database instance $\mathcal{B}(p)$ for $\mathcal{G}(p) = \langle \Psi_p, \Sigma_p \rangle$, i.e., a set of facts (also *tuples*) whose predicate symbols are

---

[2] We model universally quantified constraints, such as Keys, Functional Dependencies, and Exclusion Dependencies. Dealing with existentially quantified constraints can be done by techniques in [29].

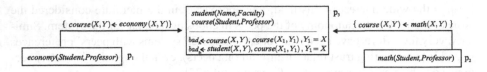

Fig. 1. The *P2P* system $\mathcal{P}^o$ in Example 3

in Ψ_p. We say that $\mathcal{B}$ *satisfies* $\mathcal{P}$ if for each peer p, *(1)* $\mathcal{B}(p) \cup \Sigma_p \not\models \mathtt{bad}_p$, and *(2)* the stable model of $\bigcup_{q \in \mathcal{N}(p)}(\mathcal{M}_p(q) \cup \mathcal{B}(q))$ is contained in $\mathcal{B}(p)$. Intuitively, $\mathcal{B}$ *satisfies* $\mathcal{P}$ if there are no violations of integrity constraints, and if the mappings correctly evaluate the tuples that are 'retrieved' from the neighbors.

Example 3 (contd.) Consider the global database $\mathcal{B}^o$ for the system $\mathcal{P}^o$ such that: $\mathcal{B}^o(p_1) = \{economy(Albert, Bill)\}$, $\mathcal{B}^o(p_2) = \{math(John, Mary), math(Mary, Tom)\}$ and $\mathcal{B}^o(p_3) = \{student(Mary, ComputerScience)\}$.

Then, $\mathcal{B}^o$ does not satisfy $\mathcal{P}^o$, because of the violation of the mapping assertions; indeed, according to the mappings reported in Figure 1, $\mathcal{B}^o(p_3)$ must contain the tuples $course(Albert, Bill)$ (imported from p_1), $course(John, Mary)$, $course(Mary, Tom)$ (imported from p_2), beside the tuple $student(Mary, ComputerScience)$ (already in p_3).

By the way, if we assume that all the above tuples are materialized in $\mathcal{B}^o(p_3)$, we yet have that $\mathcal{B}^o$ does not satisfy $\mathcal{P}^o$, because of the violation of the integrity constraints over p_3, since *Mary* will result to be both a student and a professor.   ◁

4.2   An Abductive Approach to Repair *P2P* Systems

Given that a *P2P* system may be not satisfied by a global database, a "repair" for the *P2P* system has to be computed [26, 27]. Roughly speaking, repairs may be viewed as insertions or deletions of tuples at the peers that are able to lead the system to a consistent state. We next propose an abductive approach to repair *P2P* systems based on joint explanations.

Let $\mathcal{P} = \langle P, \mathcal{G}, \mathcal{N}, \mathcal{M} \rangle$ be a *P2P* system, with $P = \{p_1, ..., p_n\}$, and $\mathcal{B}$ be a global database for $\mathcal{P}$. The *abductive scenario* $\mathcal{S}(\mathcal{P}, \mathcal{B})$ associated to $\mathcal{P}$ and $\mathcal{B}$ is the tuple $\langle \mathcal{A}(\mathcal{P}, \mathcal{B}), \mathcal{O}(\mathcal{P}, \mathcal{B}) \rangle$, where $\mathcal{A}(\mathcal{P}, \mathcal{B}) = \{p_1, ..., p_n\}$ and $\mathcal{O}(\mathcal{P}, \mathcal{B}) = \{\mathtt{ok}_{p_1}, ..., \mathtt{ok}_{p_n}\}$, s.t. for each agent $p_i \in \mathcal{A}(\mathcal{P})$:

- $H(p_i) = H_1(p_i) \cup H_2(p_i) \cup H_3(p_i)$, where
  - $H_1(p_i) = \{+\mathtt{r}(\mathtt{t}), -\mathtt{r}(\mathtt{t}) \mid r \text{ is a predicate in } \mathcal{B}(p_i) \wedge t \text{ is a tuple of constants}\}$
  - $H_2(p_i) = \{\mathtt{r}^\bullet(\mathtt{t}) \mid r \text{ is a predicate in } \mathcal{B}(p_j) \cap \mathcal{M}_{p_j}(p_i), p_i \neq p_j \wedge t \text{ is a tuple of constants}\}$
  - $H_3(p_i) = \{\mathtt{r}^\bullet(\mathtt{t}) \mid r \text{ is a predicate in } \mathcal{B}(p_i) \cap \mathcal{M}_{p_i}(p_j), p_i \neq p_j \wedge t \text{ is a tuple of constants}\}$
- $T(p_i)$ is the union of $\mathcal{M}(p_i) \cup \Sigma_p \cup \{\mathtt{ok}_{p_i} \leftarrow \mathtt{not} \; \mathtt{bad}_{p_i}.\}$ with the rules:
  $$r_1 : \mathtt{r}_\mathtt{D}(\mathtt{t}).\} \quad \forall r(\mathtt{t}) \in \mathcal{B}(p_i)$$

$$
\left.
\begin{aligned}
&r_2 : \mathtt{r}(\mathtt{t}) \leftarrow +\mathtt{r}(\mathtt{t}).\\
&r_3 : \mathtt{r}(\mathtt{t}) \leftarrow \mathtt{r}^\bullet(\mathtt{t}).\\
&r_4 : \mathtt{r}(\mathtt{t}) \leftarrow \mathtt{r}_\mathtt{D}(\mathtt{t}), \mathtt{not} - \mathtt{r}(\mathtt{t}).\\
&r_5 : \mathtt{bad}_{p_i} \leftarrow \mathtt{r}(\mathtt{t}), -\mathtt{r}(\mathtt{t}).
\end{aligned}
\right\} \quad \forall r \text{ occurring in } \mathcal{B}(p_i)
$$

$$
\left.
\begin{aligned}
&r_6 : \mathtt{bad}_{p_i} \leftarrow \mathtt{r}(\mathtt{t}), \mathtt{not} \; \mathtt{r}^\bullet(\mathtt{t}).\\
&r_7 : \mathtt{bad}_{p_i} \leftarrow \mathtt{r}^\bullet(\mathtt{t}), \mathtt{not} \; \mathtt{r}(\mathtt{t}).
\end{aligned}
\right\} \quad \forall r \text{ occurring in } \bigcup_{p_j \in Neigh(p_i)} \mathcal{M}_{p_j}(p_i)
$$

We next informally describe the meaning of this abductive scenario. Let us start by commenting the notation. For each predicate symbol r belonging to the global schema of the peer p_i, the fact $+r(t)$ (resp., $-r(t)$) indicates that the tuple $r(t)$ has been added to (resp., deleted from) $\mathcal{B}(p_i)$; the fact $\mathbf{r}(t)$ indicates that the tuple $r(t)$ is actually stored in the database associated to p_i after the repair is performed; moreover, the fact $\mathbf{r^e}(t)$ denotes that the tuple $r(t)$ is exchanged (either imported or exported) with some other peer. Finally, $\mathbf{r_D}(t)$ denotes that $r(t)$ occurs in $\mathcal{B}(p_i)$, before the updates are performed.

As for the hypotheses in $H(p_i)$, we distinguished three different sets. $H_1(p_i)$ is such that its subsets (in the case where a given tuple is not added and deleted at the same time) are in one-to-one correspondence with all the possible repairs for the database $\mathcal{B}(p_i)$.

Then, $H_2(p_i)$ contains the data that p_i exports to his neighbors (indeed, r is in this case a predicate symbol in the database of some of p_i's neighbors), while the set $H_3(p_i)$ contains the data that p_i imports from other peers.

In order to get an intuition about the need of the last two sets of hypotheses, consider two peers p_i and p_j such that $p_j \in \mathit{Neigh}(p_i)$, and let r_{p_j} be a predicate symbol in the schema of $\mathcal{B}(p_j)$. Clearly, facts of the form $\mathbf{r^e_{p_j}}(t)$ occur both in the set $H_2(p_i)$ and in the set $H_3(p_j)$. Therefore, in order to satisfy condition (1) in Definition 1, any joint explanation for $\langle \mathcal{A}(\mathcal{P}, \mathcal{B}), \mathcal{O}(\mathcal{P}, \mathcal{B}) \rangle$ must be a set of "consistent" exchanges, i.e., tuples exported by p_i to p_j must coincide with the tuples imported by p_j from p_i. Importantly, this is the only information shared by pairs of peers; in particular, each peer may be completely unaware of the content of the databases stored at his neighbors. This feature fits the basic assumptions in *P2P* systems, where the mappings represent the only way for getting data from peers.

Let us now discuss the intended meaning of agent's theory $T(p_i)$. We distinguished three sets of rules, namely $\{r_1\}$, $\{r_2, ..., r_5\}$, and $\{r_6, r_7\}$. Rule r_1 is used to store into facts over the predicate symbol $\mathbf{r_D}$ all the data contained in $\mathcal{B}(p_i)$ before any update is performed. As for the second set, rules r_2, r_3 and r_4 simply compute the value of the modified database (after updates and data exchanges). In particular, the modified database will contain: *(i)* all the tuples added to $\mathcal{B}(p_i)$ (rule r_2); *(ii)* all the tuples imported by peer p_i (rules r_3); and *(iii)* all the tuples that are in $\mathcal{B}(p_i)$ and have been not deleted in the repair (rule r_4). Moreover, rule r_5 entails $\mathrm{bad_{p_i}}$ if and only if a given tuple deleted in the repair while occurring in the database after the updates are performed. Finally, rules r_6 and r_7 are used to check whether the facts exported by p_i to his neighbors (facts over the predicate symbol $\mathbf{r^e}$) actually coincide with those computed throughout the mapping assertions in $\mathcal{M}(p_i)$. If this is not the case, then $\mathrm{bad_{p_i}}$ is entailed.

To conclude the description of the encoding, recall that the rules in Σ_p entail, by definition, $\mathrm{bad_{p_i}}$ if and only if there is some constraint violation. Therefore, by summing up all the above considerations, $\mathrm{ok_{p_i}}$ is true in the model of $T(p_i)$ if and only if a database instance for $\mathcal{G}(p_i)$ has been computed which satisfies all the mapping assertions. Then, if $\mathrm{ok_{p_i}}$ is entailed for each peer p_i, the hypotheses at hand encode a model for the system.

Formally, let $S = \langle S_1, ..., S_n \rangle$ be a joint explanation for $\mathcal{O}(\mathcal{P}, \mathcal{B})$. We denote by $\mathcal{B}_S$ the global database such that, for each peer $p_i \in P$, $\mathcal{B}_S(p_i) = \{r(t) \mid \mathbf{r}(t) \in$

$S_i \wedge r$ is a predicate symbol in $\mathcal{B}(p_i)$}. Then, the following theorem shows that joint explanations represent, in fact, an elegant way for assigning semantics to *P2P* systems.

Theorem 1. *Let* $\mathcal{P} = \langle P, \mathcal{G}, \mathcal{N}, \mathcal{M} \rangle$ *be a* P2P *system, let* $\mathcal{B}$ *be a global database for* $\mathcal{P}$, *and let* $\mathcal{S}(\mathcal{P}, \mathcal{B}) = \langle \mathcal{A}(\mathcal{P}, \mathcal{B}), \mathcal{O}(\mathcal{P}, \mathcal{B}) \rangle$ *be the associated abductive scenario. Then:*

- *if S is a joint explanation for* $\mathcal{O}(\mathcal{P}, \mathcal{B})$, *then* $\mathcal{B}_S$ *satisfies* $\mathcal{P}$;
- *if* $\mathcal{B}'$ *satisfies* $\mathcal{P}$, *then there is a joint explanation S for* $\mathcal{O}(\mathcal{P}, \mathcal{B})$ *such that* $\mathcal{B}' = \mathcal{B}_S$.

In the light of the theorem above, $\mathcal{B}_S$ is said to be a *repair for* $\mathcal{P}$ *w.r.t.* $\mathcal{B}$.

Example 3 (contd.) Let $\mathcal{S}(\mathcal{P}^o, \mathcal{B}^o) = \langle \langle A(p_1), A(p_2), A(p_3) \rangle, \{ \mathsf{ok}_{p_1}, \mathsf{ok}_{p_2}, \mathsf{ok}_{p_3} \} \rangle$ be the scenario associated to $\mathcal{P}^o$ and $\mathcal{B}^o$. Then, we can single out several joint explanations.

In particular, denote by S_1^o the joint explanation which is carried out by deleting the tuple *math(Mary, Tom)* from p_2 and the tuple *student(Mary, ComputerScience)* from p_3, and by propagating to p_3 the tuples *course(Albert, Bill)* and *course(John, Mary)* from p_1 and p_2, respectively. Moreover, let us denote by S_2^o the joint explanation which is carried out by deleting *math(John, Mary)* from p_2, and propagating *course(Albert, Bill)* and *course(Mary, Tom)* from p_1 and p_2, respectively. Then, it is easy to see that both $\mathcal{B}_{S_1^o}$ and $\mathcal{B}_{S_2^o}$ satisfy $\mathcal{P}^o$.   ◁

Importantly, we can further exploit our framework to single out repairs that enjoy some nice properties. In particular, peers' preferences can be trivially dealt with by equipping peers with weighting functions and by considering the kinds of more sophisticated notion of explanations we have investigated in the paper. This is relevant, for instance, to avoid repairs where all the tuples are deleted.

In this respect, a very common scenario occurs when each agent p is equipped with the function $\gamma_p(S) = -|\{ +\mathtt{r}(\mathtt{t}) \in S_p \} \cup \{ -\mathtt{r}(\mathtt{t}) \in S_p \}|$. In this case, agents are giving preference to those repairs requiring the minimum number of updates to the original global database. We leave to the careful reader the task of checking that given these functions, S_1^o and S_2^o in the example above are the only preferred joint explanations, while S_2^o is the only pareto joint explanation.

Clearly, more complex functions may be used to model peers' assumptions about reliability of other sources, e.g., stating to prefer tuples coming from some given peer. In all the cases, the semantics for the interactions can be chosen among those proposed in the previous section, according to the application needs. For instance, one can define the notion of preferred repair as the one where no peer gets an incentive to deviate from by selecting a different repair. As a further example application, we refer the interested reader to [28], where competitive peers are considered and a notion of agreement is proposed which basically relies on the concept of safe explanation.

5   Reasoning with Joint Explanations

After that the framework for computing joint explanations has been introduced and discussed, we turn to the investigation of the computational complexity of some basic reasoning tasks emerging in this scenario. This study can help in shedding some lights into the difficulty of finding joint explanations and is a necessary step for developing effective algorithms for their computation. So, assume that an abductive scenario $\langle \mathcal{A}, \mathcal{O} \rangle$, with $\mathcal{A} = \langle A_1, ..., A_n \rangle$, is given, and consider the following problems (which generalize classical problems in standard abduction):

- Consistency: Does there exist an explanation for $\mathcal{O}$?
- Checking: Given a tuple $\langle S_1, ..., S_n \rangle$, is $\langle S_1, ..., S_n \rangle$ an explanation for $\mathcal{O}$?
- Relevance: Given a hypothesis h and an agent A_i, is there an explanation $\langle S_1, ..., S_n \rangle$ such that h is in S_i?
- Necessity: Given a hypothesis h and an agent A_i, is the case that for each explanation $\langle S_1, ..., S_n \rangle$, h is in S_i?

Specifically, we shall study the complexity of the problems above for all the kinds of explanation (joint, preferred, pareto, stable, and safe) formalized in Section 3, where agents' theories are assumed to be ground (or, equivalently, propositional) programs.

5.1   Complexity Settings and Overview of Complexity Results

In order to make the analysis rigorous, we first discuss the parameters of interest in our studies. Indeed, in a scenario involving several agents, we can single out two different sources of complexity: the number of agents which are required to agree on finding an explanation, and the size of the agents's theories and hypotheses. Clearly enough, if we deal with a *fixed* number of agents, i.e., if agents are not part of the input problem, then all the reasoning tasks can be easily reduced to the problem of abductive reasoning with penalization in the context of one agent only [11] (details omitted here).

Therefore, we assume that the number of agents is not fixed so that two kinds of complexity setting may be explored, that are: (1) *Agent Complexity*, i.e., the complexity of solving the reasoning problems by fixing, for each agent, its theory and set of hypotheses, and (2) *Combined Complexity*, i.e., the complexity measured by assuming that both the number of players, their theories and hypotheses are not a-priori fixed. Notice that the former setting is meant to capture the difficulty lying in the agents' interactions, whereas the latter accounts for both the intricacy of the interactions and the complexity of individual abductions. In particular, the agent complexity setting provides an useful bound on the cost of *communication* required in practical implementations. Indeed, an NP-hardness result in this setting entails that it is unlikely to exists a coordinating algorithm requiring a polynomial number of messages exchanges only. Conversely, a polynomial time result in the agent complexity would suggest the existence of some efficient algorithms finding joint explanations in a decentralized fashion.

A summary of all the complexity (completeness) results for the two settings is reported in Figure 2. All the programs are assumed to be propositional/ground – as usual, results for non-grounds programs are one exponential higher.

The figure shows that problems have complexities lying at the first and second level of the polynomial hierarchy. Moreover, the results for the agent complexity are a lower bound for the combined complexity case. It is evident that the more stringent conditions imposed by the notions of pareto and stable explanations came at a high computational cost. Moreover, in both the two settings, Relevance and Necessity are always in complementary complexity classes, and Consistency provides lower bounds for

130   G. Greco

problem	joint	preferred	pareto	stable	safe
Consistency	NP	NP	NP	Σ_2^P	NP
Checking	P	P	co-NP	co-NP	P
Relevance	NP	NP	Σ_2^P	Σ_2^P	NP
Necessity	co-NP	co-NP	Π_2^P	Π_2^P	co-NP

(a)

problem	joint	preferred	pareto	stable	safe
Consistency	NP	NP	NP	Σ_2^P	P^{NP}
Checking	P	co-NP	co-NP	co-NP	P
Relevance	NP	Σ_2^P	Σ_2^P	Σ_2^P	P^{NP}
Necessity	co-NP	Π_2^P	Π_2^P	Π_2^P	P^{NP}

(b)

Fig. 2. Summary of Complexity Results: (a) Agent Complexity; (b) Combined Complexity

Relevance. Actually, the complexity of Consistency and Relevance coincides for most of the considered notions but for the intriguing cases of pareto (both in (a) and (b)), and preferred explanations (in (b) only).

Interestingly, by turning from the agent complexity to the combined complexity, the notions of preferred and safe explanations are those evidencing some more structural differences. In fact, problems for safe explanations becomes in (b) complete for the polynomial time closure of the class NP, i.e., for P^{NP}; moreover, for preferred explanations we get completeness results for the second level of the polynomial hierarchy.

5.2 Proofs of Complexity Results

In this section, we provide some selected proofs for the results in Figure 2, that are representative of the techniques exploited in our investigation. In particular, due to space constraints, we focus on the Consistency problem only.

Theorem 2. Consistency *is* NP-*complete for joint, preferred and pareto explanations. (Agent Complexity)*

Proof (Sketch). Membership. To decide the existence of joint, preferred and pareto explanations it suffices to certify the existence of a joint explanation, after Proposition 1. This latter task is feasible in NP: guess a tuple $\langle S_1, ..., S_n \rangle$ and verify in polynomial time that all the conditions in Definition 1 hold. In particular, in order to verify that condition (3) holds, we can simply compute the stable model for each program of the form $T(A_i) \cup S_i$ and let O_i contain all the observations in this model.

Hardness. We establish the NP-hardness for joint explanations by exploiting a nice connection between the problem of computing joint explanations and the problem of computing (pure) Nash equilibria in graphical games.

A *graphical game* $\mathcal{G}$ is a tuple $\langle P, Neigh, Act, U \rangle$, where P is a non-empty set of distinct players and $Neigh : P \longrightarrow 2^P$ is a function such that for each $p \in P$, $Neigh(p) \subseteq P - \{p\}$ contains all neighbors of p, Act is a function returning for each player p a set of possible actions $Act(p)$, and U associates a utility function $u_p : Act(p) \times_{j \in Neigh(p)} Act(j) \to \Re$ to each player p. For a player p, p_a denotes her choice to play the action $a \in Act(p)$. Each possible p_a is called a strategy for p, and the set of all strategies for p is denoted by $St(p)$. A global strategy $\mathbf{x}$ is a set containing exactly one strategy for players in P.

A (pure) Nash equilibrium for $\mathcal{G}$ is a global strategy $\mathbf{x}$ such that for every player $p \in P$, $\nexists p_a \in St(p)$ such that $u_p(\mathbf{x}) < u_p(\mathbf{x}_{-p}[p_a])$, where $u_p(\mathbf{x})$ denote the output of u_p on the projection of $\mathbf{x}$ to the domain of u_p, and $\mathbf{x}_{-p}[p_a]$ is the global strategy obtained by replacing p's individual strategy in $\mathbf{x}$ by p_a. The problem of deciding the existence of pure Nash equilibria in graphical games has been proven to be NP-hard in [30], even if each player has three neighbors at most.

Given a game $\mathcal{G} = \langle P, Neigh, Act, U \rangle$ with $P = \{p^1, ..., p^n\}$, we build an abductive scenario $\mathcal{S}(\mathcal{G}) = \langle \mathcal{A}(\mathcal{G}), \mathcal{O}(\mathcal{G}) \rangle$ such that: (i) $\mathcal{O}(\mathcal{G}) = \{\text{ok}_{p^i} \mid p^i \in P\}$; (ii) $\mathcal{A}(\mathcal{G}) = \{A^1, ..., A^n\}$ where, for each agent A^i, $H(A^i) = \{\text{Hp}_a^j \mid p^j \in (Neigh(p^i) \cup \{p^i\}), a \in St(p^j)\}$, and $T(A^i)$ consists of the program:

$$\text{p}_a^i \leftarrow \text{Hp}_{a_1}^{n_1}, ..., \text{Hp}_{a_k}^{n_k}. \quad where \; \{p^{n_1}, ..., p^{n_k}\} = Neigh(p) \wedge \forall p^{n_j}, a_j \in St(p^{n_j}) \wedge a \in St(p^i) \wedge$$
$$u_{p^i}(\{p_a^i, p_{a_1}^{n_i}, ..., p_{ak}^{n_k}\}) = \max_{s \in St(p^i)}(\{p_s^i, p_{a_1}^{n_i}, ..., p_{ak}^{n_k}\});$$

$$\begin{array}{ll}
\text{bad} \leftarrow \text{not } \text{p}_a^i, \text{Hp}_a^i. & \forall a \in St(p^i); \\
\text{oneSelected}_{p^j} \leftarrow \text{Hp}_a^j. & \forall a \in St(p^j) \wedge p^j \in (Neigh(p^i) \cup \{p^i\}); \\
\text{twoSelected} \leftarrow \text{Hp}_a^j, \text{Hp}_b^j. & \forall a, b \in St(p^j) \text{ s.t. } a \neq b \wedge p^j \in (Neigh(p^i) \cup \{p^i\});
\end{array}$$

$$\text{ok}_{p^i} \leftarrow \text{not bad}, \text{oneSelected}_{p^i}, \text{not twoSelected},$$
$$\text{oneSelected}_{p^{n_1}}, ..., \text{oneSelected}_{p^{n_k}}.$$

Let $\mathbf{x}$ be a global strategy for $\mathcal{G}$. We denote by $e(\mathbf{x})$ the tuple $\langle e_1(\mathbf{x}), ..., e_n(\mathbf{x}) \rangle$ such that, for each $i \in \{1, ..., n\}$, it holds that: $e_i(\mathbf{x}) = \bigcup_{p^j \in (Neigh(p^i) \cup \{p^i\})} \{\text{p}_a^j \mid p_a^j \in \mathbf{x}\}$.

Then, it is not difficult to check that $\mathbf{x}$ is a Nash equilibrium $\Leftrightarrow e(\mathbf{x})$ is a joint explanation for $\mathcal{O}(\mathcal{G})$, and the NP-hardness of Consistency for joint explanations follows. Hardness for the other kinds of explanation derives from Proposition 1.   □

Theorem 3. Consistency *is* P^{NP}*-complete for safe explanations. (Combined Complexity)*

Proof (Sketch). Membership. The problem can be solved by processing agents in a sequential manner. For each agent p, we can find the maximum value of the associated function γ_p by means of a binary search, in which at each step we guess in NP an hypothesis and verify in polynomial time that condition (1) in Definition 4 holds.

After having collected the maximum values for all the agents, we conclude with a final guess to get a tuple of individual explanations, and a subsequent check that actually each agent p gets precisely this maximum value for γ_p.

Hardness. Let Φ be a boolean formula in conjunctive normal form $\Phi = C_1 \wedge ... \wedge C_m$ over the variables $X_1, ..., X_n$. Assume an ordering over the variables, say $X_i \prec X_j$ iff $i < j$, and recall that the problem of deciding whether X_n is true in the lexicographical maximum (w.r.t. $\prec$) satisfying assignment is P^{NP}-complete [31].

We build an abductive scenario $\langle \mathcal{A}(\Phi), \mathcal{O}(\Phi) \rangle$ such that: $\mathcal{O}(\Phi) = \{\text{sat}_j \mid j \in \{1, ..., m\}\}$, and $\mathcal{A}(\Phi) = \{A, E\}$. In particular, $H(A) = \{\text{HX}_i \mid X_i \text{ occurs in } \Phi\}$, and $T(A)$ consists of the following rules (for $1 \leq j \leq m$):

$$\left. \begin{array}{l}
\text{sat}_j \leftarrow \sigma(t_{j,1}). \\
\text{sat}_j \leftarrow \sigma(t_{j,2}). \\
\text{sat}_j \leftarrow \sigma(t_{j,3}).
\end{array} \right\} \quad where \; c_j = t_{j,1} \vee t_{j,2} \vee t_{j,3}, and$$

where σ is the following mapping: $\sigma(t) = \begin{cases} \text{HX}_i & \text{, if } t = X_i, 1 \leq i \leq n \\ \text{not } \text{HX}_i & \text{, if } t = \neg X_i, 1 \leq i \leq n. \end{cases}$

Finally, let $H(E)$ be the set $\{HX_n\}$, $T(E)$ be the program $\text{sat}_j \leftarrow HX_n$. $1 \leq j \leq m$, and assume that $\gamma_A(S) = \sum_{i=1}^{n} 2^{n-i}\delta(HX_i)$, where $\delta(HX_i) = 1$ iff HX_i is in the explanation S and $\delta(HX_i) = 0$ otherwise. Moreover, let $\gamma_E(\{HX_n\}) = 1$ and $\gamma_E(\{\}) = 0$.

Then, it is easy to see that the existence of a safe explanation entails that X_n is true in the lexicographical maximum satisfying assignment. Moreover, if there is no safe explanation, then X_n is not true in the lexicographical maximum satisfying assignment (either the formula is not satisfiable or X_n is false).   □

5.3 Data Complexity for *P2P* Systems

After that the intrinsic complexity of the problem of computing joint explanations has been characterized (for general abductive scenarios), it is natural to ask whether focusing on the specific scenarios associated to *P2P* systems by means of the approach described in Section 4 makes the problem any easier. The first crucial observation is that, by Theorem 1, computing a repair $\mathcal{B}_S$ for a *P2P* system $\mathcal{P}$ w.r.t. a global database $\mathcal{B}$ is *at most* hard as computing a joint explanation for a suitable abductive scenario $\mathcal{S}(\mathcal{P}, \mathcal{B})$. Moreover, the construction of $\mathcal{S}(\mathcal{P}, \mathcal{B})$ is such that if the global schema of each peer $\mathcal{P}$ is not considered to be part of the input problem (i.e., the standard *data complexity* setting is considered), then the size of $\mathcal{S}(\mathcal{P}, \mathcal{B})$, where all the agents' theories are grounded, turns out to be polynomial in the size of $\mathcal{B}$; therefore, membership results in Figure 2.(b) apply — the result holds even if the global schema is part of the input problem, but arities of the predicates are bounded by a fixed constant.

Actually, we claim that hardness results in Figure 2.(b) hold as well. Indeed, it is not difficult to associate in polynomial time a *P2P* system $\mathcal{P}[\bar{\mathcal{S}}]$ and a global database $\mathcal{B}[\bar{\mathcal{S}}]$ to any abductive scenario $\bar{\mathcal{S}}$ (where agents' theories are propositional) such that joint explanations for $\bar{\mathcal{S}}$ coincide with joint explanations for $\mathcal{S}(\mathcal{P}[\bar{\mathcal{S}}], \mathcal{B}[\bar{\mathcal{S}}])$ and such that $\mathcal{S}(\mathcal{P}[\bar{\mathcal{S}}], \mathcal{B}[\bar{\mathcal{S}}])$ can be built in polynomial time. Basically, agents in $\bar{\mathcal{S}}$ may be associated with peers in $\mathcal{P}[\bar{\mathcal{S}}]$; hypotheses and atoms in $\bar{\mathcal{S}}$ may be associated with (propositional) facts in $\mathcal{B}[\bar{\mathcal{S}}]$; mappings for $\mathcal{P}[\bar{\mathcal{S}}]$ may be defined to propagate the hypotheses shared by pairs of agents only; and integrity constraints may be used to enforce the evaluation of the individual theories and the entailment of the observations. Since $\mathcal{B}[\bar{\mathcal{S}}]$ stores propositional facts only, agents' theories in $\mathcal{S}(\mathcal{P}[\bar{\mathcal{S}}], \mathcal{B}[\bar{\mathcal{S}}])$ are propositional, and the size of $\mathcal{S}(\mathcal{P}[\bar{\mathcal{S}}], \mathcal{B}[\bar{\mathcal{S}}])$ is polynomial in the size of $\mathcal{B}[\bar{\mathcal{S}}]$ and, hence, in the size of $\bar{\mathcal{S}}$.

6   Discussions and Conclusion

We have considered a generalization of the basic abductive framework to a scenario where several autonomous agents interact with each other, and where agents' knowledge is formalized by means of stratified logic programs. A complete picture of the computational complexity of the framework has been presented: It turned out that the main reasoning problems are confined in the first two levels of the polynomial hierarchy. The complexity results are particular interesting in the light of the exploitation of joint explanations to provide semantics for integrating data in *P2P* systems. Indeed, even though the proposed setting is quite rich and expressive in taking into account both peers' autonomy and preferences, it is well-behaved under a computational point of view, since even very simple integration tasks (in a single data integration system) are known to be complete for the same classes.

It is worthwhile noting that abduction has already played a central role in the design of multi-agent systems, e.g., to provide semantics for information exchange [32], to model agent communication [33], to define mechanisms for agent negotiation based on dialogue [34] (just to cite a few applications). Our perspective here was quite different. Indeed, we investigated some foundational problems related to a scenario where several autonomous agents jointly contribute to find explanations to a bunch of observations.

Important contributions in this direction have been already appeared in the literature. In particular, [35] investigated how to coordinate the abductive reasoning of multiple agents, by developing an architecture (ALIAS) where several coordination patterns can be chosen; in ALIAS, the agent behavior is expressed in the LAILA language [36], which allows the agents, e.g., to specify whether they desire to find explanations for a given goal either in a collaborative or in a competitive way. In both cases, agents are assumed to share the same set of hypotheses and to independently look for individual sets of hypotheses explaining the same set of observations: In the collaborative scenario, agents simply merge their individual explanations provided these are consistent with each other, whereas in the competitive scenario, one individual solution wins the competition and is eventually selected.

In this paper, we extended the above framework by removing some of its simplifying assumptions and introducing the notion of *joint explanation*. First of all, we did not require the existence of a unique centralized knowledge base containing the set of all the hypotheses for the agents and, more importantly, we allowed each agent to partially contribute to the whole explanation, i.e., to explain a proper subset of all the observations only. This is particularly useful whenever each agent does not have enough knowledge for explaining all the observations by itself and coordination with other agents is required. It is worthwhile noting that while computing joint explanations, agents are not mutually independent and solutions to abductive problems can in general be obtained neither by testing for consistency of individual solutions nor by just selecting a representative one.

Closest to our setting is perhaps [37], where diagnostic agents that need to reach a common diagnosis have been considered. Diagnosis is carried out in a completely decentralized fashion, so that each agent is responsible for monitoring part of the system. Then, a consensus component provides algorithms to achieve a consensus among agents if their diagnosis results differ, which are based on the comparison of the individual diagnosis and on a majority voting protocol. Clearly enough, while our approach shares with [37] the idea of exploiting a distributed abductive process, the proposed solutions strongly differ. Indeed, rather than being based on voting mechanisms, joint explanations emerge as kind of *compromise* between the various individual explanations. In this respect, our solution approach may be seen as a generalization to the abductive framework of the basic approach proposed in [9]. Indeed, [9] firstly introduced a semantics for defining compromises among agents, where rather than joining the theories and considering the model of this single program, the computation of the joint fixpoints for the separate theories is advocated.

However, besides the focus on abductive problems, our approach significantly extends [9] also in the investigation of those scenarios where agents assign possible contrasting preferences to set of hypotheses, and where some kinds of desirable explanation has to be computed. A related approach to allow agents to express preferences in their abductive reasoning has been proposed by [13], which mentioned as possible line of further research the possibility of exploiting ideas from multi-criteria decision theory to solve conflicts among agents. This is precisely the way pursed in this paper. And, as a result, our setting turned out to be a generalization of abduction with penalization (see, e.g., [10, 38, 11]) to multiple and competitive agents, where the semantics for the multi-objective optimization is given in terms of some game-theoretic arguments.

We believe that our investigation paves the way for further (more comprehensive) logic-based formalizations and for effective implementations. Indeed, an avenue of further research may lead to enrich the proposed setting by further kinds of preferences criteria, by replacing or complementing the weighting functions proposed in this paper. In this respect, it may be investigated whether current approaches for qualitatively handling preferences (in multi-agents systems), e.g., [41-45], may be adapted to cope with our distributed abductive scenario — notably, for some approaches (cf. [44, 45]), the ability of modelling strategic interactions in games has been already investigated.

Another avenue of further research is to investigate whether answer set engines (e.g., Smodels [39] and DLV [40]) may be used to support joint explanations and, more generally, agreements for selfish interested agents. Moreover, techniques for approximate abductive reasoning (e.g., [12]) might also be explored.

Acknowledgments. The author thanks the anonymous referees for their useful comments and suggestions.

References

1. Peirce, C.S.: Abduction and induction. In Philosophical Writings of Peirce (1955) 150–156
2. Denecker, M., Kakas, A.C.: Abduction in logic programming. Computational Logic: Logic Programming and Beyond (2002) 402–436
3. Kakas, A.C., Mancarella, P.: Generalized stable models: A semantics for abduction. In: Proc. of ECAI'90. (1990) 385–391
4. Lin, F., You, J.H.: Abduction in logic programming: A new definition and an abductive procedure based on rewriting. In: Proc. of IJCAI'01. (2001) 655–666
5. Console, L., Dupre, D.T., Torasso, P.: On the relationship between abduction and deduction. Journal of Logic and Computation 1 (1991) 661–690
6. Kakas, A.C., Nuffelen, B.V., Denecker, M.: A-system: Problem solving through abduction. In: Proc. of IJCAI'01. (2001) 591–596
7. Dung, P.M.: Negation as hypotheses: An abductive foundation for logic programming. In: Proc. of ICLP'91. (1991) 3–17
8. Endriss, U., Mancarella, P., Sadri, F., Terreni, G., Toni, F.: The ciff proof procedure for abductive logic programming with constraints. In: Proc. of JELIA'04. (2004) 31–43
9. Buccafurri, F., Gottlob, G.: Multiagent compromises, joint fixpoints, and stable models. In: Computational Logic: Logic Programming and Beyond, Essays in Honour of Robert A. Kowalski, Part I, London, UK, Springer-Verlag (2002) 561–585
10. Eiter, T., Gottlob, G.: The complexity of logic-based abduction. J. ACM 42 (1995) 3–42

11. Perri, S., Scarcello, F., Leone, N.: Abductive logic programs with penalization: Semantics, complexity and implementation. TPLP **5** (2005) 123–159
12. Abdelbar, A.M.: Approximating cost-based abduction is np-hard. Artificial Intelligence **159** (2004) 231–239
13. Kakas, A., Moraitis, P.: Argumentation based decision making for autonomous agents. In: Proc of AAMAS '03. (2003) 883–890
14. Arieli, O., Denecker, M., Nuffelen, B.V., Bruynooghe, M.: Coherent integration of databases by abductive logic programming. J. Artif. Intell. Research **21** (2004) 245–286
15. Pivkina, I., Pontelli, E., Son, T.C.: Revising knowledge in multi-agent systems using revision programming with preferences. In: Proc. of CLIMA IV. (2004) 134–158
16. Mura, P.L., Shoham, Y.: Conditional, hierarchical, multi-agent preferences. In: Proc. of TARK'98. (1998) 215–224
17. Faratin, P., de Walle, B.V.: Agent preference relations: Strict, indifferent, and incomparable. In: Proc. of AAMAS'02. (2002) 1317–1324
18. Bracciali, A., Mancarella, P., Stathis, K., Toni, F.: On modelling multi-agent systems declaratively. In: Proc. of DALT'04. (2004) 53–68
19. Nuffelen, B.V., Corts-Calabuig, A., Denecker, M., Arieli, O., Bruynooghe, M.: Data integration using id-logic. In: Proc. of CAiSE'04. (2004) 67–81
20. Greco, G., Greco, S., Zumpano, E.: A logic programming approach to the integration, repairing and querying of inconsistent databases. In: Proc. of ICLP'01. (2001) 348–364
21. Calì, A., Lembo, D., Rosati, R.: Query rewriting and answering under constraints in data integration systems. In: Proc. of IJCAI'03. (2003) 16–21
22. Bravo, L., Bertossi, L.: Logic programming for consistently querying data integration systems. In: Proc. of IJCAI'03. (2003) 10–15
23. Calvanese, D., Giacomo, G.D., Lenzerini, M., Rosati, R.: Logical foundations of peer-to-peer data integration. In: Proc. of PODS'04. (2004) 241–251
24. Schlosser, M.T., Sintek, M., Decker, S., Nejdl, W.: Hypercup - hypercubes, ontologies, and efficient search on peer-to-peer networks. In: Proc. of AP2PC'02. (2002) 112–124
25. Franconi, E., Kuper, G., Lopatenko, A., Serafini, L.: A robust logical and computational characterisation of peer-to-peer database systems. In: Proc od DBISP2P'03. (2003) 64–76
26. Bertossi, L.E., Bravo, L.: Query answering in peer-to-peer data exchange systems. In: Proc. of EDBT Workshops 2004. (2004) 476–485
27. Lenzerini, M.: Principles of p2p data integration. In: Proc. of DiWeb'04. (2004) 7–21
28. Greco, G. and Scarcello, F. On the Complexity of Computing Peer Agreements for Consistent Query Answering in Peer-to-Peer Data Integration Systems, In: Proc. of CIKM'05. (2005)
29. Calì, A., Lembo, D., Rosati, R.: On the decidability and complexity of query answering over inconsistent and incomplete databases. In: Proc. of PODS'03. (2003) 260–271
30. Gottlob, G., Greco, G., Scarcello, F.: Pure nash equilibria: hard and easy games. Journal of Artificial Intelligence Research **24** (2005) 357–406
31. Krentel, M.W.: The Complexity of Optimization Problems. Journal of Computer and System Sciences 36(3)(1988). 490-509
32. Gavanelli, M., Lamma, E., Mello, P., Torroni, P.: An abductive framework for information exchange in multi-agent systems. In: Proc. of CLIMA IV. (2004) 34–52
33. Hindriks, K.V., de Boer, F.S., van der Hoek, W., Meyer, J.J.C.: Semantics of communicating agents based on deduction and abduction. Issues in Agent Communication (2000) 63–79
34. Sadri, F., Toni, F., Torroni, P.: An abductive logic programming architecture for negotiating agents. In: Proc. of JELIA'02. (2002) 419–431
35. Ciampolini, A., Lamma, E., Mello, P., Toni, F., Torroni, P.: Cooperation and competition in alias: A logic framework for agents that negotiate. Ann. Math. Artif. Intell. **37** (2003) 65–91
36. Ciampolini, A., Lamma, E., Mello, P., Torroni, P.: Laila: a language for coordinating abductive reasoning among logic agents. Comput. Lang. **27** (2001) 137–161

37. Schroeder, M., de Almeida Mora, I., Pereirat, L.M.: A deliberative and reactive diagnosis agent based on logic programming. In: Proc of ICTAI'96. (1996) 436
38. Charniak, E., Shimony, S.: Cost-based abduction and map explanations. Artificial Intelligence **66** (1994) 345–374
39. Syrjänen, T., Niemelä, I.: The smodels system. In: Proc. of LPNMR'01. (2001) 434–438
40. Leone, N., Pfeifer, G., Faber, W., Calimeri, F., Dell'Armi, T., Eiter, T., Gottlob, G., Ianni, G., Ielpa, G., Koch, C., Perri, S., Polleres, A.: The DLV System. In: Proc of JELIA'02. (2002) 537–540
41. Brewka, G., Niemelä, I., Truszczynski, M.: Answer set optimization. In: Proc. of IJCAI'03. (2003) 867–872
42. Van Nieuwenborgh, D., Heymans, S., Vermeir, D.: On Programs with Linearly Ordered Multiple Preferences. In: Proc of ICLP'04. (2004) 180–194
43. De Vos, M., Crick, T., Padget, J., Brain, M., Cliffe, O., Needham, J.: Multi-agent Platform using Ordered Choice Logic Programming. In: Proc of DALT'05. (2005) 67–83
44. De Vos, M., Vermeir, D.: Choice Logic Programs and Nash Equilibria in Strategic Games. In: Proc of CSL'99. (1999) 266–276
45. Apt, K.R., Rossi, F., Venable, K.B.: CP-nets and Nash equilibria. In: Proc. of CIRAS'05. (2005)

The Nested List Normal Form
for Functional and Multivalued Dependencies

Sven Hartmann and Sebastian Link*

Information Science Research Centre, Dept of Information Systems,
Massey University, Palmerston North, New Zealand
{s.hartmann, s.link}@massey.ac.nz

Abstract. The Nested List Normal Form is proposed as a syntactic
normal form for semantically well-designed database schemata obtained
from any arbitrary finite nesting of records and lists. The Nested List
Normal Form is defined in terms of functional and multivalued dependencies, independent from any specific data model, and generalises the
well-known Fourth Normal Form from relational databases in order to
capture more application domains.

1   Introduction

An important issue associated with the use of any databases is the correct structure or design of data to be used. Several criteria, referred to as *normal forms*,
have been proposed as conditions for database schemata that a database design
should satisfy to ensure an absence of processing difficulties with the database.
These normal forms give a database designer unambiguous guidelines in deciding
which databases are good in the quest to avoid bad designs that have redundancy problems and update anomalies. Such normal forms have already been
introduced in [12] by Codd himself. In general, they are dependent on the type
of integrity constraints or rules which apply to data items within the database.
Important classes of integrity constraints are *functional dependencies* (FDs) [12]
and *Multivalued dependencies* (MVDs) [14]. FDs and MVDs cause difficulties
such as redundancy in the representation of data and update anomalies. The
Boyce-Codd normal form (BCNF) was proposed to overcome these difficulties
with FDs [13], and Fagin introduced the Fourth Normal form (4NF) to deal with
the more general class of FDs and MVDs [14]. Later on, after the notions of redundancy and update anomaly had been clarified and formalised, it was shown
that BCNF (4NF) precisely captures those relation schemata that are free from
redundancies and update anomalies in terms of FDs (FDs and MVDs) [7, 15, 35].
Normalisation has been studied in the context of other data models as well. There
are several normal form proposals for the nested relational data model, and a
detailed comparison can be found in [30]. Recently, the issue of normalisation
has been revived in the context of XML [2, 3, 36, 37]. XNF is defined in terms
of FDs that are based on a path-like notion in DTDs and do not enjoy a finite

---

* Sebastian Link was supported by Marsden Funding, Royal Society of New Zealand.

ground axiomatisation [2]. In [3] techniques from information theory are used to provide justifications for several normal forms and normalisation algorithms. MVDs have also been introduced into the context of XML and an extension of 4NF has been proposed [37]. Apart from [18, 31], who consider set equality in the nested relational data model, all previous approaches to defining constraints in advanced data formats do not consider equality on complex objects such as lists, sets or multisets, and are therefore unable to express important semantic information that occurs in many applications.

Several researchers have remarked that classical database design problems need to be revisited in new data formats [3, 32, 34, 35]. Biskup [8, 9] has listed two particular challenges for database design theory: finding a unifying framework and extending achievements to deal with advanced database features such as complex object types. We propose to classify data models according to the type constructors they support. Thus, the relational data model can be captured by a single application of the record type, arbitrary nesting of record and set constructor covers aggregation and grouping which are fundamental to many semantic data models as well as the nested relational data model [1, 24]. The Entity-Relationship Model and its extensions require record, set and (disjoint) union constructor [11, 33]. A minimal set of type constructors supported by any object-oriented data model includes records, lists, sets and multisets (bags) [5]. Genomic sequence data models call for support of records, lists and sets [28]. Finally, XML requires at least record (concatenation), list (Kleene Closure), union (optionality), and reference constructor [10].

In this paper we study database design in the presence of record and list constructor with respect to functional and multivalued dependencies. It is our goal to achieve an adequate extension of 4NF from relational databases, and to actually demonstrate what this extension achieves in terms of characterising the absence of adequate extensions of the notions of redundancies and update anomalies. Our studies will be based on an abstract data model that defines a database schema as an arbitrarily nested attribute where nesting applies record and list constructor. It is our intention not to focus on any specific data model in order to place emphasis on the type constructors themselves. Dependencies are defined in terms of subschemata of the underlying database schema. This approach provides a mathematically well-founded framework that is sufficiently flexible and powerful to study design problems for different classes of constraints. The fact that the set of all subschemata of some fixed database schema carries the structure of a Brouwerian algebra turns out to precisely accommodate the needs of multivalued dependencies.

Throughout the article we will apply the theory to an example from image processing that we introduce now. Digital halftoning plays a key role in almost every discipline that involves printing and displaying. All newspapers, magazines, and books are printed with digital halftoning. One method to perform digital halftoning is error diffusion [16, 25, 26]: once a pixel has been quantised, thus introducing some error, this error should affect the quantisation of the pixels in the region of its neighbours. Digital halftoning is an application of the matrix rounding problem [4]. The problem is to convert a continuous-tone image into a binary one that looks similar. The input matrix A represents a digital (gray) image, where a_{ij} represents the brightness level of the (i,j)-pixel in the $n \times n$ pixel grid. Typically, n is between 256 and 4096, and a_{ij} is an integral multiple of $\frac{1}{256}$: this means that we use 256 brightness levels. If we want to send an image using fax or print it out by a dot or ink-jet printer, brightness levels available are limited. Instead, we replace the input matrix A by an integral matrix B so that each pixel uses only two brightness levels. Here, it is important that B looks similar to A; in other words, B should be an approximation of A. In this sense, an approximation of input matrix A is a $\{0,1\}$-matrix B that minimises

the distance $\left| \sum\limits_{(i,j)\in R} a_{i,j} - \sum\limits_{(i,j)\in R} b_{i,j} \right|$ for all $R \in \mathcal{R}$. In this formulae $\mathcal{R}$ denotes the set of regions of neighbors, for instance the set of all pairs of indices that denote $2 \times 1, 1 \times 2$ and 2×2 submatrices. A region has therefore one of the following forms

$$
\begin{array}{|c|c|} \hline a & b \\ \hline c & d \\ \hline \end{array}
\qquad
\begin{array}{|c|c|} \hline a & b \\ \hline \end{array}
\qquad
\begin{array}{|c|} \hline a \\ \hline b \\ \hline \end{array} \,,
$$

and can be represented as a list of either two or four elements. The regions may have all different kinds of shapes in practice. In order to make the example more illustrative, we assume from now on that the input matrix has entries in $\{0, \frac{1}{2}, 1\}$, i.e., uses three brightness levels. Input regions can be best approximated by a number of different output regions. All inputs with overall brightness $\frac{1}{2}$ and length two, i.e. $[0, \frac{1}{2}]$ or $[\frac{1}{2}, 0]$, could be mapped to any of $[0,1]$, $[1,0]$ or $[0,0]$, each of which has distance $\frac{1}{2}$. In this sense, the set of input sequences $(\{[0, \frac{1}{2}], [\frac{1}{2}, 0]\})$ is determined by the overall brightness of the input sequences $(\frac{1}{2})$ and the length of the input sequence (2), independently of the set of output sequences $(\{[0,1], [1,0], [0,0]\})$. This is true for any inputs and outputs, e.g., all inputs with overall brightness $\frac{3}{2}$ and length four such as $[0, 0, 1, \frac{1}{2}]$ can be mapped to any of $[0,0,0,1]$, $[0,0,1,0]$, $[0,1,0,0]$, $[1,0,0,0]$, $[0,0,1,1]$, $[0,1,0,1]$, $[1,0,0,1]$, $[0,1,1,0]$, $[1,0,1,0]$, $[1,1,0,0]$.

Consider a database which stores input and output sequences together with the overall brightness of the input sequence. It is then desirable to find a $\{0,1\}$-matrix B that has for every of the possible regions of input matrix A a corresponding output region that are stored together as an entry in the database. The input matrix $A = \begin{pmatrix} 0 & 0 \\ \frac{1}{2} & \frac{1}{2} \end{pmatrix}$ has for instance the approximation $B = \begin{pmatrix} 0 & 0 \\ 0 & 1 \end{pmatrix}$. Every 2×2 matrix has five input sequences and the mappings that produce B from A are as follows: $[0,0] \mapsto [0,0]$, $[\frac{1}{2}, \frac{1}{2}] \mapsto [0,1]$, $[0, \frac{1}{2}] \mapsto [0,0]$ (left column), $[0, \frac{1}{2}] \mapsto [0,1]$ (right column) and $[0, 0, \frac{1}{2}, \frac{1}{2}] \mapsto [0,0,0,1]$.

The matrix $\begin{pmatrix} 0 & 1 \\ 0 & 0 \end{pmatrix}$, however, is not an approximation of A as the sequence $[\frac{1}{2}, \frac{1}{2}]$ should not be mapped to $[0,0]$.

Constraints that a database designer may choose to specify for this application are the following:

1. The length of the input sequence determines the length of the output sequence, and vice versa.
2. The overall brightness and length of the input sequence together determine the set of all input sequences independently from the set of the output sequences.

The example illustrates a typical scenario where list equality occurs in a constraint specification. We will formalise all parts of this example during the course of the paper, and see whether the suggested design is appropriate with respect to the constraints specified.

2   A Summary of Previous Work

2.1   The Complex-Value Data Model

This section introduces a data model based on the nesting of attributes and sub-typing. It may be used to provide a framework for the study of type constructors such as records, lists, sets, multisets, unions and references. This article, however, focuses on records and lists. In terms of XML the reader may notice that we deal with a slightly extended fragment of DTDs in which only concatenation and Kleene closure are allowed. However, the expressiveness of our constraints is different from previous approaches as we are particularly interested in list equality.

We start with the definition of flat attributes and values for them. A *universe* is a finite set $\mathcal{U}$ together with domains (, i.e., sets of values) $dom(A)$ for all $A \in \mathcal{U}$. The elements of $\mathcal{U}$ are called *flat attributes*. Flat attributes will be denoted by upper-case characters from the start of the alphabet such as A, B, C etc.

In the following we will use a set $\mathcal{L}$ of labels, and assume that the symbol λ is neither a flat attribute nor a label, i.e., $\lambda \notin \mathcal{U} \cup \mathcal{L}$. Moreover, flat attributes are not labels and vice versa, i.e., $\mathcal{U} \cap \mathcal{L} = \emptyset$.

Database schemata in our data model will be given in form of nested attributes. Let $\mathcal{U}$ be a universe and $\mathcal{L}$ a set of labels. The set $\mathcal{NA}(\mathcal{U}, \mathcal{L})$ of *nested attributes over $\mathcal{U}$ and $\mathcal{L}$* is the smallest set satisfying the following conditions: $\lambda \in \mathcal{NA}(\mathcal{U}, \mathcal{L})$, $\mathcal{U} \subseteq \mathcal{NA}(\mathcal{U}, \mathcal{L})$, for $L \in \mathcal{L}$ and $N_1, \ldots, N_k \in \mathcal{NA}(\mathcal{U}, \mathcal{L})$ with $k \geq 1$ we have $L(N_1, \ldots, N_k) \in \mathcal{NA}(\mathcal{U}, \mathcal{L})$, for $L \in \mathcal{L}$ and $N \in \mathcal{NA}(\mathcal{U}, \mathcal{L})$ we have $L[N] \in \mathcal{NA}(\mathcal{U}, \mathcal{L})$. We call λ *null attribute*, $L(N_1, \ldots, N_k)$ *record-valued attribute* and $L[N]$ *list-valued attribute*. We will use upper-case letters from the middle of the alphabet such as $N, M, etc.$ to refer to nested attributes. From now on, we assume that a set $\mathcal{U}$ of attribute names, and a set $\mathcal{L}$ of labels is fixed, and write $\mathcal{NA}$ instead of $\mathcal{NA}(\mathcal{U}, \mathcal{L})$. We may use the nested attribute

HALFTONING(Brightness,INPUT[Level],OUTPUT[Bit])

as a database schema for instances of the digital halftoning database described in the introduction. Labels are HALFTONING, INPUT and OUTPUT, and flat attribute names are Brightness, Level and Bit. The domain of the flat attribute Level is $\{0, \frac{1}{2}, 1\}$ and the domain of the flat attribute Bit is $\{0, 1\}$.

In general, we can extend the mapping dom from flat attributes to nested attributes, i.e., we define a set $dom(N)$ of values for every nested attribute $N \in \mathcal{N}A$. For a nested attribute $N \in \mathcal{N}A$ we define the *domain* $dom(N)$ as follows: $dom(\lambda) = \{ok\}$, $dom(L(N_1, \ldots, N_k)) = \{(v_1, \ldots, v_k) \mid v_i \in dom(N_i) \text{ for } i = 1, \ldots, k\}$, i.e., the set of all k-tuples $(v_1, \ldots, v_k)$ with $v_i \in dom(N_i)$ for all $i = 1, \ldots, k$, and $dom(L[N]) = \{[v_1, \ldots, v_n] \mid v_i \in dom(N) \text{ for } i = 1, \ldots, n\}$, i.e., the set of all finite lists with elements in $dom(N)$. The empty list is denoted by [ ]. For instance, the domain of INPUT$[\lambda]$ is the set of all finite lists consisting of elements ok, i.e., $\{[\,], [ok], [ok, ok], \ldots\}$. The nested attribute INPUT$[\lambda]$ therefore still tells us how long the lists over INPUT[Level] are. The value ok can be interpreted as the null value "some information exists, but is currently omitted".

The replacement of attributes by the null attribute λ decreases the amount of information modelled. This fact allows one to introduce an order between nested attributes. The *subattribute relation* $\leq$ on the set of nested attributes $\mathcal{N}A$ over $\mathcal{U}$ and $\mathcal{L}$ is defined by the following rules, and the following rules only: $N \leq N$, $\lambda \leq A$ for all flat attributes $A \in \mathcal{U}$, $\lambda \leq N$ for all list-valued attributes N, $L(N_1, \ldots, N_k) \leq L(M_1, \ldots, M_k)$ whenever $N_i \leq M_i$ for all $i = 1, \ldots, k$, and $L[N] \leq L[M]$ whenever $N \leq M$. For N, M we say that M is a *subattribute* of N if and only if $M \leq N$ holds. We write $M \not\leq N$ if M is not a subattribute of N, and $M < N$ in case $M \leq N$ and $M \neq N$.

Lemma 1 ([22]). *The subattribute relation is a partial order on nested attributes.* □

Informally, M is a subattribute of N if and only if M comprises at most as much information as N does. The informal description of the subattribute relation is formally documented by the existence of a projection function $\pi_M^N : dom(N) \to dom(M)$ in case $M \leq N$ holds. For $M \leq N$ the *projection function* $\pi_M^N : dom(N) \to dom(M)$ is defined as follows:

- if $N = M$, then $\pi_M^N = id_{dom(N)}$ is the identity on $dom(N)$,
- if $M = \lambda$, then $\pi_\lambda^N : dom(N) \to \{ok\}$ is the constant function that maps every $v \in dom(N)$ to ok,
- if $N = L(N_1, \ldots, N_k)$ and $M = L(M_1, \ldots, M_k)$, then $\pi_M^N = \pi_{M_1}^{N_1} \times \cdots \times \pi_{M_k}^{N_k}$ which maps every tuple $(v_1, \ldots, v_k) \in dom(N)$ to $(\pi_{M_1}^{N_1}(v_1), \ldots, \pi_{M_k}^{N_k}(v_k)) \in dom(M)$, and
- if $N = L[N']$ and $M = L[M']$, then $\pi_M^N : dom(N) \to dom(M)$ maps every list $[v_1, \ldots, v_n] \in dom(N)$ to the list $[\pi_{M'}^{N'}(v_1), \ldots, \pi_{M'}^{N'}(v_n)] \in dom(M)$.

The set $Sub(N)$ of *subattributes* of N is $Sub(N) = \{M \mid M \leq N\}$. Note that $Sub(N)$ is always finite. Lemma 1 shows that the restriction of $\leq$ to $Sub(N)$ is a partial order on $Sub(N)$. We study the algebraic structure of $Sub(N)$. A *Brouwerian algebra* [29] is a lattice $(L, \sqsubseteq, \sqcup, \sqcap, \dot{-}, 1)$ with top element 1 and a binary operation $\dot{-}$ which satisfies $a \dot{-} b \sqsubseteq c$ iff $a \sqsubseteq b \sqcup c$ for all $c \in L$. In this case, the operation $\dot{-}$ is called the *pseudo-difference*. The *Brouwerian complement* $\neg a$ of $a \in L$ is then defined by $\neg a = 1 \dot{-} a$. A Brouwerian algebra is also called a co-Heyting algebra or a dual Heyting algebra. The system of all closed subsets of a topological space is a well-known Brouwerian algebra, see [29]. The *join* $X \sqcup_N Y$, *meet* $X \sqcap_N Y$ and *pseudo-difference* $X \mathbin{\dot{-}}_N Y$ of X and Y in $Sub(N)$ are completely determined by the subattribute order $\leq$. We use $Y_N^{\mathcal{C}} = N \mathbin{\dot{-}} Y$ to denote the *Brouwerian complement* of Y in $Sub(N)$. The following theorem generalises the fact that $(\mathcal{P}(R), \subseteq, \cup, \cap, -, \emptyset, R)$ is a Boolean algebra for a relation schema R.

Theorem 1 ([22]). $(Sub(N), \leq, \sqcup_N, \sqcap_N, \mathbin{\dot{-}}_N, N)$ *forms a Brouwerian algebra for every* $N \in \mathcal{N}A$. □

In order to simplify notation, occurrences of λ in a record-valued attribute are usually omitted if this does not cause any ambiguities. That is, the subattribute $L(M_1, \ldots, M_k) \leq L(N_1, \ldots, N_k)$ is abbreviated by $L(M_{i_1}, \ldots, M_{i_l})$ where $\{M_{i_1}, \ldots, M_{i_l}\} = \{M_j : M_j \neq \lambda_{N_j} \text{ and } 1 \leq j \leq k\}$ and $i_1 < \cdots < i_l$. If $M_j = \lambda_{N_j}$ for all $j = 1, \ldots, k$, then we use λ instead of $L(M_1, \ldots, M_k)$. The subattribute HALFTONING(λ,INPUT[λ],λ) is abbreviated by HALFTONING(INPUT[λ]). However, the subattribute $L(A, \lambda)$ of $L(A, A)$ cannot be abbreviated by $L(A)$ since this may also refer to $L(\lambda, A)$. If the context allows, we omit the index N from the operations $\sqcup_N, \sqcap_N, \mathbin{\dot{-}}_N, (\cdot)_N^{\mathcal{C}}$ and from λ_N. The Brouwerian algebra for HALFTONING(Brightness,INPUT[Level],OUTPUT[Bit]) is illustrated in Figure 1.

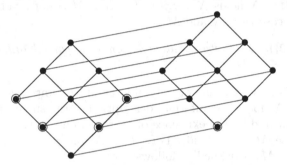

Fig. 1. Brouwerian algebra of HALFTONING(Brightness,INPUT[Level],OUTPUT[Bit])

Fundamental to lists is the following fact: if $\pi_X^N(t_1) = \pi_X^N(t_2)$ and $\pi_Y^N(t_1) = \pi_Y^N(t_2)$, then also $\pi_{X \sqcup Y}^N(t_1) = \pi_{X \sqcup Y}^N(t_2)$ for any $t_1, t_2 \in dom(N)$ [22]. This suggests to focus on join-irreducible elements of $(Sub(N), \leq, \sqcup, \sqcap, \lambda_N)$. Recall that an element a of a lattice with bottom element 0 is called *join-irreducible* if and only if $a \neq 0$ and if $a = b \sqcup c$ holds for any elements b and c, then $a = b$ or $a = c$. Let $\mathcal{B}(N)$ denote the set of join-irreducible elements of $(Sub(N), \leq, \sqcup, \sqcap, \mathbin{\dot{-}}, N)$, and $\mathcal{B}_{\mathcal{M}}(N)$ the maximal elements of $\mathcal{B}(N)$ with respect to $\leq$. The join-irreducibles of HALFTONING(Brightness,INPUT[Level],OUTPUT[Bit]) are circled in Figure 1.

2.2 An Axiomatisation for FDs and MVDs

In this section we repeat previous definitions and results [20, 22]. The data model allows us to introduce a natural extension of the notion of FDs and MVDs from the relational data model.

A *functional dependency (FD)* on the nested attribute N is an expression of the form $X \rightarrow Y$ where $X, Y \in Sub(N)$. A set $r \subseteq dom(N)$ *satisfies* the functional dependency $X \to Y$ on N, denoted by $\models_r X \to Y$, if and only if $\pi_Y^N(t_1) = \pi_Y^N(t_2)$ whenever $\pi_X^N(t_1) = \pi_X^N(t_2)$ for any $t_1, t_2 \in r$ holds. A *multivalued dependency (MVD)* on N is an expression of the form $X \twoheadrightarrow Y$ where $X, Y \in Sub(N)$. A set $r \subseteq dom(N)$ *satisfies* the multivalued dependency $X \twoheadrightarrow Y$ on N if and only if for all values $t_1, t_2 \in r$ with $\pi_X^N(t_1) = \pi_X^N(t_2)$ there is a value $t \in r$ with $\pi_{X \sqcup Y}^N(t) = \pi_{X \sqcup Y}^N(t_1)$ and $\pi_{X \sqcup Y^C}^N(t) = \pi_{X \sqcup Y^C}^N(t_2)$.

The constraints on $\textsc{Halftoning}(\text{Brightness}, \textsc{Input}[\text{Level}], \textsc{Output}[\text{Bit}])$, informally described in the introduction, can now be formalised (using abbreviations) as:

$$\textsc{Halftoning}(\textsc{Input}[\lambda]) \to \textsc{Halftoning}(\textsc{Output}[\lambda]),$$
$$\textsc{Halftoning}(\textsc{Output}[\lambda]) \to \textsc{Halftoning}(\textsc{Input}[\lambda]), \text{ and}$$
$$\textsc{Halftoning}(\text{Brightness}, \textsc{Input}[\lambda]) \twoheadrightarrow \textsc{Halftoning}(\textsc{Input}[\text{Level}]).$$

Fagin proves [14] that relational MVDs "provide a necessary and sufficient condition for a relation to be decomposable into two of its projections without loss of information (in the sense that the original relation is guaranteed to be the join of the two projections)." Let $N \in \mathcal{N}A$ and $X, Y \in Sub(N)$. Let $r_1 \subseteq dom(X)$ and $r_2 \subseteq dom(Y)$. Then $r_1 \bowtie r_2 = \{ t \in dom(X \sqcup Y) \mid$ there are $t_1 \in r_1, t_2 \in r_2$ with $\pi_X^{X \sqcup Y}(t) = t_1$ and $\pi_Y^{X \sqcup Y}(t) = t_2 \}$ is called the *generalised join* $r_1 \bowtie r_2$ of r_1 and r_2. The *projection* $\pi_X(r)$ of $r \subseteq dom(N)$ on $X \in Sub(N)$ is defined as $\{ \pi_X^N(t) \mid t \in r \}$.

Theorem 2 ([22]). *Let $N \in \mathcal{N}A$, and $r \subseteq dom(N)$. Then is $X \twoheadrightarrow Y$ satisfied by r if and only if $r = \pi_{X \sqcup Y}(r) \bowtie \pi_{X \sqcup Y^C}(r)$. If r satisfies the FD $X \to Y$, then $r = \pi_{X \sqcup Y}(r) \bowtie \pi_{X \sqcup Y^C}(r)$.* □

The notions of *implication* ($\models$) and *derivability* ($\vdash_{\mathfrak{R}}$) with respect to a set $\mathfrak{R}$ of inference rules for a class $\mathcal{C}$ of dependencies can be defined analogously to the notions in relational databases [1–pp. 164-168]. Note that finite and unrestricted implication coincide for functional and multivalued dependencies, even in the presence of lists [22]. The notions of *soundness* and *completeness* for a set $\mathfrak{R}$ of inference rules carry over as well. A dependency σ on some nested attribute N is called *trivial* if and only if $\models_r \sigma$ for every $r \subseteq dom(N)$. An FD $X \to Y$ on N is trivial iff $Y \leq X$ holds, and an MVD $X \twoheadrightarrow Y$ on N is trivial iff $Y \leq X$ or $X \sqcup Y = N$ holds. Note that $X \sqcup Y = N$ iff $Y^C \leq X$. A complete set of inference rules is said to be *minimal* if and only if none of its rules can be omitted without losing completeness.

Theorem 3 ([20, 22]). *The following inference rules*

$$\frac{}{X \to Y} \, Y \leq X$$
(reflexivity axiom)

$$\frac{X \to Y}{X \to X \sqcup Y}$$
(extension rule)

$$\frac{X \to Y, Y \to Z}{X \to Z}$$
(transitivity rule)

$$\frac{X \to Y}{X \twoheadrightarrow Y}$$
(implication rule)

$$\frac{X \twoheadrightarrow Y, Y \to Z}{X \to (Z \dot{-} Y)}$$
(mixed pseudo-transitivity rule)

$$\frac{X \twoheadrightarrow Y}{X \to Y \sqcap Y^C}$$
(mixed meet rule)

$$\frac{X \twoheadrightarrow Y, Y \twoheadrightarrow Z}{X \twoheadrightarrow (Z \dot{-} Y)}$$
(pseudo-transitivity rule)

$$\frac{X \twoheadrightarrow Y}{X \twoheadrightarrow Y^C}$$
(Brouwerian complement rule)

$$\frac{X \twoheadrightarrow Y, X \twoheadrightarrow Z}{X \twoheadrightarrow (Y \sqcup Z)}$$
(multivalued join rule)

are minimal, sound and complete for the implication of FDs and MVDs in the presence of records and lists. $\square$

Keys play a central role in the retrieval of information because they provide a method by which an element of a database may be identified.

Definition 1. Let $N \in \mathcal{N}A$ be a nested attribute and Σ a set of FDs and MVDs on N. A subattribute $X \in Sub(N)$ is called a *superkey for* N with respect to Σ if and only if $\Sigma \models X \rightarrow N$ holds. In case there is not any proper subattribute $X' < X$ which is also a superkey for N with respect to Σ, we call X a *minimal key for* N with respect to Σ. $\square$

$X \in Sub(N)$ is a superkey if and only if $X \rightarrow N \in \Sigma^+$ by Theorem 3. If $\models_r \Sigma$ for some $r \subseteq dom(N)$ and X is a superkey for N, then $t_1 = t_2$ whenever $\pi_X^N(t_1) = \pi_X^N(t_2)$ for any $t_1, t_2 \in r$. Furthermore, X is a superkey for N if and only if $X^+ = N$ where $X^+ = \bigsqcup \{Z \mid X \rightarrow Z \in \Sigma^+\}$. An FD $X \rightarrow N \in \Sigma^*$ is called a *key dependency* on N with respect to Σ if and only if X is a minimal key for N with respect to Σ. The set of all key dependencies is denoted by Σ_{key}.

3   The Nested List Normal Form

In this section we will investigate normalisation issues in the presence of records and lists in terms of FDs and MVDs. This will extend previous work in which only FDs have been considered [21].

3.1   Three Notions of Redundancy

In relational databases the definition of redundancy is based on viewing FDs and MVDs not only as integrity constraints on a relation, but also as representing the fundamental units of information for retrieving and updating the data in a relation. This interpretation of the semantics of the information stored in a relation was implicit in the original study of normalisation by Codd [12], and has since been used in many aspects of database theory. A relation schema is defined to be redundant with respect to a given set of FDs and MVDs if there exists a relation over the schema which satisfies all these FDs and MVDs and which has at least two tuples which are identical on a fact. If we formalise this notion of redundancy [6] in the framework of nested attributes, then we obtain the following definition. A nested attribute N is *redundant with respect to a* set Σ of FDs and MVDs on N if and only if there is some $r \subseteq dom(N)$ with $\models_r \Sigma$ and there are some $t_1, t_2 \in r$ with $t_1 \neq t_2$ and $\pi_{X \sqcup Y}^N(t_1) = \pi_{X \sqcup Y}^N(t_2)$ for some $X \rightarrow Y \in \Sigma$ or some $X \twoheadrightarrow Y \in \Sigma$ which is not trivial. Intuitively, this notion of redundancy seems to make perfect sense. Take a look at the FD HALFTONING(INPUT[λ]) $\rightarrow$ HALFTONING(OUTPUT[λ]). This is a non-trivial FD. The elements $(\frac{1}{2}, [0, \frac{1}{2}], [0, 0])$ and $(1, [0, 1], [0, 1])$ coincide on HALFTONING(INPUT[λ],OUTPUT[λ]), i.e., the FD causes some redundancy according to the definition above. This example shows that our current definition of redundancy is not really appropriate anymore. That is, the FD HALFTONING(INPUT[λ]) $\rightarrow$

HALFTONING(OUTPUT[Bit]) is not satisfied by the two elements above and, consequently, redundancy would need to be defined in terms of the non-maximal join-irreducible HALFTONING(OUTPUT[λ]). This, however, appears to be impossible as the information in HALFTONING(OUTPUT[λ]) will always be contained in HALFTONING(OUTPUT[Bit]). The point here is that the information in a non-maximal join-irreducible Y cannot be separated from the information in any maximal join-irreducible Z with $Y \leq Z$. We will see further evidence for this in Section 4. This motivates the following definition.

Definition 2. Let $N \in \mathcal{N}A$ be a nested attribute and Σ a set of FDs and MVDs on N. Let $\Sigma_{\text{inev}} \subseteq \Sigma^+$ denote the union of all $X \to Y \in \Sigma^+$ where $Y \leq X$ or $Y \in \mathcal{B}(N) - \mathcal{B}_{\mathcal{M}}(N)$ holds, and all $X \twoheadrightarrow Y \in \Sigma^+$ where $Y \leq X$ or $Y^C \leq X$ or $Y \in \mathcal{B}(N) - \mathcal{B}_{\mathcal{M}}(N)$ holds. The FDs (MVDs) of the closure Σ_{inev}^+ of Σ_{inev} under inference with respect to the inference rules from Theorem 3 are called *inevitable FDs (MVDs)* on N with respect to Σ. □

The following lemma characterises inevitable dependencies which are derivable from a given set of FDs and MVDs.

Lemma 2. *Let $N \in \mathcal{N}A$, Σ a set of FDs and MVDs on N. If $X \to Y \in \Sigma^+$, then $X \to Y \in \Sigma_{inev}^+$ if and only if $Y^{CC} \leq X$. If $X \twoheadrightarrow Y \in \Sigma^+$, then $X \twoheadrightarrow Y \in \Sigma_{inev}^+$ if and only if $Y^{CC} \leq X$ or $Y^C \leq X$.*

Proof (Sketch). In order to show that $X \to Y \in \Sigma_{\text{inev}}^+$ implies $Y^{CC} \leq X$, and that $X \twoheadrightarrow Y \in \Sigma_{\text{inev}}^+$ implies $Y^{CC} \leq X$ or $Y^C \leq X$, one can proceed by induction on the inference length using the inference rules from Theorem 3. The remaining direction is a matter of applying some of these inference rules as well. □

Thus, trivial FDs and MVDs are always inevitable on N with respect to any Σ, but not vice versa. We are now prepared to define a better notion of redundancy for nested attributes *in terms of FDs and MVDs*.

Definition 3. Let Σ be a set of FDs and MVDs on the nested attribute N. We call N *type-1 redundant with respect to* Σ if and only if there is some $r \subseteq dom(N)$ with $\models_r \Sigma$ and there are some distinct $t_1, t_2 \in r$ with $\pi_{X \sqcup Y}^N(t_1) = \pi_{X \sqcup Y}^N(t_2)$ for some FD $X \to Y \in \Sigma$ that is not inevitable on N with respect to Σ or some MVD $X \twoheadrightarrow Y \in \Sigma$ that is not inevitable on N with respect to Σ. □

Definition 3 allows the set of facts to be the subattributes in all the FDs and MVDs which are not inevitable in a user-supplied set of dependencies Σ. However, one may also recognise the symmetrical nature of MVDs and so allow the subattributes in any MVD that can be derived from any MVD in Σ and successive applications of the Brouwerian-complement rule to also be a fact. Finally, the last possibility is to include inferred dependencies and allow subattributes in any FD or MVD that is not inevitable and implied by Σ to be a fact. Intuitively, one would expect that the notion of redundancy is independent of which of these facts is chosen but in general this is not the case, and the proof is by no means immediate. Let Σ' denote the smallest set with the following properties: $\Sigma \subseteq \Sigma'$ and $X \twoheadrightarrow Y^C \in \Sigma'$ whenever $X \twoheadrightarrow Y \in \Sigma'$.

Definition 4. Let N be a nested attribute and Σ a set of FDs and MVDs on N. We call N *type-2(3) redundant with respect to* Σ if and only if there is some $r \subseteq dom(N)$ with $\models_r \Sigma$ and there are some distinct $t_1, t_2 \in r$ with $\pi^N_{X \sqcup Y}(t_1) = \pi^N_{X \sqcup Y}(t_2)$ for some FD $X \rightarrow Y \in \Sigma'(\Sigma^+)$ which is not inevitable on N with respect to Σ or some MVD $X \twoheadrightarrow Y \in \Sigma'(\Sigma^+)$ which is not inevitable on N with respect to Σ. □

Let $N = L(A, B, C)$ and $\Sigma = \{L(A) \twoheadrightarrow L(B), L(B) \rightarrow L(A, C)\}$. The only minimal key is $L(B)$. From $L(A) \twoheadrightarrow L(B)$ and the Brouwerian-complement rule follows that $L(A) \twoheadrightarrow L(A, C)$ is in Σ' and therefore also in Σ^+. The instance $r = \{(a, b_1, c), (a, b_2, c)\}$ with distinct $b_1, b_2 \in dom(B)$ satisfies Σ and the projections of both elements on $L(A, C)$ are identical. It follows that N is type-2 and type-3 redundant with respect to Σ. However, N is not type-1 redundant as every dependency in Σ contains the subattribute $L(B)$ and no instance over N can have duplicates on a dependency in Σ.

3.2   The Proposal

Fourth normal form (4NF,[14]) has been introduced as an extension of Boyce-Codd normal form (BCNF) and has been intensely studied. A relation schema R is in 4NF with respect to a set Σ of FDs and MVDs defined on R if and only if every $X \twoheadrightarrow Y \in \Sigma^*$ is trivial or X is a superkey for R with respect to Σ.

Definition 5. Let Σ be a set of FDs and MVDs on the nested attribute N. We say that N is in *Nested List Normal Form (NLNF)* with respect to Σ if and only if every $X \twoheadrightarrow Y \in \Sigma^+$ is an inevitable dependency on N with respect to Σ or X is a superkey for N with respect to Σ. □

Note that NLNF generalises 4NF from relational databases. In fact, every inevitable dependency on the record-valued attribute $R(A_1, \ldots, A_n)$ must be trivial since the join-irreducibles of $R(A_1, \ldots, A_n)$ form an anti-chain with respect to $\leq$.

One may define N to be in *Nested List Fourth Normal Form (NL4NF)* with respect to Σ if and only if every $X \twoheadrightarrow Y \in \Sigma^+$ is a trivial dependency on N with respect to Σ or X is a superkey for N with respect to Σ. In this case, NL4NF also extends 4NF from relational databases and implies NLNF as every trivial dependency is also inevitable. However, NLNF is strictly weaker than NL4NF as there are, in general, inevitable dependencies which are not trivial. NLNF for FDs and MVDs subsumes the NLNF for FDs only [21].

HALFTONING(Brightness,INPUT[Level],OUTPUT[Bit]) is not in NLNF with respect to the set Σ of dependencies previously specified. The MVD HALFTONING(Brightness,INPUT[λ]) $\twoheadrightarrow$ HALFTONING(INPUT[Level]) is neither inevitable nor is HALFTONING(Brightness,INPUT[λ]) a superkey with respect to Σ.

3.3   Characterising NLNF

Given some nested attribute N and some set Σ of FDs and MVDs on N, how can we verify that N is in NLNF? By Definition 5 one needs to inspect every

$X \twoheadrightarrow Y$ derivable from Σ, i.e., every $X \twoheadrightarrow Y$ in Σ^+ must be inevitable or X must be a superkey. However, an inspection of every dependency in Σ suffices.

Theorem 4. *Let Σ be a set of FDs and MVDs on the nested attribute N. N is in NLNF with respect to Σ if and only if for every $X \to Y \in \Sigma$ or $X \twoheadrightarrow Y \in \Sigma$ which is not inevitable on N with respect to Σ, the left-hand side X is a superkey for N with respect to Σ.*

Proof (Sketch). One can show the following result using the inference rules from Theorem 3. For any $X \twoheadrightarrow W$ or $X \to W$ in Σ^+ which is not inevitable on N with respect to Σ, there is some $X' \twoheadrightarrow Y$ or $X' \to Y$ in Σ with $X' \leq X \sqcup \bigsqcup \{Z \sqcap Z^C \mid X \twoheadrightarrow Z \in \Sigma^+\}$ which is not inevitable on N with respect to Σ.

Suppose N is not in NLNF with respect to Σ. Then there is some $X \twoheadrightarrow Y \in \Sigma^+$ which is not inevitable and where X is not a superkey for N with respect to Σ. The result above shows that there is some $X' \to Y'$ or $X' \twoheadrightarrow Y'$ in Σ which is not inevitable and where $X' \leq X \sqcup \bigsqcup \{Z \sqcap Z^C \mid X \twoheadrightarrow Z \in \Sigma^+\}$ holds. The mixed meet rule guarantees that $X' \leq X^+$ and therefore $X \to X' \in \Sigma^+$. Since X is not a superkey for N with respect Σ, neither can X' be.

The remaining direction is a consequence of $\Sigma \subseteq \Sigma^+$ and the implication rule.   □

We will now give another characterisation of NLNF. The result extends a classical result for relational databases [15]. In order to verify whether an instance over N in NLNF satisfies all dependencies it is sufficient to verify that all key dependencies and all inevitable dependencies are satisfied. Unlike the relational case where it is enough to look at all key dependencies for a relation schema in 4NF, one still needs to deal with all inevitable dependencies that are not trivial when a nested attribute in NLNF is given.

Theorem 5. *Let N be a nested attribute and Σ a set of FDs and MVDs on N. N is in NLNF with respect to Σ if and only if every $r \subseteq dom(N)$ with $\models_r \Sigma_{key} \cup \Sigma_{inev}^+$ implies $\models_r \Sigma$.*

Proof (Sketch). It is not difficult to see that the existence of some $r \subseteq dom(N)$ with $\models_r \Sigma_{\text{key}} \cup \Sigma_{\text{inev}}^+$ and $\not\models_r \Sigma$ implies the violation of the NLNF condition.

Suppose N is not in NLNF with respect to Σ. Let $X \twoheadrightarrow Y \in \Sigma^+$ be not inevitable and X not be a superkey for N with respect to Σ. One can show that there is some $r \subseteq dom(N)$ with $\models_r \Sigma_{\text{key}} \cup \Sigma_{\text{inev}}^+$ and $\not\models_r \Sigma$. In fact, one defines $X_{\text{inev}}^+ = \bigsqcup \{Z \mid X \to Z \in \Sigma_{\text{inev}}^+\}$ and chooses $r \subseteq dom(N)$ with $r = \{t, t'\}$ such that

$$\pi_W^N(t) = \pi_W^N(t') \qquad \text{if and only if} \qquad W \leq X_{\text{inev}}^+.$$

Note that such t, t' can always be constructed [22, Lemma 3.2].   □

3.4   Type-2 and Type-3 Redundancy

We have seen that HALFTONING(Brightness,INPUT[Level],OUTPUT[Bit]) is not in NLNF. So far, this means only that the schema does not satisfy a syntactic condition with respect to the given set of dependencies. In this section we show the equivalence of NLNF to several semantic design desiderata. This will reveal what NLNF actually achieves.

The first semantic justification of Nested List Normal Form is that a nested attribute N is in NLNF with respect to a given set Σ of FDs and MVDs precisely if N is not type-3 redundant with respect to Σ.

Theorem 6. *Let Σ be a set of FDs and MVDs on the nested attribute N. Then is N in NLNF with respect to Σ if and only if N is not type-3 redundant with respect to Σ.*

Proof (Sketch). It is not difficult too see that type-3 redundancy implies the violation of the NLNF condition. Suppose N is not type-3 redundant with respect to Σ, and $X \twoheadrightarrow Y \in \Sigma^+$ is not inevitable with respect to Σ. We sketch that X is a superkey for N with respect to Σ. As N is not type-3 redundant we have $t_1 = t_2$ for all $t_1, t_2 \in r \subseteq dom(N)$ with $\models_r \Sigma$ and $\pi^N_{X \sqcup Y}(t_1) = \pi^N_{X \sqcup Y}(t_2)$. It follows that $X \sqcup Y$ is a superkey for N with respect to Σ. One can show that $X \to Y^C \in \Sigma^+$, and $X \to Y^C$ is not inevitable with respect to Σ. It follows that $X \sqcup Y^C$ is a superkey for N with respect to Σ. Otherwise it is possible to construct some $r \subseteq dom(N), \mid r \mid \geq 2$ with $\models_r \Sigma$ and for all distinct $t_1, t_2 \in r$ we have $\pi^N_{X \sqcup Y^C}(t_1) = \pi^N_{X \sqcup Y^C}(t_2)$ contradicting the fact that N is not type-3 redundant with respect to Σ. Consequently, X is a superkey for N with respect to Σ. $\qquad\square$

The two notions of type-2 and type-3 redundancy coincide.

Theorem 7. *Let Σ be a set of FDs and MVDs on the nested attribute N. Then is N type-2 redundant with respect to Σ if and only if N is type-3 redundant with respect to Σ.*

Proof (Sketch). Type-2 redundancy implies type-3 redundancy as $\Sigma' \subseteq \Sigma^+$. We sketch that if N is not type-2 redundant, then it is also not type-3 redundant. If $X \twoheadrightarrow Y$ or $X \to Y \in \Sigma'$ is not inevitable, then one can show that $X \sqcup Y$ is a superkey for N with respect to Σ. Consider every dependency in Σ' that is not inevitable with respect to Σ:

- for an FD $X \to Y$ we infer that X must be a superkey for N with respect to Σ,
- for an MVD $X \twoheadrightarrow Y$ we also have $X \twoheadrightarrow Y^C \in \Sigma'$. Consequently, both $X \sqcup Y$ and $X \sqcup Y^C$ are superkeys for N with respect to Σ, and one can show that both $X \to Y^C, X \to Y^{CC} \in \Sigma^+$. That gives $X \to N \in \Sigma^+$, i.e., X is a superkey for N with respect to Σ.

Since $\Sigma \subseteq \Sigma'$ holds, the left-hand side of every dependency in Σ which is not inevitable with respect to Σ is a superkey for N. Theorem 4 shows that N is in NLNF with respect to Σ and we conclude that N is not type-3 redundant with respect to Σ by Theorem 6. $\qquad\square$

It follows that the notion of type-2 redundancy is invariant under different choices of equivalent sets of FDs and MVDs.

Corollary 1. *Let $N \in \mathcal{N}A$, and Σ and Θ two equivalent sets of FDs and MVDs on N. Then N is type-2 redundant with respect to Σ if and only if N is type-2 redundant with respect to Θ.* $\square$

3.5 Type-1 Redundancy

Unlike type-2 and type-3 redundancy, type-1 redundancy does depend on the choice of equivalent sets of dependencies. We will now characterise type-1 redundancy syntactically.

Theorem 8. *Let Σ be a set of FDs and MVDs on the nested attribute N. Then the following conditions are equivalent:*

1. *N is not type-1 redundant with respect to Σ,*
2. *for every $X \twoheadrightarrow Y$ and $X \to Y$ in Σ which is not inevitable on N with respect to Σ the subattribute $X \sqcup Y$ is a superkey for N with respect to Σ, and*
3. *for every $X \twoheadrightarrow Y$ and $X \to Y$ in Σ which is not inevitable on N with respect to Σ we have $X \to Y^C \in \Sigma^+$.* $\square$

Let $N = L(A, B, C)$, $\Sigma = \{L(A) \twoheadrightarrow L(B), L(B) \to L(A, C)\}$, and $\Theta = \{L(A) \twoheadrightarrow L(C), L(B) \to L(A, C)\}$. $L(B)$ is the only minimal key with respect to Σ and Θ, and Σ and Θ are equivalent sets of FDs and MVDs. N is not type-1 redundant with respect to Σ since every $X \twoheadrightarrow Y$ and every $X \to Y$ in Σ satisfies $L(B) \leq X \sqcup Y$. However, N is type-1 redundant with respect to Θ since $A \sqcup C$ is not a superkey on N with respect to Σ.

3.6 Pure MVDs

We will now provide a sufficient condition under which the different types of redundancies are equivalent. Let Σ be a set of FDs and MVDs defined on the nested attribute N. An MVD $X \twoheadrightarrow Y \in \Sigma$ is called *pure* if and only if neither $X \to Y$ nor $X \to Y^C$ are in Σ^+. Pure MVDs are not inevitable on N with respect to Σ. The set Σ is called pure if and only if every MVD in Σ is pure.

Pure MVDs reflect pure multivalued information and can therefore not be captured by FDs. A set Σ of MVDs and FDs contains at least one pure MVD if and only if Σ is not equivalent to a set of FDs.

Theorem 9. *If $N \in \mathcal{N}A$ and Σ is a pure set of FDs and MVDs on N, then N is type-1 redundant if and only if N is type-3 redundant.* $\square$

Corollary 2. *Let $N \in \mathcal{N}A$, and Σ be a pure set of FDs and MVDs on N. If N is type-1 redundant with respect to Σ, then is N type-1 redundant with respect to any set Θ of FDs and MVDs on N that is equivalent to Σ.* $\square$

3.7   Value Redundancy

The previously introduced notions of redundancy have one major deficiency in common. They all depend on the syntactic structure of FDs and MVDs making it difficult to further generalise those notions to other types of dependencies or adapting those definitions to other data models. In relational databases the notion of value redundancy has been introduced to overcome those deficiencies [35]. The occurrence of some flat attribute value in some flat relation is redundant if it can be derived from other data values in that relation and the set of dependencies which apply to that relation. A relation schema R is redundant if there exists a legal R-relation which contains an occurrence of a flat attribute value such that *any* change to this occurrence results in the violation of at least one dependency.

Care must be taken when this notion of value-redundancy is generalised to the presence of lists. Consider for instance the list-valued nested attribute OUTPUT[Bit] together with the element $[0, 1, 1, 1]$ which has projection $[ok, ok, ok, ok]$ on the subattribute OUTPUT[λ]. Then, the value occurrence $[ok, ok, ok, ok]$ cannot be changed without affecting the list $[0, 1, 1, 1]$. On the other hand, arbitrary changes to $[0, 1, 1, 1]$ may also affect the projection $[ok, ok, ok, ok]$. An alteration of the number of the elements in $[0, 1, 1, 1]$, e.g. removing the last two elements results in $[0, 1]$, and also results in a different projection, $[ok, ok]$ in this case.

In general, changes to the value $\pi_Y^N(t)$ on a join-irreducible Y will cause a change of values $\pi_Z^N(t)$ on any join-irreducible Z with $Y \leq Z$. It is therefore advisable to consider only value occurrences $\pi_M^N(t)$ on maximal join-irreducibles $M \in \mathcal{B}_\mathcal{M}(N)$. Thus, the role that flat attributes played in the definition of value redundancy in relational databases, is now taken by maximal join-irreducibles in the context of lists. Moreover, we will not consider arbitrary changes to $\pi_M^N(t)$, but only those which do not cause any changes to $\pi_W^N(t)$ for all $W < M$. We call the replacement of some data value $\pi_M^N(t)$ by $m \in dom(M)$ *admissible* if and only if $\pi_W^N(t) = \pi_W^M(m)$ for all $W < M$.

Definition 6. Let $N \in \mathcal{N}A$, $M \in \mathcal{B}_\mathcal{M}(N)$, Σ a set of dependencies on N, $r \subseteq dom(N)$ and $t \in r$. The data value occurrence $\pi_M^N(t)$ is *redundant* if and only if for every admissible replacement of $\pi_M^N(t)$ by a value m with $\pi_M^N(t) \neq m$ that results in the modified instance $r' \subseteq dom(N)$ we have $\not\models_{r'} \Sigma$.   □

The last definition extends the notion of value-redundancy from the relational case [35]. In the presence of records only, the join-irreducibles form an anti-chain with respect to Σ and, consequently, every join-irreducible is maximal and all replacements are admissible. Consider the following legal snapshot of 6 tuples

$$(\tfrac{1}{2}, [0, \tfrac{1}{2}], [0, 0]), \ (\tfrac{1}{2}, [0, \tfrac{1}{2}], [0, 1]), \ (\tfrac{1}{2}, [0, \tfrac{1}{2}], [1, 0])$$
$$(\tfrac{1}{2}, [\tfrac{1}{2}, 0], [0, 0]), \ (\tfrac{1}{2}, [\tfrac{1}{2}, 0], [0, 1]), \ (\tfrac{1}{2}, [\tfrac{1}{2}, 0], [1, 0])$$

over HALFTONING(Brightness,INPUT[Level],OUTPUT[Bit]). The admissible replacements of $[0, 0]$ are $[0, 1], [1, 0], [1, 1]$. The data value occurrence of any of these 6 tuples projected on HALFTONING(INPUT[Level]) is redundant as every admissible replacement of such an occurrence leads to a violation of the MVD

HALFTONING(Brightness,INPUT[λ]) $\twoheadrightarrow$ HALFTONING(INPUT[Level]).

The same applies to the data value occurrences of any tuple projected on HALFTONING(OUTPUT[Bit]).

The nested attribute N is in Value Redundancy Free Normal Form (VRFNF) with respect to a set Σ of dependencies defined on N if and only if there does not exist $r \subseteq dom(N)$ with $\models_r \Sigma$ which contains a data value occurrence that is redundant.

Theorem 10. *Let Σ be a set of FDs and MVDs on the nested attribute N. Then N is in VRFNF with respect to Σ if and only if N is in NLNF with respect to Σ.*

Proof (Sketch). We sketch first that if N is not in VRFNF with respect to Σ, then N is also not in NLNF with respect to Σ. If N is not in VRFNF with respect to Σ, then there is some $r \subseteq dom(N)$, some $t \in r$ and some $M \in \mathcal{B}_\mathcal{M}(N)$ such that every admissible replacement of $\pi_M^N(t)$ results in a modified instance violating Σ. That is, if $\pi_M^N(t)$ is changed to a value m' such that $m' \notin \pi_M(r) = \{\pi_M^N(s) : s \in r\}$, resulting in the new element t' and the modified instance $r' = (r - \{t\}) \cup \{t'\}$, then $\not\models_{r'} \Sigma$. One can now show that a violation of some FD or MVD in Σ by r' implies a violation of the NLNF condition. We omit the details.

If N is not in NLNF with respect to Σ, then there is some $X \twoheadrightarrow Y \in \Sigma^+$ which is not inevitable on N with respect to Σ and where X is not a superkey for N with respect to Σ. In the following let $DepB(X)$ denote the dependency basis of X with respect to Σ [22]. Let $DepB(X) = \{W_{0,1}, \ldots, W_{0,m}, W_1, \ldots, W_k\}$ with $W_{0,i} \leq X^+$ and $W_j \not\leq X^+$ for $i = 1, \ldots, m$ and $j = 1, \ldots, k$. Since X is not a superkey for N with respect to Σ, there is some $W_j \in DepB(X)$ with $W_j \not\leq X^+ < N$. One can show that there is some $r = \{t, t'\} \subseteq dom(N)$ with $\models_r \Sigma$ and $\pi_W^N(t) = \pi_W^N(t')$ iff $W \leq X^+ \sqcup \bigsqcup \{W_n : n \neq j\}$. Suppose there is some $M \in \mathcal{B}_\mathcal{M}(N)$ with $M \not\leq X$, but $M \leq X^+$. Then $\pi_M^N(t)$ and $\pi_M^N(t')$ are both redundant: changing one of these values results in a violation of $X \twoheadrightarrow X^+ \in \Sigma^+$ which is not inevitable on N with respect to Σ.

Alternatively, $(X^+)^{CC} \leq X$. Assume there is only one $W_j \in DepB(X)$ with $W_j \not\leq X^+$. $X \twoheadrightarrow Y \in \Sigma^+$ implies that Y is the join over some elements of $DepB(X)$ and as $X \twoheadrightarrow Y \in \Sigma^+$ is not inevitable on N with respect to Σ we have $Y^{CC} \not\leq X$. This leaves us with $W_j \leq Y$ and thus $N = X \sqcup W_j \leq X \sqcup Y$, i.e., $N = X \sqcup Y$. This, however, is a contradiction since $X \twoheadrightarrow Y \in \Sigma^+$ is not inevitable on N with respect to Σ. We therefore have two distinct $W_i, W_j \in DepB(X)$. Consequently, for all $M \in \mathcal{B}_\mathcal{M}(N)$ with $M \leq W_i$ (at least one such M exists) the values $\pi_M^N(t)$ and $\pi_M^N(t')$ are redundant since changing any one of these values results in the violation of $X \twoheadrightarrow W_i \in \Sigma^+$. Therefore, N is not in VRFNF with respect to Σ. $\square$

Since NLNF for FDs and MVDs reduces to NLNF for FDs if only FDs are present, Theorem 10 shows that NLNF for FDs is the exact condition required to avoid value redundancy when only FDs are present. This extends previous results [21] where value redundancy had not been considered at all.

It also follows that value redundancy is equivalent to type-2 and type-3 redundancy, and equivalent to type-1, type-2 and type-3 redundancy whenever Σ is a pure set of FDs and MVDs.

3.8   Strong Update Anomalies

In the relational model of data a relation schema in 4NF does not have any update anomalies. This is another justification why relation schemata should be in 4NF [35]. The next example reveals a surprising fact. Consider

$$\text{HALFTONING}(\text{Brightness},\text{INPUT}[\lambda],\text{OUTPUT}[\text{Bit}])$$

which is in NLNF with respect to the FD

$$\text{HALFTONING}(\text{INPUT}[\lambda]) \rightarrow \text{HALFTONING}(\text{OUTPUT}[\lambda])$$

and therefore free of any form of redundancies. Say a simple database consists of the single element $(\frac{1}{2}, [ok, ok], [0, 1])$ and the element $(\frac{1}{2}, [ok, ok], [0, 0, 0, 0])$ happens to be inserted. Then all key dependencies are trivially satisfied by the new relation, but the FD

$$\text{HALFTONING}(\text{INPUT}[\lambda]) \rightarrow \text{HALFTONING}(\text{OUTPUT}[\lambda])$$

is violated. This example shows that, in general, the absence of redundancy for a nested attribute does not imply the absence of insertion anomalies. Therefore, it cannot be expected that nested attributes in NLNF do not have update anomalies. We define, however, strong update anomalies in the context of nested attributes. The main difference to the relational case is that updated relations which define any strong anomaly do not only satisfy all key dependencies on the nested attribute, but also all inevitable dependencies.

Definition 7. Let Σ be a set of FDs and MVDs on the nested attribute N.

1. We say that N has a *strong insertion anomaly* with respect to Σ if and only if there is some $r \subseteq dom(N)$ with $\models_r \Sigma$ and some $t \notin r$ with $\models_{r \cup \{t\}} \Sigma_{\text{key}} \cup \Sigma_{\text{inev}}^+$, but $\not\models_{r \cup \{t\}} \Sigma$.
2. We say that N has a *strong deletion anomaly* with respect to Σ if and only if there is some $r \subseteq dom(N)$ with $\models_r \Sigma$ and some $t \in r$ with $\models_{r - \{t\}} \Sigma_{\text{key}} \cup \Sigma_{\text{inev}}^+$, but $\not\models_{r - \{t\}} \Sigma$.
3. We say that N has a *strong replacement anomaly*
   - *of type 1* with respect to Σ if and only if there is some $r \subseteq dom(N)$ with $\models_r \Sigma$ and some $t \in r$ and $t' \in dom(N)$ with $\pi_K^N(t) = \pi_K^N(t')$ for some minimal key K on N and $\models_{r - \{t\} \cup \{t'\}} \Sigma_{\text{key}} \cup \Sigma_{\text{inev}}^+$ and $\not\models_{r - \{t\} \cup \{t'\}} \Sigma$ hold.
   - *of type 2* with respect to Σ if and only if there is some $r \subseteq dom(N)$ with $\models_r \Sigma$ and some $t \in r$ and $t' \in dom(N)$ with $\pi_K^N(t) = \pi_K^N(t')$ for some distinguished minimal key K on N and $\models_{r - \{t\} \cup \{t'\}} \Sigma_{\text{key}} \cup \Sigma_{\text{inev}}^+$ and $\not\models_{r - \{t\} \cup \{t'\}} \Sigma$ hold.

  – *of type 3* with respect to Σ if and only if there is some $r \subseteq dom(N)$ with $\models_r \Sigma$ and some $t \in r$ and $t' \in dom(N)$ with $\pi_K^N(t) = \pi_K^N(t')$ for all minimal keys K on N and $\models_{r-\{t\}\cup\{t'\}} \Sigma_{\text{key}} \cup \Sigma_{\text{inev}}^+$ and $\not\models_{r-\{t\}\cup\{t'\}} \Sigma$ hold. $\square$

If updates only alter values on maximal join-irreducibles and keep values on non-maximal join-irreducibles fixed, then only key dependencies need to be checked. Otherwise, one also needs to check all inevitable dependencies that are not trivial. The next theorem generalises a well-known result from relational databases [15].

Theorem 11. *Let Σ be a set of FDs and MVDs on the nested attribute N. Then is N in NLNF with respect to Σ if and only if N does not have any strong insertion anomaly with respect to Σ.* $\square$

While it is relatively easy to see that a nested attribute in NLNF does not have any strong update anomalies, the absence of a strong update anomaly does not necessarily imply NLNF. For strong deletion anomalies we obtain the following result.

Theorem 12. *Let Σ be a set of FDs and MVDs on the nested attribute N. N does not have any strong deletion anomaly with respect to Σ if and only if Σ is equivalent to a set of FDs.*

Proof (Sketch). If Σ is equivalent to a set $\Sigma_{\mathcal{F}}$ of FDs, then no deletion anomaly with respect to $\Sigma_{\mathcal{F}}$ can occur. Since strong deletion anomalies are invariant under different choices of covers, it follows that no deletion anomaly with respect to Σ can occur. Hence, N does not have any strong deletion anomaly with respect to Σ.

  If N is not equivalent to a set of FDs one can show that there is some pure MVD $X \twoheadrightarrow Y$ in Σ. Then there is some $r \subseteq dom(N)$ with $\models_r \Sigma$ and four distinct elements $t_1, t_2, t_3, t_4 \in r$ with $\pi_X^N(t_1) = \pi_X^N(t_2) = \pi_X^N(t_3) = \pi_X^N(t_4)$, $\pi_{W_1}^N(t_1) = \pi_{W_1}^N(t_3)$, $\pi_{W_1}^N(t_2) = \pi_{W_1}^N(t_4)$, $\pi_{W_2}^N(t_1) = \pi_{W_2}^N(t_4)$ and $\pi_{W_2}^N(t_2) = \pi_{W_2}^N(t_3)$ such that $W_1, W_2 \in DepB(X)$ are distinct and $W_i \not\leq X^+$ for $i = 1, 2$. However, r has a deletion anomaly since deleting any of the four elements from r results in a violation of $X \twoheadrightarrow W_i \in \Sigma^+$ which is not inevitable on N with respect to Σ and where X is not a superkey for N with respect to Σ. Consequently, there is some $X' \twoheadrightarrow Y'$ in Σ which is not inevitable and where $X' \leq X \sqcup \bigsqcup \{Z \sqcap Z^C \mid X \twoheadrightarrow Z \in \Sigma^+\}$ holds. The mixed meet rule guarantees that $X' \leq X^+$ and therefore $X \to X' \in \Sigma^+$. Since X is not a superkey for N with respect to Σ, neither can X' be. $\square$

It is the subject of future research to study the relationship between the various forms of strong replacement anomalies and NLNF. We conjecture that the results for strong replacement anomalies will be similar to those established for 4NF and key-based (fact-based) replacement anomalies in case of relational databases [35].

4 NLNF Decomposition

So far we have proposed the Nested List Normal Form as a desirable normal form that we aim to achieve in a database. We now tackle the problem of how to obtain

NLNF. Theorem 2 indicates that an extension of the relational decomposition approach [12, 17] can be applied to NLNF. Given some nested attribute N and a set Σ of FDs and MVDs defined on N, the decomposition approach aims at finding a set of subattributes of N each of which is in NLNF with respect to the corresponding set of all implied FDs and MVDs on that subattribute. Moreover, any instance of N that satisfies Σ is the generalised natural join of all its projections on the subattributes, i.e., every valid database on N can be decomposed without loss of information.

Definition 8. Let $N \in \mathcal{N}A$, $N_1, \ldots, N_k \in Sub(N)$, and Σ a set of FDs and MVDs defined on N. The set $\{N_1, \ldots, N_k\}$ is called a *lossless join decomposition of N with respect to Σ* if and only if $N = \bigsqcup\{N_1, \ldots, N_k\}$ and $r = \pi_{N_1}(r) \bowtie \cdots \bowtie \pi_{N_k}(r)$ holds for all $r \subseteq dom(N)$ with $\models_r \Sigma$. The set $\{N_1, \ldots, N_k\}$ is a *lossless NLNF (NL4NF) decomposition of N with respect to Σ* if and only if $\{N_1, \ldots, N_k\}$ is a lossless join decomposition of N with respect to Σ and N_i is in NLNF (NL4NF) with respect to $\pi_{N_i}(\Sigma^+)$ for every $i = 1, \ldots, k$, and where $\pi_M(\Sigma) = \{X \to Y \in \Sigma \mid X \sqcup Y \leq M\} \cup \{X \twoheadrightarrow Y \sqcap M \in \Sigma \mid X \leq M\}$. $\square$

We will now show that it is possible to obtain a lossless NLNF decomposition for any given nested attribute N and any given set of FDs and MVDs on N. Whenever an MVD in the current state of the output schema violates NLNF, the decomposition algorithm removes the cause for this violation of NLNF by replacing the offending parent subattribute by two of its proper child subattributes which can be joined losslessly to reconstruct their parent.

Algorithm 1 (Lossless NLNF decomposition)

Input: $N \in \mathcal{N}A$, set Σ of FDs and MVDs on N

Output: set $\mathcal{S} = \{(N_1, \Sigma_1), \ldots, (N_k, \Sigma_k)\}$ where Σ_i is set of FDs and MVDs on $N_i \in Sub(N)$ and $\{N_1, \ldots, N_k\}$ is lossless NLNF decomposition of N with respect to Σ

Method:

VAR $X, Y, N_1, N_2 \in Sub(N)$

DECOMPOSE(N, Σ)

(1) **BEGIN**
(2)   **IF** N in NLNF wrt Σ, **THEN** $\mathcal{S} := \{(N, \Sigma)\}$;
(3)   **ELSE**
(4)   **LET** $X \twoheadrightarrow Y \in \Sigma$ be not inevitable on N wrt Σ and $\Sigma \not\models X \to N$;
(5) $N_1 := X \sqcup Y$;
(6) $N_2 := X \sqcup Y^C$;
(7) $\mathcal{S} := $**DECOMPOSE**$(N_1, \pi_{N_1}(\Sigma^+)) \cup $**DECOMPOSE**$(N_2, \pi_{N_2}(\Sigma^+))$;
(8)   **ENDIF**;
(9)   **RETURN**$(\mathcal{S})$;
(10) **END**; $\square$

Theorem 13. *Algorithm 1 is correct.*   □

For relational databases it is well-known that any relation schema with any set of FDs and MVDs defined on it, can be decomposed into subschemata that are all in 4NF with respect to the projected sets of FDs and MVDs. In the presence of lists, however, the situation is different. The next result is further evidence that a simple extension of 4NF to NL4NF is too strong.

Theorem 14. *There are nested attributes N and sets Σ of FDs and MVDs on N for which no lossless NL4NF-decomposition exists.*

Proof. Let $N = L[A]$ and $\Sigma = \{\lambda \twoheadrightarrow L[\lambda]\}$. The MVD is not trivial and λ is not a superkey for N with respect to Σ. Consequently, N is not in NL4NF with respect to Σ. However, any decomposition of $L[A]$ must contain the nested attribute $L[A]$ itself. Therefore, no lossless NL4NF decomposition of $L[A]$ with respect to Σ exists.   □

Algorithm 1 generalises the well-known 4NF decomposition algorithm for relational databases, see for instance [27–p.270]. It follows that the NLNF decomposition algorithm causes at least as many computational problems as its relational counterpart (,e.g. running time and dependency-preservation). However, the problems do not become harder in the presence of lists. Due to lack of space we cannot go into further details. The 4NF-decomposition from [17] may indicate how to improve the NLNF decomposition according to its running time.

We continue the example from digital halftoning. The MVD

$$\textsc{Halftoning}(\text{Brightness},\textsc{Input}[\lambda]) \twoheadrightarrow \textsc{Halftoning}(\textsc{Input}[\text{Level}])$$

is neither inevitable nor is $\textsc{Halftoning}(\text{Brightness},\textsc{Input}[\lambda])$ a superkey. A first decomposition yields

$$N_1=\textsc{Halftoning}(\text{Brightness},\textsc{Input}[\text{Level}],\lambda) \text{ and}$$
$$N_2'=\textsc{Halftoning}(\text{Brightness},\textsc{Input}[\lambda],\textsc{Output}[\text{Bit}]).$$

The nested attribute N_1 is in NLNF with respect to $\pi_{N_1}(\Sigma)$. The attribute N_2' carries the inevitable FD $\textsc{Halftoning}(\textsc{Input}[\lambda]) \rightarrow \textsc{Halftoning}(\textsc{Output}[\lambda])$, and the FD $\textsc{Halftoning}(\textsc{Output}[\lambda]) \rightarrow \textsc{Halftoning}(\textsc{Input}[\lambda])$ which is not inevitable on N_2'. A further decomposition of N_2' gives

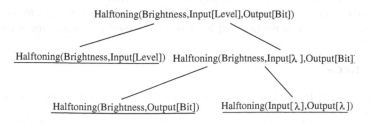

Fig. 2. Decomposition Tree for the Halftoning Example

$$N_2 = \text{HALFTONING}(\text{Brightness},\lambda,\text{OUTPUT}[\text{Bit}]) \text{ and}$$
$$N_3 = \text{HALFTONING}(\lambda,\text{INPUT}[\lambda],\text{OUTPUT}[\lambda])$$

which are in NLNF with respect to $\pi_{N_2}(\Sigma)$ and $\pi_{N_3}(\Sigma)$, respectively.

Since HALFTONING(Brightness,INPUT[Level],OUTPUT[Bit]) is not in NLNF with respect to Σ, and Σ is pure, the schema is redundant in every sense of notions that have been provided here. It also means that the schema carries strong insertion and deletion anomalies. The decomposition approach suggests to store the information in three separate schemata N_1, N_2 and N_3 which are all in NLNF with respect to the projected sets of FDs and MVDs.

5   Related and Future Work

We have proposed a suitable extension of 4NF from relational databases to the presence of lists. The results demonstrate that 4NF together with all its beautiful semantic properties can be generalised to arbitrary finite nesting of records and lists (slightly extended framework of DTDs in which only concatenation and Kleene closure are allowed). In particular, NLNF was semantically justified in several ways by showing the equivalence to the absence of appropriate extensions of different notions of redundancy and insertion anomalies. The data model has put emphasis on the record and list constructor, but will be extended to other data constructors in the future. The feature of lists to model order makes it possible to focus on join-irreducible subattributes and makes it possible to extend well-known results allowing to capture new application domains. Since set and multiset constructor neglect the order of their elements an extension of these results will be challenging.

The minimal axiomatisation from Theorem 3 follows directly from previous work [20, 22]. The complete set of inference rules is used in several proof arguments. NLNF for FDs themselves has been studied previously [21]. While [3] define information-theoretic measures to address the problem of well-designed data in data formats different from the relational data model (such as XML) we have offered an algebraic approach to database design which classifies data models according to the data type constructors supported.

Unlike our work, none of the XML FDs and MVDs [2, 37] take list equality into consideration. We believe that list equality is natural and common in real applications and should be included in defining data dependencies. An alternative way to define object equality in the context of XML can be based on homomorphisms in XML graphs [19, 23]. Notably both [18, 31] have considered set equality in their definitions of FDs in the nested relational data model.

References

1. S. Abiteboul, R. Hull, and V. Vianu. *Foundations of Databases*. Addison-Wesley, 1995.
2. M. Arenas and L. Libkin. A normal form for XML documents. *TODS*, 29(1):195–232, 2004.

3. M. Arenas and L. Libkin. An information-theoretic approach to normal forms for relational and XML data. *J.ACM*, 52(2):246–283, 2005.
4. T. Asano, N. Katoh, K. Obokata, and T. Tokuyama. Matrix rounding under the L_p-discrepancy measure and its application to digital halftoning. *SIAM Journal on Computing*, 32(6):1423–1435, 2003.
5. M. Atkinson, F. Bancilhon, D. DeWitt, K. Dittrich, D. Maier, and S. Zdonik. The object-oriented database system manifesto. In *Proceedings of the International Conference on Deductive and Object-Oriented Databases*, pages 40–57, 1989.
6. C. Beeri, P. A. Bernstein, and N. Goodman. A sophisticate's introduction to database normalization theory. In *VLDB*, pages 113–124, 1978.
7. P. A. Bernstein and N. Goodman. What does Boyce-Codd normal form do? In *VLDB*, pages 245–259, 1980.
8. J. Biskup. Database schema design theory: achievements and challenges. In *Information Systems and Data Management*, number 1066 in LNCS, pages 14–44. Springer, 1995.
9. J. Biskup. Achievements of relational database schema design theory revisited. In *Semantics in databases*, number 1358 in LNCS, pages 29–54. Springer, 1998.
10. T. Bray, J. Paoli, C. M. Sperberg-McQueen, E. Maler, and F. Yergeau. Extensible markup language (XML) 1.0 (third edition) W3C recommendation 04 February 2004. http://www.w3.org/TR/2004/REC-xml-20040204/, 2004.
11. P. P. Chen. The entity-relationship model: Towards a unified view of data. *TODS*, 1:9–36, 1976.
12. E. F. Codd. Further normalization of the database relational model. In *Courant Computer Science Symposia 6: Data Base Systems*, pages 33–64. Prentice-Hall, 1972.
13. E. F. Codd. Recent investigations in relational database system. In *Proceedings of the IFIP Conference*, pages 1017–1021, 1974.
14. R. Fagin. Multivalued dependencies and a new normal form for relational databases. *TODS*, 2(3):262–278, 1977.
15. R. Fagin. A normal form for relational databases that is based on domains and keys. *TODS*, 6(3):387–415, 1981.
16. R. W. Floyd and L. Steinberg. An adaptive algorithm for spatial grey scale. In *Proceedings of the Society of Information Display*, pages 75–77, 1976.
17. G. Grahne and K. P. Räihä. Database decomposition into 4NF. In *VLDB*, pages 186–196, 1983.
18. C. S. Hara and S. B. Davidson. Reasoning about nested functional dependencies. In *PODS*, pages 91–100, 1999.
19. S. Hartmann and S. Link. More functional dependencies for XML. In *ADBIS*, number 2798 in LNCS, pages 355–369. Springer, 2003.
20. S. Hartmann and S. Link. Multi-valued dependencies in the presence of lists. In *PODS*, pages 330–341, 2004.
21. S. Hartmann and S. Link. Normalisation in the presence of lists. In *ADC*, volume 27 of *CRPIT*, pages 53–64, 2004.
22. S. Hartmann, S. Link, and K.-D. Schewe. Functional and multivalued dependencies in nested databases generated by record and list constructor. accepted for Annals of Mathematics and Artificial Intelligence, 2006.
23. S. Hartmann and T. Trinh. Axiomatising functional dependencies for XML with frequencies. In *FoIKS*, this volume of LNCS. Springer, 2006.
24. R. Hull and R. King. Semantic database modeling: Survey, applications and research issues. *ACM Computing Surveys*, 19(3), 1987.

25. J. F. Jarvis, C. N. Judice, and W. H. Ninke. A survey of techniques for the display of continuous-tone pictures on bilevel display. *Computer Graphics Image Processing*, 5:13–40, 1976.

26. I. Katsavounidis and C.-C. J. Kuo. A multiscale error diffusion technique for digital halftoning. *Transactions on Image Processing*, 6(3):483–490, 1997.

27. M. Levene and G. Loizou. *A Guided Tour of relational databases and beyond.* Springer, 1999.

28. J. Li, S. Ng, and L. Wong. Bioinformatics adventures in database research. In *ICDT*, number 2572 in LNCS, pages 31–46. Springer, 2002.

29. J. C. C. McKinsey and A. Tarski. On closed elements in closure algebras. *Annals of Mathematics*, 47:122–146, 1946.

30. W. Y. Mok. A comparative study of various nested normal forms. *IEEE Transactions on Knowledge & Data Engineering*, 14(2):369–385, 2002.

31. M. A. Roth, H. F. Korth, and A. Silberschatz. Extended algebra and calculus for nested relational databases. *TODS*, 13(4):389–417, 1988.

32. D. Suciu. On database theory and XML. *SIGMOD Record*, 30(3):39–45, 2001.

33. B. Thalheim. *Entity-Relationship Modeling: Foundations of Database Technology.* Springer, 2000.

34. V. Vianu. A web odyssey: from Codd to XML. In *PODS*, pages 1–15, 2001.

35. M. Vincent. Semantic foundation of 4NF in relational database design. *Acta Informatica*, 36:1–41, 1999.

36. M. Vincent and J. Liu. Functional dependencies for XML. In *APWEB*, number 2642 in LNCS, pages 22–34. Springer, 2003.

37. M. Vincent, J. Liu, and C. Liu. A redundancy free 4NF for XML. In *XML Database Symposium*, number 2824 in LNCS, pages 254–266. Springer, 2003.

Axiomatising Functional Dependencies for XML with Frequencies

Sven Hartmann and Thu Trinh[*]

Information Science Research Centre,
Massey University, Palmerston North, New Zealand
{s.hartmann, t.trinh}@massey.ac.nz

Abstract. We provide a finite axiomatisation for a class of functional dependencies for XML data that are defined in the context of a simple XML tree model reflecting the permitted parent-child relationships together with their frequencies.

1 Introduction

The question of how to represent and efficiently manage complex application data is one of the major challenges database research faces today. XML (the Extensible Markup Language) has gained popularity as a standard for exchanging data on the web. The flexibility of XML and its wide acceptance as a standard make it also a good choice for modelling heterogenous and highly structured data from various application domains. As a consequence, XML databases (in the form of data-centric XML documents) have attracted a great deal of interest.

As for the relational data model (RDM), integrity constraints are needed to capture more of the semantics of the data stored in an XML database. Several types of integrity constraints have been studied in the context of XML, with a focus on various key constraints, functional dependencies, inclusion constraints, and path constraints. For relational databases, functional dependencies have been vital in the investigation of how to design "good" database schemas to avoid or minimise problems relating to data redundancy and data inconsistency. The same problems can be shown to exist in poorly designed XML databases. Not surprisingly, functional dependencies for XML (often referred to as XFDs) have recently gained much attention.

An important problem involving XFDs is that of logical implication, i.e., deciding whether a new XFD holds, given a set of existing XFDs. This is important for minimising the cost of checking that a database satisfies a set of XFDs, and may also be helpful when XFDs are propagated to view definitions. One approach to solve this problem is to develop a sound and complete set of inference rules for generating symbolic proofs of logical implication.

For the RDM, the implication problem for functional dependencies is decidable in linear time, and the Armstrong system of inference rules is sound and

---

[*] Thu Trinh's research was supported by a Lovell and Berys Clark and a William Georgetti masterate scholarship.

J. Dix and S.J. Hegner (Eds.): FoIKS 2006, LNCS 3861, pp. 159–178, 2006.
© Springer-Verlag Berlin Heidelberg 2006

complete. For XML the story is more complicated. Before studying XFDs and actually using them in database design, they have to be formally defined. In the literature [3, 4, 9-11, 13, 18-20, 17, 21, 22], several generalisations of functional dependencies to XML have been proposed, and they do not always reflect the same kind of dependencies in an XML database. The difficulty with XML data is that its nested structure is more complex than the rigid structure of relational data, and thus may well observe a larger variety of data dependencies.

In this paper we use an approach that considers XFDs in the context of a simple tree model: XFDs are defined over a schema tree that reflects the permitted parent-child relationships, and apply to the almost-copies of the schema tree that can be found in an XML data tree under inspection. XML schema trees capture information on the frequency of parent-child relationships, that is, they show whether a child is optional or required, and whether it is unique or may occur multiple times. This approach to XFDs has been suggested in [10], and comes close to the approach taken in [3, 4] where XFDs are defined on the basis of paths evolving from the root element in an XML document. This idea goes back to earlier studies of functional dependencies in semantic and object-oriented data models [15, 23]. It should be noted that XFDs may well interact in a non-trivial way with chosen specifications (like DTDs) as demonstrated, e.g., in [2, 4].

This paper is organised as follows. In Section 2, we provide preliminary notions like XML schema trees and data trees. In Section 3, we present an example that illustrates our approach, while we formally define XFDs in Section 4. In Section 5, we assemble sound inference rules, which are then shown to be complete in Section 6. In Section 7, we extend our investigation to account for ID-attributes that are widely used in XML databases. Finally, Section 8 gives an overview of related work and discusses similarities and differences.

2 Preliminary Notations

We start with reviewing basic features of a simple XML tree model. Within this paper, all graphs considered are directed, without parallel arcs and finite unless stated otherwise. For every graph G, let V_G denote its set of vertices and A_G its set of arcs. A *rooted graph* is a graph G with one distinguished vertex r_G, called the *root* of G, such that there is a directed path from r_G to every other vertex in V_G. A *rooted tree* is a rooted graph T without any (non-directed) cycles. A graph G is *empty* if A_G is empty. Specifically, G is an *empty rooted graph* if it consists of a single vertex r_G. For every vertex v, let $Succ_G(v)$ denote its (possibly empty) set of successors, called *children*, in G. A non-isolated vertex without children is a *leaf* of G. Let L_G denote the set of all leaves of G.

Definition 1. *Given a vertex $v \in V_G$ and a subset $W \subseteq L_G$ of leaves, a v-subgraph of G is the graph union of all directed walks from v to some $w \in W$. A v-walk of G is a directed walk from v to a single leaf w of G. Every v-walk or v-subgraph of a rooted tree is again a rooted tree.*

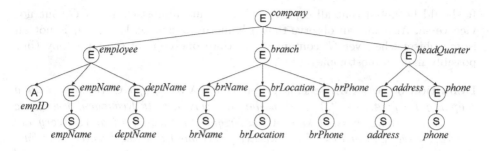

Fig. 1. An XML tree showing the names and kinds of vertices

Let *ENames* and *ANames* be fixed sets of element names and attribute names, respectively. Also let the symbols E, A and S reflect whether a vertex represents an element, attribute or text data respectively.

Definition 2. *An XML graph is a rooted graph G together with the mappings* $name : V_G \rightarrow ENames \cup ANames$ *and* $kind : V_G \rightarrow \{E, A, S\}$ *assigning every vertex its name and kind, respectively. If G is a rooted tree, then we speak of an* XML tree.

In this paper, all XML trees are assumed to be unordered trees. Let V_G^E, V_G^A and V_G^S consist of all vertices in V_G of kind E, A and S, respectively. We suppose that in an XML graph, vertices of kind A and S are always leaves and conversely all leaves are either of kind A or S, that is, $L_G = V_G^A \cup V_G^S$. Thus, within this paper, we do not consider empty elements unless G is empty.

Definition 3. *Let G' and G be two XML graphs, and consider a mapping* $\phi : V_{G'} \rightarrow V_G$. *$\phi$ is said to be* kind-preserving *if the image of a vertex is of the same kind as the vertex itself, that is, $kind(v') = kind(\phi(v'))$ for all $v' \in V_{G'}$. Further, ϕ is* name-preserving *if the image of a vertex carries the same name as the vertex itself, that is, $name(v') = name(\phi(v'))$ for all $v' \in V_{G'}$. The mapping ϕ is a* homomorphism *between G' and G if all of the following conditions hold:*

1. *the root of G' is mapped to the root of G, that is, $\phi(r_{G'}) = r_G$*
2. *every arc of G' is mapped to an arc of G, that is, $(u', v') \in A_{G'}$ implies $(\phi(u'), \phi(v')) \in A_G$*
3. *ϕ is kind-preserving and name-preserving.*

Definition 4. *A homomorphism $\phi : V_{G'} \rightarrow V_G$ is an* isomorphism *if ϕ is bijective and ϕ^{-1} is a homomorphism. Whenever such an isomorphism exists, G' is said to be* isomorphic *to G, denoted by $G' \cong G$. We also call G' a* copy *of G.*

Definition 5. *A subgraph H' of G' is a* copy *of a subgraph H of G if the restriction of $\phi : V_{G'} \rightarrow V_G$ to H' and H is an isomorphism between H' and H. An $r_{G'}$-subgraph H' of G' is a* subcopy *of G if it is a copy of some r_G-subgraph H of G. A* maximal subcopy *of G is a subcopy of G which is not an r_G-subgraph of any other subcopy of G. A maximal subcopy of G is called an* almost-copy *of G.*

It should be noted that all copies of G in G' are almost-copies of G, but not vice versa. Also we can observe that a homomorphism $\phi : V_{G'} \to V_G$ is not an isomorphism whenever G' contains more than one copy of G or no copy (but possibly many almost-copies) of G.

Definition 6. *An XML schema graph is an XML graph G together with a mapping $freq : A_G \to \{?, 1, +, *\}$ assigning every arc its frequency. Every arc $a = (v, w)$ where w is of kind A has frequency $freq(a) =?$ or 1. Every arc $a = (v, w)$ where $kind(v) = E$ and $kind(w) = S$ has frequency $freq(a) = 1$. Further, we assume no vertex in V_G has two successors with the same name and the same kind. If G is more specifically an XML tree, then we speak of an XML schema tree.*

We use "f-arc" to refer to an arc of frequency f and "f/g-arc" to refer to an arc of frequency f or g. For example, a ?-arc refers to an arc of frequency ?, while a $*/+$-arc refers to an arc of frequency $*$ or $+$. For an XML schema graph G, let $G_{\leq 1}$ be the graph union of all $?/1$-arcs in A_G, and $G_{\geq 1}$ be the graph union of all $1/+$-arcs in A_G. Note that $G_{\leq 1}$ and $G_{\geq 1}$ may not be r_G-subgraphs of G.

Definition 7. *An XML data tree is an XML tree T' together with an evaluation $val : L_{T'} \to STRING$ assigning every leaf v a (possibly empty) string $val(v)$.*

Definition 8. *Let G be an XML schema graph. An XML data tree T' is compatible with G, denoted by $T' \rhd G$, if there is a homomorphism $\phi : V_{T'} \to V_G$ between T' and G such that for each vertex v' of T' and each arc $a = (\phi(v'), w)$ of G, the number of arcs $a' = (v', w'_i)$ mapped to a is at most 1 if $freq(a) =?$, exactly 1 if $freq(a) = 1$, at least 1 if $freq(a) = +$, and arbitrarily many if $freq(a) = *$. Due to the definition of a schema graph, this homomorphism is unique if it exists.*

An XML schema graph may be developed by a database designer similar to a (rather simple) database schema, or it can be derived from other specifications (such as DTDs or XSDs). Alternatively, an XML schema graph may also be derived from an XML document itself, cf. [10]. At this point a short remark is called for. Given an XML data tree T', there is usually more than just a single XML schema graph G such that T' is compatible with G. For example, G may well be extended by adjoining new vertices and arcs, or by (partially) unfolding it. Recall that, in our definition of an XML schema graph, we did neither claim the vertices of kind E to have mutually distinct names, nor those of kind A. It is well-known that every rooted graph G may be uniquely transformed into a rooted tree T_G by completely unfolding it, cf. [8].

Let T'_1 be any almost-copy of T in an XML data tree $T' \rhd T$. It is possible that T'_1 does not contain a copy of some r_T-walk which contains an ?-arc or *-arc. Note that this flexibility is one of the desirable features of XML to adequately represent heterogenous data. We say that T'_1 is *missing* a copy of an r_T-walk C of T if T'_1 does not contain a copy of C, otherwise T'_1 is said to be *not missing* a copy of C. Similarly the data tree T' is said to be *missing* a copy of C if it does not contain a copy of C, and *not missing* a copy of C otherwise.

For two r_G-subgraphs X and Y of some graph G, we may use $X \subseteq Y$ to denote that X is an r_G-subgraph of Y in G, and more specifically $X \in Y$ to denote that X is an r_G-walk of Y in G.

Next we briefly discuss operators to construct new trees from given ones. Let G be an XML graph, and X, Y be subgraphs of G. The *union* of X and Y, denoted by $X \cup Y$, is the restriction of the graph union of X and Y to its maximal r_G-subgraph of G. For convenience, we sometimes omit the union symbol and write XY instead of $X \cup Y$. The *intersection* of X and Y, denoted by $X \cap Y$, is the union of all r_G-walks that belong to both X and Y. The *difference* between X and Y, denoted by $X - Y$, is the union of all r_G-walks belonging to X but not to Y. In particular, $X \cap Y$ and $X - Y$ are r_G-subgraphs of G.

The intersection operator is associative but the union and difference operators are not associative. The union and intersection operators are commutative but the difference operator is not. In the absence of parentheses, we suppose that the union and intersection operators bind tighter than the difference operator. For example, by $X \cup Y - Z$ we mean $(X \cup Y) - Z$.

Let G' and G be two XML graphs, and $\phi : V_{G'} \to V_G$ be a homomorphism between them. Given an r_G-subgraph H in G, the *projection* of G' to the subgraph H in G, denoted by $G'|_H$, is the union of all the subcopies of H in G'. The projection $G'|_H$ is an $r_{G'}$-subgraph of G'.

3   A Motivating Example

Our example describes information stored about a product development company and its employees. The company has one head quarter office contactable by postal mail or phone. Furthermore, the company operates various departments and has multiple branches.

In our example here, we use attributes rather than text elements purely to end up with more compact XML graphs. Of course, each attribute in an XML graph G with name n may alternatively be modelled by a vertex v of kind E with name n and a child $w \in Succ_G(v)$ of kind S and name n. Here, we are not concerned with the question of whether some information are better modelled as an attribute or text element.

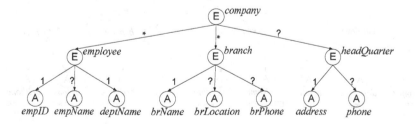

Fig. 2. The XML schema tree PD

For the ease of presentation, we have chosen examples where the leaf names are unique. In this paper, we may therefore refer to an r_G-walk to some leaf carrying the name "B" simply as $[[B]]$, e.g., $[[brPhone]]$. Further, we may refer to an r_G-subgraph X by listing the names of all leaves in X separated by white spaces, e.g., $[[brName\ brPhone]]$.

4   Functional Dependencies for XML

Two isomorphic XML data trees T' and T are said to be *value-equal*, denoted by $T' = T$, if the isomorphism $\phi : V_{T'} \to V_T$ between T' and T is evaluation-preserving, that is, $val(\phi(v')) = val(v')$ holds for every $v' \in L_{T'}$. We are now ready to present our definition of functional dependencies for XML.

Definition 9. *Given an XML schema graph T, a functional dependency (or XFD for short) on T is an expression $X \to Y$ where X and Y are non-empty r_T-subgraphs in T. Let T' be an XML data tree which is compatible with T and let $\phi : V_{T'} \to V_T$ be the unique homomorphism between T' and T. Then T' satisfies the XFD $X \to Y$, written as $\models_{T'} X \to Y$, if and only if for any two almost-copies T_1' and T_2' of T in T' the projections $T_1'|_Y$ and $T_2'|_Y$ are value-equal whenever the projections $T_1'|_X$ and $T_2'|_X$ are value-equal and copies of X, i.e., $T_1'|_Y = T_2'|_Y$ whenever $T_1'|_X = T_2'|_X \cong X$.*

Example 10. Suppose each department of our product development company is located at a single branch. Branches have unique phone numbers and are located in unique locations. Employees are assigned unique employee IDs. We use the following XFDs to model the PD company information:

$$
\begin{array}{lll}
(PD\_XFD1) & [[brName]] & \to [[brLocation]] \\
(PD\_XFD2) & [[brName]] & \to [[brPhone]] \\
(PD\_XFD3) & [[brLocation]] & \to [[brName]] \\
(PD\_XFD4) & [[deptName]] & \to [[brName]] \\
(PD\_XFD5) & [[empID]] & \to [[empName]] \\
\end{array}
$$

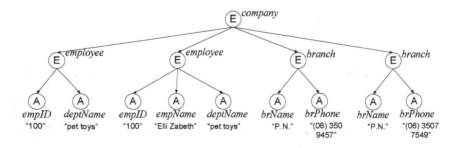

Fig. 3. An XML data tree PD$'$ compatible with PD

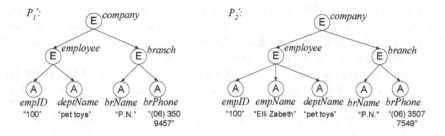

Fig. 4. Two of the four almost-copies of PD contained in PD′

The XML data tree PD′ in Figure 3 contains four almost-copies of PD, two of which are shown in Figure 4. The two remaining almost-copies can be obtained as follows:

$$P'_3 = P'_1|_{[[\,empID\ deptName\,]]} \cup P'_2|_{[[\,brName\ brPhone\,]]}$$
$$P'_4 = P'_2|_{[[\,empID\ empName\ deptName\,]]} \cup P'_1|_{[[\,brName\ brPhone\,]]}$$

The almost-copies P'_1 and P'_2 contain value-equal copies of $[[\,brName\,]]$, but they differ in their copies of $[[\,brPhone\,]]$. Therefore PD′ does not satisfy PD_XFD2. Moreover, PD′ does not satisfy PD_XFD5 since P'_1 and P'_2 contain value-equal copies of $[[\,empID\,]]$, but only P'_2 is not missing a copy of $[[\,empName\,]]$.

On the other hand, PD_XFD3 is trivially satisfied because PD′ is missing a copy of $[[\,brLocation\,]]$. P'_1 and P'_2 contain value-equal copies of $[[\,brName\,]]$ and are both missing a copy of $[[\,brLocation\,]]$. Since P'_3, P'_4 are constructed from P'_1 and P'_2, it is easy to see that any two almost-copies of PD in PD′ have a value-equal copy of $[[\,brName\,]]$ and are missing a copy of $[[\,brLocation\,]]$. Hence PD′ satisfies PD_XFD1. Both P'_1 and P'_2 contain value-equal copies of $[[\,deptName\,]]$ and $[[\,brName\,]]$. Again, we have that any two almost-copies of PD in PD′ will contain $P'_1|_{[[\,deptName\ brName\,]]}$ or $P'_2|_{[[\,deptName\ brName\,]]}$. Therefore any two almost-copies of PD in PD′ will contain value-equal copies of $[[\,deptName\,]]$ and $[[\,brName\,]]$, so that PD_XFD4 is satisfied. □

As for the RDM, we say that an XML data tree T' satisfies a given set Σ of XFDs, denoted by $\models_{T'} \Sigma$, if T' satisfies each XFD in Σ. Satisfaction of a given set of XFDs by an XML data tree usually implies the satisfaction of other XFDs. The notions of implication and derivability (with respect to a rule system $\mathcal{R}$) are defined analogously to similar notions in the RDM.

Let Σ be a set of XFDs and $X \to Y$ a single XFD. If $X \to Y$ is satisfied in every XML data tree which satisfies Σ, then Σ *implies* $X \to Y$, written as $\Sigma \models X \to Y$. The *semantic closure of* Σ, denoted by Σ^*, is the set of all XFDs which are implied by Σ, that is, $\Sigma^* = \{X \to Y \mid \Sigma \models X \to Y\}$.

Given a rule system $\mathcal{R}$, we call an XFD $X \to Y$ *derivable* from Σ by $\mathcal{R}$, denoted by $\Sigma \vdash_{\mathcal{R}} X \to Y$, if there is a finite sequence of XFDs, whose last element is $X \to Y$, such that each XFD in the sequence is in Σ or can be obtained from Σ by applying one of the inference rules in $\mathcal{R}$ to a finite number of previous

XFDs in the sequence. The *syntactic closure of Σ with respect to the rule system* $\mathcal{R}$, denoted $\Sigma_{\mathcal{R}}^{+}$, is the set of all XFDs which are derivable from Σ by means of inference rules in $\mathcal{R}$, that is, $\Sigma_{\mathcal{R}}^{+} = \{X \rightarrow Y \mid \Sigma \vdash_{\mathcal{R}} X \rightarrow Y\}$. Whenever the rule system is clearly understood, we may omit $\mathcal{R}$.

An inference rule is called *sound* if for any given set Σ of XFDs, every XFD which may be derived from Σ due to that rule is also implied by Σ. A rule system $\mathcal{R}$ is *sound* if all inference rules in $\mathcal{R}$ are sound. In other words, $\mathcal{R}$ is sound if every XFD which is derivable from Σ by $\mathcal{R}$ is also implied by Σ (i.e. $\Sigma_{\mathcal{R}}^{+} \subseteq \Sigma^{*}$). A rule system is said to be *complete* if it is possible to derive every XFD which is implied by Σ (i.e. $\Sigma^{*} \subseteq \Sigma_{\mathcal{R}}^{+}$).

5   Sound Inference Rules

In this section, we assemble sound inference rules that yield a sound and complete rule system as we will demonstrate later on.

Lemma 11. *Let T be an XML schema tree, and let X, Y, W, Z be r_T-subgraphs of T. The following inference rules for XFDs are sound:*

$$\text{(union rule)} \quad \frac{X \rightarrow Y, X \rightarrow Z}{X \rightarrow Y \cup Z}$$

$$\text{(reflexivity axiom)} \quad \frac{}{X \rightarrow Y} \quad Y \text{ is an } r_T\text{-subgraph of } X$$

$$\text{(subtree rule)} \quad \frac{X \rightarrow Y}{X \rightarrow Z} \quad Z \text{ is an } r_T\text{-subgraph of } Y$$

$$\text{(supertree rule)} \quad \frac{W \rightarrow Y}{X \rightarrow Y} \quad W \text{ is an } r_T\text{-subgraph of } X$$

For an XML schema tree T, let R_T denote the union of all r_T-walks of $T_{\leq 1}$.

Lemma 12. *Let T be an XML schema tree, and let X be an r_T-subgraph of T. The following inference rule is sound for XFDs:*

$$\text{(root axiom)} \quad \frac{}{X \rightarrow R_T}$$

Surprisingly, the transitivity rule from the RDM does not hold for XML in the presence of frequencies. Consider the XFDs $X \rightarrow Y$ and $Y \rightarrow Z$ defined on some XML schema tree T. For an XML data tree $T' \rhd T$, if any two almost-copies of T are missing a copy of some r_T-walk of Y, then T' trivially satisfies $Y \rightarrow Z$. Therefore it would be possible for two almost-copies to be not value-equal on Z while being value-equal on and not missing a copy of X, that is, $X \rightarrow Z$ can be violated.

However we can define a restricted form of the transitivity rule which is sound for the derivation of XFDs. The main idea behind such an inference rule is to use frequencies to ensure that two almost-copies are not missing a copy of every

r_T-walk of the middle term Y whenever they are not missing a copy of every r_T-walk of X or Z. The notion of Y being X, Z-compliant in the following definition accomplishes this.

Definition 13. *Let X, Y, Z be r_T-subgraphs in an XML schema tree T. We say Y is X, Z-compliant if and only if $Y \subseteq (X \cup C) \cup T_{\geq 1}$ for each r_T-walk C of Z.*

Example 14. In the XML schema tree PD in Figure 2, it is easy to see that $[[brName]] \subseteq ([[deptName]] \cup [[brLocation]]) \cup PD_{\geq 1}$ holds, that is, $[[brName]]$ is $[[deptName]], [[brLocation]]$-compliant. ☐

Lemma 15. *Let T be an XML schema tree, and let X, Y, Z be r_T-subgraphs of T. The following inference rule is sound for XFDs:*

$$\text{(restricted-transitivity rule)} \quad \frac{X \to Y, Y \to Z}{X \to Z} \quad Y \text{ is } X, Z\text{-compliant}$$

Example 16. Recall the XML schema tree PD in Figure 2 and all XFDs specified in Example 10. There are two r_{PD}-walks in $PD_{\leq 1}$, yielding $R_{PD} = [[address\ phone]]$. Using the *root axiom* we derive XFDs like $[[empID]] \to [[address\ phone]]$ (let this be denoted by PD_XFD6) and $[[empName\ brName]] \to [[address\ phone]]$.

The *supertree rule* enables us to derive from PD_XFD6 the XFD $[[empID\ empName\ brName]] \to [[address\ phone]]$. Applying the *subtree rule* to PD_XFD6 gives us the XFDs $[[empID]] \to [[address]]$ and $[[empID]] \to [[phone]]$. Using the *reflexivity axiom*, we can derive the XFD $[[empID\ empName\ brName]] \to [[empID\ empName]]$. From PD_XFD1 and PD_XFD2 and an application of the *union rule* we obtain the XFD $[[brName]] \to [[brLocation\ brPhone]]$.

Since $[[brName]]$ is $[[deptName]], [[brLocation]]$-compliant, an application of the *restricted-transitivity rule* to PD_XFD4 and PD_XFD1 yields the XFD $[[deptName]] \to [[brLocation]]$. ☐

Next, we define the notion of a unit of some r_T-walk which is needed for the final inference rule presented in this section.

Definition 17. *Let B be an r_T-walk of some XML schema tree T. The unit of B, denoted by U_B, is the union of all r_T-walks sharing some */+-arc with B.*

We continue with some useful observations about the unit of an r_T-walk. For one, it is the case that $U_C = U_B$ for any r_T-walk $C \in U_B$. Furthermore, in any data tree $T' \rhd T$, every almost-copy of $T - U_B$ together with any almost-copy of U_B form an almost-copy of T in T'. In particular, for any two almost-copies T_1', T_2' of T in T', it is the case that $T_1'|_{T-U_B} \cup T_2'|_{U_B}$ and $T_2'|_{T-U_B} \cup T_1'|_{U_B}$ are also almost-copies of T in T'. The mix-and-match approach is only possible because $T_2'|_{U_B}$ shares with $T_1'|_{T-U_B}$ exactly those arcs (and vertices) which $T_2'|_{U_B}$ shares with $T_2'|_{T-U_B}$, and likewise $T_1'|_{U_B}$ shares with $T_1'|_{T-U_B}$ exactly those arcs which $T_1'|_{U_B}$ shares with $T_1'|_{T-U_B}$.

Lemma 18. *Let T be an XML schema tree, let X be an r_T-subgraph of T, and let B an r_T-walk of T. The following inference rule is sound for XFDs:*

$$\text{(noname rule)} \qquad \frac{((X \cup B) \cup T_{\geq 1} - U_B) \cup X \to B}{X \to B}$$

Example 19. The noname rule allows us to derive XFDs which have not been derivable using the other derivation rules only, e.g., the new XFD $[[\,empName\,]] \to [[\,brName\,]]$ for our example above. To see this, we first find $([[\,empName\,]] \cup [[\,brName\,]]) \cup \mathrm{PD}_{\geq 1}$ as the r_{PD}-subgraph $[[\,empID\ empName\ deptName\ brName\,]]$ of PD, and the unit $U_{[[\,brName\,]]}$ as the r_{PD}-subgraph $[[\,brName\ brLocation\ brPhone\,]]$. This amounts to the premise of the noname rule being the XFD $[[\,empID\ empName\ deptName\,]] \to [[\,brName\,]]$, which can be derived from $[[\,deptName\,]] \to [[\,brName\,]]$ (that is, PD_XFD4) by means of the supertree rule. □

6   A Sound and Complete Rule System

In this section, we observe that the inference rules assembled above form a complete rule system for XFDs in the presence of frequencies. Let the $\mathcal{F}$-*rule system* consist of the following inference rules: *reflexivity axiom, root axiom, subtree rule, supertree rule, union rule, restricted-transitivity rule* and *noname rule*.

We take the usual approach to verifying completeness. Consider an XML schema tree T and a set Σ of XFDs on T. If $X \to Y$ cannot be derived from Σ by means of the inference rules, then we show that there is an XML data tree $T' \rhd T$ such that $\models_{T'} \Sigma$ but $\not\models_{T'} X \to Y$. Because of the union rule, this means that there is some r_T-walk $B \in Y$ such that $X \to B$ is not derivable from Σ and T' does not satisfy $X \to B$. Therefore T' must contain two almost-copies T_1', T_2' of T such that $T_1'|_X = T_2'|_X \cong X$ and $T_1'|_B \neq T_2'|_B$. In the sequel, we will outline a *general construction* for such a *counterexample data tree T'*.

Without frequencies, we can construct a counterexample data tree from the arc-disjoint union of exactly two copies T_a', T_b' of $X \cup B$ that are value-equal only on X. Particularly, T_a', T_b' are both missing a copy of every r_T-walk not in $X \cup B$. In the presence of frequencies, however, we face the additional complication that at least one almost-copy of T in T' must contain a copy of $(X \cup B) \cup T_{\geq 1}$. This means, in addition to $X \cup B$, we also need to determine whether or not T_1' and T_2' should be value-equal on any of the remaining r_T-walks in $(X \cup B) \cup T_{\geq 1}$, keeping in mind that T' must still satisfy Σ.

We first define the analogous of the closure of a set of attributes in the RDM.

Definition 20. *Let T be an XML schema tree, X be an r_T-subgraph, and Σ be a set of XFDs on T. Further let $\mathcal{R}$ be a rule system. The pre-closure $X_{\mathcal{R}}^+$ of X with respect to Σ and $\mathcal{R}$ is the following r_T-subgraph of T:*

$$X_{\mathcal{R}}^+ = \bigcup \{Y \mid X \to Y \in \Sigma_{\mathcal{R}}^+\}$$

As pointed out above, we need two almost-copies T_1', T_2' of T such that $T_1'|_X = T_2'|_X \cong X$ and $T_1'|_B \neq T_2'|_B$. To ensure $\models_{T'} \Sigma$ we must further guarantee that $T_1'|_{X_{\mathcal{F}}^+} = T_2'|_{X_{\mathcal{F}}^+}$. For that, the two almost-copies of T must not be missing value-equal copies of $X_{\mathcal{F}}^+ \cap ((X \cup B) \cup T_{\geq 1})$. A peculiar situation is encountered: there may be XFDs that are not implied by Σ (and hence not derivable by the $\mathcal{F}$-rule system), but need to be non-trivially satisfied in this situation because at least one of T_1', T_2' is not missing a copy of $(X \cup B) \cup T_{\geq 1}$. The restricted-transitivity rule yields $(X_{\mathcal{F}}^+)_{\mathcal{F}}^+ \supseteq X_{\mathcal{F}}^+$, but $X_{\mathcal{F}}^+$ will in general not be a closure as its counterpart attribute closure in the RDM.

Example 21. Our previous observation is illustrated by the XML schema tree S in Figure 5(a) and the set Σ of XFDs defined on S. We find that $[[U]] \not\subseteq ([[X]] \cup [[D]]) \cup S_{\geq 1}$, that is, $[[U]]$ is not $[[X]], [[D]]$-compliant. Therefore, we cannot use the restricted-transitivity rule to derive $[[X]] \rightarrow [[D]]$. In fact, the $\mathcal{F}$-rule does not allow us to derive $[[X]] \rightarrow [[D]]$. If an XML data tree compatible with S should satisfy Σ and violate $[[X]] \rightarrow [[D]]$ it must simply be missing a copy of $[[U]]$, see Figure 5(b).

However, if an XML data tree compatible with S is not missing a copy of $[[B]]$, then it will not be missing a copy of $[[U]]$ either. Consider the data tree S'' in Figure 5(c). It contains exactly two almost-copies of S, which we denote by S_1', S_2'. Since $\models_{S''} \Sigma$, we have $\models_{S''} [[X]] \rightarrow [[U]]$ and $\models_{S''} [[U]] \rightarrow [[D]]$. It follows from $\models_{S''} [[X]] \rightarrow [[U]]$ and $S_1'|_{[[X]]} = S_2'|_{[[X]]} \cong [[X]]$ that $S_1'|_{[[U]]} = S_2'|_{[[U]]}$. Moreover, since $[[U]] \in ([[X]] \cup [[B]]) \cup S_{\geq 1}''$, neither of S_1', S_2' is missing a copy of $[[U]]$. This means we have $S_1'|_{[[U]]} = S_2'|_{[[U]]} \cong [[U]]$, and hence $S_1'|_{[[D]]} = S_2'|_{[[D]]}$ due to $\models_{S''} [[U]] \rightarrow [[D]]$.   □

Consequently, it is insufficient to stop after having considered only $X_{\mathcal{F}}^+$. Since T_1', T_2' are value-equal on and not missing a copy of $X_{\mathcal{F}}^+ \cap ((X \cup B) \cup T_{\geq 1})$, it follows from $\models_{T'} \Sigma$ that T_1', T_2' must be value-equal on and not missing a copy of $(X_{\mathcal{F}}^+ \cap ((X \cup B) \cup T_{\geq 1}))_{\mathcal{F}}^+ \cap ((X \cup B) \cup T_{\geq 1})$. This then

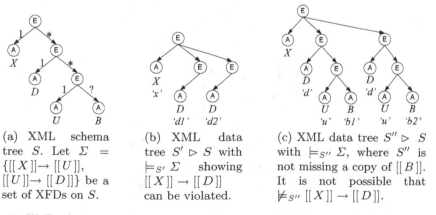

(a) XML schema tree S. Let $\Sigma = \{[[X]] \rightarrow [[U]], [[U]] \rightarrow [[D]]\}$ be a set of XFDs on S.

(b) XML data tree $S' \triangleright S$ with $\models_{S'} \Sigma$ showing $[[X]] \rightarrow [[D]]$ can be violated.

(c) XML data tree $S'' \triangleright S$ with $\models_{S''} \Sigma$, where S'' is not missing a copy of $[[B]]$. It is not possible that $\not\models_{S''} [[X]] \rightarrow [[D]]$.

Fig. 5. XML schema tree and data trees illustrating that there can be an XFD which is not derivable but which is satisfied whenever there occur copies of certain r_S-walks

means that T_1', T_2' must be value-equal on and not missing a copy of $((X_{\mathcal{F}}^+ \cap ((X \cup B) \cup T_{\geq 1}))_{\mathcal{F}}^+ \cap ((X \cup B) \cup T_{\geq 1}))_{\mathcal{F}}^+ \cap ((X \cup B) \cup T_{\geq 1})$, and so on. In the process, we obtain a sequence of pre-closures restricted to $(X \cup B) \cup T_{\geq 1}$. Eventually there is some fix-point X_n since XML schema trees are finite so that there are only finitely many r_T-walks in $(X \cup B) \cup T_{\geq 1}$. In summary, T_1', T_2' must be value-equal on and not missing a copy of the r_T-subgraph X_n in order for $\models_{T'} \Sigma$.

This outlines our general approach for constructing a counterexample data tree, though the actual proof of completeness uses some additional considerations. Recall that $T_1'|_{T-U_B} \cup T_2'|_{U_B}$ and $T_2'|_{T-U_B} \cup T_1'|_{U_B}$ are possibly further almost-copies of T in T'. In particular, if $T_1'|_{T-U_B} \neq T_2'|_{T-U_B}$ and $T_1'|_{U_B} \neq T_2'|_{U_B}$ then there are at least four almost-copies of T in T'. To simplify the discussion it is desirable to have only two almost-copies of T in the counterexample data tree under construction. To ensure this, we can force T' to contain only one copy of $T|_{T-U_B} \cap ((X \cup B) \cup T_{\geq 1})$. Actually, $T|_{T-U_B} \cap ((X \cup B) \cup T_{\geq 1})$ equals $(X \cup B) \cup T_{\geq 1} - U_B$. Therefore, we want to have $T_1'|_{(X \cup B) \cup T_{\geq 1} - U_B} = T_2'|_{(X \cup B) \cup T_{\geq 1} - U_B}$.

However, there might be some r_T-walk in $(X \cup B) \cup T_{\geq 1} - U_B$ that is not in the r_T-subgraph X_n introduced above. This is rectified by actually computing the sequence of restricted pre-closures starting with $X_0 = ((X \cup B) \cup T_{\geq 1} - U_B) \cup X$ rather than just X. The noname rule guarantees that each restricted pre-closure in the sequence remains unaffected except for the additional r_T-subgraph $X_0 - X$.

Example 22. We demonstrate the approach described thus far with an example. Consider the XML schema tree Q shown in Figure 6. (Note that we left out some vertex names in Q for convenience.) Let $\Sigma = \{[[DB]] \to [[C]], [[X]] \to [[B]]\}$ be a set of XFDs given on Q. It is easy to check that $[[XF]] \to [[W]]$ is not derivable from Σ by the $\mathcal{F}$-rule system.

Suppose we want to construct a counterexample data tree $Q' \rhd Q$ such that $\models_{Q'} \Sigma$ but $\not\models_{Q'} [[XF]] \to [[W]]$. The r_Q-subgraph $([[XF]] \cup [[W]]) \cup Q_{\geq 1}$ is

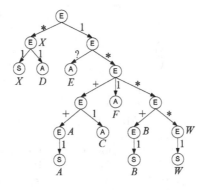

Fig. 6. An XML schema tree Q

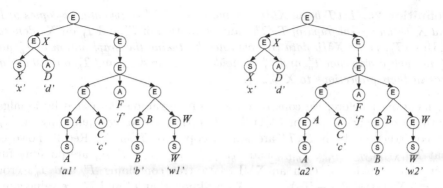

Fig. 7. Two almost-copies Q'_1, Q'_2 of the XML schema tree Q

the schema tree Q without the r_Q-walk $[[E]]$. Thus, each restricted pre-closure must not contain $[[E]]$. The sequence of restricted pre-closures is as follows:

$$[[XF]]_0 = \left(\left(\left([[XF]] \cup [[W]]\right) \cup Q_{\geq 1} - U_{[[W]]}\right) \cup [[XF]]\right)$$
$$= \left(\left[[XFABCDW]\right] - \left[[ABCFW]\right]\right) \cup [[XF]]$$
$$= [[XD]] \cup [[XF]] = [[XFD]]$$

$$[[XF]]_1 = \left([[XFD]]\right)_{\mathcal{F}}^{+} \cap \left(\left([[XF]] \cup [[W]]\right) \cup Q_{\geq 1}\right) = [[XFDB]]$$

$$[[XF]]_2 = \left([[XFDB]]\right)_{\mathcal{F}}^{+} \cap \left(\left([[XF]] \cup [[W]]\right) \cup Q_{\geq 1}\right) = [[XFDBC]]$$
$$= [[XF]]_3 = [[XF]]_4 = \dots$$

Therefore, Q' should contain two almost-copies Q'_1, Q'_2 of Q which are value-equal on and not missing a copy of the r_Q-subgraph $[[XFDBC]]$. Two such almost-copies of Q are depicted in Figure 7.   □

As discussed above, we construct T' in general by taking the union of two copies of $(X \cup B) \cup T_{\geq 1}$ that are value-equal only on X_n. Of course, T' should be compatible with T and thus conform to the frequencies specified on T. Consequently, in T' the two copies may need to share certain arcs or even copies of entire r_T-walks of T. We next provide a notion for describing that two almost-copies of some schema tree T share the same copy of some r_T-walk of T in a compatible data tree T'.

Definition 23. *Let T be an XML schema tree. Two almost-copies T'_1, T'_2 of T in an XML data tree $T' \rhd T$ are said to* coincide *on an r_T-walk B of T if and only if $T'_1|_B$ and $T'_2|_B$ are graph unions of exactly the same set of arcs in T'. Two almost-copies of T coincide on an r_T-subgraph Y of T if and only if they coincide on each r_T-walk of Y.*

Example 24. Recall the XML data tree PD$'$ in Figure 3 containing almost-copies P'_1 to P'_4 of PD, cf. Figure 4. P'_1 and P'_3 coincide on the r_{PD}-subgraph $[[empID\ deptName]]$ but do not coincide on any other r_{PD}-walk.   □

The following amalgamation operator allows us to specify how two almost-copies of an XML schema tree can be combined to construct an XML data tree.

Definition 25. *Let T be an XML schema tree, T'_a, T'_b be two almost-copies of T, and X be an r_T-subgraph of T. The* amalgamation *of T'_a and T'_b on X, denoted $T'_a \amalg_{[X]} T'_b$, is the XML data tree obtained by taking the graph union of T'_a and T'_b in such a way that T'_a and T'_b coincide on X, and T'_a and T'_b only share all arcs in their projections to $X \cup T_{\leq 1}$.*

As mentioned before, we construct the counterexample data tree by amalgamating two copies T'_a, T'_b of $(X \cup B) \cup T_{\geq 1}$ that are value-equal only on X_n. By construction, both T'_a, T'_b are almost-copies of T in T'. Recall, however, that $T'_a|_{(X \cup B) \cup T_{\geq 1} - U_B} \cup T'_b|_{U_B}$ and $T'_b|_{(X \cup B) \cup T_{\geq 1} - U_B} \cup T'_a|_{U_B}$ are possibly further almost-copies of T in an XML data tree containing T'_a and T'_b. From $X_0 = ((X \cup B) \cup T_{\geq 1} - U_B) \cup X \subseteq X_n$ we know that T'_a and T'_b are value-equal on $(X \cup B) \cup T_{\geq 1} - U_B$. In order for the constructed data tree T' to contain only two almost-copies of T (namely T'_a and T'_b), we require that T'_a and T'_b are amalgamated on $(X \cup B) \cup T_{\geq 1} - U_B$. The resulting XML data tree T' is in fact the desired counterexample data tree we are looking for.

As a consequence, we conclude the completeness of the $\mathcal{F}$-rule system.

Theorem 26. *The $\mathcal{F}$-rule system is sound and complete for XFDs in the presence of frequencies.*

Example 27. Let us continue with Example 22. To construct a counterexample data tree $Q' \rhd Q$ such that $\models_{Q'} \Sigma$ but $\not\models_{Q'} [[XF]] \to [[W]]$, we use two copies Q'_a, Q'_b of $([[XF]] \cup [[W]]) \cup Q_{\geq 1} = [[XFABCDW]]$ that are value-equal on the r_Q-subgraph $[[XFDBC]]$. Incidentally, these can be the two almost-copies Q'_1, Q'_2 of Q shown in Figure 7.

From Example 22, we have that $([[XF]] \cup [[W]]) \cup Q_{\geq 1} - U_{[[W]]} = [[XD]]$. As suggested, we amalgamate Q'_a and Q'_b on $[[XD]]$. The resulting XML data tree $Q'_a \amalg_{[[XD]]} Q'_b$, denoted Q' for short, is shown in Figure 8. Note that Q' contains only the two almost-copies Q'_1, Q'_2 of Q. Though we have left out the names of some vertices for convenience, the homomorphism between Q' and

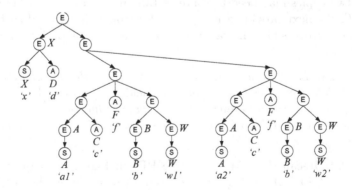

Fig. 8. The XML data tree $Q' = Q'_a \amalg_{[XD]} Q'_b$ where $Q' \rhd Q$. Here Q'_a and Q'_b correspond to the almost-copies Q'_1 and Q'_2 in Figure 7.

Q is the name-preserving mapping which maps every vertex from Q' to the vertex carrying the same name in Q. By examining the frequencies given on Q in Figure 6, it can easily be verified that $Q' \rhd Q$. Thus Q' is the desired counterexample data tree. $\square$

7   The Impact of Identifiers

Elements in an XML document can be identified using ID-attributes. The values of ID-attributes have to be unique throughout the document. This feature is borrowed from object-oriented data models [15, 23] and is supported, e.g., by DTDs. (XSDs also support ID-attributes, but provide further opportunities to declare unique values, too.) In this section, we discuss XFDs in the presence of XML schema trees where attributes may be declared as ID-attributes. We first revise some earlier definitions to account for the presence of ID-attributes.

We use "I" to differentiate an ID-attribute from an ordinary attribute. The kind assignment is thus extended to $kind : V_G \rightarrow \{E, A, I, S\}$. The possible frequencies for an arc $a = (v, w)$ where $kind(w) = I$ is the same as for $kind(w) = A$, that is, $freq(a) =?$ or 1. Furthermore, we call an r_T-walk to some ID-attribute an *identifier*. Let V_G^I consist of all vertices in V_G of kind I. Leaves are now vertices of kind A, I or S, that is, $L_G = V_G^A \cup V_G^S \cup V_G^I$. It is not necessary to modify the definitions of XML data tree and equivalence of XML data trees.

Values assigned to ID-attributes need to be unique in an XML data tree T', that is, for every leaf $v \in V_{T'}^I$ of the data tree T' there does not exist any other leaf $u \in L_{T'}$ such that $val(v) = val(u)$. We refer to this as the *unique identifier value constraint*. Now, the revised version of Definition 8 [compatible] simply requires that T' also satisfies the unique identifier value constraint.

It is easy to see that all inference rules which are sound in the presence of frequencies are also sound in the presence of frequencies and identifiers. However, a few more inference rules are required to obtain a sound and complete rule system again. The first new rule stems from the uniqueness of values taken by ID-attributes. For an r_T-subgraph X, let X_{ID} be the union of all identifiers in X, that is, the r_T-subgraph of X whose leaves are just the leaves of kind I of X.

Lemma 28. *Let T be an XML schema graph, and X be an r_T-subgraph of T with X_{ID} being non-empty. The following inference rule is sound for XFDs:*

$$\text{(identifier axiom)} \qquad \frac{}{X_{ID} \rightarrow X_{ID} \cup T_{\leq 1}}$$

If we allow the left hand side of an XFD to be an empty r_T-subgraph, then we can rewrite the root axiom from Lemma 12 such that the identifier axiom is a straightforward generalisation of the rewritten root axiom.

We next extend the notion of a unit of some r_T-walk to account for the presence of identifiers. This will then provide us a generalisation of the noname rule from Lemma 18.

Definition 29. *Let X be an r_T-subgraph, and B be an r_T-walk of an XML schema graph T. The* unit *of B relative to X, denoted by U_B^X, is the union of all r_T-walks sharing some $*/+$-arc a with B where a is not an arc of X_{ID}.*

Analogous to U_B, for any r_T-walk $C \in U_B^X$, we find that $U_C^X = U_B^X$. It remains to check whether for any two almost-copies T_1', T_2' of T in an XML data tree $T' \rhd T$, it is still true that $T_1'|_{T-U_B^X} \cup T_2'|_{U_B^X}$ and $T_2'|_{T-U_B^X} \cup T_1'|_{U_B^X}$ are almost-copies of T in T'.

If X_{ID} is empty, that is, there is no identifier in X, then $U_B^X = U_B$. Obviously in this case, $T_1'|_{T-U_B^X} \cup T_2'|_{U_B^X}$ and $T_2'|_{T-U_B^X} \cup T_1'|_{U_B^X}$ are almost-copies of T from the observation we made before. Suppose instead that X_{ID} is non-empty. The above observation is not true unless an additional condition is satisfied. Let $I \in X_{ID}$ be an identifier which shares with each r_T-walk of $U_B - U_B^X$ the last $*/+$-arc which that r_T-walk shares with U_B^X. For any two almost-copies T_1', T_2' of T such that $T_1'|_I = T_2'|_I \cong I$, we can then observe that $T_1'|_{T-U_B^X} \cup T_2'|_{U_B^X}$ and $T_2'|_{T-U_B^X} \cup T_1'|_{U_B^X}$ are also almost-copies of T. Similar to previously, this mix-and-match is possible because the unique identifier value constraint guarantees that $T_2'|_{U_B^X}$ shares with $T_1'|_{T-U_B^X}$ exactly those arcs (and vertices) which $T_2'|_{U_B^X}$ shares with $T_2'|_{T-U_B^X}$, while $T_1'|_{U_B^X}$ shares with $T_2'|_{T-U_B^X}$ exactly those arcs which $T_1'|_{U_B^X}$ shares with $T_1'|_{T-U_B^X}$.

Lemma 30. *Let T be an XML schema graph, X be an r_T-subgraph of T, and B be an r_T-walk of T. The following inference rule is sound for XFDs:*

$$\text{(generalised noname rule)} \qquad \frac{((X \cup B) \cup T_{\geq 1} - U_B^X) \cup X \to B}{X \to B}$$

Note that, if X_{ID} is a empty, the generalised noname rule reduces to the noname rule from Lemma 18.

We may finally record the main result of this paper.

Theorem 31. *The $\mathcal{I}$-rule system consisting of the reflexivity axiom, root axiom, subtree rule, supertree rule, union rule, restricted-transitivity rule, identifier axiom and generalised noname rule is sound and complete for XFDs in the presence of frequencies and identifiers.*

8 Discussion

In this section, we assemble some remarks on related work, and point out similarities and differences.

8.1 Relational FDs in the Presence of Null Values

The observation that the transitivity rule is no longer sound for XML corresponds to a similar observation for relational databases in the presence of missing information, cf. Lien [12], who also provides an axiomatisation for functional dependencies in relational databases with null values. Later Atzeni and Morfuni [6] generalised this axiomatisation to functional dependencies in the presence of null values (NFDs) and existence constraints (ECs) over flat relations. An existence constraint is an expression of the form $e : X \vdash Y$ which is satisfied by a relation if each tuple t which is X-total (i.e. $t[X]$ does not contain any null values) is also Y-total. Like frequencies, ECs offer a means to control the presence of missing values. In particular, ECs are required for defining a restricted form of the transitivity rule which is sound in the presence of null values. ECs may express some, but not all, frequencies and conversely frequencies may express some, but not all, ECs. It should be noted that the Atzeni-Morfuni rule system contains additional inference rules for the derivation of ECs which are not required for frequencies. More importantly, the $\mathcal{F}$-rule system contains two additional inference rules to the Atzeni-Morfuni rule system, namely the root axiom and the noname rule, which arise from the nested structure of XML data.

8.2   XML Schema Trees and DTDs

The currently most popular approach towards functional dependencies for XML is due to Arenas and Libkin [3, 4], who define XFDs in the presence of a DTD. In practice, XML documents do not always possess a DTD, but may be developed either from scratch or using some other specification (such as an XSD). Therefore, we use schema graphs rather than DTDs for the definition of functional dependencies. However, if a DTD is available it may be used to derive an XML schema tree, cf. [9].

It should be noted that DTDs and XML schema graphs differ in their expressiveness. DTDs use regular expressions to specify the element type definition associated with a particular element name, that is, the permitted combinations of names for the children of an element of that name. In our approach, we do not consider disjunctions. If a DTD contains a disjunctive expression, we first try to rewrite it without using disjunction. For example, $(A|B|C)^*$ may be rewritten (A^*, B^*, C^*). Otherwise, we treat the disjunction as a sequential expression with each element name in the disjunction being assigned a frequency of $*$ or $?$. For example, if $(A|B|C^+)$ is a regular expression in a given DTD, then this is treated as $(A^?, B^?, C^*)$. This is similar to the idea of simplifying a DTD [14], and to considering simple regular expressions [4]. Recent empirical studies show that the vast majority of DTDs do not contain disjunctive expressions [7]. A regular expression s is *simple* if there is a non-disjunctive regular expression s' such that every word w in the language represented by s is a permutation of a word in the language represented by s', and vice versa. In [4], a DTD is called *simple* if it contains only simple regular expressions as element type definitions.

Secondly, regular expressions in DTDs may be recursive and thus give rise to an XML schema graph with cycles, or alternatively, a rational XML schema tree. In this paper, we restrict ourselves to finite schema trees which in turn correspond to a finite number of unfoldings of the recursive expression, cf. [10]. In [4], a DTD is called *non-recursive* if it contains no recursive regular expressions as element type definitions.

Thirdly, XML schema trees may contain different vertices with the same name, but different sets of children. DTDs do not allow element type definitions associated with a particular element name to be overwritten. This is a feature that XML schema trees share with tree automata, XSDs and Relax NG specifications of XML data. For a discussion, see [7].

8.3   XFDs Studied by Arenas and Libkin

For their definition of XFDs, Arenas and Libkin [3, 4] introduce the notion of a "tree tuple", where a tree tuple is a finite partial evaluation of the paths in the underlying DTD. An XFD which consists of a set of paths on the left hand side and a single path on the right hand side is said to be satisfied if whenever two tree tuples agree on the paths on the left hand side of an XFD they must also agree on the path on the right hand side of the XFD. Note that XFDs in [3, 4] may contain paths that end in an internal vertex, rather than in a leaf. In our approach, identifier-attributes are used instead to refer to a particular vertex. This is more in line with the original intentions of XML and provides more flexibility: identifier-attributes are specified if desired, but not compulsory.

If an XML schema tree T is obtained from a DTD, the almost-copies of T in an XML data tree T' conforming to T are just the tree representations of the maximal tree tuples over the DTD that are subsumed by T'. As a consequence of Theorem 26, we find that the $\mathcal{I}$-rule system is sound and complete for XFDs in the presence of a simple non-recursive DTD. While Arenas and Libkin do not give inference rules for their XFDs, they show that XFDs in the presence of a DTD cannot be finitely axiomatised. The proof given in [4] suggests that the usage of non-simple disjunctive expressions in DTDs causes XFDs not to be finitely axiomatisable. This is one reason for considering XFDs over XML schema trees without disjunctions. Furthermore, [4] also shows the implication problem for XFDs in the presence of a simple non-recursive DTD to be solvable in quadratic time. Note that finite axiomatisability is a stronger property than the existence of an algorithm as the former implies the latter but not the other way around [1].

8.4   XFDs Studied by Vincent and Liu

More recently, Vincent and Liu (partly together with co-authors) studied another kind of functional dependencies for XML [13, 18, 19, 20, 17, 21]. Their definition of XFDs (called *strong XFDs*) does not assume a DTD, but does also not consider frequencies. A major distinguishing feature of the approach taken by Vincent and Liu is the concept of strong satisfaction, cf. [5]. Similar to relations with null values, they suggest to think of an XML tree with missing information as representing a collection of "complete" XML trees with non-missing information. An XML data tree then strongly satisfies an XFD if every possible completion satisfies the XFD. Translated into our approach, an XML data tree containing at least one almost-copy which is not a copy of the given XML schema tree is regarded as being incomplete. That is, Vincent and Liu use the "unknown" interpretation (values exist but not currently known) for information deemed to be missing. We remark that the "no information" interpretation is more general than the "unknown" interpretation as it also includes the "does not exist" interpretation (values do not exist). Due to the flexibility of XML data modelling, we argue that there is good reason for using the "no information" or "does not exist" interpretation of missing information, as done by our approach, and also by Arenas and Libkin [3, 4].

Vincent and Liu [19] provide a system of sound inference rules for strong XFDs which, however, is only proven to be complete for strong XFDs with a single path on the left hand side (called *unary XFDs*). Due to the different interpretation of null values, their system of inference rules looks quite different from ours, and also from the Armstrong rule system for the RDM. We also emphasise that the transitivity rule is sound for strong XFDs, but not sound for XFDs in our approach.

Finally, Liu, Vincent and Liu [13] also studied XFDs that use only paths ending in leaves (called *primary XFDs*) which is similar to our approach to XFDs, but without frequencies and identifiers. This paper does not address axiomatisation, but is focussed on the design of XML documents.

8.5   Further Approaches to XFDs

There is a number of further works that discuss functional dependencies for XML. Lee, Ling and Low [11] study XFDs for designing XML databases that are similar to the approach of Arenas and Libkin [3, 4], but they do not give a precise definition of XFDs. More recently, Wang and Topor [22] have defined XFDs that may use upward paths and are thus more expressive than other approaches, but do not address axiomatisation.

In [9, 10] we discuss several variations of XFDs in the presence of XML schema graphs. In particular, we introduce XFDs whose definition uses pre-images of v-subgraphs instead of almost-copies, and motivate their application by a number of examples. Moreover, we suggest to investigate axiomatisations for XFDs in the presence of XML schema graphs.

9   Conclusion

In this paper, we studied XFDs that are defined in the presence of an XML schema tree. An XFD is satisfied by an XML data tree if whenever two almost-copies of the schema tree coincide and are complete on the left hand side of the XFD then they must also coincide on the right hand side of the XFD. Our approach is similar to the approach by Arenas and Libkin [3, 4], but uses XML schema trees. This allows us to provide a sound and complete system of inference rules that may be used to solve the implication problem for the class of XFDs under discussion. Further, we extended our result to XFDs defined on XML schema trees that may contain ID-attributes. It should be mentioned, that our axiomatisation may be used to conclude a normal form for XML data that avoids redundancies caused by this class of XFDs, cf. [16].

References

1. ABITEBOUL, S., HULL, R., AND VIANU, V. *Foundations of Databases.* Addison-Wesley, 1995.
2. ARENAS, M., FAN, W., AND LIBKIN, L. What's hard about XML schema constraints? In *Database and Expert Systems Applications - DEXA* (2002), vol. 2453 of *LNCS*, Springer, pp. 269–278.
3. ARENAS, M., AND LIBKIN, L. A normal form for XML documents. In *Principles of Database Systems - PODS* (2002), ACM, pp. 85–96.
4. ARENAS, M., AND LIBKIN, L. A normal form for XML documents. *ACM Trans. Database Syst. 29* (2004), 195–232.
5. ATZENI, P., AND DE ANTONELLIS, V. *Relational database theory.* Benjamin-Cummings, 1993.
6. ATZENI, P., AND MORFUNI, N. M. Functional dependencies and constraints on null values in database relations. *Inform. and Control 70* (1986), 1–31.
7. BEX, G. J., NEVEN, F., AND VAN DEN BUSSCHE, J. DTDs versus XML schema: A practical study. In *Web and Databases - WebDB* (2004), pp. 79–84.
8. COURCELLE, B. Fundamental properties of infinite trees. *Theoret. Comput. Sci. 25* (1983), 95–169.
9. HARTMANN, S., AND LINK, S. More functional dependencies for XML. In *Advances in Databases and Information Systems - ADBIS* (2003), vol. 2798 of *LNCS*, Springer, pp. 355–369.
10. HARTMANN, S., LINK, S., AND KIRCHBERG, M. A subgraph-based approach towards functional dependencies for XML. In *Systemics, Cybernetics and Informatics - SCI* (2003), IIIS, pp. 200–205.
11. LEE, M.-L., LING, T. W., AND LOW, W. L. Designing functional dependencies for XML. In *Advances in Database Technology - EDBT* (2002), vol. 2287 of *LNCS*, Springer, pp. 124–141.
12. LIEN, Y. E. On the equivalence of database models. *J. ACM 29* (1982), 333–362.
13. LIU, J., VINCENT, M., AND LIU, C. Functional dependencies, from relational to XML. In *Perspectives of System Informatics - PSI* (2003), pp. 531–538.
14. SHANMUGASUNDARAM, J., TUFTE, K., ZHANG, C., HE, G., DEWITT, D. J., AND NAUGHTON, J. F. Relational databases for querying XML documents: Limitations and opportunities. In *Very Large Databases - VLDB* (1999), pp. 302–314.
15. THALHEIM, B. *Entity-relationship modeling.* Springer, Berlin, 2000.
16. TRINH, T. Functional dependencies for XML: Axiomatisation and normal form in the presence of frequencies and identifiers. MSc thesis, Massey University, 2004.
17. VINCENT, M., AND LIU, J. Completeness and decidability properties for functional dependencies in XML. *CoRR cs.DB/0301017* (2003), 1017.
18. VINCENT, M., AND LIU, J. Functional dependencies for XML. In *Asia-Pacific Web Conference - APWeb* (2003), vol. 2642 of *LNCS*, Springer, pp. 22–34.
19. VINCENT, M., AND LIU, J. Strong functional dependencies and a redundancy free normal form for XML. In *Systemics, Cybernetics and Informatics - SCI* (2003), IIIS, pp. 218–223.
20. VINCENT, M., LIU, J., AND LIU, C. Redundancy free mappings from relations to XML. In *Web-Age Information Management - WAIM* (2003), pp. 55–67.
21. VINCENT, M., LIU, J., AND LIU, C. Strong functional dependencies and their application to normal forms in XML. *ACM Trans. Database Syst. 29* (2004), 445–462.
22. WANG, J., AND TOPOR, R. W. Removing xml data redundancies using functional and equality-generating dependencies. In *Database Technologies - ADC* (2005), vol. 39 of *CRPIT*, Australian Computer Society, pp. 65–74.
23. WEDDELL, G. E. Reasoning about functional dependencies generalized for semantic data models. *ACM Trans. Database Syst. 17* (1992), 32–64.

Guarded Open Answer Set Programming with Generalized Literals

Stijn Heymans, Davy Van Nieuwenborgh*, and Dirk Vermeir

Dept. of Computer Science,
Vrije Universiteit Brussel, VUB,
Pleinlaan 2, B1050 Brussels, Belgium
{sheymans, dvnieuwe, dvermeir}@vub.ac.be

Abstract. We extend the open answer set semantics for programs with generalized literals. Such *extended programs (EPs)* have interesting properties, e.g. the ability to express infinity axioms - EPs that have but infinite answer sets. However, reasoning under the open answer set semantics, in particular satisfiability checking of a predicate w.r.t. a program, is already undecidable for programs without generalized literals. In order to regain decidability, we restrict the syntax of EPs such that both rules and generalized literals are *guarded*. Via a translation to guarded fixed point logic (μGF), in which satisfiability checking is 2-EXPTIME-complete, we deduce 2-EXPTIME-completeness of satisfiability checking in such *guarded EPs* (GEPs). *Bound GEPs* are restricted GEPs with EXPTIME-complete satisfiability checking, but still sufficiently expressive to optimally simulate *computation tree logic* (CTL). We translate Datalog LITE programs to GEPs, establishing equivalence of GEPs under an open answer set semantics, alternation-free μGF, and Datalog LITE. Finally, we discuss ω-restricted logic programs under an open answer set semantics.

1 Introduction

In closed answer set programming (ASP) [6], a program consisting of a rule $p(X) \leftarrow not\ q(X)$ and a fact $q(a)$ is grounded with the program's constant a, yielding $p(a) \leftarrow not\ q(a)$ and $q(a)$. This program has one answer set $\{q(a)\}$ such that one concludes that the predicate p is not satisfiable, i.e. there is no answer set of the program that contains a literal with predicate p. Adding more constants to the program could make p satisfiable, e.g., in the absence of a deducible $q(b)$, one has $p(b)$. However, in the context of conceptual modeling, such as designing database schema constraints, this implicit dependence on constants in the program in order to reach sensible conclusions, i.e. the closedness of reasoning, is infeasible. One wants to be able to test satisfiability of a predicate p in a schema independent of any associated data.

For answer set programming, this problem was solved in [7], where k-belief sets are the answer sets of a program that is extended with k extra constants. We extended this idea, e.g. in [13], by allowing for arbitrary, thus possibly infinite, universes. *Open answer sets* are pairs (U, M) with M an answer set of the program grounded with U. The above program has an open answer set $(\{x, a\}, \{q(a), p(x)\})$ where p is satisfiable.

---

* Supported by the FWO.

J. Dix and S.J. Hegner (Eds.): FoIKS 2006, LNCS 3861, pp. 179–200, 2006.
© Springer-Verlag Berlin Heidelberg 2006

In this paper, we extend programs with generalized literals, resulting in *extended programs (EPs)*. A generalized literal is a first-order formula of the form $\forall Y \cdot \phi \Rightarrow \psi$ where Y is a sequence of variables, ϕ is a finite boolean formula and ψ is an atom. Intuitively, such a generalized literal is true in an open interpretation (U, M) if for all substitutions $[Y \mid y]$, y in U, such that $\phi[Y \mid y]$ is true in M, $\psi[Y \mid y]$ is true in M.

Generalized literals $\forall Y \cdot \phi \Rightarrow \psi$, with ϕ an atom instead of a boolean formula, were introduced in Datalog[1] with the language Datalog LITE [8]: stratified Datalog with generalized literals, where rules are monadic or guarded, and under an appropriate extension of the least fixed point semantics. In open answer set programming (OASP), we define a reduct that removes the generalized literals. E.g., a rule $r : ok \leftarrow \forall X \cdot critical(X) \Rightarrow work(X)$ expresses that a system is OK if all critical devices are functioning: the *GLi-reduct (generalized literal reduct)* of such a rule for an open interpretation $(\{x_0, \ldots\}, M)$ where M contains $critical(x_i)$ for even i, contains a rule $r' : ok \leftarrow work(x_0), work(x_2), \ldots$, indicating that the system is OK if the critical devices $x_0, x_2, \ldots$ are working. The GLi-reduct does not contain generalized literals and one can apply the normal answer set semantics, modified to take into account the infinite body.

Just like it is not feasible to introduce all relevant constants in a program to ensure correct conceptual reasoning, it is not feasible, not even possible, to write knowledge directly as in r' for it has an infinite body. Furthermore, even in the presence of a finite universe, generalized literals allow for a more robust representation of knowledge than would be possible without them. E.g., with critical devices y_1 and y_2, a rule $s : ok \leftarrow work(y_1), work(y_2)$ does the job as good as r (and in fact s is the GLi-reduct of r), but adding new critical devices, implies revisiting s and replacing it by a rule that reflects the updated situation. Not only is this cumbersome, it may well be impossible as s contains no explicit reference to critical devices, and the knowledge engineer may not have a clue as to which rules to modify.

Characteristic about (O)ASP is its treatment of negation as failure (naf): one guesses an interpretation for a program, removes naf by computing the GL-reduct, calculates the iterated fixed point of this reduct, and checks whether this fixed point equals the initial interpretation. In [14], these external manipulations, i.e. not expressible in the language of programs itself, were compiled into fixed point logic (FPL) [11], i.e. into an extension of first-order logic with fixed point formulas. We will show how to modify the FPL translation to take into account generalized literals.

Satisfiability checking w.r.t. arbitrary EPs, even without generalized literals, under the open answer set semantics is undecidable (e.g. the domino problem can be reduced to it), and satisfiability checking in FPL is as well, as it is an extension of the undecidable first-order logic. Thus, with the FPL translation, we have a mapping from one undecidable framework into another undecidable framework. This is interesting in its own right, as it provides a characterization of an answer set semantics in FPL. But more interesting, is the deployment of the translation in order to identify decidable subclasses of EPs: if the FPL translation of a class of EPs falls into a decidable fragment of FPL, this class of EPs is decidable.

---

[1] The extension of logic programming syntax with first-order formulas dates back to [17].

Guarded fixed point logic (μGF) [11] is such a decidable fragment of FPL that is able to express fixed point formulas. It restricts the use of quantified variables by demanding that they are guarded by an atom. We restrict EPs, resulting in *guarded EPs (GEPs)*, such that all variables in a rule appear in an atom in the positive body and all generalized literals are guarded, where a generalized literal is guarded, basically, if it can be written as a guarded formula in μGF. The FPL translation of GEPs then falls into the μGF fragment, yielding a 2-EXPTIME upper complexity bound for satisfiability checking. Together with the 2-EXPTIME-completeness of guarded programs without generalized literals from [14], this establishes 2-EXPTIME-completeness for satisfiability checking w.r.t. GEPs. As a consequence, adding generalized literals to a guarded program does not increase the complexity of reasoning. We further illustrate the expressiveness of (bound) GEPs by simulating reasoning in *computational tree logic (CTL)* [4], a logic for expressing temporal knowledge.

Finally, we reduce Datalog LITE reasoning, without monadic rules, to reasoning with GEPs. In particular, we prove a generalization of the well-known result from [6] that the unique answer set of a stratified program coincides with its least fixed point model: for a universe U, the unique open answer set (U, M) of a stratified Datalog program with generalized literals is identical[2] to its least fixed point model with input structure $id(U)$, the identity relation on U. Furthermore, the Datalog LITE simulation, together with the reduction of GEPs to alternation-free[3] μGF, as well as the equivalence of alternation-free μGF and Datalog LITE [8], lead to the conclusion that alternation-free μGF, Datalog LITE, and OASP with GEPs, are equivalent, i.e. their satisfiability checking problems can be polynomially reduced to one another.

GEPs are just as expressive as Datalog LITE, however, from a knowledge representation viewpoint, GEPs allow for a compact expression of circular knowledge. E.g., the omni-present construction with rules $a(X) \leftarrow not\ b(X)$ and $b(X) \leftarrow not\ a(X)$ is not stratified and cannot be (directly) expressed in Datalog LITE. The reduction to Datalog LITE does indicate that negation as failure under the (open) answer set semantics is not that special regarding expressiveness, but can be regarded as convenient semantic sugar.

The remainder of the paper is organized as follows. After extending the open answer set semantics to support generalized literals in Section 2, we give the FPL translation in Section 3. Section 4 defines GEPs, proves a 2-EXPTIME complexity upper bound for satisfiability checking, and concludes with a CTL simulation. Section 5 describes a simulation of Datalog LITE, without monadic rules, yielding equivalence of alternation-free μGF, Datalog LITE, and GEPs. Section 6 describes the relationship with ω-restricted programs. Section 7 contains conclusions and directions for further research. Due to space restrictions, proofs and further related work have been omitted; the former can be found in http://tinf2.vub.ac.be/ sheymans/tech/goasp-gl.ps.gz, for the latter we refer to [14] and the references therein.

2   Open Answer Set Programming with Generalized Literals

A *term t* is a constant or a variable, where the former is denoted with $a, b, \ldots$ and the latter with $X, Y, \ldots$ A *k-ary atom* is of the form $p(t)$ for a sequence of terms $t =$

---

[2] Modulo equality atoms, which are implicit in OASP, but explicit in Datalog LITE.

[3] μGF without nested fixed point variables in alternating least and greatest fixed point formulas.

$t_1, \ldots, t_k$, $0 \leq k < \omega^4$, and a k-ary predicate symbol p. A *literal* is an atom $p(t)$ or a *naf-atom* $not\ p(t)$ for an atom $p(t)$.[5] The *positive part* of a set of literals α is $\alpha^+ = \{p(t) \mid p(t) \in \alpha\}$ and the *negative part* of α is $\alpha^- = \{p(t) \mid not\ p(t) \in \alpha\}$, i.e. the positive part of a set of literals are the atoms, the negative part are the naf-atoms without the *not* symbol. We assume the existence of binary predicates $=$ and $\neq$, where $t = s$ is considered as an atom and $t \neq s$ as $not\ t = s$. E.g. for $\alpha = \{X \neq Y, Y = Z\}$, we have $\alpha^+ = \{Y = Z\}$ and $\alpha^- = \{X = Y\}$. A *regular* atom is an atom that is not an equality atom. For a set X of atoms, $not\ X = \{not\ l \mid l \in X\}$.

A *generalized literal* is a first-order formula of the form $\forall Y \cdot \phi \Rightarrow \psi$, where ϕ is a finite boolean formula of atoms (i.e. using $\neg$, $\vee$, and $\wedge$) and ψ is an atom; we call ϕ the *antecedent* and ψ the *consequent*. We refer to literals and generalized literals as *extended literals*. For a set of extended literals α, $\alpha^x \equiv \{l \mid l$ generalized literal in $\alpha\}$, the set of generalized literals in α. We extend α^+ and α^- for extended literals as follows: $\alpha^+ = (\alpha \backslash \alpha^x)^+$ and $\alpha^- = (\alpha \backslash \alpha^x)^-$; thus $\alpha = \alpha^+ \cup not\ \alpha^- \cup \alpha^x$.

An *extended program (EP)* is a countable set of *rules* $\alpha \leftarrow \beta$, where α is a finite set of literals, $|\alpha^+| \leq 1$, β is a countable[6] set of extended literals, and $\forall t, s \cdot t = s \notin \alpha^+$, i.e. α contains at most one positive atom, and this atom cannot be an equality atom. The set α is the *head* of the rule and represents a disjunction[7] of literals, while β is called the *body* and represents a conjunction of extended literals. If $\alpha = \emptyset$, the rule is called a *constraint*. *Free rules* are rules of the form $q(t) \vee not\ q(t) \leftarrow$ for a tuple t of terms; they enable a choice for the inclusion of atoms. We call a predicate p free if there is a free rule $p(t) \vee not\ p(t) \leftarrow$. Literals are *ground* if they do not contain variables, generalized literals are ground if they do not contain free variables, and rules and EPs are ground if all extended literals in it are ground.

For an EP P, let $cts(P)$ be the constants in P, and $preds(P)$ its predicates. For a (generalized) literal l, we define $vars(l)$ as the (free) variables in l. For a rule r, we define $vars(r) \equiv \cup\{vars(l) \mid l$ extended literal in $r\}$. Let $\mathcal{B}_P$ be the set of regular ground atoms that can be formed from an EP P. An *interpretation* I of P is then any subset of $\mathcal{B}_P$. For a ground regular atom $p(t)$, we write $I \models p(t)$ if $p(t) \in I$; for an equality atom $p(t) \equiv t = s$, we have $I \models p(t)$ if s and t are equal terms. We have $I \models not\ p(t)$ if $I \not\models p(t)$. We further extend this, by induction, for any boolean formula of ground atoms. For such ground boolean formulas ϕ and ψ, we have $I \models \phi \wedge \psi$ iff $I \models \phi$ and $I \models \psi$, $I \models \phi \vee \psi$ iff $I \models \phi$ or $I \models \psi$, and $I \models \neg \phi$ iff $I \not\models \phi$. For a set of ground literals X, we have $I \models X$ iff $I \models x$ for every $x \in X$. A ground rule $r : \alpha \leftarrow \beta$, not containing generalized literals, is *satisfied* w.r.t. I, denoted $I \models r$, if $I \models l$ for some $l \in \alpha$ whenever $I \models \beta$, i.e. r is *applied* whenever it is *applicable*. A ground constraint $\leftarrow \beta$ is satisfied w.r.t. I if $I \not\models \beta$. For a ground program P without

---

[4] We thus allow for 0-ary predicates, i.e. propositions.

[5] We have no classical negation $\neg$, however, programs with $\neg$ can be reduced to programs without it, see e.g. [16]. To be precise, we should then refer to stable models instead of answer sets.

[6] Thus the rules may have an infinite body.

[7] The condition $|\alpha^+| \leq 1$ ensures that the GL-reduct is non-disjunctive. This allows for the definition of an immediate consequence operator, on which we rely in our proofs to make the correspondence with FPL. In the presence of positive disjunction, the currently defined operator does not suffice and it is not clear whether this can be fixed (and how).

not and without generalized literals, an interpretation I of P is a *model* of P if I satisfies every rule in P; it is an *answer set* of P if I is subset minimal, i.e. there is no model I' of P with $I' \subset I$. For ground EPs P containing *not* but still without generalized literals, the *GL-reduct* [6] w.r.t. I is defined as P^I, where P^I contains $\alpha^+ \leftarrow \beta^+$ for $\alpha \leftarrow \beta$ in P if $I \models not\ \beta^-$ and $I \models \alpha^-$. I is an *answer set* of a ground P without generalized literals if I is an answer set of P^I.

Example 1. Take the program P with rules $p(a) \leftarrow not\ q(a)$ and $q(a) \leftarrow not\ p(a)$. Then P has 4 interpretations $\emptyset, \{p(a)\}, \{q(a)\}$, and $\{p(a), q(a)\}$. The GL-reduct of P w.r.t. $\emptyset$ is $\{p(a) \leftarrow; q(a) \leftarrow\}$ which has $\{p(a), q(a)\}$ as its minimal model, and thus $\emptyset$ is not an answer set. The GL-reduct of P w.r.t. $\{p(a), q(a)\}$ is $\emptyset$ which has $\emptyset$ as its minimal model, and thus $\{p(a), q(a)\}$ is not an answer set. The GL-reduct of P w.r.t. $\{p(a)\}$ is $\{p(a) \leftarrow\}$ which has $\{p(a)\}$ as its minimal model, making $\{p(a)\}$ an answer set. Similarly, one can deduce that $\{q(a)\}$ is an answer set.

A *universe* U for an EP P is a non-empty countable superset of the constants in P: $cts(P) \subseteq U$. Let $\mathcal{B}_P^U$ be the set of regular ground atoms that can be formed from an EP P and the terms in a universe U for P. An *open interpretation* of an EP P is a pair (U, I) where U is a universe for P and I is any subset of $\mathcal{B}_P^U$.

For ground EPs P the *GLi-reduct* $P^{X(U,I)}$ w.r.t. an open interpretation (U, I) removes the generalized literals from the program: $P^{X(U,I)}$ contains the rules[8]

$$\alpha \leftarrow \beta \backslash \beta^X, \quad \bigcup_{\forall \boldsymbol{Y} \cdot \phi \Rightarrow \psi \in \beta^X} \{\psi[\boldsymbol{Y}|\boldsymbol{y}] \mid \boldsymbol{y} \subseteq U, I \models \phi[\boldsymbol{Y}|\boldsymbol{y}]\}, \qquad (1)$$

for $\alpha \leftarrow \beta$ in P. Intuitively, a generalized literal $\forall \boldsymbol{Y} \cdot \phi \Rightarrow \psi$ is replaced by those $\psi[\boldsymbol{Y}|\boldsymbol{y}]$ for which $\phi[\boldsymbol{Y}|\boldsymbol{y}]$ is true, such that[9], e.g., $p(a) \leftarrow [\forall X \cdot q(X) \Rightarrow r(X)]$ means that in order to deduce $p(a)$ one needs to deduce $r(x)$ for all x where $q(x)$ holds. If only $q(x_1)$ and $q(x_2)$ hold, then the GLi-reduct contains $p(a) \leftarrow r(x_1), r(x_2)$. With an infinite universe and a condition ϕ that holds for an infinite number of elements in the universe, one can thus have a rule with an infinite body in the GLi-reduct. An open interpretation (U, I) is an *open answer set* of a ground P if I is an answer set of $P^{X(U,I)}$.

We call P_U the ground EP obtained from an EP P by substituting every (free) variable in a rule in P by every element in U. In the following, an EP is assumed to be a finite set of rules; infinite EPs only appear as byproducts of grounding a finite program with an infinite universe, or, by taking the GLi-reduct w.r.t. an infinite universe. An *open answer set* of P is an open interpretation (U, M) of P with (U, M) an open answer set of P_U. An n-ary predicate p in P is *satisfiable* if there is an open answer set (U, M) of P and a $\boldsymbol{x} \in U^n$ such that $p(\boldsymbol{x}) \in M$. We assume, basically for technical reasons (see Example 4), that when satisfiability checking a predicate p, p is always non-free, i.e. there are no free rules with p in the head. Note that satisfiability checking of a free n-ary predicate p w.r.t. P can always be linearly reduced to satisfiability checking of a new non-free n-ary predicate p' w.r.t. $P \cup \{p'(\boldsymbol{X}) \leftarrow p(\boldsymbol{X})\}$.

---

[8] We denote the substitution of $\boldsymbol{Y} = Y_1, \ldots, Y_d$ with $\boldsymbol{y} = y_1, \ldots, y_d$ in an expression (be it an atom, set of atoms, boolean formula, or rule) X as $X[\boldsymbol{Y}|\boldsymbol{y}]$. If the substitution is clear from the context we write $X[]$.

[9] We put square brackets around generalized literals for clarity.

Example 2. Take an EP P

$$p(X) \leftarrow [\forall Y \cdot q(Y) \Rightarrow r(Y)] \qquad r(X) \leftarrow q(X)$$
$$q(X) \vee not \ q(X) \leftarrow$$

and an open interpretation $(\{x, y\}, \{p(x), r(x), q(x), p(y)\})$. Intuitively, the first rule says that $p(X)$ holds if for every Y where $q(Y)$ holds, $r(Y)$ holds (thus $p(X)$ also holds if $q(Y)$ does not hold for any Y). The GLi-reduct of $P_{\{x,y\}}$ is

$$p(x) \leftarrow r(x) \qquad p(y) \leftarrow r(x)$$
$$r(x) \leftarrow q(x) \qquad r(y) \leftarrow q(y)$$
$$q(x) \leftarrow$$

which has exactly $\{p(x), r(x), q(x), p(y)\}$ as its minimal model such that the open interpretation $(\{x, y\}, \{p(x), r(x), q(x), p(y)\})$ is indeed an open answer set.

There are EPs, not containing (in)equality atoms, for which predicates are only satisfiable by infinite open answer sets.

Example 3. Take the program P, the open answer set variant of the classical infinity axiom in guarded fixed point logic from [11]:

$$r_1 : \qquad\qquad q(X) \leftarrow f(X, Y)$$
$$r_2 : \qquad\qquad\qquad \leftarrow f(X, Y), not \ q(Y)$$
$$r_3 : \qquad\qquad\qquad \leftarrow f(X, Y), not \ well(Y)$$
$$r_4 : \qquad\qquad well(Y) \leftarrow q(Y), [\forall X \cdot f(X, Y) \Rightarrow well(X)]$$
$$r_5 : f(X, Y) \vee not \ f(X, Y) \leftarrow$$

In order to satisfy q with some x, one needs to apply r_1, which enforces an f-successor y. The second rule ensures that also for this y an f-successor must exist, etc. The third rule makes sure that every f-successor is on a well-founded f-chain. The well-foundedness itself is defined by r_4 which says that y is on a well-founded chain of elements where q holds if all f-predecessors of y satisfy the same property.

For example, take an infinite open answer set (U, M) with $U = \{x_0, x_1, \ldots\}$ and $M = \{q(x_0), well(x_0), f(x_0, x_1), q(x_1), well(x_1), f(x_1, x_2), \ldots\}$. P_U contains the following grounding of r_4:

$$r_4^0 : well(x_0) \leftarrow q(x_0), [\forall X \cdot f(X, x_0) \Rightarrow well(X)]$$
$$r_4^1 : well(x_1) \leftarrow q(x_1), [\forall X \cdot f(X, x_1) \Rightarrow well(X)]$$
$$\vdots$$

Since, for r_4^0, there is no $f(y, x_0)$ in M, the body of the corresponding rule in the GLi-reduct w.r.t. (U, M) contains only $q(x_0)$. For r_4^1, we have that $f(x_0, x_1) \in M$ such that we include $well(x_0)$ in the body:

$$well(x_0) \leftarrow q(x_0)$$
$$well(x_1) \leftarrow q(x_1), well(x_0)$$
$$\vdots$$

Thus, (U, M) is an open answer set of the EP, satisfying q.

Moreover, no finite open answer set can satisfy q. First, note that an open answer set (U, M) of P cannot contain loops, i.e. $\{f(x_0, x_1), \ldots, f(x_n, x_0)\} \subseteq M$ is not possible. Assume the contrary. By rule r_3, we need $well(x_0) \in M$. However, the GLi-reduct of P_U contains rules:

$$well(x_0) \leftarrow q(x_0), well(x_n), \ldots$$
$$well(x_n) \leftarrow q(x_n), well(x_{n-1}), \ldots$$
$$\vdots$$
$$well(x_1) \leftarrow q(x_1), well(x_0), \ldots$$

such that $well(x_0)$ cannot be in any open answer set: we have a circular dependency and cannot use these rules to motivate $well(x_0)$, i.e. $well(x_0)$ is unfounded. Thus an open answer set cannot contain loops.

Assume that q is satisfied in an open answer set (U, M) with $q(x_0) \in M$. Then, by rule r_1, we need some X such that $f(x_0, X) \in M$. Since M cannot contain loops X must be different from x_0 and we need some new x_1. By rule r_2, $q(x_1) \in M$, such that by rule r_1, we again need an X such that $f(x_1, X)$. Using x_0 or x_1 for X results in a loop, such that we need a new x_2. This process continues infinitely, such that there are only infinite open answer sets that make q satisfiable w.r.t. P.

3 Open Answer Set Programming with EPs Via Fixed Point Logic

We assume first-order logic (FOL) interpretations have the same form as open interpretations: a pair (U, M) corresponds with the FOL interpretation M over the domain U. Furthermore, we consider FOL with equality such that equality is always interpreted as the identity relation over U.

We define *Fixed Point Logic (FPL)* along the lines of [11], i.e. as an extension of first-order logic, where formulas may additionally be *fixed point formulas* of the form

$$[\text{LFP } W\boldsymbol{X}.\psi(W, \boldsymbol{X})](\boldsymbol{X}) \quad \text{or} \quad [\text{GFP } W\boldsymbol{X}.\psi(W, \boldsymbol{X})](\boldsymbol{X}) , \quad (2)$$

where W is an n-ary predicate variable, $\boldsymbol{X}$ is an n-ary sequence of distinct variables, $\psi(W, \boldsymbol{X})$ is a formula with all free variables contained in $\boldsymbol{X}$ and W appears only positively in $\psi(W, \boldsymbol{X})$.

For an interpretation (U, M) and a valuation χ of the free predicate variables, except W, in ψ, we define the operator $\psi^{(U,M),\chi} : 2^{U^n} \to 2^{U^n}$ on sets S of n-ary tuples

$$\psi^{(U,M),\chi}(S) \equiv \{\boldsymbol{x} \in U^n \mid (U, M), \chi \cup \{W \to S\} \models \psi(W, \boldsymbol{x})\} , \quad (3)$$

where $\chi \cup \{W \to S\}$ is the valuation χ extended such that W is assigned to S. If $\psi(W, \boldsymbol{X})$ contains only the predicate variable W, we often omit the valuation χ and write just $\psi^{(U,M)}$. By definition, W appears only positively in ψ such that $\psi^{(U,M),\chi}$ is monotonic on sets of n-ary U-tuples and thus has a least and greatest fixed point, which we denote by $\text{LFP}(\psi^{(U,M),\chi})$ and $\text{GFP}(\psi^{(U,M),\chi})$ respectively. Finally, we have that

$$(U, M), \chi \models [\text{LFP } W\boldsymbol{X}.\psi(W, \boldsymbol{X})](\boldsymbol{x}) \iff \boldsymbol{x} \in \text{LFP}(\psi^{(U,M),\chi}) , \quad (4)$$

and similarly for greatest fixed point formulas. As in [8], we call an FPL *sentence* (i.e. an FPL formula without free variables) *alternation-free* if it does not contain subformulas $\psi \equiv [\text{LFP } TX.\varphi](X)$ and $\theta \equiv [\text{GFP } SY.\eta](Y)$ such that T occurs in η and θ is a subformula of φ, or S occurs in φ and ψ is a subformula of η. We can eliminate greatest fixed point formulas from a formula, by the equivalence: $[\text{GFP } WX.\psi] \equiv \neg[\text{LFP } WX.\neg\psi[W|\neg W]]$, where $\neg\psi[W|\neg W]$ is $\neg\psi$ with W replaced by $\neg W$. If we thus remove greatest fixed point predicates, and if negations appear only in front of atoms or least fixed point formulas, then a formula is alternation-free iff no fixed point variable W appears in the scope of a negation.

First, we rewrite an arbitrary EP as an EP containing only one designated predicate p and (in)equality; this makes sure that when calculating a fixed point of the predicate variable p, it constitutes a fixed point of the whole program. We assume without loss of generality that the set of constants and the set of predicates in an EP are disjoint and that each predicate q has one associated arity, e.g. $q(x)$ and $q(x, y)$ are not allowed. An EP P is a *p-EP* if p is the only predicate in P different from the (in)equality predicate. In [14], we showed how to rewrite any program P (without generalized literals) as an equivalent p-program P_p. We adapt that transformation to cope with generalized literals as well. For an EP P, let $in(Y) \equiv \cup\{Y \neq a \mid a \in preds(P) \cup \{0\}\}$, i.e. a set of inequalities between the variable Y and the predicates in P as well as a new constant 0. For a sequence of variables Y, we have $in(Y) \equiv \cup_{Y \in Y} in(Y)$. For a set of extended literals α, we construct α_p in two stages:

1. replace every regular m-ary atom $q(t)$ appearing in α (either in atoms, naf-atoms, or generalized literals) by $p(t, 0, q)$ where p has arity n, with n the maximum of the arities of predicates in P augmented by 1, 0 a sequence of new constants 0 of length $n - m - 1$, and q a new constant with the same name as the original predicate,
2. in the set thus obtained, replace every generalized literal $\forall Y \cdot \phi \Rightarrow \psi$ by $\forall Y \cdot \phi \wedge \bigwedge in(Y) \Rightarrow \psi$, where $Y \neq t$ in $in(Y)$ stands for $\neg(Y = t)$ (we defined generalized literals in function of boolean formulas of atoms).

The p-EP P_p is the program P with all non-free rules $r : \alpha \leftarrow \beta$ replaced by $r_p : \alpha_p \leftarrow \beta_p, in(X)$ where $vars(r) = X$. Note that P and P_p have the same free rules.

Example 4. Let P be the EP:

$$q(X) \leftarrow [\forall Y \cdot r(Y) \Rightarrow f(X, Y)]$$
$$r(a) \leftarrow$$
$$f(X, Y) \vee not\, f(X, Y) \leftarrow$$

Then q is satisfiable $(\{a, x\}, \{f(x, a), r(a), q(x)\})$. The p-EP P_p is

$$p(X, 0, q) \leftarrow [\forall Y \cdot p(Y, 0, r) \wedge \bigwedge in(Y) \Rightarrow p(X, Y, f)], in(X)$$
$$p(a, 0, r) \leftarrow$$
$$p(X, Y, f) \vee not\, p(X, Y, f) \leftarrow$$

where $in(X) = \{X \neq f, X \neq q, X \neq r, X \neq 0\}$. The corresponding open answer set for this program is $(\{a, x, f, r, q, 0\}, \{p(x, a, f), p(a, 0, r), p(x, 0, q)\})$. Note that the free rule in P_p may introduce unwanted literals $p(q, x, f)$, i.e. where X is grounded with a predicate q from P. Those unwanted literals will, however, never make non-free rules applicable since the latter have $X \neq q$ in the body, and hence the assumption that we only check satisfiability of non-free predicates.

Proposition 1. *Let P be an EP and q a predicate in P. q is satisfiable w.r.t. P iff there is an open answer set (U', M') of the p-EP P_p with $p(\boldsymbol{x}, \boldsymbol{0}, q) \in M'$.*

Note that the size of P_p is polynomial in the size of P.

In [3], a similar motivation drives the reduction of Horn clauses to clauses consisting of only one defined predicate. Their encoding does not introduce new constants to identify old predicates and depends entirely on the use of (in)equality.

As was shown in [14], we can reduce a p-program P (without generalized literals) to an equivalent FPL formula. We extend this translation for EPs, i.e. we take into account generalized literals. The *completion* $\mathtt{comp}(P)$ of an EP P consists of formulas that demand that different constants in P are interpreted as different elements:

$$a \neq b \tag{5}$$

for every pair of different constants a and b in P, and where $a \neq b \equiv \neg(a = b)$. $\mathtt{comp}(P)$ contains formulas ensuring the existence of at least one element in the domain of an interpretation:

$$\exists X \cdot \textbf{true} \tag{6}$$

Besides these technical requirements matching FOL interpretations with open interpretations, $\mathtt{comp}(P)$ contains the formulas in $\mathtt{fix}(P) = \mathtt{sat}(P) \cup \mathtt{gl}(P) \cup \mathtt{gli}(P) \cup \mathtt{fpf}(P)$, which can be intuitively categorized as follows: $\mathtt{sat}(P)$ ensures that a model of $\mathtt{fix}(P)$ satisfies all rules in P, $\mathtt{gl}(P)$ is an auxiliary component defining atoms that indicate when a rule in P belongs to the GL-reduct, $\mathtt{gli}(P)$ indicates when the antecedent of generalized literals are true, and finally $\mathtt{fpf}(P)$ ensures that every model of $\mathtt{fix}(P)$ is a minimal model of the GL-reduct of the GLi-reduct of P; it uses the atoms defined in $\mathtt{gl}(P)$ to select, for the calculation of the fixed point, only those rules in P that are in the GL-reduct of the GLi-reduct of P; the atoms defined in $\mathtt{gli}(P)$ ensure that the generalized literals are interpreted correctly.

We interpret a naf-atom *not a* in a FOL formula as the literal $\neg a$. Moreover, we assume that, if a set X is empty, $\bigwedge X = \textbf{true}$ and $\bigvee X = \textbf{false}$. In the following, we assume that the arity of p, the only predicate in a p-EP is n.

Definition 1. *Let P be a p-EP. The fixed point translation of P is $\mathtt{fix}(P) \equiv \mathtt{sat}(P) \cup \mathtt{gli}(P) \cup \mathtt{gl}(P) \cup \mathtt{fpf}(P)$, where*

1. $\mathtt{sat}(P)$ contains formulas

$$\forall \boldsymbol{Y} \cdot \bigwedge \beta \Rightarrow \bigvee \alpha \tag{7}$$

for rules $r : \alpha \leftarrow \beta \in P$ with $vars(r) = \boldsymbol{Y}$,
2. $\mathtt{gl}(P)$ contains the formulas

$$\forall \boldsymbol{Y} \cdot r(\boldsymbol{Y}) \Leftrightarrow \bigwedge \alpha^- \wedge \bigwedge \neg \beta^- \tag{8}$$

for rules $r : \alpha \leftarrow \beta \in P^{10}$ with $vars(r) = \boldsymbol{Y}$,

---

[10] We assume that rules are uniquely named.

3. $\texttt{gli}(P)$ *contains the formulas*

$$\forall \boldsymbol{Z} \cdot g(\boldsymbol{Z}) \Leftrightarrow \phi \tag{9}$$

for generalized literals $g : \forall \boldsymbol{Y} \cdot \phi \Rightarrow \psi \in P^{11}$ *where* ϕ *contains the variables* $\boldsymbol{Z}$,
4. $\texttt{fpf}(P)$ *contains the formula*

$$\forall \boldsymbol{X} \cdot p(\boldsymbol{X}) \Rightarrow [\text{LFP } W\boldsymbol{X}.\phi(W, \boldsymbol{X})](\boldsymbol{X}) \tag{10}$$

with

$$\phi(W, \boldsymbol{X}) \equiv W(\boldsymbol{X}) \vee \bigvee_{r:p(t)\vee\alpha\leftarrow\beta\in P} E(r) \tag{11}$$

and

$$E(r) \equiv \exists \boldsymbol{Y} \cdot X_1 = t_1 \wedge \ldots \wedge X_n = t_n \wedge \bigwedge \beta^+[p \mid W] \wedge \bigwedge \gamma \wedge r(\boldsymbol{Y}) \tag{12}$$

where $\boldsymbol{X} = X_1, \ldots, X_n$ *are* n *new variables,* $vars(r) = \boldsymbol{Y}$, W *is a new (second-order) variable,* $\beta^+[p \mid W]$ *is* β^+ *with* p *replaced by* W, *and* γ *is* β^x *with*
 – *every generalized literal* $g : \forall \boldsymbol{Y} \cdot \phi \Rightarrow \psi$ *replaced by* $\forall \boldsymbol{Y} \cdot g(\boldsymbol{Z}) \Rightarrow \psi$, $\boldsymbol{Z}$ *the variables of* ϕ, *and, subsequently,*
 – *every* p *replaced by* W.

The completion *is* $\texttt{comp}(P) \equiv \texttt{fix}(P) \cup \{a \neq b \mid a, b$ *different in* $cts(P)\} \cup \{\exists X \cdot \boldsymbol{true}\}$.

The predicate W appears only positively in $\phi(W, \boldsymbol{X})$ such that the fixed point formula in (10) is well-defined. Note that the predicate p is replaced by the fixed point variable W in $E(r)$ except in the antecedents of generalized literals, which were replaced by g-atoms, and the negative part of r, which were replaced by r-atoms, thus respectively encoding the GLi-reduct and the GL-reduct.

Example 5. We rewrite the program from Example 3 as the p-EP P.

$r_1 :$ $p(X, 0, q) \leftarrow p(X, Y, f), in(X), in(Y)$
$r_2 :$ $\leftarrow p(X, Y, f), not\ p(Y, 0, q), in(X), in(Y)$
$r_3 :$ $\leftarrow p(X, Y, f), not\ p(Y, 0, well), in(X), in(Y)$
$r_4 :$ $p(Y, 0, well) \leftarrow p(Y, 0, q), in(Y),$
                             $[\forall X \cdot p(X, Y, f) \wedge \bigwedge in(X) \Rightarrow p(X, 0, well)]$
$r_5 : p(X, Y, f) \vee not\ p(X, Y, f) \leftarrow$

where $in(X)$ and $in(Y)$ are shorthand for the inequalities with the new constants. $\texttt{sat}(P)$ consists of the sentences

 – $\forall X, Y \cdot p(X, Y, f) \wedge \bigwedge in(X) \wedge \bigwedge in(Y) \Rightarrow p(X, 0, q)$,
 – $\forall X, Y \cdot p(X, Y, f) \wedge \neg p(Y, 0, q) \wedge \bigwedge in(X) \wedge \bigwedge in(Y) \Rightarrow \boldsymbol{false}$,
 – $\forall X, Y \cdot p(X, Y, f) \wedge \neg p(Y, 0, well) \wedge \bigwedge in(X) \wedge \bigwedge in(Y) \Rightarrow \boldsymbol{false}$, and
 – $\forall Y \cdot p(Y, 0, q) \wedge \bigwedge in(Y) \wedge (\forall X \cdot p(X, Y, f) \wedge \bigwedge in(X) \Rightarrow p(X, 0, well))$
    $\Rightarrow p(Y, 0, well)$,
 – $\forall X, Y \cdot \boldsymbol{true} \Rightarrow p(X, Y, f) \vee \neg p(X, Y, f)$.

---

[11] We assume that generalized literals are named.

$gl(P)$ contains the sentences

- $\forall X, Y \cdot r_1(X, Y) \Leftrightarrow in(X) \wedge in(Y)$,
- $\forall X, Y \cdot r_2(X, Y) \Leftrightarrow \neg p(Y, 0, q) \wedge in(X) \wedge in(Y)$,
- $\forall X, Y \cdot r_3(X, Y) \Leftrightarrow \neg p(Y, 0, well) \wedge in(X) \wedge in(Y)$,
- $\forall Y \cdot r_4(Y) \Leftrightarrow in(Y)$, and
- $\forall X, Y \cdot r_5(X, Y) \Leftrightarrow p(X, Y, f)$.

$gli(P)$ contains the sentence $\forall X, Y \cdot g(X, Y) \Leftrightarrow p(X, Y, f) \wedge \bigwedge in(X)$, and $fpf(P)$ is constructed with

- $E(r_1) \equiv \exists X, Y \cdot X_1 = X \wedge X_2 = 0 \wedge X_3 = q \wedge W(X, Y, f) \wedge r_1(X, Y)$,
- $E(r_4) \equiv \exists Y \cdot X_1 = Y \wedge X_2 = 0 \wedge X_3 = well \wedge W(Y, 0, q) \wedge$
  $(\forall X \cdot g(X, Y) \Rightarrow W(X, 0, well)) \wedge r_4(Y)$.
- $E(r_5) \equiv \exists X, Y \cdot X_1 = X \wedge X_2 = Y \wedge X_3 = f \wedge r_5(X, Y)$.

Take an infinite FOL interpretation (U, M) with $U = \{q, f, well, 0, x_0, x_1, \ldots\}$ and[12]

$$M = \{p(x_0, 0, q), p(x_0, 0, well), p(x_0, x_1, f),$$
$$p(x_1, 0, q), p(x_1, 0, well), p(x_1, x_2, f), \ldots$$
$$r_1(x_0, x_0), r_1(x_0, x_1), \ldots, r_1(x_1, x_0), \ldots, r_4(x_0), r_4(x_1), \ldots$$
$$r_5(x_0, x_1), r_5(x_1, x_2), \ldots, g(x_0, x_1), g(x_1, x_2), \ldots\}) \ .$$

$sat(P), gl(P)$, and $gli(P)$ are satisfied. We check that $fpf(P)$ is satisfied by M. We first construct the fixed point of $\phi^{(U,M)}$ where $\phi(W, X_1, X_2, X_3) \equiv W(X_1, X_2, X_3) \vee E(r_1) \vee E(r_4) \vee E(r_5)$ as in [9], i.e. in stages starting from $W^0 = \emptyset$. We have that

- $W^1 = \phi^{(U,M)}(W^0) = \{(x_0, x_1, f), (x_1, x_2, f), \ldots\}$, where the (x_i, x_{i+1}, f) are introduced by $E(r_5)$,
- $W^2 = \phi^{(U,M)}(W^1) = W_1 \cup \{(x_0, 0, q), (x_1, 0, q), \ldots\}$, where the $(x_i, 0, q)$ are introduced by $E(r_1)$,
- $W^3 = \phi^{(U,M)}(W^2) = W_2 \cup \{(x_0, 0, well)\}$, where $(x_0, 0, well)$ is introduced by $E(r_4)$,
- $W^4 = \phi^{(U,M)}(W^3) = W_3 \cup \{(x_1, 0, well)\}$,
- $\ldots$

The least fixed point $\text{LFP}(\phi^{(U,M)})$ is then $\cup_{\alpha < \omega} W^\alpha$ [9]. The sentence $fpf(P)$ is then satisfied since every p-literal in M is also in this least fixed point. (U, M) is thus a model of $comp(P)$, and it corresponds to an open answer set of P.

Proposition 2. *Let P be a p-EP. Then, (U, M) is an open answer set of P iff $(U, M \cup R \cup G)$ is a model of $comp(P)$, where $R \equiv \{r(\boldsymbol{y}) \mid r[\boldsymbol{Y} \mid \boldsymbol{y}] : \alpha[] \leftarrow \beta[] \in P_U, M \models \alpha[]^- \cup not\ \beta[]^-, vars(r) = \boldsymbol{Y}\}$, i.e. the atoms corresponding to rules for which the GLi-reduct version will be in the GL-reduct, and $G \equiv \{g(\boldsymbol{z}) \mid g : \forall \boldsymbol{Y} \cdot \phi \Rightarrow \psi \in P, vars(\phi) = \boldsymbol{Z}, M \models \phi[\boldsymbol{Z} \mid \boldsymbol{z}]\}$, i.e. the atoms corresponding to true antecedents of generalized literals in P.*

---

[12] We interpret the constants in $comp(P)$ by universe elements of the same name.

Using Propositions 1 and 2, we can reduce satisfiability checking in OASP to satisfiability checking in FPL. Moreover, since $\text{comp}(P)$ contains only one fixed point predicate, the translation falls in the alternation-free fragment of FPL. If the number of constants in a program P is c, then the number of formulas $a \neq b$ is $\frac{1}{2}c(c-1)$; since the rest of $\text{comp}(P)$ is linear in P, this yields a quadratic bound for the size of $\text{comp}(P)$.

Theorem 1. *Let P be an EP and q an n-ary predicate in P. q is satisfiable w.r.t. P iff $\exists X \cdot p(X, 0, q) \land \bigwedge \text{comp}(P_p)$ is satisfiable. Moreover, this reduction is polynomial.*

4   Open Answer Set Programming with Guarded Extended Programs

We repeat the definitions of the *guarded fragment* [2] of first-order logic as in [11]: *The guarded fragment GF of first-order logic is defined inductively as follows:*

(1) *Every relational atomic formula belongs to GF.*
(2) *GF is closed under propositional connectives $\neg$, $\land$, $\lor$, $\Rightarrow$, and $\Leftrightarrow$.*
(3) *If X, Y are tuples of variables, $\alpha(X, Y)$ is an atomic formula, and $\psi(X, Y)$ is a formula in GF such that $free(\psi) \subseteq free(\alpha) = X \cup Y$, then the formulas*

$$\exists Y \cdot \alpha(X, Y) \land \psi(X, Y)$$
$$\forall Y \cdot \alpha(X, Y) \Rightarrow \psi(X, Y)$$

belong to GF, (where $free(\psi)$ are the free variables of ψ). $\alpha(X, Y)$ is the guard of the formula.

The *guarded fixed point logic μGF* is GF extended with fixed point formulas (2) where $\psi(W, X)$ is a formula such that W does not appear in guards.

Definition 2. *A generalized literal $\forall Y \cdot \phi \Rightarrow \psi$ is guarded if ϕ is of the form $\gamma \land \phi'$ with γ an atom, and $vars(Y) \cup vars(\phi') \cup vars(\psi) \subseteq vars(\gamma)$; we call γ the guard of the generalized literal. A rule $r : \alpha \leftarrow \beta$ is guarded if every generalized literal in r is guarded, and there is an atom $\gamma_b \in \beta^+$ such that $vars(r) \subseteq vars(\gamma_b)$; we call γ_b a body guard of r. It is fully guarded if it is guarded and there is a $\gamma_h \subseteq \alpha^-$ such that $vars(r) \subseteq vars(\gamma_h)$; γ_h is called a head guard of r.*

An EP P is a (fully) guarded EP ((F)GEP) if every non-free rule in P is (fully) guarded.

Example 6. Reconsider the EP from Example 3. r_1, r_2, and r_3 are guarded with guard $f(X, Y)$. The generalized literal in r_4 is guarded by $f(X, Y)$, and r_4 itself is guarded by $q(Y)$. Note that r_5 does not influence the guardedness as it is a free rule.

Every fully guarded EP is guarded. Vice versa, we can transform every guarded EP into an equivalent fully guarded one.

Example 7. Take the guarded EP consisting of the rules r_1 and r_5 from Example 3. We rewrite r_1 as the fully guarded rule $q(X) \lor not\ f(X, Y) \leftarrow f(X, Y)$, i.e. take the body guard and write it negated in the head, where it serves as head guard. Intuitively, rules in the original EP where the body guard cannot be satisfied are removed in the GL-reduct of the new EP; if the body guard is true then the GL-reduct removes the head guard from the head. The effect is in both cases the same.

For a GEP P, P^f is P with the non-free rules $\alpha \leftarrow \beta$ replaced by $\alpha \cup not\ \gamma_b \leftarrow \beta$ for the body guard γ_b of $\alpha \leftarrow \beta$. For a GEP P, we have that P^f is a FGEP, where the head guard of each non-free rule is equal to the body guard. Moreover, the size of P^f is linear in the size of P.

Proposition 3. *Let P be a GEP. An open interpretation (U, M) of P is an open answer set of P iff (U, M) is an open answer set of P^f.*

We have that the construction of a p-EP retains the guardedness properties.

Proposition 4. *Let P be an EP. Then, P is a (F)GEP iff P_p is a (F)GEP.*

For a fully guarded p-EP P, we can rewrite $\text{comp}(P)$ as the equivalent μGF formulas $\text{gcomp}(P)$. For a guarded generalized literal $\xi \equiv \forall Y \cdot \phi \Rightarrow \psi$, define $\xi^g = \forall Y \cdot \gamma \Rightarrow \psi \vee \neg\phi'$, where, since the generalized literal is guarded, $\phi = \gamma \wedge \phi'$, and $vars(Y) \cup vars(\phi') \cup vars(\psi) \subseteq vars(\gamma)$, making formula ξ^g a guarded formula. The extension of this operator for sets (or boolean formulas) of generalized literals is as usual.

$\text{gcomp}(P)$ is $\text{comp}(P)$ with the following modifications.

- Formula $\exists X \cdot \textbf{true}$ is replaced by

$$\exists X \cdot X = X \ , \tag{13}$$

such that it is guarded by $X = X$.
- Formula (7) is removed if $r : \alpha \leftarrow \beta$ is free or otherwise replaced by

$$\forall Y \cdot \gamma_b \Rightarrow \bigvee \alpha \vee \bigvee \neg(\beta^+ \setminus \{\gamma_b\}) \vee \bigvee \beta^- \vee \bigvee \neg(\beta^X)^g \ , \tag{14}$$

where γ_b is a body guard of r, thus we have logically rewritten the formula such that it is guarded. If r is a free rule of the form $q(t) \vee not\ q(t) \leftarrow$ we have $\forall Y \cdot \textbf{true} \Rightarrow q(t) \vee \neg q(t)$ which is always true and can thus be removed from $\text{comp}(P)$.
- Formula (8) is replaced by the formulas

$$\forall Y \cdot r(Y) \Rightarrow \bigwedge \alpha^- \wedge \bigwedge \neg\beta^- \tag{15}$$

and

$$\forall Y \cdot \gamma_h \Rightarrow r(Y) \vee \bigvee \beta^- \vee \bigvee \neg(\alpha^- \setminus \{\gamma_h\}) \ , \tag{16}$$

where γ_h is a head guard of $\alpha \leftarrow \beta$. We thus rewrite an equivalence as two implications where the first implication is guarded by $r(Y)$ and the second one is guarded by the head guard of the rule - hence the need for a fully guarded program, instead of just a guarded one.
- Formula (9) is replaced by the formulas

$$\forall Z \cdot g(Z) \Rightarrow \phi \tag{17}$$

and

$$\forall Z \cdot \gamma \Rightarrow g(Z) \vee \neg\phi' \tag{18}$$

where $\phi = \gamma \wedge \psi$ by the guardedness of the generalized literal $\forall Y \cdot \phi \Rightarrow \psi$. We thus rewrite an equivalence as two implications where the first one is guarded by $g(Z)$ ($vars(\phi) = Z$ by definition of g), and the second one is guarded by γ ($vars(g(Z) \vee \neg\phi') = vars(Z) = vars(\gamma)$).

– For every $E(r)$ in (10), replace $E(r)$ by

$$E'(r) \equiv \bigwedge_{t_i \notin \mathbf{Y}} X_i = t_i \wedge \exists \mathbf{Z} \cdot (\bigwedge \beta^+[p|W] \wedge \bigwedge \gamma \wedge r(\mathbf{Y}))[t_i \in \mathbf{Y}|X_i] , \quad (19)$$

with $\mathbf{Z} = \mathbf{Y} \setminus \{t_i \mid t_i \in \mathbf{Y}\}$, i.e. move all $X_i = t_i$ where t_i is constant out of the scope of the quantifier, and remove the others by substituting each t_i in $\bigwedge \beta^+[p|W] \wedge \bigwedge \gamma \wedge r(\mathbf{Y})$ by X_i. This rewriting makes sure that every (free) variable in the quantified part of $E'(R)$ is guarded by $r(\mathbf{Y})[t_i \in \mathbf{Y}|X_i]$.

Example 8. The rule $r : p(X) \vee not\ p(X) \leftarrow p(X), [\forall Y \cdot p(Y) \wedge p(b) \Rightarrow p(a)]$ constitutes a fully guarded p-EP P. The generalized literal is guarded by $p(Y)$ and the rule by head and body guard $p(X)$. $\mathtt{sat}(P)$ contains the formula $\forall X \cdot p(X) \wedge (\forall Y \cdot p(Y) \wedge p(b) \Rightarrow p(a)) \Rightarrow p(X) \vee \neg p(X)$, $\mathtt{gl}(P)$ consists of $\forall X \cdot r(X) \Leftrightarrow p(X)$, $\mathtt{gli}(P)$ is the formula $\forall Y \cdot g(Y) \Leftrightarrow p(Y) \wedge p(b)$ and $E(r) \equiv \exists X \cdot X_1 = X \wedge W(X) \wedge (\forall Y \cdot g(Y) \Rightarrow W(a)) \wedge r(X)$.

$\mathtt{gcomp}(P)$ consists then of the corresponding guarded formulas:

– $\forall X \cdot p(X) \Rightarrow p(X) \vee \neg p(X) \vee \neg(\forall Y \cdot p(Y) \Rightarrow p(a) \vee \neg p(b))$,
– $\forall X \cdot r(X) \Rightarrow p(X)$,
– $\forall X \cdot p(X) \Rightarrow r(X)$,
– $\forall Y \cdot g(Y) \Rightarrow p(Y) \wedge p(b)$,
– $\forall Y \cdot p(Y) \Rightarrow g(Y) \vee \neg p(b)$, and
– $E'(r) \equiv W(X_1) \wedge (\forall Y \cdot g(Y) \Rightarrow W(a)) \wedge r(X_1)$.

As $\mathtt{gcomp}(P)$ is basically a linear logical rewriting of $\mathtt{comp}(P)$, they are equivalent. Moreover, $\bigwedge \mathtt{gcomp}(P)$ is an alternation-free μGF formula.

Proposition 5. *Let P be a fully guarded p-EP. (U, M) is a model of $\mathtt{comp}(P)$ iff (U, M) is a model of $\mathtt{gcomp}(P)$.*

Proposition 6. *Let P be a fully guarded p-EP. Then, $\bigwedge \mathtt{gcomp}(P)$ is an alternation-free μGF formula.*

For a GEP P, we have that P^f is a FGEP. By Proposition 4, we have that $(P^f)_p$ is a fully guarded p-EP, thus the formula $\mathtt{gcomp}((P^f)_p)$ is defined. By Proposition 3, q is satisfiable w.r.t. P iff q is satisfiable w.r.t. P^f. By Theorem 1, we have that q is satisfiable w.r.t. P^f iff $\exists X \cdot p(X, 0, q) \wedge \bigwedge \mathtt{comp}((P^f)_p)$ is satisfiable. Finally, Proposition 5 yields that q is satisfiable w.r.t. P iff $\exists X \cdot p(X, 0, q) \wedge \bigwedge \mathtt{gcomp}((P^f)_p)$ is satisfiable.

The polynomial reduction in Theorem 1 is the worst reduction used, thus yielding the upper bound for the overall reduction.

Theorem 2. *Let P be a GEP and q an n-ary predicate in P. q is satisfiable w.r.t. P iff $\exists X \cdot p(X, 0, q) \wedge \bigwedge \mathtt{gcomp}((P^f)_p)$ is satisfiable. Moreover, this reduction is polynomial.*

For a GEP P, we have, by Proposition 6, that $\bigwedge \mathtt{gcomp}((P^f)_p)$ is an alternation-free μGF formula such that $\exists X \cdot p(X, 0, q) \wedge \bigwedge \mathtt{gcomp}((P^f)_p)$ is a μGF sentence.

Corollary 1. *Satisfiability checking w.r.t. GEPs can be polynomially reduced to satisfiability checking of alternation-free μGF-formulas.*

Since satisfiability checking of μGF formulas is 2-EXPTIME-complete (Proposition [1.1] in [11]), satisfiability checking w.r.t. GEPs is, by Corollary 1, in 2-EXPTIME.

Corollary 2. *Satisfiability checking w.r.t. GEPs is in 2-EXPTIME.*

Thus, adding generalized literals to guarded programs does not come at the cost of increased complexity of reasoning, as also for guarded programs without generalized literals, reasoning is in 2-EXPTIME [14]. In [14], we established 2-EXPTIME-completeness for satisfiability checking w.r.t. guarded programs (without generalized literals). Since every guarded program is a GEP, 2-EXPTIME-hardness w.r.t. GEPs follows.

Theorem 3. *Satisfiability checking w.r.t. GEPs is 2-EXPTIME-complete.*

To conclude this section, we illustrate the use of open answer set programming with GEPs as a general purpose knowledge representation formalism by simulating satisfiability checking of *computation tree logic (CTL)* [4, 5] formulas. Let AP be the finite set of available proposition symbols. Computation tree logic (CTL) formulas are defined as follows[13]: every proposition symbol $P \in AP$ is a formula, if p and q are formulas, so are $p \wedge q$ and $\neg p$, if p and q are formulas, then EGp, E$(p \cup q)$, and EXp are formulas. The semantics of a CTL formula is given by *(temporal) structures*. A structure K is a tuple (S, R, L) with S a countable set of states, $R \subseteq S \times S$ a total relation on S, i.e. $\forall s \in S \cdot \exists t \in S \cdot (s, t) \in R$, and $L : S \rightarrow 2^{AP}$ a function labeling states with propositions. Intuitively, R indicates the permitted transitions between states and L indicates which propositions are true at certain states.

A path π in K is an infinite sequence of states $(s_0, s_1, \ldots)$ such that $(s_{i-1}, s_i) \in R$ for each $i > 0$. For a path $\pi = (s_0, s_1, \ldots)$, we denote the element s_i with π_i. For a structure $K = (S, R, L)$, a state $s \in S$, and a formula p, we inductively define when K is a *model* of p at s, denoted $K, s \models p$:

- $K, s \models P$ iff $P \in L(s)$ for $P \in AP$,
- $K, s \models \neg p$ iff not $K, s \models p$.
- $K, s \models p \wedge q$ iff $K, s \models p$ and $K, s \models q$,
- $K, s \models$ EGp iff there exists a path π in K with $\pi_0 = s$ and $\forall k \geq 0 \cdot K, \pi_k \models p$,
- $K, s \models$ E$(p \cup q)$ iff there exists a path π in K with $\pi_0 = s$ and $\exists k \geq 0 \cdot (K, \pi_k \models q \wedge \forall j < k \cdot K, \pi_j \models p)$,
- $K, s \models$ EXp iff there is a $(s, t) \in R$ and $K, t \models p$.

The expression $K, s \models$ EGp can be read as "there is some path from s along which p holds Globally (everywhere)", $K, s \models$ EXp as "there is some neXt state where p holds", and $K, s \models$ E$(p \cup q)$ as "there is some path from s along which p holds Until q holds (and q eventually holds)". A structure $K = (S, R, L)$ *satisfies* a CTL formula p if there is a state $s \in S$ such that $K, s \models p$; we also call K a *model* of p. A CTL formula p is *satisfiable* iff there is a model of p.

For a CTL formula p, let $clos(p)$ be the *closure* of p: the set of subformulas of p. We construct a GEP $G \cup D_p$ consisting of a generating part G and a defining part D_p. The

---

[13] In order to make the treatment as simple as possible, we do not include formulas involving the path quantifier A. However, as indicated in [15], the defined constructs are adequate, i.e. every CTL formula can be rewritten using only those, while preserving satisfiability.

guarded program G contains free rules (g_1) for every proposition $P \in AP$, free rules (g_2) allowing for state transitions, and rules (g_3) that ensure that the transition relation is total:

$$[P](S) \lor not\ [P](S) \leftarrow \qquad\qquad\qquad (g_1)$$

$$next(S, N) \lor not\ next(S, N) \leftarrow \qquad\qquad\qquad (g_2)$$

$$succ(S) \leftarrow next(S, N) \qquad \leftarrow S = S, not\ succ(S) \quad (g_3)$$

where $[P]$ is the predicate corresponding to the proposition P. The $S = S$ is necessary merely for having guarded rules; note that any rule containing only one (free) variable can be made guarded by adding such an equality.

The GEP D_p introduces for every non-propositional CTL formula in $clos(p)$ the following rules (we write $[q]$ for the predicate corresponding to the CTL formula $q \in clos(p)$); as noted before we tacitly assume that rules containing only one (free) variable S are guarded by $S = S$:

$$[\neg q](S) \leftarrow not\ [q](S) \qquad\qquad\qquad (d_1)$$

$$[q \land r](S) \leftarrow [q](S), [r](S) \qquad\qquad\qquad (d_2)$$

$$[\mathsf{EG}q](S) \leftarrow not\ [\mathsf{AF}\neg q](S) \qquad\qquad\qquad (d_3^1)$$

$$[\mathsf{AF}\neg q](S) \leftarrow not\ [q](S) \qquad\qquad\qquad (d_3^2)$$

$$[\mathsf{AF}\neg q](S) \leftarrow \forall N \cdot next(S, N) \Rightarrow [\mathsf{AF}\neg q](N) \qquad (d_3^3)$$

$$[\mathsf{E}(q\ \mathsf{U}\ r)](S) \leftarrow [r](S) \qquad\qquad\qquad (d_4)$$

$$[\mathsf{E}(q\ \mathsf{U}\ r)](S) \leftarrow [q](S), next(S, N), [\mathsf{E}(q\ \mathsf{U}\ r)](N) \qquad (d_5)$$

$$[\mathsf{EX}q](S) \leftarrow next(S, N), [q](N) \qquad\qquad\qquad (d_6)$$

The rules $(d_{\{1,2,6\}})$ are direct translations of the CTL semantics.

Rules (d_3^2) and (d_3^3) ensure that if $[\mathsf{AF}\neg q](s)$ holds, then on all paths from s we eventually reach a state where q does not hold. In particular this is true if q does not hold in the current state (d_3^2), or if it holds for all successors (d_3^3); minimality of open answer sets ensures that after a finite time a state where q does not hold is reached. Rule (d_3^1) then defines $[\mathsf{EG}q]$ as the negation of $[\mathsf{AF}\neg q]$.

Rules (d_4) and (d_5) are in accordance with the characterization $\mathsf{E}(q\ \mathsf{U}\ r) \equiv r \lor (q \land \mathsf{EXE}(q\ \mathsf{U}\ r))$ [4], and make implicit use of the minimality of answer sets to eventually ensure realization of r.

Theorem 4. *Let p be a CTL formula. p is satisfiable iff $[p]$ is satisfiable w.r.t. the GEP $G \cup D_p$.*

Since CTL satisfiability checking is EXPTIME-complete [4] and satisfiability checking w.r.t. GEPs is 2-EXPTIME-complete (Theorem 3), the reduction from CTL to GEPs does not seem to be optimal. However, we can show that the particular GEP $G \cup D_p$ is a *bound* GEP for which reasoning is indeed EXPTIME-complete and thus optimal.

Define the *width* of a formula ψ as the maximal number of free variables in its subformulas [10]. We define *bound* programs by looking at their first order form and the arity of its predicates.

Definition 3. *Let P be an EP. Then, P is* bound *if every formula in* $\mathrm{sat}(P)$ *is of bounded width and the predicates in P have a bounded arity.*

For a CTL formula p, one has that $G \cup D_p$ is a bound GEP. Indeed, every subformula of formulas in $\mathrm{sat}(G \cup D_p)$ contains at most 2 free variables and the maximum arity of the predicates is 2 as well.

Let P be a bound GEP. We have that $(P^f)_p$ is bound and one can check that $\exists \boldsymbol{X} \cdot p(\boldsymbol{X}, \boldsymbol{0}, q) \wedge \bigwedge \mathrm{gcomp}((P^f)_p)$ is of bounded width.

Using Theorem 2, one can reduce satisfiability checking of a bound GEP to satisfiability of a μGF-formula with bounded width. The latter can be done in EXPTIME by Theorem 1.2 in [11], such that satisfiability checking w.r.t. bound GEPs is in EXPTIME.

The EXPTIME-hardness follows from Theorem 4 and the EXPTIME-hardness of CTL satisfiability checking [4].

Theorem 5. *Satisfiability checking w.r.t. bound GEPs is* EXPTIME-*complete.*

5   Equivalence with Datalog LITE

We define *Datalog* LITE as in [8]. A *Datalog rule* is a rule $\alpha \leftarrow \beta$ where $\alpha = \{a\}$ for some atom a and β does not contain generalized literals. A *basic Datalog program* is a finite set of Datalog rules such that no head predicate appears in negative bodies of rules. Predicates that appear only in the body of rules are *extensional* or *input* predicates. Note that equality is, by the definition of rules, never a head predicate and thus always extensional. The semantics of a basic Datalog program P, given a relational input structure $\mathcal{U}$ defined over extensional predicates of P^{14}, is given by the unique (subset) minimal model whose restriction to the extensional predicates yields $\mathcal{U}$. We refer to [1] for more details.

For a query (P, q), where P is a basic Datalog program and q is a n-ary predicate, we write $\boldsymbol{a} \in (P, q)(\mathcal{U})$ if the minimal model M of Σ_P with input $\mathcal{U}$ contains $q(\boldsymbol{a})$, where Σ_P are the first-order clauses corresponding to P, see [1]. We call (P, q) satisfiable if there exists a $\mathcal{U}$ and an $\boldsymbol{a}$ such that $\boldsymbol{a} \in (P, q)(\mathcal{U})$.

A program P is a *stratified Datalog program* if it can be written as a union of basic Datalog programs $(P_1, \ldots, P_n)$, so-called *strata*, such that each of the head predicates in P is a head predicate in exactly one stratum P_i. Furthermore, if a head predicate in P_i is an extensional predicate in P_j, then $i < j$. This definition entails that head predicates in the positive body of rules are head predicates in the same or a lower stratum, and head predicates in the negative body are head predicates in a lower stratum. The semantics of stratified Datalog programs is defined stratum per stratum, starting from the lowest stratum and defining the extensional predicates on the way up. For an input structure $\mathcal{U}$ and a stratified program $P = (P_1, \ldots, P_n)$, define as in [1]:

$$\mathcal{U}_0 \equiv \mathcal{U}$$
$$\mathcal{U}_i \equiv \mathcal{U}_{i-1} \cup P_i(\mathcal{U}_{i-1} | edb(P_i))$$

---

[14] We assume that an input structure always defines equality, and that it does so as the identity relation.

where $S_i \equiv P_i(\mathcal{U}_{i-1}|edb(P_i))$ is the minimal model of Σ_{P_i} among those models of Σ_{P_i} whose restriction to the extensional predicates of P_i is equal to $\mathcal{U}_{i-1}|edb(P_i)$. The least fixed point model of P is per definition $\mathcal{U}_n$.

A *Datalog* LITE *generalized literal* is a generalized literal $\forall Y \cdot a \Rightarrow b$ where a and b are atoms and $vars(b) \subseteq vars(a)$. Note that Datalog LITE generalized literals $\forall Y \cdot a \Rightarrow b$ can be replaced by the equivalent $\forall Z \cdot a \Rightarrow b$ where $Z \equiv Y \setminus \{Y \mid Y \notin vars(a)\}$, i.e. with the variables that are not present in the formula $a \Rightarrow b$ removed from the quantifier. After such a rewriting, Datalog LITE generalized literals are guarded according to Definition 2.

A *Datalog* LITE *program* is a stratified Datalog program, possibly containing Datalog LITE generalized literals in the positive body, where each rule is *monadic* or *guarded*. A rule is monadic if each of its (generalized) literals contains only one (free) variable; it is guarded if there exists an atom in the positive body that contains all variables (free variables in the case of generalized literals) of the rule. The definition of stratified is adapted for generalized literals: for a $\forall Y \cdot a \Rightarrow b$ in the body of a rule where the underlying predicate of a is a head predicate, this head predicate must be a head predicate in a lower stratum (i.e. a is treated as a naf-atom) and a head predicate underlying b must be in the same or a lower stratum (i.e. b is treated as an atom). The semantics can be adapted accordingly since a is completely defined in a lower stratum, as in [8]: every generalized literal $\forall Y \cdot a \Rightarrow b$ is instantiated (for any x grounding the free variables X in the generalized literal) by $\bigwedge\{b[X \mid x][Y \mid y] \mid a[X \mid x][Y \mid y] \text{ is true}\}$, which is well-defined since a is defined in a lower stratum than the rule where the generalized literal appears.

For stratified Datalog programs, least fixed point models with as input the identity relation on a universe U coincide with open answer sets with universe U.

Proposition 7. *Let $P = (P_1, \ldots, P_n)$ be a stratified Datalog program, possibly with generalized literals, U a universe for P, and l a literal. For the least fixed point model $\mathcal{U}_n$ of P with input $\mathcal{U} = \{id(U)\}$, we have $\mathcal{U}_n \models l$ iff there exists an open answer set (U, M) of P such that $M \models l$.*

Moreover, for any open answer set (U, M) of P, we have that $M = \mathcal{U}_n \setminus id(U)$.

From Proposition 7, we obtain a generalization of Corollary 2 in [6] (*If Π is stratified, then its unique stable model is identical to its fixed point model.*) for stratified programs with generalized literals and an open answer set semantics.

Corollary 3. *Let P be a stratified Datalog program, possibly with generalized literals, and U a universe for P. The unique open answer set (U, M) of P is identical to its least fixed point model (minus the equality atoms) with input structure $id(U)$.*

We generalize Proposition 7, to take into account arbitrary input structures $\mathcal{U}$. For a stratified Datalog program P, possibly with generalized literals, define $F_P \equiv \{q(X) \vee not\ q(X) \leftarrow \mid q \text{ extensional (but not =) in } P\}$.

Proposition 8. *Let $P = (P_1, \ldots, P_n)$ be a stratified Datalog program, possibly with generalized literals, and l a literal. There exists an input structure $\mathcal{U}$ for P with least fixed point model $\mathcal{U}_n$ such that $\mathcal{U}_n \models l$ iff there exists an open answer set (U, M) of $P \cup F_P$ such that $M \models l$.*

The set of free rules F_P ensures a free choice for extensional predicates, a behavior that corresponds to the free choice of an input structure for a Datalog program P. Note that $P \cup F_P$ is not a Datalog program anymore, due to the presence of naf in the heads of F_P.

Define a Datalog LITEM program as a Datalog LITE program where all rules are guarded. As we will see below this is not a restriction. As F_P contains only free rules, $P \cup F_P$ is a GEP if P is a Datalog LITEM program. Furthermore, the size of the GEP $P \cup F_P$ is linear in the size of P.

Proposition 9. *Let P be a Datalog LITEM program. Then, $P \cup F_P$ is a GEP whose size is linear in the size of P.*

By Propositions 8 and 9, satisfiability checking of Datalog LITEM queries can be reduced to satisfiability checking w.r.t. GEPs.

Theorem 6. *Let (P, q) be a Datalog LITEM query. Then, (P, q) is satisfiable iff q is satisfiable w.r.t. $P \cup F_P$. Moreover, this reduction is linear.*

With a similar reasoning as in [14], one can show that the opposite direction holds as well. In [8], Theorem 8.5., a Datalog LITEM query (π_φ, q_φ) was defined for an alternation-free μGF sentence φ such that $(U, M) \models \varphi$ iff $(\pi_\varphi, q_\varphi)(M \cup id(U))$ evaluates to true, where the latter means that q_φ is in the least fixed point model of $(\pi_\varphi, q_\varphi)(M \cup id(U))$. For the formal details of this reduction, we refer to [8]. Satisfiability checking w.r.t. GEPs can then be polynomially reduced to satisfiability checking in Datalog LITEM. Indeed, by Theorem 2, we have that q is satisfiable w.r.t. a GEP P iff $\varphi \equiv \exists \boldsymbol{X} \cdot p(\boldsymbol{X}, \boldsymbol{0}, q) \wedge \mathrm{gcomp}((P^f)_p)$ is satisfiable. Since φ is an alternation-free μGF sentence, we have that φ is satisfiable iff (π_φ, q_φ) is satisfiable. By Theorem 2, the translation of P to φ is polynomial in the size of P and the query (π_φ, q_φ) is quadratic in φ [8], resulting in a polynomial reduction.

Theorem 7. *Let P be a GEP, q an n-ary predicate in P and φ the μGF sentence $\exists \boldsymbol{X} \cdot p(\boldsymbol{X}, \boldsymbol{0}, q) \wedge \mathrm{gcomp}((P^f)_p)$. q is satisfiable w.r.t. P iff (π_φ, q_φ) is satisfiable. Moreover, this reduction is polynomial.*

Theorems 6 and 7 lead to the conclusion that Datalog LITEM and open ASP with GEPs are equivalent (i.e. satisfiability checking in either one of the formalisms can be polynomially reduced to satisfiability checking in the other). Furthermore, since Datalog LITEM, Datalog LITE, and alternation-free μGF are equivalent as well [8], we have the following concluding result.

Theorem 8. *Datalog LITE, alternation-free μGF, and open ASP with GEPs are equivalent.*

6 ω-Restricted Logic Programs

A class of logic programs with function symbols are the ω-*restricted programs* from [19]. The Herbrand Universe of ω-restricted programs is possibly infinite (in the presence of function symbols), however, answer sets are guaranteed to be finite, exactly by the structure of ω-restricted programs. Informally, an ω-restricted program consists of a stratified part and a part that cannot be stratified (the ω-*stratum*), where every rule is such that every variable in a rule is "guarded" by an atom of which the predicate is defined in a strictly lower stratum. The answer sets of ω-restricted programs can then be computed by instantiating the strata from the bottom up. We refer to [19] for precise definitions.

We extend the definition of universe for programs that contain function symbols. A *universe U* for a program P is a non-empty countable superset of the Herbrand Universe $\mathcal{H}_P$ of P. Thus, a universe U is equal to $\mathcal{H}_P \cup X$ for some countable X; as usual, we call the elements from $U \setminus \mathcal{H}_P$ *anonymous*.

For ω-restricted programs, the open answer set semantics coincides with the normal answer set semantics.

Theorem 9. *Let P be an ω-restricted program and U a universe for P. (U, M) is an open answer set of P iff M is an answer set of P.*

Since checking whether there exists an answer set of an ω-restricted program is in general 2-NEXPTIME-complete [19], we have, with Theorem 9, 2-NEXPTIME-completeness for consistency checking under the open answer set semantics for ω-restricted programs, where consistency checking involves checking whether there exists an open answer set of a program.

Theorem 10. *Consistency checking w.r.t. ω-restricted programs is 2-NEXPTIME-complete.*

Furthermore, since reasoning with ω-restricted programs is implemented in the SMOD-ELS reasoner [18], Theorem 9 implies an implementation of the open answer set semantics for ω-restricted programs as well.

In [20], ω-restricted programs allow for *cardinality constraints* and *conditional literals*. Conditional literals have the form $X.L : A$ where X is a set of variables, A is an atom (the condition) and L is an atom or a naf-atom. Intuitively, conditional literals correspond to generalized literals $\forall X \cdot A \Rightarrow L$, i.e., the defined reducts add instantiations of L to the body if the corresponding instantiation of A is true. However, conditional literals appear only in cardinality constraints $Card(b, S)$[15] where S is a set of literals (possibly conditional), such that a *for all* effect such as with generalized literals cannot be obtained with conditional literals.

Take, for example, the rule $q \leftarrow [\forall X \cdot b(X) \Rightarrow a(X)]$ and a universe $U = \{x_1, x_2\}$ with an interpretation containing $b(x_1)$ and $b(x_2)$. The reduct will contain a rule $q \leftarrow a(x_1), a(x_2)$ such that, effectively, q holds only if a holds everywhere where b holds. The equivalent rule rewritten with a conditional literal would be something like $q \leftarrow Card(n, \{X.a(X) : b(X)\})$, resulting in a rule $q \leftarrow Card(n, \{a(x_1), a(x_2)\})$. In order to have the *for all* effect, we have that n must be 2. However, we cannot know this n in advance, making it impossible to express a *for all* restriction.

Further note that consistent ω-restricted programs (with cardinality constraints and conditional literals) always have finite answer sets, which makes a reduction from GEPs (in which infinity axioms can be expressed) to ω-restricted programs non-trivial.

---

[15] $Card(b, S)$ is true if at least b elements from S are true.

7   Conclusions and Directions for Further Research

We defined GEPs, guarded programs with generalized literals, under an open answer set semantics, and showed 2-EXPTIME-completeness of satisfiability checking by a reduction to μGF. Furthermore, we translated Datalog LITEM programs to GEPs, and generalized the result that the unique answer set of a stratified program is identical to its least fixed point.

We plan to extend GEPs to loosely guarded EPs, where a guard may be a set of atoms; a reduction to the loosely guarded fixed point logic should then provide for decidability. More liberal generalized literals, with the consequent a conjunction of atoms and naf-atoms instead of just an atom, does not affect the definition of the GLi-reduct, but the FPL translation requires modification to ensure no fixed point variable appears negatively.

We plan to look into the correspondence with Datalog and use decidability results for Datalog satisfiability checking, as, e.g., in [12], to search for decidable fragments under an open answer set semantics.

Although adding generalized literals to guarded programs does not increase the complexity of reasoning, it does seem to increase expressivity: one can, for example, express infinity axioms. Given the close relation with Datalog LITE and the fact that Datalog LITE without generalized literals cannot express well-founded statements, it seems unlikely that guarded programs without generalized literals can express infinity axioms; this is subject to further research.

We only considered generalized literals in the positive body. If the antecedents in generalized literals are atoms, it seems intuitive to allow also generalized literals in the negative body. E.g., take a rule $\alpha \leftarrow \beta, not\ [\forall X \cdot b(X) \Rightarrow a(X)]$; it seems natural to treat $not\ [\forall X \cdot b(X) \Rightarrow a(X)]$ as $\exists X \cdot b(X) \wedge \neg a(X)$ such that the rule becomes $\alpha \leftarrow \beta, b(X), not\ a(X)$. A rule like $[\forall X \cdot b(X) \Rightarrow a(X)] \vee \alpha \leftarrow \beta$ is more involved and it seems that the generalized literal can only be intuitively removed by a modified GLi-reduct.

We established the equivalence of open ASP with GEPs, alternation-free μGF, and Datalog LITE. Intuitively, Datalog LITE is not expressive enough to simulate normal μGF since such μGF formulas could contain negated fixed point variables, which would result in a non-stratified program when translating to Datalog LITE [8]. Open ASP with GEPs does not seem to be sufficiently expressive either: fixed point predicates would need to appear under negation as failure, however, the GL-reduct removes naf-literals, such that, intuitively, there is no real recursion through naf-literals. Note that it is unlikely (but still open) whether alternation-free μGF and normal μGF are equivalent, i.e., whether the alternation hierarchy can always be collapsed.

References

1. S. Abiteboul, R. Hull, and V. Vianu. *Foundations of Databases*. Addison-Wesley, 1995.
2. J. Van Benthem. Dynamic Bits and Pieces. In *ILLC research report*. University of Amsterdam, 1997.
3. A. K. Chandra and D. Harel. Horn Clauses and the Fixpoint Query Hierarchy. In *Proc. of PODS '82*, pages 158–163. ACM Press, 1982.

4. E. A. Emerson. Temporal and Modal Logic. In J. van Leeuwen, editor, *Handbook of Theoretical Computer Science*, pages 995–1072. Elsevier Science Publishers B.V., 1990.
5. E. A. Emerson and E. M. Clarke. Using Branching Time Temporal Logic to Synthesize Synchronization Skeletons. *Sciene of Computer Programming*, 2(3):241–266, 1982.
6. M. Gelfond and V. Lifschitz. The Stable Model Semantics for Logic Programming. In *Proc. of ICLP'88*, pages 1070–1080, Cambridge, Massachusetts, 1988. MIT Press.
7. M. Gelfond and H. Przymusinska. Reasoning in Open Domains. In *Logic Programming and Non-Monotonic Reasoning*, pages 397–413. MIT Press, 1993.
8. G. Gottlob, E. Grädel, and H. Veith. Datalog LITE: A deductive query language with linear time model checking. *ACM Transactions on Computational Logic*, 3(1):1–35, 2002.
9. E. Grädel. Guarded Fixed Point Logic and the Monadic Theory of Trees. *Theoretical Computer Science*, 288:129–152, 2002.
10. E. Grädel. Model Checking Games. In *Proceedings of WOLLIC 02*, volume 67 of *Electronic Notes in Theoretical Computer Science*. Elsevier, 2002.
11. E. Grädel and I. Walukiewicz. Guarded Fixed Point Logic. In *Proc. of LICS '99*, pages 45–54. IEEE Computer Society, 1999.
12. A. Halevy, I. Mumick, Y. Sagiv, and O. Shmueli. Static Analysis in Datalog Extensions. *Journal of the ACM*, 48(5):971–1012, 2001.
13. S. Heymans, D. Van Nieuwenborgh, and D. Vermeir. Nonmonotonic Ontological and Rule-Based Reasoning with Extended Conceptual Logic Programs. In *Proc. of ESWC 2005*, number 3532 in LNCS, pages 392–407. Springer, 2005.
14. S. Heymans, D. Van Nieuwenborgh, and D. Vermeir. Guarded Open Answer Set Programming. In *8th International Conference on Logic Programming and Nonmonotonic Reasoning (LPNMR 2005)*, number 3662 in LNAI, pages 92–104. Springer, 2005.
15. M. R. A. Huth and Mark Ryan. *Logic in Computer Science: Modelling and Reasoning about Systems*. Cambridge University Press, 2000.
16. V. Lifschitz, D. Pearce, and A. Valverde. Strongly Equivalent Logic Programs. *ACM Transactions on Computational Logic*, 2(4):526–541, 2001.
17. J. Lloyd and R. Topor. Making Prolog More Expressive. *J. Log. Program.*, 1(3):225–240, 1984.
18. P. Simons. SMODELS homepage. http://www.tcs.hut.fi/Software/smodels/.
19. T. Syrjänen. Omega-restricted Logic Programs. In *Proc. of LPNMR*, volume 2173 of *LNAI*, pages 267–279. Springer, 2001.
20. T. Syrjänen. Cardinality Constraint Programs. In *Proc. of JELIA'04*, pages 187–200. Springer, 2004.

Reasoning Support for Expressive Ontology Languages Using a Theorem Prover

Ian Horrocks and Andrei Voronkov*

The University of Manchester
{horrocks, voronkov}@cs.man.ac.uk

Abstract. It is claimed in [45] that first-order theorem provers are not efficient for reasoning with ontologies based on description logics compared to specialised description logic reasoners. However, the development of more expressive ontology languages requires the use of theorem provers able to reason with full first-order logic and even its extensions. So far, theorem provers have extensively been used for running experiments over TPTP containing mainly problems with relatively small axiomatisations. A question arises whether such theorem provers can be used to reason in real time with large axiomatisations used in expressive ontologies such as SUMO. In this paper we answer this question affirmatively by showing that a carefully engineered theorem prover can answer queries to ontologies having over 15,000 first-order axioms with equality. Ontologies used in our experiments are based on the language KIF, whose expressive power goes far beyond the description logic based languages currently used in the Semantic Web.

State-of-the-art theorem provers for first-order logic (FOL) are highly sophisticated and efficient programs. Moreover, they are very flexible tools and can be tuned to a number of applications. For example, VAMPIRE [35] provides a large collection of parameters that can be used to give better performance for various classes of applications. In addition, VAMPIRE implements a number of literal selection functions and internally contains a library for defining such functions in a simple way; this makes it possible to simulate various proof-search algorithms and even provide decision procedures for decidable classes of first-order logic (see, e.g., [23]).

However, there was a common belief that provers like VAMPIRE cannot directly be used for efficient reasoning with very large ontologies using expressive languages such as KIF [14] for two reasons. Firstly, these provers are optimised for reasoning with relatively small axiomatisations. Secondly, they do not support some extensions of FOL required in KIF.

In this paper we describe an adaptation of VAMPIRE to support reasoning for expressive ontology languages and present experimental results which show that it can be used for efficient reasoning with large ontologies using extensions of the first-order language.

---

* The authors are partially supported by grants from EPSRC.

J. Dix and S.J. Hegner (Eds.): FoIKS 2006, LNCS 3861, pp. 201–218, 2006.
© Springer-Verlag Berlin Heidelberg 2006

This paper is structured as follows. In Section 1 we briefly overview expressive languages for ontologies, including the language KIF, and FO provers. In Section 2 we describe the adaptation of VAMPIRE for reasoning with large ontologies. For our experiments we selected ontologies implemented in KIF since they offer a high degree of sophistication compared to ontologies using Description Logic (DL) based languages, and there are publicly available KIF-based ontologies containing thousands of FO formulas with equality. However, our adaptation is quite general, and could be used for other expressive ontology languages.

Note that there is no way to compare VAMPIRE with description logic provers on these ontologies, since their subsets corresponding to description logic languages and hardly interesting and not representative.

Section 3 contains a summary and a description of future work. Finally, in the appendix we demonstrate the efficiency and advanced features of VAMPIRE by showing two (out of a large number of) inconsistency proofs found by it in SUMO and a terrorism ontology.

1   Introduction

Expressive Languages for Ontologies. Ontologies play a major role in the Semantic Web where they are widely used in, e.g., bio-informatics, medical terminologies and other knowledge management applications [5, 44, 47, 41, 34, 40]. They are also of increasing importance in the Grid, where they may be used, e.g., to support semantic based discovery, execution and monitoring of Grid services [9, 46, 11].

State of the art ontology languages, such as DAML+OIL [48] and OWL [3], are based on expressive description logics (DLs). This establishes a firm formal foundation for the language, e.g., by providing well-defined semantics and a broad understanding of the computational properties of key inference problems; it also allows applications to exploit the reasoning services provided by highly optimised DL reasoners such as FaCT and Racer [19, 15, 13]. It is widely recognised, however, that the expressive power of such languages is inadequate in some applications, and in particular applications related to the discovery and composition of Web and Grid services. This has led to efforts to develop languages based on more expressive logics up to and including full first-order predicate logic [17, 38, 6, 4].

Motivation. The availability of efficient reasoners has proved to be important in both the design and deployment of ontologies. Designing ontologies is an extremely complex task, and modern ontology design tools typically use reasoners to support the ontologist by highlighting inconsistencies in the design and allowing them to compare their intuitions about implicit subsumption relationships between classes with those computed by the reasoner [2, 31]. This kind of reasoning support is, for example, provided by both OilEd and the ProtégéOWL plugin, tools which are increasingly used for ontology development in e-Science. Applications of ontologies typically involve querying, and this again means using a reasoner, e.g., to determine when an individual (or a tuple of individuals) satisfies a query expression, or to retrieve all individuals (or tuples) satisfying a given query [22, 10]. For example, a biologist may want to answer queries about gene product data annotated with terms from the Gene Ontology, with a reasoner being used to determine which gene products are instances of complex descriptions that also use terms from the ontology [16].

The known decidability of key reasoning problems (such as satisfiability, subsumption and instance retrieval), and the availability of efficient reasoners based on highly optimised tableau decision procedures, were crucial factors in motivating the DL based design of DAML+OIL and OWL [19, 20]. Decidability comes, however, at a cost in terms of restricted expressive power. In particular, while such languages are generally equipped with a relatively rich set of constructors for use with classes (unary predicates), they only provide a very limited set of constructors for use with properties (binary predicates). These limitations can be onerous in some applications, in particular those where aggregation plays a prominent role. For example, in complex physically structured domains such as biology and medicine it is often important to describe structures that are exactly equivalent to the aggregation of their parts, and to have properties of the component parts transfer to the whole (a femur with a fractured shaft is a fractured femur) [32]. The importance of this kind of knowledge can be gauged from the fact that it can invariably be expressed in ontology languages designed specifically for medicine, even those that are otherwise relatively weak [33, 40]; various "work-arounds" have also been described for use with ontology languages that cannot express this kind of knowledge directly [37].

Similarly, in grid and web services applications, it may be necessary to describe composite processes in terms of their component parts, and to express relationships between the properties of the various components and those of the composite process. For example, in a sequential composition of processes it may be useful to express a relationship between the inputs and outputs of the composite and those of the first and last component respectively, as well as relationships between the outputs and inputs of successive components [44].

These limitations can be overcome to some extent by extending the DL language with a so-called *role-box* [21], but in order to maintain decidability it is necessary to impose severe restrictions on what can be expressed. For example, this framework would not allow the expression of simple family relationships such as the fact that "uncle" is equivalent to the composition of "parent" and "brother".

In addition to these problems with domain ontologies, many richly axiomatised foundational ontologies, such as SUMO[1] and DOLCE [29], are based on full FOL with relations of arbitrary arity, and even on extensions of FOL using relations with variable arities. This makes it impossible for DL based tools to exploit these foundational ontologies in order to structure or validate domain ontologies, and to improve interoperability between ontologies.

A recognition of the limitations of DL based ontology languages, in particular in web services applications, has led to proposals to extend them with, e.g.,

---

[1] See http://suo.ieee.org/

Horn-clause axioms [17, 18], or even axioms supporting arbitrary use of first order quantification [6, 4, 38]. These extended languages are based on larger fragments (than the DL fragment) of FOL, and may even be equivalent to full FOL; as a consequence, computing class consistency and subsumption is no longer decidable in general.

The utility of such languages, and the applications that use them, will crucially depend on the provision of reasoning support: there is little point in building complex models of web services without any means of manipulating or querying them.

First-Order Theorem Provers. First-order theorem provers have traditionally been used for the same purpose as DL-based ontology reasoners: providing reasoning services. They have a long history: indeed, some first-order theorem provers had already been implemented in the 1960s. There are, however, many important differences between FO provers and DL reasoners which explain why FO provers have not yet achieved widespread use in the Semantic Web.

FO provers deal with an undecidable logic. They are highly optimised for general-purpose reasoning, and are especially optimised for reasoning with equality. For example, they can often find very complex combinatorial proofs of identities in algebras. FO provers are based on a highly advanced theory of saturation algorithms with redundancy. This theory is very flexible—for example, completeness theorems in it have been proven for inference systems using arbitrary literal selection functions that can simulate various proof-search strategies, such as bottom-up or top-down reasoning.

Recently, there have been papers showing how FO provers can be used to reason in theories with a rich definitional structure [8, 12]. Experiments into their use for classifying DL-based ontologies, using a naive translation of DL formulas into first-order formulas, have also shown encouraging results [45].

Nonetheless, [45] also shows that on DL-based ontologies using simple languages DL reasoners are much faster than a straightforwardly used FO prover. However, the use of DL reasoners for more expressive languages faces a number of obstacles. For example, different languages need different reasoning algorithms, and an efficient implementation of a new inference algorithm may require the re-implementation of data structures supporting efficient inference procedures. In contrast, FO provers use a well-established uniform inference mechanism with thoroughly investigated implementation techniques and data structures (see, e.g., [39, 35, 36]); tuning them for new applications usually requires only implementation of new preprocessing algorithms, and finding the best settings for a wide range of already available parameters.

Traditionally, FO provers have been used for proving theorems in mathematics, and for software and hardware verification. For these applications the axiomatisation is normally relatively small, and also has a small number of function and predicate symbols. In contrast, ontologies may contain a very large number of axioms and predicate symbols. Moreover, axiomatisations of different theories in FOL offer a great variety of different constructs, while ontologies typically contain many similarly-structured "definitions".

The experimental results in [49] show that the inverse method (a non-tableau method based on a saturation algorithm) can be implemented just as efficiently as tableau-based DL provers. The implementation reported in this paper required less than one second to answer queries to the SUMO ontology, which contains about 5,000 first-order axioms with equality. Moreover, it took only a few seconds to to find a large number of non-trivial inconsistencies in various versions of SUMO and in an even larger terrorism ontology.

The KIF Language. KIF is a language for expressing knowledge that contains full first-order logic, and extends it with several features. First, it supports some datatypes, for example real numbers, and evaluable functions on these datatypes. Second, it can use arbitrary terms, including variables, as function and predicate symbols. Third, it has row variables which range over sequences of terms of arbitrary finite lengths. As a consequence, KIF allows for functions and relations of variable arities. The semantics of KIF is described in [14].

Support for datatypes seems to be crucial for many applications, and there are numerous proposals to include datatypes in Semantic Web languages. Full support of datatypes in first-order logic is impossible: having a datatype of integers with simple operations on it means that one can express arithmetic, and therefore there is no hope for the automation of reasoning in first-order logic with datatypes.

The second feature of KIF—variables as function and predicate symbols—is not well-understood. Indeed, it was believed that one could translate this feature in first-order logic by adding a (meta) predicate $holds$, and replacing formulas like $x(t)$ by $holds(x, t)$, but [25] have shown that this is not so.

Concerning row variables, [14] note that they are not first-order, but their full expressive power should be further investigated. The following theorem shows that one can express arithmetic in first-order logic with row variables.

Theorem 1. *There exists a polynomial-time translation τ of arithmetic into predicate logic with row variables such that for every sentence F or arithmetic, F holds in the standard model of arithmetic if and only if $\tau(F)$ is valid in predicate logic with row variables.*

This theorem shows that there is no hope for the automation of reasoning in first-order with arbitrary row variables; in particular, the set of theorems of first-order logic with row variables is not recursively enumerable.

2   Adapting Vampire for Large Ontologies

Query Answering. Query answering requires retrieving individuals which satisfy a given formula. That is, given a query $Q(\bar{x})$ with free variables $\bar{x}$, one has to find (all, or a given number of) vectors of terms $\bar{t}$ such that $Q(\bar{t})$ is a logical consequence of formulas in the ontology. To implement query answering, one needs to modify inference algorithms to return individuals satisfying the query. To implement it efficiently against large knowledge bases, one has to, in addition, be able to answer a sequence of queries without reloading the ontology.

Query answering requires retrieving individuals which satisfy a given formula. For example, one can ask the following query to find all individuals who published a paper at FoIKS

```
has_paper(X,'FoIKS')
```

To implement query answering, one needs to modify inference algorithms to return individuals satisfying the formula.

Returning Individuals. Database systems perform query answering but not theorem proving. DL reasoners were originally designed for theorem proving, but some of them can now perform restricted forms of query answering as well. For FO provers, query answering is not very different from theorem proving and can be implemented essentially at no extra cost. The standard way of providing query answering is via an answer predicate. For example, for the above query with one variable X we introduce a unary answer predicate **answer**, replace the query by the formula

```
has_paper(X,'FoIKS') -> answer(X)
```

and run a standard saturation algorithm with one difference: instead of searching for a derivation of the empty clause (to signal that a refutation is found) we search for derivations of clauses whose only predicate symbol is **answer**. For example, if one derives

```
answer('Andrei'),
```

then 'Andrei' is an answer to the query. Answer predicates are a very powerful mechanism. They can be used for finding multiple answers, disjunctive answers, or general answers with variables. For example, a derivation of

```
answer('Andrei') \/ answer('Ian'),
```

means that either 'Andrei' or 'Ian' satisfy the query (but there may be not enough information for a definite answer). Likewise, having derived **answer(Y)**, where Y is a variable, means that every object satisfies the query (which may signal that something is wrong with the ontology or with the query).

Answer predicates are implemented in a number of theorem provers, including VAMPIRE, Otter [28] and Gandalf [43]. For VAMPIRE, one can also specify that only definite answers should be considered. To implement definite answers, VAMPIRE replaces any clause $a(s_1, \ldots, s_n) \vee a(t_1, \ldots, t_n) \vee C$, where a is the answer predicate, by the clause $s_1 \neq t_1 \vee \ldots \vee s_n \neq t_n \vee a(t_1, \ldots, t_n) \vee C$. Completeness of resolution with answer predicates was studied in [42].

Answering Sequences of Queries. One essential difference between theorem provers and query answering systems is that the former are normally invoked to solve a single problem. If another problem has to be solved, the theorem prover has to be called again. The cost of activating a query answering system working

```
<!-- load the terrorism ontology -->
<kb_load kb="terrorism" syntax="kif"
        file="SemWeb/terrorism.kif" />

<!-- answer the following query to the terrorism ontology -->
<query max_answers="10" time_limit="5">
  (instance ?X ?Y)
</query>

<!-- load SUMO and disable the terrorism ontology -->
<kb_load kb="sumo" syntax="kif"
        file="SemWeb/sumo139.kif" />
<kb_status kb="terrorism" enabled="no" />

<!-- answer the following query to SUMO -->
<query max_answers="10" time_limit="5">
  (instance ?X ?Y)
</query>

<!-- enable the terrorism ontology -->
<kb_status kb="terrorism" enabled="yes" />

<!-- answer the same query using formulas from both SUMO
 and the terrorism ontology -->
<query max_answers="10" time_limit="5">
  (instance ?X ?Y)
</query>

<!-- assert two new facts about the instance relation -->
<assert syntax="kif">
  (instance a b)
  (instance b c)
</assert>

<!-- answer the same query using formulas from SUMO,
  the terrorism ontology and newly asserted facts -->
<query max_answers="10" time_limit="5">
  (instance ?X ?Y)
</query>
```

Fig. 1. Pre-compiled Knowledge Bases

with a large ontology may be prohibitive. For example, VAMPIRE implements sophisticated preprocessing algorithms for first-order formulas, and the collection of clauses obtained from them can be further processed for simplifications. It is better to have a system which can answer a sequence of queries to an ontology or collection of ontologies without restarts. VAMPIRE solves this problem by implementing *pre-compiled knowledge bases*.

Figure 1 shows a part of VAMPIRE's *bag file*[2] that uses pre-compiled knowledge bases in VAMPIRE. The command `kb_load` reads an ontology or a knowledge base from a file and compiles it. This is done by preprocessing formulas in the ontology, converting them to CNF, and applying various simplification rules to the CNF. The resulting set of clauses is then stored internally, along with some information that allows one to quickly retrieve only a part of the ontology that is relevant to a particular query. The command `kb_status` can be used to enable or disable loaded ontologies (disabled ontologies are not used for query answering but remain stored and pre-compiled). The formulas in the enabled knowledge bases can be used for query answering. Answering some queries may require information from several knowledge bases. As far as we know, VAMPIRE is the only first-order prover implementing pre-compiled knowledge bases.

Goal-Oriented Proof-Search for Query Answering. When answering queries to large ontologies, it is of paramount importance that query answering be goal-oriented. One can use the modern theory of resolution (see [1]) to make resolution proof-search goal-oriented. For example, one can use literal selection functions and term orderings that prefer literals and terms coming from the goal. Another possibility is to use the *set-of-support strategy*, in which inferences involving only clauses not derived from the query are prohibited. The use of the set of support strategy together with selection functions is incomplete, so we used the following modification of this strategy: we select *all* literals in clauses not derived from the query. Experimental results have shown that this variant of the set-of-support strategy works very well: a typical query response time for queries to SUMO falls within one second.

We believe that in future one can improve goal-oriented proof search by also providing relevance filters, which will allow one to focus on a small part of the ontology. This seems to be especially promising for ontologies, in which the majority of knowledge is represented as definitions of predicates. Such a filtering technique proved indispensable in the use of theorem provers for classifying ontologies based on the subsumption relation (see [45]).

Consistency Checking. For DL-based ontologies, consistency (of the ontology as a whole) is usually not an issue. For ontologies and knowledge bases using expressive languages, such as FOL, consistency may be a problem. Such ontologies may be created by people using the same symbols with a different meaning, people who do not know logic well, or people who understand the ontology domain differently. Our experiments with several versions of SUMO [30] have shown that all of them had numerous axioms creating inconsistency. Two examples (one of them for a terrorism ontology) are given in the appendix.

Checking consistency of ontologies is a more difficult problem than query answering. For query answering one can focus on the query and formulas derived from it. For consistency-checking, there is no query to focus on. Consistency checking in VAMPIRE was implemented by a standard saturation algorithm. How-

---

[2] A bag file may contain any kind of information for VAMPIRE, including commands, options, and queries.

ever it turned out that the standard options used for theorem proving did not perform well for large ontologies. (This may be one of the reasons for the relatively slow performance of VAMPIRE reported in [45].) By turning off some simplification rules (backward subsumption, backward demodulation, forward subsumption resolution) and fine-tuning some other options, we were able to increase the speed by a factor of 12.

Beyond First-Order Logic. Even relatively simple extensions of both DLs and FOL can be very difficult to implement or even lead to theories for which no complete algorithms exist. For example, it is not hard to achieve the full expressive power of arithmetic by adding integers as a datatype and using this datatype in an unrestricted way. Nonetheless, extensions of first-order logic may turn out to be crucial for applications, and in this case one has to find a compromise between the expressive power of such extensions and the possibility of implementing them efficiently on top of existing implementations. In this section we describe a light-weight implementation in VAMPIRE of some features taking the system beyond first-order logic. We illustrate the utility of such extensions by showing inconsistency proofs for ontologies using these features.

Support for Datatypes. Theorem provers are not able to deal with datatypes (such as integers and strings) other than via complex axiomatisations that would severely degrade performance. Typical ontologies, particularly in web services applications, will contain many datatypes and data values recording knowledge about names, addresses, prices and so on. Full support of these datatypes in a prover is impossible since first-order logic with datatypes is not recursively enumerable. However, a limited form of reasoning with datatypes can be implemented.

VAMPIRE supports three datatypes: integers, real numbers and strings. It can understand constants of these datatypes occurring in the input ontology, for example it knows that 1 is an integer constant and "one" is a string constant. It also implements a number of built-in functions and relations on these datatypes, such as comparison operators on numeric datatypes or concatenation of strings. VAMPIRE can evaluate simple expressions. For example, it can evaluate $2 + 3 < 6$ to **true**. It cannot do more complex reasoning tasks with the datatypes, for example, it cannot to derive a contradiction from the facts $c < 2$ and $3 < c$, but will be able derive a contradiction from them if the input contains the transitivity axiom for $<$.

Moreover, VAMPIRE has a mechanism that allows one to define new functions and relations on the datatypes using recursive definitions, so in a way it contains a small functional programming language inside. Such definitions can be implemented using pre-oriented equalities and pre-selected literals. The ability to define new functions and relations on datatypes can become standard in future expressive ontology languages.

Since there is no standard convention on the names for built-in functions and relations on datatypes, VAMPIRE also provides an interface for mapping the input ontology names to the internal names for these functions and relations.

Meta-Predicates. It is non-trivial to implement variables and functions as predicate symbols, as there is no proof theory for this feature of KIF. If VAMPIRE finds such variables in the input, it transforms the input using the predicate symbol *holds*, as mentioned in Section 1, and the function symbol *apply*. This style of reasoning is incomplete, since VAMPIRE does not implement *reflection*, that is, the rule replacing $holds(r, t_1, \ldots, t_n)$ by $r(t_1, \ldots, t_n)$ or vice versa. Even the current lightweight implementation using *holds* was strong enough to answer some queries related to transitive relations axiomatised by

$$transitive(u) \leftrightarrow \forall x \forall y \forall z (u(x, y) \wedge u(y, z) \rightarrow u(x, z))$$

and to discover inconsistencies in SUMO, see the appendix.

Row Variables. Theorem 1 implies the impossibility of reasoning with row variables. If all row variables are bound by essentially universal quantifiers (that is, positively occurring universal or negatively occurring existential ones), then the set of provable formulas is still recursively enumerable but one has to implement reasoning modulo associativity, see [14]. Implementing reasoning modulo associativity is very difficult, for example, two terms may have an infinite number of minimal unifiers modulo associativity. It is pointed out in [14] that one can use a weaker class of formulas with row variables in which a row variable may occur only as the last argument. In this case it is enough to implement only *sequence variables* [26].

However, one can note that the main use of row variables in SUMO is for relations or relatively small arities. For this reason, VAMPIRE only substitutes for row variable sequences of variables of a bounded length. The default upper bound on the length is 2 but it can be changed by the user.

If the input contains row variables, VAMPIRE does the following:

1. Reject formulas that contain row variables bound by essentially existential quantifiers;
2. Substitute every remaining row variable @ by sequences or ordinary variables $x_1, \ldots, x_i$, such that i ranges over $0, \ldots, n$ where n is the upper bound specified by the user. This rule is called *row variable expansion*.

Appendix A gives an example involving reasoning with row variables. The latest version of SUMO contains only two formulas rejected by VAMPIRE.

Other Issues

Proof Output. It is important that an answer to a query comes with an explanation. Likewise, when inconsistency is discovered, one needs an explanation to find a source of inconsistency. One can also require that the answers provided by an ontology reasoner could be checked by a proof checker. To this end, one needs a system able to produce proofs. Most of the currently available theorem provers produce a proof in some form. VAMPIRE can produce proofs in several formats, including XML. Our experiments on checking consistency of SUMO have shown that the proof should be understandable by humans. Indeed, when an inconsistency is discovered, one should find the axioms that cause inconsistency and repair the ontology to remove all sources of inconsistency. We have found that the proof format in which each inference is displayed separately is easier to read and understand. Moreover, an ASCII proof is not very readable, so we included an option to output proofs in LATEX. Both proofs in the appendix have been generated by VAMPIRE automatically and slightly edited to fit in the paper size.

The output proof format of VAMPIRE is still far from perfect. We believe that further research should be done to improve presentation of complex computer-generated proofs in a human-friendly form.

We are currently working on producing proofs checkable in $O(n \log n)$ time. These proofs will be more detailed than the human-readable proofs but the ability to check proofs by using a proof-checker is indispensable both in debugging the theorem prover, the use of the prover for safety-critical applications and automatic analysis of proofs.

3 Future Work

The techniques we have described greatly improve the performance of VAMPIRE when answering queries to and checking the consistency of ontologies. Like other resolution theorem provers, however, it is not very effective at proving satisfiability, and hence at proving non-subsumption. This is a problem if VAMPIRE is to be used for general purpose ontology reasoning: in typical ontologies, for example, most classes are satisfiable and most pairs of concepts are not in a subsumption relationship.

Future work will, therefore, include investigations of a number of strategies for addressing this problem. Firstly, we will investigate improved literal selection strategies (that exploit the structure of ontology axioms) to improve the performance of VAMPIRE on satisfiable problems. Secondly, we will investigate the enhancement of existing model building methods (see, e.g., [24, 27, 7]), which are designed to prove satisfiability in FOL. The idea here is to use similar techniques to those we have already successfully employed in VAMPIRE, in particular using relevance filters to reduce the effective size of the ontology and exploiting the special structure of ontology axioms (in this case to try to minimise the problem of exponential explosion in model size). Finally, for suitable undecidable extensions of DLs (e.g., the proposed SWRL Horn-clause extension to OWL [17, 18]), we will investigate the development of model building algorithms based on existing tableaux decision procedures for DLs. Such an algorithm would still be sound for satisfiability (i.e., it would only succeed in building a model if the problem is satisfiable), but it will no longer be guaranteed to terminate.

We believe that a combination of some or all of these satisfiability testing techniques with a suitably optimised resolution prover (such as VAMPIRE) will be able to solve the vast majority of problems encountered when reasoning with ontologies.

References

1. L. Bachmair and H. Ganzinger. Resolution theorem proving. In A. Robinson and A. Voronkov, editors, *Handbook of Automated Reasoning*, volume I, chapter 2, pages 19–99. Elsevier Science, 2001.
2. S. Bechhofer, I. Horrocks, C. Goble, and R. Stevens. OilEd: A reason-able ontology editor for the semantic web. In F. Baader, G. Brewka, and T. Eiter, editors, *KI 2001: Advances in Artificial Intelligence*, volume 2174 of *Lecture Notes in Computer Science*, pages 396–408. Springer Verlag, 2001.
3. S. Bechhofer, F. van Harmelen, J. Hendler, I. Horrocks, D. L. McGuinness, P. F. Patel-Schneider, and L. A. Stein. OWL web ontology language 1.0 reference. W3C Recommendation, 10 February 2004. Available at http://www.w3.org/TR/owl-ref/.
4. D. Berardi, M. Grüninger, R. Hull, and S. McIlraith. Towards a first-order ontology for semantic web services. http://www.w3.org/2004/08/ws-cc/mci-20040904, sep 2004.
5. T. Berners-Lee, J. Hendler, and O. Lassila. The semantic Web. *Scientific American*, 284(5):34–43, 2001.
6. H. Boley, M. Dean, B. Grosof, M. Sintek, B. Spencer, S. Tabet, and G. Wagner. First-Order-Logic RuleML. http://www.ruleml.org/fol/, 2004.
7. K. Claessen and N. Sörensson. New techniques that improve mace-style model finding. In *Proceedings of the Workshop Model Computation 2003*, 2003.
8. A. Degtyarev, R. Nieuwenhuis, and A. Voronkov. Stratified resolution. *Journal of Symbolic Computations*, 36(1-2):79–99, 2003.
9. A. Emmen. The grid needs ontologies—onto-what?, 2002. http://www.hoise.com/primeur/03/articles/monthly/AE-PR-02-03-7.html.
10. R. Fikes, P. Hayes, and I. Horrocks. OWL-QL—a language for deductive query answering on the Semantic Web. *Journal of Web Semantics*, 2004. To Appear.
11. I. Foster, C. Kesselman, J. Nick, and S. Tuecke. The physiology of the grid: An open grid services architecture for distributed systems integration, 2002. http://www.globus.org/research/papers/ogsa.pdf.
12. H. Ganzinger and J. Stuber. Superposition with equivalence reasoning and delayed clause normal form transformation. In F. Baader, editor, *19th International Conference on Automated Deduction (CADE-19)*, volume 2741 of *Lecture Notes in Computer Science*, pages 335–349. Springer Verlag, 2003.
13. V. Haarslev and R. Möller. *RACER User's Guide and Reference Manual. Version 1.7.7*, Sept. 2003.
14. P. Hayes and C. Menzel. A semantics for the knowledge interchange format. In *IJCAI 2001 Workshop on the IEEE Standard Upper Ontology*, 2001.
15. I. Horrocks. Using an expressive description logic: FaCT or fiction? In A. Cohn, L. Schubert, and S. Shapiro, editors, *Principles of Knowledge Representation and Reasoning: Proceedings of the Sixth International Conference (KR'98)*, pages 636–647, San Francisco, CA, June 1998. Morgan Kaufmann.
16. I. Horrocks, L. Li, D. Turi, and S. Bechhofer. The instance store: DL reasoning with large numbers of individuals. In *Proc. of the 2004 Description Logic Workshop (DL 2004)*, pages 31–40, 2004.
17. I. Horrocks and P. F. Patel-Schneider. A proposal for an OWL rules language. In *Proc. of the Thirteenth International World Wide Web Conference (WWW 2004)*, pages 723–731. ACM, 2004.

18. I. Horrocks, P. F. Patel-Schneider, H. Boley, S. Tabet, B. Grosof, and M. Dean. SWRL: A semantic web rule language combining owl and ruleml. W3C Note, 21 May 2004. Available at http://www.w3.org/Submission/SWRL/.

19. I. Horrocks, P. F. Patel-Schneider, and F. van Harmelen. Reviewing the design of DAML+OIL: An ontology language for the semantic web. In *Proc. of the 18th National Conference on Artificial Intelligence (AAAI 2002)*, pages 792–797. AAAI Press, 2002.

20. I. Horrocks, P. F. Patel-Schneider, and F. van Harmelen. From SHIQ and RDF to OWL: The making of a web ontology language. *Journal of Web Semantics*, 1(1):7–26, 2003.

21. I. Horrocks and U. Sattler. The effect of adding complex role inclusion axioms in description logics. In *Proc. of the 18th Int. Joint Conf. on Artificial Intelligence (IJCAI 2003)*, pages 343–348. Morgan Kaufmann, 2003.

22. I. Horrocks and S. Tessaris. Querying the semantic web: a formal approach. In I. Horrocks and J. Hendler, editors, *First International Semantic Web Conference (ISWC 2002)*, number 2342 in Lecture Notes in Computer Science, pages 177–191. Springer Verlag, 2002.

23. U. Hustadt, B. Konev, A. Riazanov, and A. Voronkov. TeMP: A temporal monodic prover. In M. R. D.A. Basin, editor, *Automated Reasoning - Second International Joint Conference, IJCAR 2004*, volume 3097 of *Lecture Notes in Computer Science*, pages 326–330. Springer Verlag, 2004.

24. U. Hustadt and R. A. Schmidt. Using resolution for testing modal satisfiability and building models. In I. P. Gent, H. van Maaren, and T. Walsh, editors, *SAT 2000: Highlights of Satisfiability Research in the Year 2000*, volume 63 of *Frontiers in Artificial Intelligence and Applications*. IOS Press, Amsterdam, 2000. Also to appear in a special issue of *Journal of Automated Reasoning*.

25. P. P.-S. I. Horrocks. Three theses of representation in the semantic web. In *Proceedings of the Twelfth International World Wide Web Conference, WWW2003*, pages 39–47, Budapest, Hungary, Jan. 2003. ACM.

26. T. Kutsia. Theorem proving with sequence variables and flexible arity symbols. In M.Baaz and A. Voronkov, editors, *Logic for Programming, Artificial Intelligence, and Reasoning (LPAR 2002)*, volume 2514 of *Lecture Notes in Artificial Intelligence*, pages 278–291, Tbilisi, Georgia, 2002.

27. W. McCune. Mace4 reference manual and guide. Technical Memorandum 264, Argonne National Laboratory, Aug. 2003.

28. W. McCune. OTTER 3.3 reference manual. Technical Memorandum 263, Argonne National Laboratory, Aug. 2003.

29. P. Mika, D. Oberle, A. Gangemi, and M. Sabou. Foundations for service ontologies: Aligning OWL-S dolce. In S. Feldman, M. Uretsky, M. Najork, and C. Wills, editors, *WWW 2004, Proceedings of the 13th international conference on World Wide Web*, pages 563–572. ACM, 2004.

30. A. Pease, I. Niles, and J. Li. The Suggested Upper Merged Ontology: A large ontology for the Semantic Web and its applications. In *Working Notes of the AAAI-2002 Workshop on Ontologies and the Semantic Web*, Edmonton, Canada, 2002.

31. Protégé. http://protege.stanford.edu/, 2003.

32. A. Rector. Analysis of propagation along transitive roles: Formalisation of the galen experience with medical ontologies. In *Proc. of DL 2002*. CEUR (http://ceur-ws.org/), 2002.

33. A. Rector, S. Bechhofer, C. A. Goble, I. Horrocks, W. A. Nowlan, and W. D. Solomon. The GRAIL concept modelling language for medical terminology. *Artificial Intelligence in Medicine*, 9:139–171, 1997.
34. A. Rector and I. Horrocks. Experience building a large, re-usable medical ontology using a description logic with transitivity and concept inclusions. In *Proc. of the 13th Nat. Conf. on Artificial Intelligence (AAAI 97)*, 1997.
35. A. Riazanov and A. Voronkov. The design and implementation of Vampire. *AI Communications*, 15(2-3):91–110, 2002.
36. A. Riazanov and A. Voronkov. Limited resource strategy in resolution theorem proving. *Journal of Symbolic Computations*, 36(1-2):101–115, 2003.
37. S. Schulz and U. Hahn. Parts, locations, and holes - formal reasoning about anatomical structures. In *Proc. of AIME 2001*, volume 2101 of *Lecture Notes in Artificial Intelligence*. Springer Verlag, 2001.
38. Common Logic Standard. http://cl.tamu.edu/.
39. R. Sekar, I. Ramakrishnan, and A. Voronkov. Term indexing. In A. Robinson and A. Voronkov, editors, *Handbook of Automated Reasoning*, volume II, chapter 26, pages 1853–1964. Elsevier Science, 2001.
40. K. Spackman. Managing clinical terminology hierarchies using algorithmic calculation of subsumption: Experience with snomed-rt. *J. of the Amer. Med. Informatics Ass.*, 2000. Fall Symposium Special Issue.
41. R. Stevens, C. Goble, I. Horrocks, and S. Bechhofer. Building a bioinformatics ontology using OIL. *IEEE Transactions on Information Technology in Biomedicine*, 6(2):135–141, 2002.
42. T. Tammet. Completeness of resolution for definite answers. *Journal of Logic and Computation*, 5(4):449–471, 1995.
43. T. Tammet. Gandalf. *Journal of Automated Reasoning*, 18(2):199–204, 1997.
44. The DAML Services Coalition. DAML-S: Web service description for the semantic web. In *Proc. of the 2003 International Semantic Web Conference (ISWC 2003)*, number 2870 in Lecture Notes in Computer Science. Springer Verlag, 2003.
45. D. Tsarkov, A. Riazanov, S. Bechhofer, and I. Horrocks. Using Vampire to reason with OWL. In *Semantic Web 2004*. Springer Verlag, 2004. to appear.
46. S. Tuecke, K. Czajkowski, I. Foster, J. Frey, S. Graham, C. Kesselman, and P. Vanderbilt. Grid service specification (draft). GWD-I draft , GGF Open Grid Services Infrastructure Working Group, 2002.
47. M. Uschold, M. King, S. Moralee, and Y. Zorgios. The enterprise ontology. *Knowledge Engineering Review*, 13, 1998.
48. F. van Harmelen, P. F. Patel-Schneider, and I. Horrocks. Reference description of the DAML+OIL (March 2001) ontology markup langauge, Mar. 2001.
49. A. Voronkov. kЯ: a theorem prover for *K*. In H. Ganzinger, editor, *Automated Deduction—CADE-16. 16th International Conference on Automated Deduction*, volume 1632 of *Lecture Notes in Artificial Intelligence*, pages 383–387, Trento, Italy, July 1999.

A   Appendix

In this appendix we give two inconsistency proofs found by VAMPIRE. Each of the proofs is explained in a separate section. We have not changed the formulation of the problems but do not present them in the KIF syntax. The proofs were produced using the LaTeX output facility of VAMPIRE. We had to edit them slightly to fit in the paper size. We also renamed several functions and relations in SUMO for a better readability using the renaming feature provided by VAMPIRE. For example, we write $x : y$ instead of $instance(x, y)$. A proof consists of a sequence of inferences. Each inference infers a formula, called the *conclusion* of an inference from zero or more formulas, called the *premises* of the inference. All formulas occurring in the proof are numbered. Each inference is annotated by the numbers of the premises, the number of conclusion, and inference rules used in the inference. For example, the annotation of last inference in the first proof means that formula 9 was obtained from formulas 1,2 and 8 using the resolution and forward subsumption resolution inference rules. The symbol □ denotes the empty clause, which is logically equivalent to contradiction.

We have many examples of inconsistency proofs found by VAMPIRE in the latest versions of SUMO and the ontology of terrorism from the Sumo Web page http://ontology.teknowledge.com. A typical proof occupies at least two pages, so we cannot give them here in detail.

We give a very brief account of one (very short) proof of inconsistency of the ontology of terrorism. This ontology has about 18,000 first-order axioms. This example illustrates why most (if not all) of the previous provers would not be able to find this inconsistency. Although the proof is relatively short, VAMPIRE generates over 60,000 formulas to find it.

Essentially, the proof is based on three axioms in the ontology. Two of the axioms are facts asserting information about the number of victims of two Hamas attacks. The problem with the axiomatisation of these attacks comes from the fact that they were given the same name.

$$victimDeathCount(HAMAS\text{-}KnifeAttack\text{-}25\text{-}Jun\text{-}92, 0);$$
$$victimDeathCount(HAMAS\text{-}KnifeAttack\text{-}25\text{-}Jun\text{-}92, 2).$$

These two axioms look contradictory, but they do not contradict each other in FOL. However, the ontology also contains the following axiom

$$victimDeathCount(x_0, x_3) \supset$$
$$x_3 = CardinalityFn(KappaFn(x_4, and(patient(x_0, x_4),$$
$$holdsDuring(ImmediateFutureFn(x_0),$$
$$attribute(x_4, Dead))))).$$

This axiom looks quite complex, nonetheless it is easy to see that it implies that in the relation *victimDeathCount* the second argument is a function of the first argument. This and the two facts given above imply that $0 = 2$. Since VAMPIRE knows simple arithmetic, it immediately derives contradiction. The proof is found in 1.7 seconds on a computer with a 1GHz Intel processor and 2GB of RAM.

The proof is neither long nor very sophisticated. However, to find it one has to apply the equality rule paramodulation to rather complex terms used in the last formula. This would make it very difficult if possible at all to find it for a prover not having efficient built-in equality reasoning or a prover using a translation of logic with equality into logic without equality. In addition, knowledge of simple arithmetic is needed to derive contradiction from $0 = 2$, and as far as we know the most efficient FO provers do not have built-in arithmetic in any form.

We have a large collection of proofs of inconsistency of several versions of SUMO found by VAMPIRE; some of them, if translated into human proofs, use very refined argument, for example showing problems with a careless use in SUMO of row variables and relations of arbitrary arity. We are planning to analyse these proofs in a separate paper.

A Proof of Inconsistency of SUMO. Here we give a proof demonstrating inconsistency of the Suggested Upper Merger Ontology version of July 2004. This proof is generated by the auto-mode of VAMPIRE in 34.5 seconds. VAMPIRE can also find the inconsistency proof in 3 seconds with a time limit of 4 seconds.[3] Using the optimal settings for ontology reasoning VAMPIRE can prove inconsistency in 2.7 seconds, and also in 0.7 second with a time limit of 1 second. If incomplete strategies are used, the proof can be found in 0.2 seconds. Note that consistency checking is much harder than query answering, since there is no goal to focus on, so a theorem prover must perform a brute force non-goal-oriented search from the initial set of about 5,000 FO formulas with equality. This proof derives contradiction from a formula containing a row variable and uses the row variable expansion rule. A row variables occurs, for example, in the atomic subformula $holds(x, @_1, w)$ of input formula 5.

The proof is written in a rather condensed form. For example, the CNF transformation rule in the proof consists of a number of smaller steps, including skolemisation introducing the skolem function σ.

Proof.

[1, input]
$$ListFn : TotalValuedRelation$$

[2, input]
$$ListFn : VariableArityRelation$$

[3, input]
$$x : VariableArityRelation \supset \neg(\exists y)valence(x, y)$$

[3 → 4, cnf transformation]
$$\frac{x : VariableArityRelation \supset \neg(\exists y)valence(x, y)}{\neg valence(x, y) \vee \neg x : VariableArityRelation}$$

[5, input]
$x : TotalValuedRelation \equiv$
$\quad (\exists y)(\, x : Relation \wedge valence(x, y) \wedge$
$\quad\quad ((\forall z \forall u \forall v)(z < y \wedge domain(x, z, v) \wedge u = nth(ListFn(@_1), z) \supset u : v) \supset$
$\quad\quad (\exists w)holds(x, @_1, w)))$

---

[3] VAMPIRE may work much faster when a time limit is given, for details see [36].

[5 → 6, row variable expansion]

$x : TotalValuedRelation \equiv$
$(\exists y)(\, x : Relation \land valence(x, y) \land$
$\quad ((\forall z \forall u \forall v)(z < y \land domain(x, z, v) \land u = nth(ListFn(@_1), z) \supset u : v) \supset$
$\quad (\exists w)holds(x, @_1, w)))$

$x : TotalValuedRelation \equiv$
$(\exists y)(\, x : Relation \land valence(x, y) \land$
$\quad ((\forall z \forall u \forall v)(z < y \land domain(x, z, v) \land u = nth(ListFn(w), z) \supset u : v) \supset$
$\quad (\exists x_6)holds(x, w, x_6)))$

[6 → 7, cnf transformation]

$x : TotalValuedRelation \equiv$
$(\exists y)(\, x : Relation \land valence(x, y) \land$
$\quad ((\forall z \forall u \forall v)(z < y \land domain(x, z, v) \land u = nth(ListFn(w), z) \supset u : v) \supset$
$\quad (\exists x_6)holds(x, w, x_6)))$

$valence(y, \sigma(x, y)) \lor \neg y : TotalValuedRelation$

[4, 7 → 8, resolution]

$\neg valence(x, y) \lor \neg x : VariableArityRelation$
$valence(y, \sigma(x, y)) \lor \neg y : TotalValuedRelation$

$\neg y : VariableArityRelation \lor \neg y : TotalValuedRelation$

[1, 2, 8 → 9, resolution, forward subsumption resolution]

$ListFn : TotalValuedRelation$
$ListFn : VariableArityRelation$

$\neg y : VariableArityRelation \lor \neg y : TotalValuedRelation$

$\square$

A Proof of Inconsistency of the Terrorism Ontology.

Here we give a proof of inconsistency of the ontology of terrorism from the Sumo Web page http://ontology.teknowledge.com. This ontology has about 18,000 first-order axioms. The proof is rather short but requires knowledge of the datatype of integers. Namely, at the last inference step it uses the fact $0 \neq 2$. In addition, the proof uses built-in equality reasoning. The proof is found in 7 seconds using the standard mode and in 1.7 seconds using the optimal settings for ontology reasoning. Note that the proof uses applications of equality to large terms, so it is unlikely to be found quickly by a prover without built-in equality reasoning.

Proof.

[1, input]

$victimDeathCount(x, y) \supset$
$\quad y = CardinalityFn(KappaFn(z, and(\, patient(x, z),$
$\qquad\qquad\qquad\qquad\qquad holdsDuring(\, ImmediateFutureFn(x),$
$\qquad\qquad\qquad\qquad\qquad\qquad attribute(z, Dead)))))$

[1 → 2, cnf transformation]

$victimDeathCount(x, y) \supset$
$\quad y = CardinalityFn(KappaFn(z, and(patient(x, z),$
$\qquad\qquad\qquad\qquad\qquad holdsDuring(ImmediateFutureFn(x),$
$\qquad\qquad\qquad\qquad\qquad\qquad attribute(z, Dead)))))$

$\quad y = CardinalityFn(KappaFn(x, and(patient(z, x),$
$\qquad\qquad\qquad\qquad\qquad holdsDuring(ImmediateFutureFn(z),$
$\qquad\qquad\qquad\qquad\qquad\qquad attribute(x, Dead))))) \lor$
$\neg victimDeathCount(z, y)$

[3, input]

$\qquad victimDeathCount(HAMAS\text{-}KnifeAttack\text{-}25\text{-}Jun\text{-}92, 0)$

[2, 3 → 4, resolution]

$\quad y = CardinalityFn(KappaFn(x, and(patient(z, x),$
$\qquad\qquad\qquad\qquad\qquad holdsDuring(ImmediateFutureFn(z),$
$\qquad\qquad\qquad\qquad\qquad\qquad attribute(x, Dead))))) \lor$
$\neg victimDeathCount(z, y)$
$\qquad victimDeathCount(HAMAS\text{-}KnifeAttack\text{-}25\text{-}Jun\text{-}92, 0)$

$CardinalityFn(KappaFn(y, and(patient(HAMAS\text{-}KnifeAttack\text{-}25\text{-}Jun\text{-}92, y),$
$\qquad holdsDuring(ImmediateFutureFn(HAMAS\text{-}KnifeAttack\text{-}25\text{-}Jun\text{-}92),$
$\qquad\qquad attribute(y, Dead))))) = 0$

[5, input]

$\qquad victimDeathCount(HAMAS\text{-}KnifeAttack\text{-}25\text{-}Jun\text{-}92, 2)$

[4, 2, 5 → 6, resolution, forward demodulation, evaluation]

$CardinalityFn(KappaFn(y, and(patient(HAMAS\text{-}KnifeAttack\text{-}25\text{-}Jun\text{-}92, y),$
$\qquad holdsDuring(ImmediateFutureFn(HAMAS\text{-}KnifeAttack\text{-}25\text{-}Jun\text{-}92),$
$\qquad\qquad attribute(y, Dead))))) = 0$
$\quad y = CardinalityFn(KappaFn(x, and(patient(z, x),$
$\qquad\qquad\qquad\qquad\qquad holdsDuring(ImmediateFutureFn(z),$
$\qquad\qquad\qquad\qquad\qquad\qquad attribute(x, Dead))))) \lor$
$\neg victimDeathCount(z, y)$
$\qquad victimDeathCount(HAMAS\text{-}KnifeAttack\text{-}25\text{-}Jun\text{-}92, 2)$

$\qquad\qquad\qquad\qquad\qquad \Box$

Consistency Checking Algorithms for Restricted UML Class Diagrams

Ken Kaneiwa and Ken Satoh

National Institute of Informatics,
2-1-2 Hitotsubashi, Chiyoda-ku, Tokyo 101-8430, Japan
{kaneiwa, ksatoh}@nii.ac.jp

Abstract. Automatic debugging of UML class diagrams helps in the visual specification of software systems because users cannot detect errors in logical inconsistency easily. This paper focuses on *tractable* consistency checking of UML class diagrams. We accurately identify inconsistencies in these diagrams by translating them into first-order predicate logic generalized by counting quantifiers and classify their expressivities by eliminating some components. For class diagrams of different expressive powers, we introduce optimized algorithms that compute their respective consistencies in P, NP, PSPACE, or EXPTIME with respect to the size of a class diagram. In particular, for two cases in which class diagrams contain (i) disjointness constraints and overwriting/multiple inheritances and (ii) these components along with completeness constraints, the restriction of attribute value types decreases the complexities from EXPTIME to P and PSPACE. Additionally, we confirm the existence of a meaningful restriction of class diagrams that prevents any logical inconsistency.

1 Introduction

The Unified Modeling Language (UML) [11, 6] is a standard modeling language; it is used as a visual tool for designing software systems. However, visualized descriptions make it difficult to determine consistency in formal semantics. In order to design UML diagrams, designers check not only for syntax errors but also for *logical inconsistency*, which may be present implicitly in the diagrams. Automatic detection of errors is very helpful for designers; for example, it enables them to revise erroneous parts of UML diagrams by determining inconsistent classes or attributes. Moreover, in order to confirm the accuracy of debugging (soundness, completeness, and termination), a consistency checking algorithm should be developed computationally and theoretically.

Class diagrams, which are a type of UML diagrams, are employed to model concepts in static views. The consistency of class diagrams has been investigated as follows. Evans [5] attempted a rigorous description of UML class diagrams by using the Object Constraint Language (OCL) and treated UML reasoning. Beckert, Keller, and Schmitt [1] defined a translation of UML class diagrams with OCL into first-order predicate logic. Further, Tsiolakis and Ehrig [13] analyzed

J. Dix and S.J. Hegner (Eds.): FoIKS 2006, LNCS 3861, pp. 219–239, 2006.

the consistency of UML class and sequence diagrams by using attributed graph grammars. The OCL and other approaches provide rigorous semantics and logical reasoning on UML class diagrams; however, they do not theoretically analyze the worst-case complexity of consistency checking. On the other hand, a number of object-oriented models and their consistency [10, 12] have been considered for developing software systems, but the models do not characterize the components of UML class diagrams; for example, the semantics of attribute multiplicities is not supported.

Berardi, Calvanese, and De Giacomo presented the correspondence between UML class diagrams and description logics (DLs), which enables us to utilize DL-based systems for reasoning on UML class diagrams [2]. In fact, Franconi and Ng implemented the concept modeling system ICOM [7] using DLs. The cyclic expressions of class diagrams are represented by *general axioms* for DLs. For example, a class diagram is cyclic if a class C has an attribute and the type of the attribute value is defined by the same class. However, it is well known that reasoning on general axioms of the necessary DLs is exponential time hard [3]. Therefore, consistency checking of the class diagrams in DLs requires exponential time in the worst case.

In order to reduce the complexity, we consider restricted UML class diagrams obtained by deleting some components. A meaningful restriction of class diagrams is expected to avoid intractable reasoning, thus facilitating automatic debugging. This solution provides us with not only tractable consistency checking but also a sound family of class diagrams (i.e., its consistency is theoretically guaranteed without checking).

The aim of this paper is to present optimized algorithms for testing the consistency of restricted UML class diagrams, which are designed to be suitable for class diagrams of different expressive powers. The algorithms detect the logical inconsistency of class diagram formulation in first-order predicate logic generalized by counting quantifiers [9]. Although past approaches employ reasoning algorithms of DL and OCL, we develop consistency checking algorithms specifically for UML class diagrams. Our algorithms deal directly with the structure of UML class diagrams; hence, they enable the following:

- Easy recognition of the inconsistency triggers in the diagram structure, such as combinations of disjointness/completeness constraints, attribute multiplicities, and overwriting/multiple inheritances, and
- Refinement of the algorithms when the expressivity is changed by the presence of the inconsistency triggers.

The inconsistency triggers captured by the diagram structure are used to restrict some relevant class diagram components in order to derive a classification of UML class diagrams. Since we can theoretically prove that there arises no inconsistency of eliminated components, the algorithms will become simplified and optimized for their respective expressivity.

The contributions of this paper are as follows:

1. *Inconsistency triggers:* We accurately identify the inconsistency triggers that cause logical inconsistency among classes, attributes, and associations.

2. *Expressivity:* We classify the expressivity of UML class diagrams by deleting and adding certain inconsistency triggers.
3. *Algorithms and complexities:* We develop several consistency checking algorithms for class diagrams of different expressive powers and demonstrate that they compute the consistency of those class diagrams in P, NP, PSPACE, or EXPTIME with respect to the size of a class diagram.
4. *Tractable consistency checking in the optimized algorithms:* When the attribute value types are defined with restrictions in class diagrams, consistency checking is respectively computable in P and PSPACE for two cases in which the diagrams contain (i) disjointness constraints and overwriting/multiple inheritances and (ii) these components with completeness constraints.
5. *Consistent class diagrams:* We demonstrate that every class diagram is consistent if the expressivity is restricted by deleting disjointness constraints and overwriting/multiple inheritances (but allowing attributes multiplicities and simple inheritances). Thus, we need not test the consistency of such less expressive class diagrams ($\mathcal{D}_0^-$ and $\mathcal{D}_{com}^-$).

There are two main advantages with regard to the results of this study. First, the optimized algorithms support efficient reasoning for various expressive powers of class diagrams. In contrast, the DL formalisms do not provide optimized algorithms for the *restricted* UML class diagrams because general axioms of DLs require exponential time even if DLs are restricted [3]. Therefore, the classification of DLs does not fit into the classification of UML class diagrams[1]. Second, a meaningful restriction of UML class diagrams is analyzed. We confirm the existence of restricted class diagrams that permit attribute multiplicities but that cause no logical inconsistency.

2   Class Diagrams in FOPL with Counting Quantifiers

We define a translation of UML class diagrams into first-order predicate logic generalized by counting quantifiers. The reasons for encoding into first-order predicate logic with counting quantifiers are as follows. First, the semantics of UML class diagrams should be defined by encoding them in a logical language because consistency checking is based on the semantics of encoded formulas. In other words, no consistency checking algorithm can operate on original diagrams without formal semantics. Second, variables and quantifiers in first-order logic lead to an explicit formulation that is useful to restrict/classify the expressive powers. In contrast, DL encoding [2] conceals the quantification of variables in expressions.

The alphabet of UML class diagrams consists of a set of class names, a set of attribute names, a set of operation names, a set of association names, and a set of datatype names. Let C, C', C_i be class names, a, a' attribute names, f, f'

---

[1] Note that reasoning on general axioms becomes exponential hard even if the small DL $\mathcal{AL}$ contains no disjunction, qualified existential restriction, and number restriction.

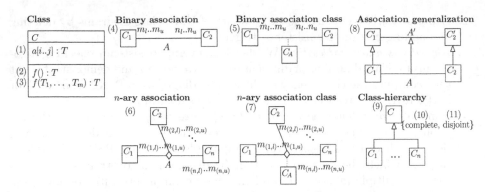

Fig. 1. Components of UML class diagrams

operation names, A, A' association names, and t, t', t_i datatype names. Let type T be either a class or a datatype. The leftmost figure in Fig.1 represents a class C with an attribute $a[i..j]\colon T$, a 0-ary operation $f()\colon T$, and an n-ary operation $f(T_1, \ldots, T_n)\colon T$, where $[i..j]$ is the attribute multiplicity and T and $T_1, \ldots, T_n$ are types. Any class C can be expressed as the unary predicate C in first-order logic. Let F_1 and F_2 be first-order formulas. We denote the implication form $F_1 \to F_2$ as the universal closure $\forall x_1 \cdots \forall x_n.(F_1 \to F_2)$ where $x_1, \ldots, x_n$ are all the free variables occurring in $F_1 \to F_2$. Let $F(x)$ denote a formula F in which the free variable x occurs. The counting quantifier formula $\exists_{\geq i} x.F(x)$ implies that at least i elements x satisfy $F(x)$, while the counting quantifier formula $\exists_{\leq i} x.F(x)$ implies that at most i elements x satisfy $F(x)$. The value type T and multiplicity $[i..j]$ of the attribute a in the class C are specified by the following implication forms:

(1) $C(x) \to (a(x, y) \to T(y))$ and $C(x) \to \exists_{\geq i} z.a(x, z) \land \exists_{\leq j} z.a(x, z)$

where a is a binary predicate and T is a unary predicate. Moreover, the 0-ary operation $f()\colon T$ of the class C is specified by the following implication forms:

(2) $C(x) \to (f(x, y) \to T(y))$ and $C(x) \to \exists_{\leq 1} z.f(x, z)$

where f is a binary predicate and T is a unary predicate. The n-ary operation $f(T_1, \ldots, T_n)\colon T$ of the class C is specified by the following implication forms:

(3) $C(x) \to (f(x, y_1, \ldots, y_n, z) \to T_1(y_1) \land \cdots \land T_n(y_n) \land T(z))$
  $C(x) \to \exists_{\leq 1} z.f(x, y_1, \ldots, y_n, z)$

where f is an $n + 2$-ary predicate and each T_i, T are unary predicates.

We next formalize associations A that imply connections among classes C_1, $\ldots, C_n$ (as in (4) and (6) of Fig.1). A binary association A between two classes C_1 and C_2 and the multiplicities $m_l..m_u$ and $n_l..n_u$ are specified by the forms:

(4) $A(x_1, x_2) \to C_1(x_1) \land C_2(x_2)$
  $C_1(x) \to \exists_{\geq n_l} x_2.A(x, x_2) \land \exists_{\leq n_u} x_2.A(x, x_2)$
  $C_2(x) \to \exists_{\geq m_l} x_1.A(x_1, x) \land \exists_{\leq m_u} x_1.A(x_1, x)$

where A is a binary predicate and C_1, C_2 are unary predicates. In addition to the formulas, if an association is represented by a class, then the association class C_A is specified by supplementing the implication forms below:

(5) $A(x_1, x_2) \rightarrow (r_0(x_1, x_2, z) \rightarrow C_A(z))$
$\quad A(x_1, x_2) \rightarrow \exists_{=1} z.r_0(x_1, x_2, z)$ and $\exists_{\leq 1} z.(r_0(x_1, x_2, z) \wedge C_A(z))$

where C_A is a unary predicate and r_0 is a ternary predicate. By extending the formulation of a binary association, the n-ary association A among classes C_1, ..., C_n and their multiplicities "$m_{(1,l)}..m_{(1,u)}$", ..., "$m_{(n,l)}..m_{(n,u)}$" (as shown in (6) of Fig.1) are specified by the following implication forms:

(6) $A(x_1, \ldots, x_n) \rightarrow C_1(x_1) \wedge \cdots \wedge C_n(x_n)$
$\quad C_k(x) \rightarrow \exists_{\geq m_{(1,l)}} x_1 \cdots \exists_{\geq m_{(k-1,l)}} x_{k-1} \exists_{\geq m_{(k+1,l)}} x_{k+1} \cdots \exists_{\geq m_{(n,l)}} x_n.A(x_1, \ldots, x_n)[x_k/x]$
$\quad C_k(x) \rightarrow \exists_{\leq m_{(1,u)}} x_1 \cdots \exists_{\leq m_{(k-1,u)}} x_{k-1} \exists_{\leq m_{(k+1,u)}} x_{k+1} \cdots \exists_{\leq m_{(n,u)}} x_n.A(x_1, \ldots, x_n)[x_k/x]$

where A is an n-ary predicate and $[x_k/x]$ is a substitution of x_k with x. In addition, the association class C_A is specified by adding the implication forms below:

(7) $A(x_1, \ldots, x_n) \rightarrow (r_0(x_1, \ldots, x_n, z) \rightarrow C_A(z))$
$\quad A(x_1, \ldots, x_n) \rightarrow \exists_{=1} z.r_0(x_1, \ldots, x_n, z)$ and $\exists_{\leq 1} z.(r_0(x_1, \ldots, x_n, z) \wedge C_A(z))$

where C_A is a unary predicate and r_0 is an $n + 1$-ary predicate. Furthermore, we treat association generalization (not discussed in [2]) such that the binary association A' between classes C_1' and C_2' generalizes the binary association A between classes C_1 and C_2 (as in (8) of Fig.1). More universally, the generalization between n-ary associations A and A' is specified by the following implication forms:

(8)' $A(x_1, \ldots, x_n) \rightarrow A'(x_1, \ldots, x_n)$ and $C_1(x) \rightarrow C_1'(x)$, ..., $C_n(x) \rightarrow C_n'(x)$

where A, A' are n-ary predicates and each C_i, C_j' are unary predicates.

We consider class hierarchies and disjointness/completeness constraints of the classes in hierarchies, as shown in (9), (10), and (11) of Fig.1. A class hierarchy (a class C generalizes classes $C_1, \ldots, C_n$) is specified by the implication forms below:

(9) $C_1(x) \rightarrow C(x)$, ..., $C_n(x) \rightarrow C(x)$

where C and $C_1, \ldots, C_n$ are unary predicates. The completeness constraint between class C and classes $C_1, \ldots, C_n$ and the disjointness constraint among classes $C_1, \ldots, C_n$ are respectively specified by the implication forms:

(10) $C(x) \rightarrow C_1(x) \vee \cdots \vee C_n(x)$

(11) $C_i(x) \rightarrow \neg C_{i+1}(x) \wedge \cdots \wedge \neg C_n(x)$ for all $i \in \{1, \ldots, n-1\}$

where C and $C_1, \ldots, C_n$ are unary predicates.

Let D be a UML class diagram. $\mathcal{G}(D)$ is called the translation of D and denotes the set of implication forms obtained by the encoding of D in first-order predicate logic with counting quantifiers (using (1)–(11)).

3   Inconsistencies in Class Diagrams

In this section, we analyze inconsistencies among classes, attributes, and associations in UML class diagrams. We first define the syntax errors of duplicate names and irrelevant attribute value types as follows.

Duplicate name errors/attribute value type errors. A UML class diagram D contains a duplicate name error if (i) two classes C_1 and C_2 appear and C_1 and C_2 have the same class name, (ii) two associations A_1 and A_2 appear and A_1 and A_2 have the same association name, or (iii) two attributes a_1 and a_2 appear in a class C and a_1 and a_2 have the same attribute name. Moreover, if two classes have the same name's attributes $a\colon T_1$ and $a\colon T_2$, such that T_1 is a class and T_2 is a datatype, then the class diagram contains an attribute value type error. Obviously, the checking of these syntax errors in a UML class diagram can be computed in linear time.

We elaborate three inconsistency triggers for the UML class diagrams. The reflexive and transitive closure of $\rightarrow$ over classes and associations are denoted by $\rightarrow^*$ such that (i) $C(x) \rightarrow^* C(x)$, (ii) $A(x_1,\ldots,x_n) \rightarrow^* A(x_1,\ldots,x_n)$, (iii) if $C(x) \rightarrow F(x)$, or $C(x) \rightarrow^* C'(x)$ and $C'(x) \rightarrow^* F(x)$, then $C(x) \rightarrow^* F(x)$, and (iv) if $A(x_1,\ldots,x_n) \rightarrow F(x_1,\ldots,x_n)$, or $A(x_1,\ldots,x_n) \rightarrow^* A'(x_1,\ldots,x_n)$ and $A'(x_1,\ldots,x_n) \rightarrow^* F(x_1,\ldots,x_n)$, then $A(x_1,\ldots,x_n) \rightarrow^* F(x_1,\ldots,x_n)$, where $F(x)$ and $F(x_1,\ldots,x_n)$ are any formulas including the free variables.

Inconsistency trigger 1 (generalization and disjointness). The first inconsistency trigger is caused by a combination of generalization and a disjointness constraint. A class diagram has an inconsistency trigger if it contains the formulas $C(x) \rightarrow^* C_k(x)$ and $C(x) \rightarrow^* \neg C_1(x) \wedge \cdots \wedge \neg C_n(x)$ where $1 \le k \le n$.

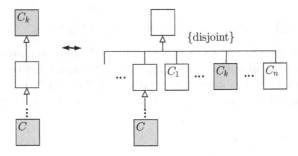

As shown in the above figure, this inconsistency appears when a class C has a superclass C_k but the classes C and C_k are defined as disjoint to each other in the constraint of a class hierarchy.

Inconsistency trigger 2 (overwriting/multiple inheritance). The second inconsistency trigger is caused by one of the following situations:

1. (a) conflict between value types T_1 and T_2 when they appear in attributes $a\colon T_1$ and $a\colon T_2$ of the same name, or (b) conflict between multiplicities $[i..j]$ and $[i'..j']$ when they appear in multiplicities $a\colon T_1$ and $a\colon T_2$ of attributes with the same names.

2. conflict between multiplicities when they appear in association and super-associations.

More formally, a class diagram has an inconsistency trigger if it contains a group of the following formulas:

1. $C_2(x) \rightarrow^* C_1(x)$, or $C(x) \rightarrow^* C_1(x)$ and $C(x) \rightarrow^* C_2(x)$, together with
   (a) **Attribute value types:** $C_1(x) \rightarrow (a(x,y) \rightarrow T_1(y))$ and $C_2(x) \rightarrow (a(x,y) \rightarrow T_2(y))$ where T_1 and T_2 are disjoint[2], or
   (b) **Attribute multiplicities:** $C_1(x) \rightarrow \exists_{\geq i} z.a(x,z) \wedge \exists_{\leq j} z.a(x,z)$ and $C_2(x) \rightarrow \exists_{\geq i'} z.a(x,z) \wedge \exists_{\leq j'} z.a(x,z)$ where $i > j'$.

2. **Association multiplicities:** $A(x_1,\ldots,x_n) \rightarrow A'(x_1,\ldots,x_n)$ with
   $C_k(x) \rightarrow \exists_{\geq m_{(1,l)}} x_1 \cdots \exists_{\geq m_{(k-1,l)}} x_{k-1} \exists_{\geq m_{(k+1,l)}} x_{k+1} \cdots \exists_{\geq m_{(n,l)}} x_n . A(x_1, \ldots, x_n)[x_k/x]$ and $C'_k(x') \rightarrow \exists_{\leq m'_{(1,u)}} x'_1 \cdots \exists_{\leq m'_{(k-1,u)}} x'_{k-1} \exists_{\leq m'_{(k+1,u)}} x'_{k+1} \cdots \exists_{\leq m'_{(n,u)}} x'_n . A'(x'_1, \ldots, x'_n)[x'_k/x']$ where $m_{(i,l)} > m'_{(i,u)}$.

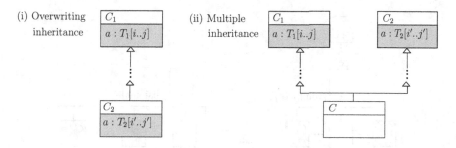

(i) Overwriting inheritance: a class C_2 with attribute $a: T_2[i'..j']$; superclass C_1 with $a: T_1[i..j]$.
(ii) Multiple inheritance: class C inherits from C_1 with $a: T_1[i..j]$ and C_2 with $a: T_2[i'..j']$.

This figure explains that (i) a class C_2 with an attribute $a: T_2[i'..j']$ inherits the same name's attribute $a: T_1[i..j]$ from a superclass C_1 and (ii) a class C inherits the two attributes $a: T_1[i..j]$ and $a: T_2[i'..j']$ of the same name from superclasses C_1 and C_2. The former is called *overwriting inheritance*; the latter, *multiple inheritance*. In these cases, if the attribute value types T_1 and T_2 are disjoint or if the multiplicities $[i..j]$ and $[i'..j']$ conflict with each other, then the attributes are determined to be inconsistent. For example, the multiplicities $[1..5]$ and $[10..*]$ cannot simultaneously hold for the same name's attributes.

Inconsistency trigger 3 (completeness and disjointness). A disjointness constraint combined with a completeness constraint can yield the third inconsistency trigger. A class diagram has an inconsistency trigger if it contains the formulas $C(x) \rightarrow^* C_1(x) \vee \cdots \vee C_n(x)$ and $C(x) \rightarrow^* \neg C'_1(x) \wedge \cdots \wedge \neg C'_m(x)$, where $\{C_1,\ldots,C_n\} \subseteq \{C'_1,\ldots,C'_m\}$. This inconsistency appears when classes C and $C_1,\ldots,C_n$ satisfy the completeness constraint in a class hierarchy and classes C and $C'_1,\ldots,C'_m$ satisfy the disjointness constraint in another class

---

[2] Types T_1 and T_2 are disjoint if they are classes C_1 and C_2 such that $C_1(x) \rightarrow^* \neg C_2 \in \mathcal{G}(D)$ or if they are datatypes t_1 and t_2 such that $t_1 \cap t_2 = \emptyset$.

hierarchy. Intuitively, any instance of class C must be an instance of one of the classes $C_1, \ldots, C_n$, but each instance of class C cannot be an instance of classes $C_1', \ldots, C_m'$. Hence, this situation is contradictory.

The third inconsistency trigger may be more complicated when the number of completeness and disjointness constraints that occur in a class diagram is increased. In other words, disjunctive expressions raised by many completeness constraints expand the search space of finding inconsistency. Let us define the relation $C(x) \rightarrow^+ C_1(x) \vee \cdots \vee C_n(x)$ as follows: (i) if $C(x) \rightarrow^* C_1(x) \vee \cdots \vee C_n(x)$, then $C(x) \rightarrow^+ C_1(x) \vee \cdots \vee C_n(x)$, and (ii) if $C(x) \rightarrow^+ C_1(x) \vee \cdots \vee C_n(x)$ and $C_1(x) \rightarrow^+ DC_1(x), \ldots, C_n(x) \rightarrow^+ DC_n(x)$, then $C(x) \rightarrow^+ DC_1(x) \vee \cdots \vee DC_n(x)$ where each DC_i denotes $C_1'(x) \vee \cdots \vee C_m'(x)$ as disjunctive classes. A class diagram has an inconsistency trigger if it contains the formulas $C(x) \rightarrow^+ C_1(x) \vee \cdots \vee C_n(x)$ and for each $i \in \{1, \ldots, n\}$, $C(x) \rightarrow^* \neg C_{(i,1)}(x) \wedge \cdots \wedge \neg C_{(i,m_i)}(x)$, where C_i is one of the classes $C_{(i,1)}, \ldots, C_{(i,m_i)}$. For example, the following figure illustrates that two completeness constraints are complicatedly inconsistent with respect to a disjointness constraint.

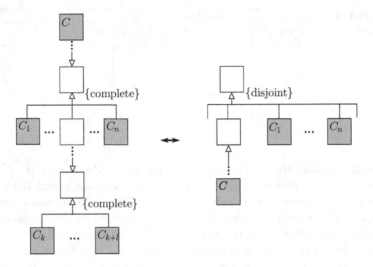

The three inconsistency triggers describe all the logical inconsistencies in UML class diagrams if they contain association generalization but not roles. In the next section, we will design a *complete* consistency checking algorithm for finding those inconsistency triggers.

We define a formal model of UML class diagrams using the semantics of FOPL with counting quantifiers. An interpretation $\mathcal{I}$ is an ordered pair (U, I) of the universe U and an interpretation function I for a first-order language.

Definition 1 (UML Class Diagram Models). *Let $\mathcal{I} = (U, I)$ be an interpretation. The interpretation $\mathcal{I}$ is a model of a UML class diagram D (called a UML-model of D) if*

1. *$I(C) \neq \emptyset$ for every class C in D and*
2. *$\mathcal{I}$ satisfies $\mathcal{G}(D)$ where $\mathcal{G}(D)$ is the translation of D.*

The first condition indicates that every class is a non-empty class (i.e., an instance of the class exists) and the second condition implies that $\mathcal{I}$ is a first-order model of the class diagram formulation $\mathcal{G}(D)$. A UML class diagram D is consistent if it has a UML-model.

Furthermore, the following class diagram is invalid because the association class C_A cannot be used for two different binary associations between classes C_1 and C_2 and between classes C_1 and C_3.

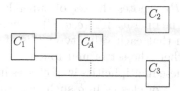

Instead of C_A, we describe a ternary association or two association classes. It appears that the EXPTIME-hardness in [2] relies on such expressions. This is because when we reduce (EXPTIME-hard) concept satisfiability in $\mathcal{ALC}$ KBs to class consistency in a UML class diagram, the $\mathcal{ALC}$ KB $\{C_1 \sqsubseteq \exists P_A.C_2,\ C_1 \sqsubseteq \exists P_A.C_3\}$ is encoded into an invalid association class. This condition is important in order to avoid the EXPTIME-hardness and therefore to derive the complexity results in Section 5. This implies that the consistency checking of some restricted UML class diagram groups is computable in P and PSPACE.

4   Consistency Checking

This section presents a consistency checking algorithm for a set of implication forms Γ_0 (corresponding to the UML class diagram formulation $\mathcal{G}(D)$). It consists of two sub-algorithms $Cons$ and $Assoc$: $Cons$ checks the consistency of a class in Γ_0 and $Assoc$ tests the consistency of association generalization in Γ_0.

4.1   Algorithm for Testing Consistency

We decompose an implication form set Γ_0 in order to apply our consistency checking algorithm to it. Let Γ_0 be a set of implication forms, C be a class, and $F_i(x)$ be any formula including a free variable x. Γ is a decomposed set of Γ_0 if the following conditions hold: (i) $\Gamma_0 \subseteq \Gamma$, (ii) if $C(x) \to F_1(x) \wedge \cdots \wedge F_n(x) \in \Gamma$, then $C(x) \to F_1(x) \in \Gamma, \ldots, C(x) \to F_n(x) \in \Gamma$, and (iii) if $C(x) \to F_1(x) \vee \cdots \vee F_n(x) \in \Gamma$, then $C(x) \to F_i(x) \in \Gamma$ for some $i \in \{1, \ldots, n\}$. We denote $\Sigma(\Gamma_0)$ as the family of decomposed sets of Γ_0.

We denote $cls(\Gamma_0)$ as the set of classes, $att(\Gamma_0)$ as the set of attributes, and $asc(\Gamma_0)$ as the set of associations that occur in the implication form set Γ_0.

Definition 2. *The following operations will be embedded as subroutines in the consistency checking algorithm:*

1. $H(C, \Gamma) = \{C' \mid C(x) \to^* C'(x) \in \Gamma\} \cup \{\neg C' \mid C(x) \to^* \neg C'(x) \in \Gamma\}$.
2. $E(\delta, a, \Gamma) = \bigcup_{C \in \delta} E(C, a, \Gamma)$ where $E(C, a, \Gamma) = \{C' \mid C(x) \to^* (a(x, y) \to C'(y)) \in \Gamma$ and $C(x) \to^* \exists_{\geq i} z.a(x, z) \in \Gamma\}$ with $i \geq 1$.
3. $N(\delta, a, \Gamma) = \bigcup_{C \in \delta} N(C, a, \Gamma)$ where $N(C, a, \Gamma) = \{\geq i \mid C(x) \to^* \exists_{\geq i} z.a(x, z) \in \Gamma\} \cup \{\leq j \mid C(x) \to^* \exists_{\leq j} z.a(x, z) \in \Gamma\}$.
4. $\mu_0(\delta, \Gamma) = \{C\}$ if for all $C' \in \mu(\delta, \Gamma)$, $C \preceq C'$ and $C \in \mu(\delta, \Gamma)$ where $\mu(\delta, \Gamma) = \{C \in \delta \mid \delta \subseteq H(C, \Gamma)\}$ and $\preceq$ is a linear order over $cls(\Gamma_0)$.

The operation $H(C, \Gamma)$ denotes the set of superclasses C' of C and disjoint classes $\neg C'$ of C in Γ. The operation $E(\delta, a, \Gamma)$ gathers the set of value types T of attribute a in Γ such that each value type T is of classes in δ. Further, the operation $N(\delta, a, \Gamma)$ gathers the set of multiplicities $\geq i$ and $\leq j$ of attribute a in Γ such that each of these multiplicities is of classes in δ. The operation $\mu(\delta, \Gamma)$ returns a set $\{C_1, \ldots, C_n\}$ of classes in δ such that the superclasses of each C_i (in Γ) subsume all the classes in δ. The operation $\mu_0(\delta, \Gamma)$ returns the singleton set $\{C\}$ of a class in $\mu(\delta, \Gamma)$ such that C is the least class in $\mu(\delta, \Gamma)$ over $\preceq$. The consistency checking algorithm *Cons* is described as follows.

Algorithm *Cons*
input set of classes δ, family of sets of classes Δ, set of implication forms Γ_0
output 1 (consistent) or 0 (inconsistent)
begin
    for $\Gamma \in \Sigma(\Gamma_0)$ **do**
        $S = \bigcup_{C \in \delta} H(C, \Gamma)$; $f_\Gamma = 0$;
        if $\{C, \neg C\} \not\subseteq S$ and $\{t_1, \ldots, t_n\} \not\subseteq S$ s.t. $t_1 \cap \cdots \cap t_n = \emptyset$ **then** $f_\Gamma = 1$;
            for $a \in att(\Gamma_0)$ **do**
                if $i > j$ s.t. $\{\geq i, \leq j\} \subseteq N(\delta, a, \Gamma)$ **then** $f_\Gamma = 0$;
                else $\delta_a = E(\delta, a, \Gamma)$;
                    if $\delta_a \neq \emptyset$ and $\delta_a, \mu_0(\delta_a, \Gamma) \not\subseteq \Delta$ **then** $f_\Gamma = Cons(\delta_a, \Delta \cup \{\delta\}, \Gamma_0)$;
            esle
            rof
        fi
        if $f_\Gamma = 1$ **then** **return** 1;
    rof
    return 0;
end;

In order to decide the consistency of the input implication form set Γ_0, we execute the algorithm $Cons(\{C\}, \emptyset, \Gamma_0)$ for every class $C \in cls(\Gamma_0)$. If C is consistent in Γ_0, it returns 1, else 0 is returned. At the first step of the algorithm, a decomposed set Γ of Γ_0 (in $\Sigma(\Gamma_0)$) is selected, which is one of all the disjunctive branches with respect to the completeness constraints in Γ_0. Subsequently, for each $\Gamma \in \Sigma(\Gamma_0)$, the following three phases are performed.

(1) For the selected Γ, the algorithm checks whether all the superclasses of classes in $\delta = \{C\}$ (obtained from $S = \bigcup_{C \in \delta} H(C, \Gamma)$) are disjoint to each other. Intuitively, it sets a dummy instance of class C and then, the dummy instance is regarded as an instance of the superclasses C' of C and of the disjoint classes $\neg C'$ of C along the implication forms $C(x) \to^* C'(x)$ and $C(x) \to^* \neg C'(x)$ in Γ. If an

inconsistent pair C_i and $\neg C_i$ possesses the dummy instance, then δ is determined to be inconsistent in Γ. For example, $\{C\}$ is inconsistent in $\Gamma_1 = \{C(x) \to C_1(x), C_1(x) \to C_2(x), C(x) \to \neg C_2(x)\}$ since the inconsistent pair C_2 and $\neg C_2$ must have the dummy instance of the class C, i.e., $H(C, \Gamma_1) = \{C, C_1, C_2, \neg C_2\}$.

(2) If phase (1) finds no inconsistency in Γ, the algorithm next checks the multiplicities of all the attributes $a \in att(\Gamma_0)$. The multiplicities of the same attribute name a are obtained by $N(\delta, a, \Gamma)$; therefore, when $N(\delta, a, \Gamma)$ contains $\{\geq i, \leq j\}$ with $i > j$, these multiplicities are inconsistent. Intuitively, similar to phase (1), the algorithm checks whether superclasses involve conflicting multiplicities along the implication form $C(x) \to^* C'(x)$ in Γ. For example, $\{C\}$ is inconsistent in $\Gamma_2 = \{C(x) \to \exists_{\geq 10} z.\, a(x,z), C(x) \to C_1(x), C_1(x) \to \exists_{\leq 5} z.a(x,z)\}$ since the counting quantifiers $\exists_{\geq 10}$ and $\exists_{\leq 5}$ cannot simultaneously hold when $N(\{C\}, a, \Gamma_2) = \{\geq 10, \leq 5\}$.

(3) Next, the disjointness of attribute value types is checked. Along the implication form $C(x) \to^* C'(x)$ in Γ, the algorithm gathers all the value types of the same name's attributes, obtained by $\delta_a = E(\delta, a, \Gamma)$ for each $a \in att(\Gamma_0)$. For example, $\Gamma_3 = \{C(x) \to C_1(x), C(x) \to C_2(x), C_1(x) \to (a(x,y) \to C_3(y)), C_2(x) \to (a(x,y) \to C_4(y))\}$ derives $\delta_a = \{C_3, C_4\}$ by $E(\{C\}, a, \Gamma_3)$ since superclasses C_1 and C_2 of C have the attributes $a: C_3$ and $a: C_4$. In other words, each value of attribute a is typed by C_3 and C_4. Hence, the algorithm needs to check the consistency of $\delta_a = \{C_3, C_4\}$. In order to accomplish this, it recursively calls $Cons(\delta_a, \Delta \cup \{\{C\}\}, \Gamma_0)$, where δ_a is consistent if 1 is returned. The second argument $\Delta \cup \{\{C\}\}$ prevents infinite looping by storing sets of classes where each set is already checked in the caller processes.

In order to find a consistent decomposed set Γ in the disjunctive branches of $\Sigma(\Gamma_0)$, if the three phases (1), (2), and (3) do not detect any inconsistency in Γ, then the algorithm sets the flag $f_\Gamma = 1$, else it sets $f_\Gamma = 0$. Thus, the flag $f_\Gamma = 1$ indicates that $\{C\}$ is consistent in the input Γ_0, i.e., $Cons(\{C\}, \emptyset, \Gamma_0) = 1$.

In addition to the algorithm $Cons$, the consistency checking of multiplicities over association generalization is processed by the following algorithm $Assoc$. If Γ_0 does not cause any inconsistency with respect to associations, $Assoc(\Gamma_0)$ returns 1, which is computable in polynomial time.

Algorithm $Assoc$
input set of implication forms Γ_0
output 1 (consistent) or 0 (inconsistent)
begin
    for $A \in asc(\Gamma_0)$ and $k \in \{1, \ldots, n\}$ s.t. $arity(A) = n$ **do**
        if $i_v > j_v$ s.t. $\{(\geq i_1, \ldots, \geq i_{k-1}, \geq i_{k+1}, \ldots, \geq i_n),$
            $(\leq j_1, \ldots, \leq j_{k-1}, \leq j_{k+1}, \ldots, \leq j_n)\} \subseteq N_k(H(A, \Gamma_0), \Gamma_0)$ **then return** 0;
    rof
    return 1;
end;

As defined below, the operations $H(A, \Gamma_0)$ and $N_k(\alpha, \Gamma_0)$ respectively return the set of super-associations A' of A and the set of $n - 1$-tuples of multiplicities of n-ary associations A in α along the implication forms $C_k(x) \to \exists_{\geq i_1} x_1 \cdots$

$\exists_{\geq i_{k-1}} x_{k-1} \exists_{\geq i_{k+1}} x_{k+1} \cdots \exists_{\geq i_n} x_n.A(x_1, \ldots, x_n)[x_k/x]$ and $C_k(x) \rightarrow \exists_{\leq j_1} x_1$ $\cdots \exists_{\leq j_{k-1}} x_{k-1} \exists_{\leq j_{k+1}} x_{k+1} \cdots \exists_{\leq j_n} x_n.A(x_1, \ldots, x_n)[x_k/x]$, respectively.

Definition 3. *The operations $H(A, \Gamma_0)$ and $N_k(\alpha, \Gamma_0)$ are defined as follows:*

1. $H(A, \Gamma_0) = \{A' \mid A(x_1, \ldots, x_n) \rightarrow^* A'(x_1, \ldots, x_n) \in \Gamma_0\}$.
2. $N_k(\alpha, \Gamma_0) = \bigcup_{A \in \alpha} N_k(A, \Gamma_0)$ *where* $N_k(A, \Gamma_0) = \{(\geq i_1, \ldots, \geq i_{k-1}, \geq i_{k+1}, \ldots, \geq i_n) \mid C_k(x) \rightarrow \exists_{\geq i_1} x_1 \cdots \exists_{\geq i_{k-1}} x_{k-1} \exists_{\geq i_{k+1}} x_{k+1} \cdots \exists_{\geq i_n} x_n.A(x_1, \ldots, x_n)[x_k/x] \in \Gamma_0\} \cup \{(\leq j_1, \ldots, \leq j_{k-1}, \leq j_{k+1}, \ldots, \leq j_n) \mid C_k(x) \rightarrow \exists_{\leq j_1} x_1 \cdots \exists_{\leq j_{k-1}} x_{k-1} \exists_{\leq j_{k+1}} x_{k+1} \cdots \exists_{\leq j_n} x_n.A(x_1, \ldots, x_n)[x_k/x] \in \Gamma_0\}$.

4.2   Soundness, Completeness, and Termination

We sketch a proof of the completeness for the algorithms *Cons* and *Assoc*. Assume that $Cons(\{C\}, \emptyset, \mathcal{G}(D))$ for all $C \in cls(\mathcal{G}(D))$ and $Assoc(\mathcal{G}(D))$ are called. We construct an implication tree of $(C, \mathcal{G}(D))$ that expresses the consistency checking proof of C in $\mathcal{G}(D)$. If $Cons(\{C\}, \emptyset, \mathcal{G}(D)) = 1$, there exists a non-closed implication tree of $(C, \mathcal{G}(D))$. In order to prove the existence of a UML-model of D, a canonical interpretation is constructed by consistent subtrees of the non-closed implication trees of $(C_1, \mathcal{G}(D)), \ldots, (C_n, \mathcal{G}(D))$ (with $cls(\mathcal{G}(D)) = \{C_1, \ldots, C_n\}$) and by $Assoc(\mathcal{G}(D)) = 1$. This proves that D is consistent.

Corresponding to calling $Cons(\delta_0, \emptyset, \Gamma_0)$, we define an implication tree of a class set δ_0 that expresses the consistency checking proof of δ_0.

Definition 4. *Let Γ_0 be a set of implication forms and let $\delta_0 \subseteq cls(\Gamma_0)$. An implication tree of (δ_0, Γ_0) is a finite and minimal tree such that (i) the root is a node labeled with δ_0, (ii) each non-leaf node is labeled with a non-empty set of classes, (iii) each leaf is labeled with 0, 1, or w, (iv) each edge is labeled with Γ or (Γ, a) where $\Gamma \in \Sigma(\Gamma_0)$ and $a \in att(\Gamma_0)$, and (v) for each node labeled with δ and each $\Gamma \in \Sigma(\Gamma_0)$, if $\bigcup_{C \in \delta} H(C, \Gamma)$ contains $\{C, \neg C\}$ or $\{t_1, \ldots, t_n\}$ with $t_1 \cap \cdots \cap t_n = \emptyset$, then there is a child of δ labeled with 0 and the edge of the nodes δ and 0 is labeled with Γ, and otherwise:*

– *if $att(\Gamma_0) = \emptyset$, then there is a child of δ labeled with 1 and the edge of the nodes δ and 1 is labeled with Γ, and*
– *for all $a \in att(\Gamma_0)$, the following conditions hold:*
    1. *if $i > j$ such that $\{\geq i, \leq j\} \in N(\delta, a, \Gamma)$, then there is a child of δ labeled with 0 and the edge of the nodes δ and 0 is labeled with (Γ, a),*
    2. *if $E(\delta, a, \Gamma) = \emptyset$, then there is a child of δ labeled with 1 and the edge of the nodes δ and 1 is labeled with (Γ, a),*
    3. *if there is an ancestor labeled with $E(\delta, a, \Gamma)$ or $\mu_0(E(\delta, a, \Gamma), \Gamma)$, then there is a child of δ labeled with w and the edge of the nodes δ and w is labeled with (Γ, a), and*
    4. *otherwise, there is a child of δ labeled with $E(\delta, a, \Gamma)$ and the edge of the nodes δ and $E(\delta, a, \Gamma)$ is labeled with (Γ, a).*

Let T be an implication tree of (δ_0, Γ_0). A node d in T is closed if (i) d is labeled with 0 or if (ii) d is labeled with δ and for every $\Gamma \in \Sigma(\Gamma_0)$, there is an edge (d, d') labeled with Γ or (Γ, a) such that d' is closed. An implication tree of (δ_0, Γ_0) is closed if the root is closed; it is non-closed otherwise. A forest of Γ_0 is a set of implication trees of $(\{C_1\}, \Gamma_0), \ldots, (\{C_n\}, \Gamma_0)$ such that $cls(\Gamma_0) = \{C_1, \ldots, C_n\}$. A forest S of Γ_0 is closed if there exists a closed implication tree T in S. The following lemma states the correspondence between the consistency checking for every $C \in cls(\Gamma_0)$ and the existence of a non-closed forest of Γ_0.

Lemma 1. *Let Γ_0 be a set of implication forms. For every class $C \in cls(\Gamma_0)$, $Cons(\{C\}, \emptyset, \Gamma_0) = 1$ if and only if there is a non-closed forest of Γ_0.*

We define a consistent subtree T' of a non-closed implication tree T such that T' is constructed by non-closed nodes in T.

Definition 5 (Consistent Subtree). *Let T be a non-closed implication tree of $(\{C_0\}, \Gamma_0)$ and d_0 be the root where Γ_0 is a set of implication forms and $C_0 \in cls(\Gamma_0)$. A tree T' is a consistent subtree of T if (i) T' is a subtree of T, (ii) every node in T' is not closed, and (iii) every non-leaf node has m children of all the attributes $a_1, \ldots, a_m \in att(\Gamma_0)$ where each child is labeled with 1, w, or a set of classes and each edge of the non-leaf node and its child is labeled with (Γ, a_i).*

We show the correspondence between the consistency of an implication form set Γ_0 and the existence of a non-closed forest of Γ_0. We extend the first-order language by adding the new constants $\bar{d}$ for all the elements $d \in U$ such that each new constant is interpreted by itself, i.e., $I(\bar{d}) = d$. In addition, we define the following operations:

1. $proj_k^n(x_1, \ldots, x_n) = x_k$ where $1 \leq k \leq n$.
2. $Max_{\geq}(X) = (Max(X_1), \ldots, Max(X_n))$ where X is a set of n-tuples and for each $v \in \{1, \ldots, n\}$, $X_v = \{ proj_v^n (i_1, \ldots, i_n) \mid (\geq i_1, \ldots, \geq i_n) \in X\}$.
3. $AC(A, \Gamma) = (C_1, \ldots, C_n)$ if $A(x_1, \ldots, x_n) \to C_1(x_1) \wedge \cdots \wedge C_n(x_n) \in \Gamma$.

A canonical interpretation of an implication form set Γ_0 is constructed by consistent subtrees of the non-closed implication trees in a forest of Γ_0, that is used to prove the completeness of the algorithm $Cons$. A class C is consistent in Γ if there exists a non-closed implication tree of $(\{C\}, \Gamma_0)$ such that the root labeled with $\{C\}$ has a non-closed child node labeled with Γ or (Γ, a).

Definition 6 (Canonical Interpretation). *Let Γ_0 be a set of implication forms such that $Assoc(\Gamma_0) = 1$ and let $S = \{T_1, \ldots, T_n\}$ be a non-closed forest of Γ_0. For every $T_i \in S$, there is a consistent subtree T_i' of T_i, and we set $S' = \{T_1', \ldots, T_n'\}$ as the set of consistent subtrees of $T_1, \ldots, T_n$ in S. An canonical interpretation of Γ_0 is a pair $\mathcal{I} = (U, I)$ such that $U_0 = \{d \mid d$ is a non-leaf node in $T_1' \cup \cdots \cup T_n'\}$, each $e_0, e_j, e_{(v,w)}$ are new individuals, and the following conditions hold:*

1. $U = U_0 \cup \bigcup_{\substack{d \in T_1' \cup \cdots \cup T_n' \\ a \in att(\Gamma_0)}} U_{d,a} \cup \bigcup_{\substack{d \in T_1' \cup \cdots \cup T_n' \\ A \in asc(\Gamma_0)}} U_{d,A}$ and $I(x) = I_0(x) \cup \bigcup_{\substack{d \in T_1' \cup \cdots \cup T_n' \\ a \in att(\Gamma_0)}} I_{d,a}(x) \cup \bigcup_{\substack{d \in T_1' \cup \cdots \cup T_n' \\ A \in asc(\Gamma_0)}} I_{d,A}(x).$

2. For each $\Gamma \in \Sigma(\Gamma_0)$,

- $d \in I_0(C)$ iff a non-leaf node d is labeled with δ where $C \in \bigcup_{C' \in \delta} H(C', \Gamma)$, and
- $(d, d') \in I_0(a)$ iff (i) d' is a non-leaf node and (d, d') is an edge labeled with (Γ, a), or (ii) a node d has a child labeled with w, there is a witness d_0 of d, and (d_0, d') is an edge labeled with (Γ, a).

3. For each edge (d, d') labeled with (Γ, a) such that the node d is labeled with δ and $Max_{\geq}(N(\delta, a, \Gamma)) = k$,

- $U_{d,a} = \{e_1, \ldots, e_{k-1}\}$,
- $(d, e_1), \ldots, (d, e_{k-1}) \in I_{d,a}(a)$ iff $(d, d') \in I_0(a)$,
- $e_1, \ldots, e_{k-1} \in I_{d,a}(C)$ iff $d' \in I_0(C)$, and
- $(e_1, d''), \ldots, (e_{k-1}, d'') \in I_{d,a}(a')$ iff $(d', d'') \in I_0(a')$.

4. For all nodes $d \in I_0(C_k)$ such that $AC(A, \Gamma_0) = (C_1, \ldots, C_k, \ldots, C_n)$ and $Max_{\geq}(N_k(H(A, \Gamma_0), \Gamma_0)) = (i_1, \ldots, i_{k-1}, i_{k+1}, \ldots, i_n)$,

- $U_{d,A} = \{e_0\} \cup \bigcup_{v \in \{1, \ldots, n\} \setminus \{k\}} \{e_{(v,1)}, \ldots, e_{(v,i_v)}\}$,
- for all $(w_1, \ldots, w_{k-1}, w_{k+1}, \ldots, w_n) \in \mathbf{N}^n$ such that $1 \leq w_v \leq i_v$, $(e_{(1,w_1)}, \ldots, e_{(k-1,w_{k-1})}, d, e_{(k+1,w_{k+1})}, \ldots, e_{(n,w_n)}) \in I_{d,A}(A)$ and $e_{(1,w_1)} \in I_{d,A}(C_1)$, $\ldots, e_{(k-1,w_{k-1})} \in I_{d,A}(C_{k-1})$, $e_{(k+1,w_{k+1})} \in I_{d,A}(C_{k+1}), \ldots, e_{(n,w_n)} \in I_{d,A}(C_n)$,
- $e_{(v,w)} \in I_{d,A}(C')$ for all $C' \in H(C_v, \Gamma')$ iff $e_{(v,w)} \in I_{d,A}(C_v)$ and C_v is consistent in Γ',
- $(u_1, \ldots, u_n) \in I_{d,A}(A')$ for all $A' \in H(A, \Gamma_0)$ iff $(u_1, \ldots, u_n) \in I_{d,A}(A)^3$,
- $(e_{(v,w)}, d'') \in I_{d,A}(a)$ and $e_{(v,w)} \in I_{d,A}(C_v)$ iff $(d', d'') \in I_0(a)$ and $d' \in I_0(C_v)$, and
- for all $(w_1, \ldots, w_{k-1}, w_{k+1}, \ldots, w_n) \in \mathbf{N}^n$ such that $1 \leq w_v \leq i_v$, and $(e_{(1,w_1)}, \ldots, e_{(k-1,w_{k-1})}, e, e_{(k+1,w_{k+1})}, \ldots, e_{(n,w_n)}) \in I_{d,A}(A)$ iff $e \in I(C_k)$ where e is e_0, e_j, or $e_{(x,y)}$.

5. For all $A \in asc(\Gamma_0)$,

- $(u_1, \ldots, u_n, e_0) \in I_{d,A}(r_0)$ and $e_0 \in I_{d,A}(C_A)$ iff $(u_1, \ldots, u_n) \in I_{d,A}(A)$,
- $e_0 \in I_{d,A}(C)$ for all $C \in H(C_A, \Gamma')$ iff $e_0 \in I_{d,A}(C_A)$ and C_v is consistent in Γ', and
- $(e_0, d'') \in I_{d,A}(a)$ and $e_0 \in I_{d,A}(C_A)$ iff $(d', d'') \in I_0(a)$ and $d' \in I_0(C_A)$.

Lemma 2. Let Γ_0 be a set of implication forms. There exists an interpretation $\mathcal{I}$ such that for every $C_0 \in cls(\Gamma_0)$, $\mathcal{I} \models \exists x. C_0(x)$ if and only if (i) there exists a non-closed forest of Γ_0 and (ii) $Assoc(\Gamma_0) = 1$.

The correctness for the algorithms *Cons* and *Assoc* is obtained as follows:

Theorem 1 (Soundness and Completeness). *Let D be a UML class diagram with association generalization and without roles, and let $\mathcal{G}(D)$ be the translation of D into a set of implication forms. D is consistent if and only if $Cons(\{C\}, \emptyset, \mathcal{G}(D)) = 1$ for all $C \in cls(\mathcal{G}(D))$ and $Assoc(\mathcal{G}(D)) = 1$.*

---

[3] Note that d, d', d'', d_0 are nodes, $e_0, e_j, e_{(v,w)}$ are new constants, and u, u_j are nodes or new constants.

Theorem 2 (Termination). *The consistency checking algorithm Cons termi-nates.*

5 Algorithms and Complexities for Various Expressivities

The proposed consistency checking algorithm terminates; however, *Cons* still exhibits a double-exponential complexity in the worst case (and *Assoc* exhibits polynomial time complexity). In this section, we will present optimized consis-tency checking algorithms for class diagrams of different expressive powers.

5.1 Restriction of Inconsistency Triggers

We denote the set of UML class diagrams with association generalization and without roles as $\mathcal{D}_{ful}^-$. By deleting certain inconsistency triggers, we classify UML class diagrams that are less expressive than $\mathcal{D}_{ful}^-$. The least set $\mathcal{D}_0^-$ of class dia-grams is obtained by deleting disjointness/completeness constraints and over-writing/multiple inheritances. We define $\mathcal{D}_{dis}^-$, $\mathcal{D}_{com}^-$, and $\mathcal{D}_{inh}^-$ as extensions of $\mathcal{D}_0^-$ by adding disjointness constraints, completeness constraints, and over-writing/multiple inheritances, respectively. We denote $\mathcal{D}_{dis+com}^-$, $\mathcal{D}_{dis+inh}^-$, and $\mathcal{D}_{inh+com}^-$ as the unions of $\mathcal{D}_{dis}^-$ and $\mathcal{D}_{com}^-$, $\mathcal{D}_{dis}^-$ and $\mathcal{D}_{inh}^-$, and $\mathcal{D}_{inh}^-$ and $\mathcal{D}_{com}^-$, respectively. In order to design algorithms suitable for these expressivities, we divide the class diagrams into the five groups, as shown in Fig.2.

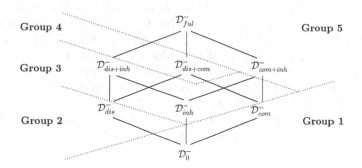

Fig. 2. Classification of UML class diagrams

The least expressive Group 1 is the set of class diagrams obtained by delet-ing disjointness constraints and overwriting/multiple inheritances (but allowing attribute multiplicities). Groups 2 and 3 prohibit $C_1(x) \vee \cdots \vee C_m(x)$ as dis-junctive classes by deleting completeness constraints, and furthermore, Group 2 contains no overwriting/multiple inheritances. Group 4 is restricted by elimi-nating overwriting/multiple inheritances (but allowing disjointness constraints, completeness constraints, and attribute multiplicities).

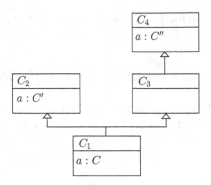

Fig. 3. Attribute value types in overwriting/multiple inheritances

5.2 Restriction of Attribute Value Types

Apart from the restriction of inconsistency triggers, we naturally restrict attribute value types in the overwriting/multiple inheritances. Consider the class hierarchy in Fig.3. Class C_1 with attribute a: C inherits attributes a: C' and a: C'' from superclasses C_2 and C_4. In this case, if the value type C is a subclass of all the other value types C' and C'' of the same name's attributes in the class hierarchy, then the consistency checking of the value types C, C', and C'' can be guaranteed by the consistency checking of only the value type C.

Let $C \in cls(\Gamma_0)$ and $\Gamma \in \Sigma(\Gamma_0)$. The value types of attributes in class C are said to be *restrictedly defined in* Γ when if the superclasses $C_1, \ldots, C_n$ of C (i.e., $H(C, \Gamma) = \{C_1, \ldots, C_n\}$) have the same name's attributes and the value types are classes $C'_1, \ldots, C'_m$, then a value type C'_i is a subclass of the other value types $\{C'_1, \ldots, C'_m\} \backslash \{C'_i\}$ (i.e., $\{C'_1, \ldots, C'_m\} \subseteq H(C'_i, \Gamma)$). Every attribute value type is restrictedly defined if the value types of attributes in any class $C \in cls(\Gamma_0)$ are restrictedly defined in any $\Gamma \in \Sigma(\Gamma_0)$. For example, as shown in Fig.3, the value types C, C', and C'' of attribute a in class C_1 are restrictedly defined in $\Gamma_1 = \{C_1(x) \rightarrow C_2(x), C_1(x) \rightarrow C_3(x), C_3(x) \rightarrow C_4(x), C_1(x) \rightarrow (a(x, y) \rightarrow C(y)), C_2(x) \rightarrow (a(x, y) \rightarrow C'(y)), C_4(x) \rightarrow (a(x, y) \rightarrow C''(y)), \ldots\}$ if $\{C, C', C''\} \subseteq H(C_0, \Gamma_1)$, where C_0 is C, C', or C''.

5.3 Optimized Algorithms

We show that Group 1 does not cause any inconsistency and we devise consistency checking algorithms suitable for Groups 2–5. The following algorithm $Cons1$ computes the consistency of class diagrams in $\mathcal{D}^-_{dis+inh}$, $\mathcal{D}^-_{inh}$, and $\mathcal{D}^-_{dis}$ if we call $Cons1(\{C_0\}, \emptyset, \Gamma_0)$ for every class $C_0 \in cls(\Gamma_0)$. Let X be a set and Y be a family of sets. Then, we define $ADD(X, Y) = \{X_i \in Y \mid X_i \not\subset X\} \cup \{X\}$ such that X is added to Y and all $X_i \subset X$ are removed from Y. Since $\mathcal{D}^-_{dis+inh}$, $\mathcal{D}^-_{inh}$, and $\mathcal{D}^-_{dis}$ do not contain any completeness constraints, there is a unique decomposed set of Γ_0, namely, $\Sigma(\Gamma_0) = \{\Gamma\}$. Instead of recursive calls, $Cons1$ performs looping of consistency checking for each element of variable P that stores unchecked sets of classes.

Algorithm $Cons1$ for $\mathcal{D}_{dis+inh}^{-}$, $\mathcal{D}_{inh}^{-}$, and $\mathcal{D}_{dis}^{-}$
input set of classes δ, family of sets of classes Δ, set of implication forms Γ_0
output 1 (consistent) or 0 (inconsistent)
begin
  $P = \{\delta\}; G = \Delta;$
  while $P \neq \emptyset$ **do**
    $\delta \in P; P = P - \{\delta\}; \Gamma \in \Sigma(\Gamma_0); S = \bigcup_{C \in \delta} H(C, \Gamma);$
    if $\{C, \neg C\} \subseteq S$ or $\{t_1, \ldots, t_n\} \subseteq S$ s.t. $t_1 \cap \cdots \cap t_n = \emptyset$ **then return** 0;
    else $G = ADD(\delta, G);$
      for $a \in att(\Gamma_0)$ **do**
        if $i > j$ s.t. $\{\geq i, \leq j\} \subseteq N(\delta, a, \Gamma)$ **then return** 0;
        else $\delta_a = E(\delta, a, \Gamma);$
          if $\delta_a \neq \emptyset$ and $\delta_a, \mu_0(\delta_a, \Gamma) \not\subseteq \delta'$ for all $\delta' \in G$ **then**
            if $\mu(\delta_a, \Gamma) \neq \emptyset$ **then** $\delta_a = \mu_0(\delta_a, \Gamma);$
            $P = ADD(\delta_a, P);$
        fi
      esle
      rof
    esle
  elihw
  return 1;
end;

The algorithm $Cons2$ is simply designed for testing the consistency of an input class C_0 in every $\Gamma \in \Sigma(\Gamma_0)$, where the multiplicities of attributes in C_0 are checked but the disjointness of its attribute value types are not. This is because $\mathcal{D}_{dis+com}^{-}$ involves no overwriting/multiple inheritances, i.e., each attribute value is uniquely typed and if type T is a class (in $cls(\Gamma_0)$), the consistency of T can be checked in another call $Cons2(T, \Gamma_0)$. This algorithm computes the consistency of $\mathcal{D}_{dis+com}^{-}$ if $Cons2(C_0, \Gamma_0)$ is called for every class $C_0 \in cls(\Gamma_0)$.

Algorithm $Cons2$ for $\mathcal{D}_{dis+com}^{-}$
input class C_0, set of implication forms Γ_0
output 1 (consistent) or 0 (inconsistent)
begin
  for $\Gamma \in \Sigma(\Gamma_0)$ **do**
    $S = H(C_0, \Gamma);$
    if $\{C, \neg C\} \not\subseteq S$ and $\{t_1, \ldots, t_n\} \not\subseteq S$ s.t. $t_1 \cap \cdots \cap t_n = \emptyset$ **then**
      for $a \in att(\Gamma_0)$ **do**
        if $i > j$ s.t. $\{\geq i, \leq j\} \subseteq N(C_0, a, \Gamma)$ **then return** 0;
      return 1;
    fi
  rof
  return 0;
end;

It should be noted that the algorithm $Cons$ requires double exponential time in the worst case. We develop the optimized algorithm $Cons3$ as the single exponential version by skipping the sets of classes that are already checked as consistent or inconsistent in any former routine (but $Cons$ limits the skipping to the set Δ stored in the caller processes). It computes the consistency of $\mathcal{D}^-_{com+inh}$ and $\mathcal{D}^-_{ful}$ if we call $Cons3(\{C_0\}, \emptyset, \Gamma_0)$ for every class $C_0 \in cls(\Gamma_0)$.

Algorithm $Cons3$ for $\mathcal{D}^-_{com+inh}$ and $\mathcal{D}^-_{ful}$
input set of classes δ, family of sets of classes Δ, set of implication forms Γ_0
output 1 (consistent) or 0 (inconsistent)
global variables $G = \emptyset$, $NG = \emptyset$
begin
    for $\Gamma \in \Sigma(\Gamma_0)$ s.t. $(\delta, \Gamma) \notin NG$ **do**
        $S = \bigcup_{C \in \delta} H(C, \Gamma)$; $f_\Gamma = 0$;
        if $\{C, \neg C\} \not\subseteq S$ and $\{t_1, \ldots, t_n\} \not\subseteq S$ s.t. $t_1 \cap \cdots \cap t_n = \emptyset$ **then** $f_\Gamma = 1$;
            for $a \in att(\Gamma_0)$ **do**
                if $i > j$ s.t. $\{\geq i, \leq j\} \subseteq N(\delta, a, \Gamma)$ **then** $f_\Gamma = 0$;
                else $\delta_a = E(\delta, a, \Gamma)$;
                        if $\delta_a \neq \emptyset$ and $\delta_a, \mu_0(\delta_a, \Gamma) \not\subseteq \delta'$ for all $\delta' \in \Delta \cup G$ **then**
                              if $\mu(\delta_a, \Gamma) \neq \emptyset$ **then** $\delta_a = \mu_0(\delta_a, \Gamma)$;
                              $f_\Gamma = Cons3(\delta_a, \Delta, \Gamma_0)$;
                fi
            esle
            rof
        fi
        if $f_\Gamma = 1$ **then** $G = ADD(\delta, G)$; **return** 1;
        else $NG = ADD((\delta, \Gamma), NG)$;
    rof
    return 0;
end;

The optimization method of using good and no good variables G and NG is based on the EXPTIME tableau algorithm in [4]. In $Cons1$ and $Cons3$, the good variable $G = \{\delta_1, \ldots, \delta_n\}$ contains sets of classes such that each set δ_i is consistent in a decomposed set of Γ_0 (in $\Sigma(\Gamma_0)$). In $Cons3$, the no good variable NG contains pairs of a set δ of classes and a decomposed set Γ of Γ_0 such that δ is inconsistent in Γ. Each element in NG exactly indicates the inconsistency of δ in the set Γ by storing the pair (δ, Γ), so that it is never checked again. In addition to this method, we consider that further elements can be skipped by the condition "$\delta_a, \mu_0(\delta_a, \Gamma) \not\subseteq \delta'$ for all $\delta' \in \Delta \cup G$." This implies that $Cons1$ and $Cons3$ skip the consistency checking of the target set δ_a if a superset δ' of either δ_a or $\mu_0(\delta_a, a, \Gamma)$ is already checked in former processes (i.e., $\delta' \in \Delta \cup G$). With regard to the skipping condition, the following lemma guarantees that if $\mu(\delta, \Gamma) \neq \emptyset$, then all the classes $C_1, \ldots, C_n$ in δ and the sole class C in $\mu_0(\delta, \Gamma)$ $(= \{C\})$ have the same superclasses. In other words, the consistency checking of δ can be replaced with the consistency checking of $\mu_0(\delta, \Gamma)$. Therefore, the computational steps can be decreased by skipping the target set δ_a since this set can be replaced by an already checked superset of the singleton $\mu_0(\delta_a, a, \Gamma)$.

Lemma 3. *Let Γ_0 be a set of implication forms and let $\Gamma \in \Sigma(\Gamma_0)$. For all $\delta \subseteq cls(\Gamma_0)$ and $a \in att(\Gamma_0)$, if $\mu(\delta, \Gamma) \neq \emptyset$, then $\bigcup_{C \in \delta} H(C, \Gamma) = \bigcup_{C \in \mu_0(\delta, \Gamma)} H(C, \Gamma)$, $N(\delta, a, \Gamma) = N(\mu_0(\delta, \Gamma), a, \Gamma)$, and $E(\delta, a, \Gamma) = E(\mu_0(\delta, \Gamma), a, \Gamma)$.*

We adjust the algorithm $Cons3$ to class diagrams in which every attribute value type is restrictedly defined. The algorithm $Cons4$ is shown below; as indicated by the underlined text, this algorithm is improved by only storing sets of classes in NG (similar to G). The restriction of value types leads to $\mu(\delta_a, \Gamma) \neq \emptyset$; therefore, the size of NG is limited to a set of singletons of classes. In other words, $Cons4$ can be adjusted to decrease the space complexity (i.e., NG) to polynomial space by using the property of Lemma 3. Unfortunately, this adjustment does not yield a single exponential algorithm if attribute value types are *unrestrictedly* defined. Hence, we need both $Cons3$ and $Cons4$ for the case where attribute value types are restrictedly defined or not.

Algorithm $Cons4$ for $\mathcal{D}^-_{com+inh}$ and $\mathcal{D}^-_{ful}$
input set of classes δ, family of sets of classes Δ, set of implication forms Γ_0
output 1 (consistent) or 0 (inconsistent)
global variables $G = \emptyset$, $NG = \emptyset$
begin
   for $\Gamma \in \Sigma(\Gamma_0)$ **do**
     $S = \bigcup_{C \in \delta} H(C, \Gamma)$; $f_\Gamma = 0$;
     if $\{C, \neg C\} \not\subseteq S$ and $\{t_1, \ldots, t_n\} \not\subseteq S$ s.t. $t_1 \cap \cdots \cap t_n = \emptyset$ **then** $f_\Gamma = 1$;
      for $a \in att(\Gamma_0)$ **do**
        if $i > j$ s.t. $\{\geq i, \leq j\} \subseteq N(\delta, a, \Gamma)$ **then** $f_\Gamma = 0$;
        else $\delta_a = E(\delta, a, \Gamma)$;
          if $\delta_a \neq \emptyset$ and $\delta_a, \mu_0(\delta_a, \Gamma) \not\subseteq \delta'$ for all $\delta' \in \Delta \cup G$ **then**
            if $\mu(\delta_a, \Gamma) \neq \emptyset$ **then** $\delta_a = \mu_0(\delta_a, \Gamma)$;
            <u>**if** $\delta_a \in NG$ **then** $f_\Gamma = 0$;</u>
            <u>**else** $f_\Gamma = Cons4(\delta_a, \Delta, \Gamma_0)$;</u>
        fi
       esle
      rof
    fi
    if $f_\Gamma = 1$ **then** $G = ADD(\delta, G)$; **return** 1;
   rof
   <u>$NG = ADD(\delta, NG)$</u>; **return** 0;
end;

Without losing the completeness of consistency checking (see Appendix in [8]), these algorithms have the following computational properties for each class diagram group (as shown in Table 1). We believe that the complexity classes 0, P, NP, and PSPACE less than EXPTIME are suitable for us to implement the algorithms for different expressive powers of class diagram groups. For all the class diagram groups, complexity1 in Table 1 arranges the complexities of algorithms $Cons1$, $Cons2$, and $Cons3$ with respect to the size of a class diagram. Every class diagram in $\mathcal{D}^-_0$ and $\mathcal{D}^-_{com}$ is consistent; therefore, the complexity is zero (i.e., we do not need to check consistency). $Cons1$ computes the consistency of $\mathcal{D}^-_{dis}$ in

Table 1. Upper-bound complexities of algorithms for testing consistency

UML group	complexity1	algorithm	complexity2	algorithm
$\mathcal{D}_0^-$	**0**		**0**	
$\mathcal{D}_{com}^-$	**0**		**0**	
$\mathcal{D}_{dis}^-$	**P**	$Cons1$	**P**	$Cons1$
$\mathcal{D}_{inh}^-$	EXPTIME		**P**	
$\mathcal{D}_{dis+inh}^-$	EXPTIME		**P**	
$\mathcal{D}_{dis+com}^-$	**NP**	$Cons2$	**NP**	$Cons2$
$\mathcal{D}_{com+inh}^-$	EXPTIME	$Cons3$	**PSPACE**	$Cons4$
$\mathcal{D}_{ful}^-$	EXPTIME		**PSPACE**	

P (polynomial time) and that of $\mathcal{D}_{inh}^-$ and $\mathcal{D}_{dis+inh}^-$ in EXPTIME (exponential time). $Cons2$ computes the consistency of $\mathcal{D}_{dis+com}^-$ in NP (non-deterministic polynomial time), and $Cons3$ computes the consistency of $\mathcal{D}_{com+inh}^-$ and $\mathcal{D}_{ful}^-$ in EXPTIME.

Moreover, complexity2 in Table 1 shows the complexities of the algorithms $Cons1$, $Cons2$, and $Cons4$ for the case in which every attribute value type is restrictedly defined. In particular, $Cons1$ computes the consistency of $\mathcal{D}_{inh}^-$ and $\mathcal{D}_{dis+inh}^-$ in P, and $Cons4$ computes the consistency of $\mathcal{D}_{com+inh}^-$ and $\mathcal{D}_{ful}^-$ in PSPACE (polynomial space). Therefore, by Lemma 3 and by the skipping method of consistency checking, the complexities of $Cons1$ and $Cons4$ are respectively reduced from EXPTIME to P and PSPACE. Due to spatial constraints, detailed proofs of the lemmas and theorems have been omitted (see [8]).

6   Conclusion

We introduced the restriction of UML class diagrams based on

(i) inconsistency triggers (disjointness constraints, completeness constraints, and overwriting/multiple inheritances) and
(ii) attribute value types defined with restrictions in overwriting/multiple inheritances.

Inconsistency triggers are employed to classify the expressivity of class diagrams, and their combination with the attribute value types results in tractable consistency checking of the restricted class diagrams. First, we presented a complete algorithm for testing the consistency of class diagrams including any inconsistency triggers. Second, the algorithm was suitably refined in order to develop optimized algorithms for different expressive powers of class diagrams obtained by deleting some inconsistency triggers. Our algorithms were easily modified depending on the presence of diagram components. The algorithms clarified that every class diagram in $\mathcal{D}_0^-$ and $\mathcal{D}_{com}^-$ must have a UML-model (i.e., consistency is guaranteed) and when every attribute value type is restrictedly defined, the complexities of class diagrams in $\mathcal{D}_{inh}^-$ and $\mathcal{D}_{dis+inh}^-$ and in $\mathcal{D}_{com+inh}^-$ and $\mathcal{D}_{ful}^-$

are essentially decreased from EXPTIME to P and PSPACE, respectively. Restricted/classified UML class diagrams and their optimized algorithms are new results; however, the translation into first-order logic is similar to and based on the study of [1, 2].

Our future research is concerned with the complexity in terms of the depth of class hierarchies and the average-case complexity for consistency checking. Furthermore, an experimental evaluation should be performed in order to ascertain the applicability of optimized consistency algorithms.

References

1. B. Beckert, U. Keller, and P. H. Schmitt. Translating the object constraint language into first-order predicate logic. In *Proceedings of VERIFY, Workshop at Federated Logic Conferences (FLoC)*, 2002.
2. D. Berardi, A. Cali, D. Calvanese, and G. De Giacomo. Reasoning on UML class diagrams. *Artificial Intelligence*, 168(1-2):70–118, 2005.
3. F. M. Donini. Complexity of reasoning. In *Description Logic Handbook*, pages 96–136, 2003.
4. F. M. Donini and F. Massacci. EXPTIME tableaux for ALC. *Artificial Intelligence*, 124(1):87–138, 2000.
5. A. S. Evans. Reasoning with UML class diagrams. In *Second IEEE Workshop on Industrial Strength Formal Specification Techniques, WIFT'98, USA*, 1998.
6. M. Fowler. *UML Distilled: A Brief Guide to the Standard Modeling Object Language*. Object Technology Series. Addison-Wesley, third edition, September 2003.
7. E. Franconi and G. Ng. The i.com tool for intelligent conceptual modeling. In *KRDB*, pages 45–53, 2000.
8. K. Kaneiwa and K. Satoh. Consistency checking algorithms for restricted UML class diagrams. NII Technical Report, NII-2005-013E, National Institute of Informatics, 2005, http://research.nii.ac.jp/TechReports/05-013E.html.
9. P. G. Kolaitis and J. A. Väänänen. Generalized quantifiers and pebble games on finite structures. *Annals of Pure and Applied Logic*, 74(1):23–75, 1995.
10. H. Mannila and K.-J. Räihä. On the complexity of inferring functional dependencies. *Discrete Applied Mathematics*, 40(2):237–243, 1992.
11. J. Rumbaugh, I. Jacobson, and G. Booch. *The Unified Modeling Language Reference Manual*. Addison-Wesley, Reading, Massachusetts, USA, 1st edition, 1999.
12. K.-D. Schewe and B. Thalheim. Fundamental concepts of object oriented databases. *Acta Cybern*, 11(1-2):49–84, 1993.
13. A. Tsiolakis and H. Ehrig. Consistency analysis between UML class and sequence diagrams using attributed graph gammars. In *Proceedings of joint APPLIGRAPH/ GETGRATS Workshop on Graph Transformation Systems*, pages 77–86, 2000.

Some Contributions to the Minimum Representation Problem of Key Systems

Gyula O.H. Katona[1,*] and Krisztián Tichler[2,**]

[1] Alfréd Rényi Institute of Mathematics, 1053 Budapest, Hungary
ohkatona@renyi.hu
[2] Technische Universität Berlin, Strasse des 17. Juni 135,
10623 Berlin, Germany
krisz@renyi.hu

Abstract. Some new and improved results on the minimum representation problem for key systems will be presented. By improving a lemma of the second author we obtain better or new results on badly representable key systems, such as showing the most badly representable key system known, namely of size

$$2^{n(1-c \cdot \log\log n/\log n)},$$

where n is the number of attributes. We also make an observation on a theorem of J. Demetrovics, Z. Füredi and the first author and give some new well representable key systems as well.

Keywords: Labelled directed tree, relational database, minimum matrix representation, extremal problems.

1 Introduction

Consider a relation name R in the *relational database model*. Let $\Omega(R)$ be a finite set, its elements are called the *attributes* of R. If $|\Omega| = n$ we say that the *arity* of R is n. The relation name and the set of attributes together are called the *relation schema* and denoted by $R[\Omega]$. Suppose, that there is given a (countably) infinite set **dom**. An *n-tuple* over the relation schema $R[\Omega]$ is a total mapping u from Ω to **dom**. A *relation instance* over a relation schema $R[\Omega]$ is a (possibly empty) finite (multi)set I of n-tuples over Ω.

The value of the n-tuple u on an attribute A is denoted by $u(A)$. If $X \subseteq \Omega$ then $\pi_X(u)$ is an $|X|$-tuple v over X, such that $v(A) = u(A)$ for all $A \in X$. A relation instance I over $R[\Omega]$ satisfies $K \to \Omega$, denoted by $I \vDash K \to \Omega$, if for each pair u and v of tuples in I, $\pi_K(u) = \pi_K(v)$ implies $\pi_{\Omega \setminus K}(u) = \pi_{\Omega \setminus K}(v)$. $K \to \Omega$ is a *key dependency* where $K \subseteq \Omega$ is called a *key*. (Less formally, $K \subseteq \Omega$ is a key, if the values in K of an n-tuple determine the whole n-tuple.)

* The work was supported by the Hungarian National Foundation for Scientific Research, grant numbers T037846 and T034702.
** The work was supported by the Young Researcher Fellowship of the Alfréd Rényi Institute and the COMBSTRU program (Marie Curie Fellowship of the Europian Union) at Universität Bielefeld and Technische Universität Berlin.

J. Dix and S.J. Hegner (Eds.): FoIKS 2006, LNCS 3861, pp. 240–257, 2006.

A key is called *minimal key*, if it does not contain any other key as a proper subset. Since the set of keys and minimal keys determine each other (each subset of the attributes that contain a key is also a key), it is natural to investigate the system of minimal keys, which usually contains fewer members, smaller in size. Keys can be widely applied in database management, see [1].

A family $\mathcal{F}$ of subsets of a finite set is called a *Sperner system*, if for $F_1, F_2 \in \mathcal{F}$ the property $F_1 \not\subseteq F_2$ holds. The system of minimal keys is clearly a non-empty Sperner system. For a Sperner system $\mathcal{K}$ let us introduce the notation

$$I(\mathcal{K}) = \{I | I \vDash K \to \Omega \text{ if and only if } \exists K', K' \subseteq K, K' \in \mathcal{K}\}.$$

We call an element of $I(\mathcal{K})$ a *representation* of $\mathcal{K}$. The following basic theorem of W.W. Armstrong and J. Demetrovics states, that for every non-empty Sperner system, there is always a representation of it, i.e., there exists a relation, in which the system of minimal keys is exactly the given family of sets.

Theorem 1.1. *[2, 4] If $\mathcal{K}$ is non-empty, then $I(\mathcal{K}) \neq \emptyset$.*

In view of Theorem 1.1 it is natural to ask for the minimum size of a relation, that represents a given system of keys. Formally let $s(\mathcal{K}) = \min\{|I| \mid I \in I(\mathcal{K})\}$ denote this minimum.

Suppose, that little a priori information is known about the structure of a given database instance. If a theorem ensures the validity of an inequality among the parameters of a database and we have information on the actual values of a part of these parameters then a statement may be deduced for the rest of the parameters of the given instance. In our case, we have a theorem for the following three parameters: number of attributes, system of minimal keys (this is not a number!) and the size of the relation. So if the size of the instance is less than the size of this minimal sample database, then the system of minimal keys can not be this one, our hypothesis on the system of keys can be rejected. This argument is trying to justify the investigation of the quantity $s(\mathcal{K})$. The goal of the present paper is to extend our knowledge on the minimum representation problem of key systems. In addition to its importance they usually raise interesting and sometimes challenging mathematical problems.

Let us start with presenting some earlier results on minimum representation.

$A \subseteq \Omega$ is an *antikey* if it is not a key. An antikey is called a *maximal antikey*, if other antikeys do not include it. If $\mathcal{K}$ is the system of minimal keys, denote the system of maximal antikeys by $\mathcal{K}^{-1}$. There is a strong connection between $s(\mathcal{K})$ and $|\mathcal{K}^{-1}|$, namely the magnitude of $s(\mathcal{K})$ is between $|\mathcal{K}^{-1}|$ and its square root.

Theorem 1.2. *[5] If $\mathcal{K} \neq \emptyset$ is a Sperner system, then the following two inequalities hold,*

$$|\mathcal{K}^{-1}| \leq \binom{s(\mathcal{K})}{2} \quad and \quad s(\mathcal{K}) \leq 1 + |\mathcal{K}^{-1}|. \tag{1}$$

Informally, we say that a Sperner system $\mathcal{K}$ is well/badly representable if $s(\mathcal{K})$ is close to the lower/upper bound of Theorem 1.2. It is easy to see, that a minimal representation have the following two basic properties.

Proposition 1.1. *Suppose, that $I \in I(\mathcal{K}), |I| = s(\mathcal{K})$ is a minimal representation of the Sperner system $\mathcal{K}$, then the following properties hold,*

(i) *for every $A \in \mathcal{K}^{-1}$ there exist two tuples, u and v in I, such that $\pi_A(u) = \pi_A(v)$,*
(ii) *there are no two different tuples u and v in I, such that $\pi_K(u) = \pi_K(v)$ holds for some $K \in \mathcal{K}$.*

Let us mention two results from the 1980's. $\mathcal{K}_k^n$ denotes the family of all k-element subsets of the n-element Ω.

Theorem 1.3. *[5]* $2 \leq k < n \Rightarrow \exists c_1 = c_1(k), c_2 = c_2(k)$

$$c_1 n^{(k-1)/2} < s(\mathcal{K}_k^n) < c_2 n^{(k-1)/2}.$$

Theorem 1.4. *[8]* $n > n_0(k), k \geq 1 \Rightarrow \exists c_3 = c_3(k), c_4 = c_4(k)$

$$c_3 n^{(2k+1)/3} < s(\mathcal{K}_{n-k}^n) < c_4 n^k, \qquad \frac{1}{12} n^2 < s(\mathcal{K}_{n-2}^n) < \frac{1}{2} n^2.$$

It has been proved in [7] that there is a Sperner system $\mathcal{K}$ such that

$$s(\mathcal{K}) > \frac{1}{n^2} \binom{n}{\lfloor \frac{n}{2} \rfloor}.$$

Its proof is, however probabilistic. It does not give a construction. No constructed Sperner system $\mathcal{K}$ exists in the literature with exponential $s(\mathcal{K})$. We will show such a construction in section 3. More precisely, $s(\mathcal{K})$ will have an exponent nearly n. The method of proving a lower estimate on $s(\mathcal{K})$ is the same as that of [11]. The method was based on a lemma on labelled (by subsets) trees. Section 2 improves the statement of this lemma, giving a sharp estimate replacing the estimate of [11]. This improvement makes us able to prove the exponential lower estimate.

In section 4 we return to Theorem 1.3. First (subsection 4.1) the upper estimate is improved (it becomes independent of k). Subsection 4 is a small observation showing that the method of [5] can be used to prove a good upper estimate for other (non-uniform) $\mathcal{K}$s. In section 5 we summarize the related questions to be done.

2   An Extremal Problem on Labelled Directed Trees Revisited: A Tool for Deriving Results on Badly Representable Key Systems

A tree F is called a *directed tree,* if there is a direction on the edges, so that a vertex v_0 *(root)* has only out-neighbours, and an arbitrary vertex $v \neq v_0$ has a uniquely determined in-neighbour $n(v)$. $N(v)$ denotes the out-neighbourhood of v. The set of the leaves of a tree F is denoted by $\ell(F)$. Let U be a (finite) set.

A tree $F = F(U)$ is called *labelled*, if a subset $A(v)$ of U is associated with each vertex v of F.

For fixed integers $k \geq 1$, $\ell \geq 2$ and $U = \{1, 2, ..., m\}$ consider the family of labelled directed trees $\mathcal{F}_{k,\ell}^{(m)}$, for which the vertices of each tree $F \in \mathcal{F}_{k,\ell}^{(m)}$ are labelled as follows. The label of the root v_0 of F is $A(v_0) = U$. For an arbitrary vertex v of F there is a disjoint partition $N(v) = \bigcup_{i=1}^{\ell} N_i(v)$ of its out-neighbourhood satisfying the following properties.

$$A(v) \subseteq A(n(v)) \quad (v \neq v_0), \tag{2}$$

$$|A(v)| \geq k + 1, \tag{3}$$

$$w_1, w_2 \in N_i(v) \Rightarrow A(w_1) \cap A(w_2) = \emptyset \quad (1 \leq i \leq \ell), \tag{4}$$

$$w_1 \in N_i(v), w_2 \in N_j(v) \Rightarrow |A(w_1) \cap A(w_2)| \leq k \quad (1 \leq i < j \leq \ell). \tag{5}$$

Introduce the notation $T_{k,\ell}(m) = \max\{|\ell(F)| \mid F \in \mathcal{F}_{k,\ell}^{(m)}\}$. If $k = 1$, we simply write $\mathcal{F}_\ell^{(m)}$ for $\mathcal{F}_{k,\ell}^{(m)}$ and $T_\ell(m)$ for $T_{k,\ell}(m)$.

Throughout the rest of the paper we write simply log for $\log_2$. We have for $\mathcal{F}_\ell$ the following:

Theorem 2.1.

$$T_2(m) \leq \frac{1}{2} m \log m. \tag{6}$$

and equality holds if and only if m is a power of 2.

Theorem 2.2.

$$T_\ell(m) = \Theta_\ell(m \log^\alpha m) \tag{7}$$

for $\ell \geq 3$. Where $\alpha = \alpha(\ell) = \log \ell$.

In [11], the magnitude of $T_{k,\ell}$ was determined.

2.1   Case $\ell = 2$

We will use the concept of *entropy* [3] in the proof. Entropy is a measure of a random variable X:

$$H(X) = -\sum_i p_i \log p_i, \tag{8}$$

where $\text{Prob}(X = i) = p_i$. It is known, that

$$H((X, Y)) \leq H(X) + H(Y). \tag{9}$$

The proof is by induction on m. (6) holds for $m = 2$. Suppose that the statement holds for every integer smaller than m.

Let $F \in \mathcal{F}_2^{(m)}$ be a tree with $|\ell(F)| = T(m)$. Furthermore let $N(v_0) = \{v_1, \ldots, v_s, w_1, \ldots, w_t\}$, $N_1(v_0) = \{v_1, \ldots, v_s\}$, $N_2(v_0) = \{w_1, \ldots, w_t\}$. Let us use the short notations $A_i = A(v_i)$, $a_i = |A_i|$, $(1 \leq i \leq s)$, $B_i = A(w_i)$, $b_i = |B_i|$, $(1 \leq i \leq t)$. The subtree of F of root v_i (w_j) is denoted by F_i (F_{s+j}), $1 \leq i \leq s$ $(1 \leq j \leq t)$.

By the induction hypothesis

$$T(a_i) \le \frac{1}{2} a_i \log a_i, \ (1 \le i \le s), \quad \text{and} \quad T(b_i) \le \frac{1}{2} b_i \log b_i, \ (1 \le i \le t)$$

holds. So it is enough to prove that

$$\sum_{i=1}^{s} a_i \log a_i + \sum_{i=1}^{t} b_i \log b_i \le m \log m, \tag{10}$$

since then

$$T(m) = |\ell(F)| = \sum_{i=1}^{s+t} |\ell(F_i)| \le \sum_{i=1}^{s} T(a_i) + \sum_{i=1}^{t} T(b_i)$$

$$\le \sum_{i=1}^{s} \frac{1}{2} a_i \log a_i + \sum_{i=1}^{t} \frac{1}{2} b_i \log b_i \le \frac{1}{2} m \log m.$$

Let $s_1 = m - \sum_{i=1}^{s} a_i$ and $s' = s + s_1$. Add s_1 disjoint sets $A_{s+1}, \ldots, A_{s'}$ of cardinality 1, such that $\{1, 2, \ldots, m\} = \bigcup_{i=1}^{s'} A_i$. We define t_1, t' and the sets $B_{t+1}, \ldots, B_{t'}$ analogously. The sets A_i $(1 \le i \le s')$ and B_j $(1 \le j \le t')$ have the following properties:

$$\{A_i, 1 \le i \le s'\} \text{ is a partition of } \{1, 2, \ldots, m\}, \tag{11}$$
$$\{B_j, 1 \le j \le t'\} \text{ is a partition of } \{1, 2, \ldots, m\}, \tag{12}$$
$$|A_i \cap B_j| \le 1, 1 \le i \le s', 1 \le j \le t'. \tag{13}$$

Let $\Omega_X = \{1, 2, \ldots, m\}$ be the event space of the random variable X. Furthermore, let $X(\omega) = \omega$, $\omega \in \Omega_X$ and $\text{Prob}(X = \omega) = 1/m$. Let us define another two random variables, $Y(X \in A_i) = i$, $1 \le i \le s'$ and $Z(X \in B_j) = j$, $1 \le j \le t'$. Then

$$\text{Prob}(Y = i) = \frac{a_i}{m} \ (1 \le i \le s') \quad \text{and} \quad \text{Prob}(Z = j) = \frac{b_j}{m} \ (1 \le j \le t').$$

The random variables Y and Z are well defined by (11) and (12). Furthermore, by (13) we get

$$\text{Prob}((Y, Z) = (i, j)) = \begin{cases} \text{Prob}(X = k) = 1/m & \text{if} \quad A_i \cap B_j = \{k\}, \\ 0 & \text{if} \quad A_i \cap B_j = \emptyset. \end{cases}$$

So we have for the entropies of Y, Z and (Y, Z):

$$H(Y) = \sum_{i=1}^{s'} \frac{a_i}{m} \log \frac{m}{a_i}, \qquad H(Z) = \sum_{j=1}^{t'} \frac{b_j}{m} \log \frac{m}{b_j},$$

$$H((Y, Z)) = -\sum_{i=1}^{s'} \sum_{j=1}^{t'} \text{Prob}((Y, Z) = (i, j)) \log \text{Prob}((Y, Z) = (i, j)) =$$

$$\sum_{i=1}^{m} \frac{1}{m} \log m = \log m.$$

Therefore, by (9) we get

$$\log m \le \sum_{i=1}^{s'} \frac{a_i}{m} \log \frac{m}{a_i} + \sum_{j=1}^{t'} \frac{b_j}{m} \log \frac{m}{b_j},$$

which is equivalent to (10).

2.2   Case $\ell \ge 3$

To prove Theorem 2.2 we need somewhat more counting. We could not find a straightforward way to generalize the concept of entropy for this case, altough the main idea (see equation (16)) of the proof comes from the proof of the case $\ell = 2$.

In this section we will use the following notations. If $\mathbf{u} = (u_1, \dots, u_t)$ for some t, then let $P_i(\mathbf{u}) = \sum_{j=1}^{t} u_j^i$ and $\sigma_2(\mathbf{u}) = \sum_{1 \le i < j \le t} u_i u_j$. Let $A(\mathbf{u}) = (\sum_{j=1}^{t} u_j)/t$ and $G(\mathbf{u}) = (\prod_{j=1}^{t} u_j)^{1/t}$ denote the arithmetic and geometric mean, respectively. We will use the notations $A_2(\mathbf{u}) = A(\sigma_2(\mathbf{u}))$ and $G_2(\mathbf{u}) = G(\sigma_2(\mathbf{u}))$ as well.

Lower estimation of $T_\ell(m)$. Let $\mathcal{H}_\ell^q$, $(\ell \le q)$ be a partial affine plane of order q, i.e., a set of ℓq lines on q^2 points, such that the lines form ℓ parallel classes and lines from different parallel classes intersect in 1.

Suppose, that we have for m square

$$T_\ell(\sqrt{m}) \ge C_\ell \cdot \sqrt{m} \log^\alpha \sqrt{m}, \tag{14}$$

and let $F \in \mathcal{F}_\ell^{(\sqrt{m})}$ be a tree with $|\ell(F)| = T_\ell(\sqrt{m})$. Suppose furthermore, that there exists a partial affine plane, $\mathcal{H}_\ell^{\sqrt{m}}$.

Let $T \in \mathcal{F}_\ell^{(m)}$ be the following tree. The root has $\ell\sqrt{m}$ out-neighbours. Each of them are labelled by one of the $\ell\sqrt{m}$ members of the partial affine plane, $\mathcal{H}_\ell^{\sqrt{m}}$. These out-neighbours are roots of one copy of F each. Then

$$T_\ell(m) \ge |\ell(T)| = \ell\sqrt{m}|\ell(F)| \ge \ell\sqrt{m}(C_\ell\sqrt{m}\log^\alpha \sqrt{m}) = C_\ell \cdot m \log^\alpha m.$$

It is known, that there exist a partial affine plane of order q with ℓ parallel classes if and only if there exist ℓ pairwise orthogonal latin squares of order q.

There are only partial results known about the existence of pairwise orthogonal latin squares, [10]. Of course, partial affine planes exist if affine planes exist, i.e., for prime powers.

The statement follows from the fact that prime powers occur densly, see (39).

Upper estimation of $T_\ell(m)$. Let $\ell \ge 3$ arbitrary. We prove by induction on m. We have to prove, that

$$T_\ell(m) \le cm \log^\alpha m \tag{15}$$

holds for some $c = c(\ell)$ to be chosen later.

For small m (15) is true if c is large enough. Suppose that the statement is true for every integer smaller than m.

Let $F \in \mathcal{F}_\ell^{(m)}$ be a tree with $|\ell(F)| = T(m)$. Let $N(v_0) = \{v_1, \ldots, v_t\}$. The number of the leaves can be maximal only if for the subtrees F_i of F of root v_i, $|\ell(F_i)| = T(m_i)$ holds, where $m_i = |A(v_i)|$. Furthermore let us introduce the short notation $N_j = N_j(v_0)$, $1 \le j \le \ell$.

Case A. $\exists 1 \le j \le t$, such that $m_j \ge m/\log^{1/2} m$.

We need the following observation:

Proposition 2.1. $T_\ell(m) \le T_\ell(m-1) + m - 1$.

Proof. Induction on m. The statement is true for $m = 1$. Suppose that the statement is true for every integer smaller than m. Let $F \in \mathcal{F}_\ell^{(m)}$ be a tree with $|\ell(F)| = T(m)$. Let $N(v_0) = \{v_1, \ldots, v_t\}$. The number of the leaves can be maximal only if for the subtrees F_i of F of root v_i, $|\ell(F_i)| = T(m_i)$ holds, where $m_i = |A(v_i)|$. Let $m_i' = |A(v_i)\backslash\{m\}|$. Consider the following tree $F' \in \mathcal{F}_\ell^{(m-1)}$. Let $N(v_0') \subseteq \{v_1', \ldots, v_t'\}$, $v_i' \in N(v_0') \Leftrightarrow m_i' \ge 2$. $A(v_i') = A(v_i)\backslash\{m\}$. The subtree F_i' of F' of root v_i' is a tree with $|\ell(F_i')| = m_i'$. Then

$$T(m) = \sum_{i=1}^t T(m_i) = \sum_{i=1}^t T(m_i') + \sum_{i=1}^t (T(m_i) - T(m_i')) = |\ell(F')| +$$

$$\sum_{m \in A(v_i)} (T(m_i) - T(m_i - 1)) \le T(m-1) + \sum_{m \in A(v_i)} (m_i - 1) \le T(m-1) + m - 1.$$

The last inequality holds by (5), the previous one by the induction hypothesis. □

Let $m_j = \beta m$. It is enough to prove, that $\sum_{i=1}^t T(m_i) \le cm \log^\alpha m$.

$$\sum_{i=1}^t T(m_i) = T(m_j) + T(m - m_j) + \sum_{i \ne j : A(v_i) \cap A(v_j) \ne \emptyset} (T(m_i) - T(m_i - 1)) \le$$

$$c\beta m \log^\alpha \beta m + c(1-\beta)m \log^\alpha(1-\beta)m + \ell(m - m_j) = cm \log^\alpha m +$$

$$c\beta m(\log^\alpha \beta m - \log^\alpha m) + c(1-\beta)m(\log^\alpha(1-\beta)m - \log^\alpha m) + \ell(1-\beta)m \le$$

$$cm \log^\alpha m - 2\frac{c\alpha\beta(1-\beta)m}{\ln 2} \log^{\alpha-1} m + \ell(1-\beta)m \le cm \log^\alpha m +$$

$$(1 - \beta)m(\ell - \frac{2\alpha c}{\ln 2} \log^{1/20} m) \le cm \log^\alpha m,$$

if $\log^{1/20} m > (\ell \ln 2)/(2\alpha c)$, which is true for every m if c is large enough. We have used only (4)-(5) (at the first inequality), Proposition 2.1, and the fact that the function $x \mapsto \log^\alpha x$ is concave.

Case B. $\forall 1 \le j \le t$, $m_j < m/\log^{1/2} m$.

Let us introduce the notation $N_j' = N_j \cup \{u_j^{(h)} : 1 \le h \le m, \nexists v_i \in N_j : h \in A(v_i)\}$. Let $A(u_j^{(h)}) = \{h\}$, $m_i = A(w)$ for $w \in N_j'$. Then the following notation is well defined. Let $m_j^{(h)} = |A(w)|$, where $w \in N_j'$, $h \in A(w)$.

We will prove that the following inequality holds for every $1 \leq h \leq m$ and $m > M(\ell)$:

$$\log^\alpha m - \sum_{i=1}^\ell \log^\alpha m_i^{(h)} \geq \frac{\alpha}{\ln 2} \log^{\alpha-1} m (1 - \frac{\sum_{1 \leq j < k \leq \ell} m_j^{(h)} m_k^{(h)}}{\binom{\ell}{2} m}). \tag{16}$$

Using this inequality we get for $m > M$:

$$m \log^\alpha m - \sum_{j=1}^\ell \sum_{i:i \in N_j'} m_i \log^\alpha m_i = \sum_{h=1}^m (\log^\alpha m - \sum_{j=1}^\ell \log^\alpha m_j^{(h)}) \geq$$

$$\sum_{h=1}^m \frac{\alpha}{\ln 2} \log^{\alpha-1} m (1 - \frac{\sum_{1 \leq j < k \leq \ell} m_j^{(h)} m_k^{(h)}}{\binom{\ell}{2} m}) =$$

$$\frac{\alpha \log^{\alpha-1} m}{(\ln 2)\binom{\ell}{2} m} \sum_{h=1}^m \sum_{1 \leq j < k \leq \ell} (m - m_j^{(h)} m_k^{(h)}) =$$

$$\frac{\alpha \log^{\alpha-1} m}{(\ln 2)\binom{\ell}{2} m} \sum_{1 \leq j < k \leq \ell} \sum_{h=1}^m (m - m_j^{(h)} m_k^{(h)}) \geq$$

$$\frac{\alpha \log^{\alpha-1} m}{(\ln 2)\binom{\ell}{2} m} \sum_{1 \leq j < k \leq \ell} (m^2 - (\sum_{w \in N_j'} |A(w)|)(\sum_{w \in N_k'} |A(w)|)) = 0.$$

From this, (15) straightforwardly follows in *Case B*, as well. So, we only have to prove, that inequality (16) holds in the case when m is large enough and none of the m_js are too large. Let us fix h, we will omit the upper index of $m_j^{(h)}$ in the rest of the proof, so from now on, we suppose, that $1 \leq m_j \leq m/\log^{1/2} m$ ($1 \leq j \leq \ell$) holds.

Case B.1. $\forall 1 \leq i \leq \ell : m_i \leq \sqrt{m} \log^5 m$.
Case B.1.1. $\forall 1 \leq j \leq \ell : m_j = \sqrt{m} + \delta_j, |\delta_j| \geq (1/3)\sqrt{m}$.

Proposition 2.2.

$$(\log^\alpha)^{(i)}(x) = \sum_{k=1}^i C_k^{(i)} \frac{1}{\ln^k 2} \frac{\alpha!}{(\alpha - k)!} x^{-k} \log^{\alpha-k} x \tag{17}$$

and the constants $C_k^{(i)} \in \mathbb{Z}$ have the following properties:

(i) $C_k^{(i)} = -(i-1)C_k^{(i-1)} + C_{k-1}^{(i-1)}$,
(ii) $C_1^{(i)} = (-1)^{i-1}(i-1)!$,
(iii) $|C_k^{(i)}| \leq \frac{i!}{k!} 2^i$

Proof. Easy induction on i. □

Proposition 2.3.

$$\sum_{k=0}^{\infty} \left| \binom{\alpha}{k} \right| < \infty \tag{18}$$

holds for $\alpha > 1$.

Proof.

$$\sum_{k=0}^{\infty} \left| \binom{\alpha}{k} \right| = \sum_{k=0}^{\lceil\alpha\rceil-1} \left| \binom{\alpha}{k} \right| + \sum_{i=1}^{\infty} \sum_{i\lceil\alpha\rceil}^{(i+1)\lceil\alpha\rceil-1} \left| \binom{\alpha}{k} \right| \leq$$

$$\sum_{k=0}^{\lceil\alpha\rceil-1} \left| \binom{\alpha}{k} \right| + \sum_{i=1}^{\infty} \lceil\alpha\rceil \frac{1}{\binom{i\lceil\alpha\rceil}{\lceil\alpha\rceil}} \leq \sum_{k=0}^{\lceil\alpha\rceil-1} \left| \binom{\alpha}{k} \right| + \lceil\alpha\rceil \sum_{i=1}^{\infty} \frac{1}{i^{\alpha}} < \infty. \quad \square$$

Let $a_2(\alpha) = \sum_{k=2}^{\infty} |\binom{\alpha}{k}|$, and $a_{\max}(\alpha) = \max_{k=0}^{\infty} |\binom{\alpha}{k}|$. Note, that $a_{\max}(\alpha) \leq \ell$ and $a_2(\alpha) \leq \ell + 2\alpha + 2$.

Let $\delta = (\delta_1, \delta_2, \ldots, \delta_\ell)$. Then we have to prove the following inequality:

$$\log^{\alpha} m - \sum_{i=1}^{\ell} \log^{\alpha}(\sqrt{m} + \delta_i) \geq -\frac{\alpha}{\ln 2}(\log^{\alpha-1} m)\left(\frac{2P_1(\delta)}{\ell\sqrt{m}} + \frac{\sigma_2(\delta)}{\binom{\ell}{2}m}\right). \tag{19}$$

If δ is small enough we know, that

$$\log^{\alpha}(x + \delta) = \sum_{i=0}^{\infty} \frac{(\log^{\alpha})^{(i)}(x)\delta^i}{i!} \tag{20}$$

holds, so using (17) we get for every $1 \leq j \leq \ell$

$$\log^{\alpha}(\sqrt{m} + \delta_j) = \log^{\alpha}\sqrt{m} + \frac{\alpha}{\ln 2}\frac{\log^{\alpha-1}\sqrt{m}}{\sqrt{m}}\delta_j - \frac{\alpha}{\ln 2}\frac{\log^{\alpha-1}\sqrt{m}}{(\sqrt{m})^2}\delta_j^2 +$$

$$\frac{\alpha(\alpha-1)}{\ln^2 2}\frac{\log^{\alpha-2}\sqrt{m}}{(\sqrt{m})^2}\delta_j^2 + \sum_{i=3}^{\infty}\frac{\delta_j^i}{i!(\sqrt{m})^i}\sum_{k=1}^{i}\frac{\alpha!}{(\alpha-k)!}\frac{C_k^{(i)}}{\ln^k 2}\log^{\alpha-k}\sqrt{m}.$$

Using Proposition 2.2 we get

$$\left|\sum_{i=3}^{\infty}\frac{\delta_j^i}{i!(\sqrt{m})^i}\sum_{k=2}^{i}\frac{\alpha!}{(\alpha-k)!}\frac{C_k^{(i)}}{\ln^k 2}\log^{\alpha-k}\sqrt{m}\right| \leq$$

$$\sum_{i=3}^{\infty}\frac{|\delta_j|^i}{i!(\sqrt{m})^i\ln^i 2}\sum_{k=2}^{i}\left|\binom{\alpha}{k}\right||k!\ln^{i-k}2|C_k^{(i)}|\log^{\alpha-k}\sqrt{m} \leq$$

$$\sum_{i=3}^{\infty} \frac{|\delta_j|^i}{i!(\sqrt{m})^i \ln^i 2} a_{max}(\alpha) i! 2^i \frac{1}{1-\ln 2} \log^{\alpha-2} \sqrt{m} =$$

$$\frac{1}{1-\ln 2} a_{max}(\alpha) \log^{\alpha-2} \sqrt{m} \sum_{i=3}^{\infty} (\frac{2|\delta_j|}{\sqrt{m} \ln 2})^i <$$

$$2100 a_{max}(\alpha) (\frac{|\delta_j|}{\sqrt{m}})^3 \log^{\alpha-2} \sqrt{m}.$$

Using Proposition 2.2 and substituting into (19) we get that it is enough to prove the following inequality:

$$\frac{\alpha}{\ln 2} \log^{\alpha-1} \sqrt{m} (\frac{P_1^2(\delta) + (2\ell - 3)P_2(\delta)}{(2\ell - 2)m} - \sum_{i=3}^{\infty} \frac{(-1)^{i-1}}{i} \frac{P_i(\delta)}{(\sqrt{m})^i} -$$

$$\frac{1}{\log \sqrt{m}} (\frac{P_2(\delta)}{m} \frac{\alpha - 1}{\ln 2} + 2100 a_{max}(\alpha) \frac{P_3(\delta^+)}{(\sqrt{m})^3})) \geq 0, \quad (21)$$

where $\delta^+ = (|\delta_1|, |\delta_2|, \ldots, |\delta_\ell|)$. The LHS of the previous inequality can be underestimated by

$$\frac{\alpha}{\ln 2} \log^{\alpha-1} \sqrt{m} (\frac{P_1^2(\delta) + (2\ell - 3)P_2(\delta^+)}{(2\ell - 2)m} - \frac{1}{3} \frac{P_3(\delta^+)}{(\sqrt{m})^3} -$$

$$\frac{1}{\log \sqrt{m}} (\frac{P_2(\delta^+)}{m} \frac{\alpha - 1}{\ln 2} + 2100 a_{max}(\alpha) \frac{P_3(\delta^+)}{(\sqrt{m})^3})) \geq \frac{\alpha}{\ln 2} \log^{\alpha-1} \sqrt{m} (\frac{P_1^2(\delta)}{(2\ell - 2)m} +$$

$$\frac{P_2(\delta^+)}{m} (\frac{2\ell - 3}{2\ell - 2} - \frac{1}{9} - \frac{1}{\log \sqrt{m}} (\frac{\alpha - 1}{\ln 2} + 700 a_{max}(\alpha)))), \quad (22)$$

which is nonnegative if $m > \ell^5 2^{2200\ell - 5}$. (We used in the last estimation that $|\delta_i| < (1/3)\sqrt{m}, 1 \leq i \leq \ell$.)

Case B.1.2. $\exists 1 \leq j \leq \ell : |m_j - \sqrt{m}| > (1/3)\sqrt{m}$.

Proposition 2.4. Let $\mathbf{u} = (u_1, u_2, \ldots, u_\ell)$, $\ell \geq 3$, $u_i > 0, 1 \leq i \leq \ell$, $u_\ell \notin [2/3, 4/3]$. Then there exist a constant $C(\ell) > 0$, such that the following inequality holds.

$$A_2(\mathbf{u}) - 1 - \ln G_2(\mathbf{u}) \geq C(\ell).$$

Proof. The LHS of (24) is nonnegative, since $A_2(\mathbf{u}) \geq G_2(\mathbf{u})$ holds by the arithmetic-geometric inequality and $G_2(\mathbf{u}) - 1 \geq \ln G_2(\mathbf{u})$ is true as well due to the fact $x - 1 \geq \ln x$, $x > 0$. In both inequalities equality can hold for infinitely many $\mathbf{u}$'s under the conditions of the lemma, but the statement says that equality cannot hold simultaneously.

First, suppose that $|G_2(\mathbf{u}) - 1| \geq 1/10$. In this case, $G_2(\mathbf{u}) - 1 - \ln G_2(\mathbf{u}) \geq \min\{0.1 - \ln 1.1, -0.1 - \ln 0.9\} > 1/214$.

So we can assume, that $|G_2(\mathbf{u}) - 1| < 1/10$. Let $\mathbf{u}^* = (u_1, u_2, \ldots, u_{\ell-1})$ and $\mathbf{u}' = (u_1, u_2, \ldots, u_{\ell-1}, v)$, $v(\neq u_\ell) > 0$ to be chosen later. So we have

$$A_2(\mathbf{u}) - G_2(\mathbf{u}) = (u_\ell - v)\frac{2}{\ell}A(\mathbf{u}^*) + A_2(\mathbf{u}') - G_2(\mathbf{u}) \geq (u_\ell - v)\frac{2}{\ell}G(\mathbf{u}^*) +$$

$$G_2(\mathbf{u}') - G_2(\mathbf{u}) = (u_\ell - v)(\frac{2}{\ell}G(\mathbf{u}^*) - \frac{u_\ell^{2/\ell} - v_\ell^{2/\ell}}{u_\ell - v}G(\mathbf{u}^*)^{\frac{2(\ell-1)}{\ell}}) \geq$$

$$(u_\ell - v)G(\mathbf{u}^*)\frac{2}{\ell}(1 - v^{2/\ell-1}G(\mathbf{u}^*)^{\frac{\ell-2}{\ell}}) =$$

$$u_\ell^{1-\frac{1}{\ell-1}}(1 - \frac{v}{u_\ell})G(\mathbf{u})^{\frac{\ell}{\ell-1}}\frac{2}{\ell}(1 - (\frac{G(\mathbf{u}^*)}{v})^{\frac{\ell-2}{\ell}}), \quad (23)$$

in the first inequality we used the arithmetic-geometric inequality twice, while in the second ineqality the fact that $x \mapsto x^{2/\ell}$ is a concave function.

If $u_\ell > 4/3$, then $G(\mathbf{u}^*) < G(\mathbf{u}) = \sqrt{G_2(\mathbf{u})}$, so if we choose $v = 1.2$, we have a lower estimate of $1/(37\ell)$ for the RHS of (23) (minimizing the product term by term).

On the other hand, if $u_\ell < 2/3$, then $G(\mathbf{u}^*) > G(\mathbf{u}) = \sqrt{G_2(\mathbf{u})}$, so if we choose $v = 0.8$, we get a lower estimate of $1/(72\ell)$.

So the statement holds if we choose

$$C(\ell) = 1/(72\ell). \quad (24)$$

$\square$

Let $a_i = u_i\sqrt{m}$, $u_i \leq \log^5 m, 1 \leq i \leq \ell, u_\ell \notin [2/3, 4/3]$. So in this case (16) has the following form:

$$\log^\alpha m - \sum_{i=1}^{\ell} \log^\alpha(u_i\sqrt{m}) \geq \frac{\alpha}{\ln 2}\log^{\alpha-1} m(1 - \frac{\sum_{1\leq j<k\leq \ell} u_j u_k}{\binom{\ell}{2}}). \quad (25)$$

By the generalized binomial theorem we have

$$(\log u_i + \log \sqrt{m})^\alpha = \sum_{j=0}^{\infty} \binom{\alpha}{j}(\log^j u_i)(\log^{\alpha-j}\sqrt{m}) \geq$$

$$\log^\alpha \sqrt{m} + \alpha(\log u_i)\log^{\alpha-1}\sqrt{m} + (\ell - \alpha - 1)(\log^2 u_i)\log^{\alpha-2}\sqrt{m}, \quad (26)$$

using the fact, that $\ell = 2^\alpha = \sum_{j=0}^{\infty}\binom{\alpha}{j}$. If m is large enough (by some counting it can be checked that $\log m > 10^{12}\ell^6$ is sufficient) the following inequality holds,

$$\log^2 u_i < \frac{\alpha}{144(\ln 2)\ell^2}\log\sqrt{m}. \quad (27)$$

Substituting (27) into (26), and (26) into (25) we get the following inequality to prove after simplifications,

$$-\frac{2}{\ell}(\sum_{i=1}^{\ell}\ln u_i) - \frac{\ell - \alpha - 1}{72\ell^2} \geq 1 - \frac{\sum_{1\leq j<k\leq \ell} u_j u_k}{\binom{\ell}{2}}.$$

Let $\mathbf{u} = (u_1, u_2, \ldots, u_\ell)$, then we can write the previous inequality in the following form,

$$A_2(\mathbf{u}) - 1 - \ln G_2(\mathbf{u}) \geq \frac{\ell - \alpha - 1}{72\ell^2},$$

which is true by Proposition 2.4 (more precisely, by (24)).

Case B.2. $\exists 1 \leq i \leq \ell : m_i > \sqrt{m} \log^5 m$.
Case B.2.1. $\exists i \neq j \; m_i m_j \geq m \log^2 m$
In this case we have

$$\text{RHS}(16) \leq \frac{\alpha}{\ln 2} \log^{\alpha - 1} (1 - \frac{\log^2 m}{\binom{\ell}{2}}) \tag{28}$$

and

$$\text{LHS}(16) \geq -(\ell - 1) \log^\alpha m, \tag{29}$$

so by (28) and (29), (16) holds if m is large enough. (Say, $\log m > (3\ell^3 \ln 2)/(4\alpha)$ is a good choice.)

Case B.2.2. $\forall i \neq j \; m_i m_j < m \log^2 m$
We would like to minimize the LHS of (16), for such m_is satisfying the conditions

$$\sqrt{m} \log^5 m \leq m_1 \leq m/\log^{1/2} m \quad \text{and} \quad m_1 m_i \leq m \log^2 m (2 \leq i \leq \ell). \tag{*}$$

The value of the LHS of (16) will not increase if we replace all m_is by $(m \log^2 m)/m_1$. So

$$\min_{(*)} \text{LHS}(16) \geq \min\{\sqrt{m} \log^5 m \leq x \leq m/\log^{1/2} m |$$

$$\log^\alpha m - \log^\alpha x - (\ell - 1) \log^\alpha \frac{m \log^2 m}{x}\}.$$

Let $f(x) = \log^\alpha m - \log^\alpha x - (\ell - 1) \log^\alpha (m \log^2 m/x)$. This function is monotonically decreasing for $x > (\ell - 1)^{1/(2(\alpha - 1))} \sqrt{m} \log m$, consider the derivative of $f(x)$, which is smaller than the lower bound for m_1 if $m > 2$. So the minimum is achieved when $x = m/\log^{1/2} m$.
Substituting into (16) we get

$$\min_{(*)} \text{LHS}(16) \geq \frac{\alpha}{2} \log^{\alpha - 1} m \log \log m - \frac{a_2(\alpha)}{4} \log^{\alpha - 2} m \log^2 \log m -$$

$$(\ell - 1) \left(\frac{5}{2}\right)^\alpha \log^\alpha \log m \geq (\alpha/8) \log^{\alpha - 1} m \log \log m. \tag{30}$$

In the first inequality we used the generalized binomial theorem. The second inequality holds if m is large enough. Say, $m > \max\{2^{((\ell + 2\alpha + 2)/\alpha)^2}, \ell 2^{10^4}\}$. On the other hand,

$$\text{RHS}(16) \leq \frac{\alpha}{\ln 2} \log^{\alpha - 1} m. \tag{31}$$

If $m > 2^{3100}$ then (16) holds by (30) and (31). $\qquad\square$

Remark 2.1. (16) holds for $m > M$, where

$$M = \max\{\frac{\ell \ln 2}{2\alpha}, \ell^5 2^{2200\ell - 5}, 2^{10^{12}\ell^6}, 2^{(3\ell^3 \ln 2)/(4\alpha)}, 3, 2^{((\ell + 2\alpha + 2)/\alpha)^2}, \ell 2^{10^4}, 2^{3100}\},$$

so the constant in Theorem 2.2 is at most

$$\binom{M}{2} < 2^{2 \cdot 10^{12} \ell^6}.$$

$\square$

3   Construction of a Badly Representable Key System

In [7], it was shown that there exist badly representable Sperner systems, namely of size

$$s(\mathcal{K}) > \frac{1}{n^2} \binom{n}{\lfloor \frac{n}{2} \rfloor}. \tag{32}$$

The proof of this theorem is not constructive. L. Rónyai's observation is that the number of Sperner families that can be represented by a matrix of at most r rows is quite limited, and so r should be at least as big as in (32) to get a representation even for all antichains at the middle level of the Boolean lattice. In the following, we show an explicit badly representable Sperner system close to the middle level, no worse is known up to now.

Remember, that if $\mathcal{K}$ is a Sperner system, $\mathcal{K}^{-1}$ denotes the set of maximal elements, that are not contained in any element of $\mathcal{K}$. $\mathcal{K}_k^n$ denotes the complete k-uniform hypergraph and $K + L$ is the disjoint union of the hypergraphs K and L on the union of the vertices.

Theorem 3.1. *[11] Let* $n = n_1 + n_2 + \ldots + n_t$, $n_i \leq N (1 \leq i \leq t)$. *Let* $\mathcal{K}_n = \mathcal{K}_k^{n_1} + \mathcal{K}_k^{n_2} + \ldots + \mathcal{K}_k^{n_t}$. *Then*

$$|\mathcal{K}_n^{-1}| \leq T_\ell(s(\mathcal{K}_n)) \tag{33}$$

holds for $\ell = \binom{N}{k-1}$.

Proof (For details, see [11]). Suppose, that $\mathcal{K}_n$ is represented by a relation I of size $s(\mathcal{K}_n)$. We can recursively construct a labelled directed tree, $F \in \mathcal{F}_\ell^{(s(\mathcal{K}_n))}$ having the property, that there is an injection from $\mathcal{K}_n^{-1}$ to the leaves of F. A maximal antikey contains exactly $k - 1$ elements from each clique, so there are $\binom{n_i}{k-1}$ possibilities for the intersection of a maximal antikey and the ith clique. The key observation is that if A_1 and A_2 are two maximal antikeys and $A_j^i = A_j \cap V(\mathcal{K}_k^{n_i})$, $(j = 1, 2)$, then there is no u and v satisfying both $\pi_{A_1^i}(u) = \pi_{A_1^i}(v)$ and $\pi_{A_2^i}(u) = \pi_{A_2^i}(v)$. (By Proposition 1.1 we know, that a representation of a Sperner family $\mathcal{K}$ should contain two rows, for each $A \in \mathcal{K}^{-1}$, that are equal in A, but should not contain two rows that are equal in an element of $\mathcal{K}$.)

The labels of the vertices of the tree are subsets of I. Let the label of the root be I, the whole relation. The out-neighborhood of the root can be devided into $\binom{n_1}{k-1}$ classes, each corresponding to a $(k-1)$-subset of $V(K_k^{n_1})$. For a $(k-1)$-subset S consider $\pi_S(u)$ for all $u \in I$. By the equality of $\pi_S(u)$'s we get a partition of I. For each element of the partition of size at least two, add a new vertex to the tree, and label it by its elements. By the above remark labels of vertices from the same class are disjoint, while labels of vertices from distinct class can intersect in at most one.

We can continue building up the tree the similar way. The only difference is that instead of considering subsets of I we consider subsets of the label of the actual vertex. It is easy to see, that we get a tree of $\mathcal{F}_\ell^{(s(\mathcal{K}_n))}$, and by Proposition 1.1 the above mentioned correspondence is really an injection.

Corollary 3.1. *There exists a sequence of Sperner systems $\mathcal{K}_n$, such that*

$$s(\mathcal{K}_n) > 2^{n(1-(26/3)\log\log n/\log n)} \tag{34}$$

holds for n large enough.

Proof. We would like to apply Theorem 3.1, let $n_i = s - 1$ or $n_i = s$, $1 \leq i \leq \lceil n/s \rceil$. So our task remains to choose s and k. Let $k = g(n)+1$ and $s = 2g(n)+1$, $g(n)$ to be chosen later.

$$\log |\mathcal{K}_n^{-1}| \geq \frac{n}{s} \log \binom{s-1}{k-1} > \frac{n}{2g(n)} \log \frac{2^{2g(n)}}{2\sqrt{2g(n)}} \geq n(1 - \frac{1}{3}\frac{\log g(n)}{g(n)}),$$

if $g(n) \geq 2^9$. So by Theorem 2.2 and Theorem 3.1,

$$n(1 - \frac{1}{3}\frac{\log g(n)}{g(n)}) < \log C(2^{2g(n)}) + \log s(\mathcal{K}_n) + 2g(n)\log\log s(\mathcal{K}_n). \tag{35}$$

where $C(x) = 2^{2\cdot10^{12}x^6}$.

$$2g(n)\log\log s(\mathcal{K}_n) < \frac{\log g(n)}{6g(n)}\log s(\mathcal{K}_n) \tag{36}$$

holds if, say, $g(n) < n^{1/4}/7$ (using only, that $s(\mathcal{K}_n) > 2^{n/4}$, n is large enough).
On the other hand,

$$\log C(2^{2g(n)}) < \frac{\log g(n)}{6g(n)}\log s(\mathcal{K}_n), \tag{37}$$

if $g(n) \leq \log n/13 - 4$. By (35)-(37), we get

$$\log s(\mathcal{K}_n) > n(1 - \frac{1}{3}\frac{\log g(n)}{g(n)})/(1 + \frac{1}{3}\frac{\log g(n)}{g(n)}) > n(1 - \frac{2}{3}\frac{\log g(n)}{g(n)}) \tag{38}$$

By (38), the statement holds if we choose $g(n) = \lfloor \log n/13 \rfloor - 4$.   □

4   Well Representable Key Systems

4.1   Improving the Upper Bound on Complete k-Uniform Key Systems

In Theorem 1.3 $c_2(k)$ depends on k exponentially. We can replace this by a constant 2.

Theorem 4.1.
$$s(\mathcal{K}_k^n) < 2n^{(k-1)/2} + o(n^{(k-1)/2}).$$

Proof (For details, see [5]). Let p be a prime. The original proof defines polynomials of degree at most k over the finite field $GF(p)$. It's coefficients are chosen from a difference set D. D is called a difference set, if $D - D = GF(p)$, i.e., each element of $GF(p)$ can be written as a difference of two elements from D. The jth coordinate of the ith vector of the representing relation instance I is $P_i(j)$ over $GF(p)$, where P_i is the ith polynomial. The size of a difference set is about $2\sqrt{p}$. Our first observation is one can take 2 classes of polynomials. One of the classes contains polynomials of degree $k-1$ coefficients from D_1 (except for the coefficient of x^{k-1}, which is 1). The other one contains polynomials of degree $k-2$ and coefficients from D_2.

One can choose D_1 and D_2 so, that each of them have size around $\sqrt{p}$ and each element of $GF(p)$ is of the form $d_1 - d_2$, where $d_i \in D_i(i = 1, 2)$. Let $D_1 = \{0, \lceil\sqrt{p}\rceil, \ldots, (\lceil\sqrt{p}\rceil - 1)\lceil\sqrt{p}\rceil\}$, $D_2 = \{0, 1, \ldots, \lceil\sqrt{p}\rceil\}$.

One can easily check, that the constructed instance represents $\mathcal{K}_k^p$. Two tuples can not have k equal coordinates. That would mean that the difference of the corresponding polynomials have k roots, but it is a polynomial of degree at most $k-1$. On the other hand, the polynomial

$$w(x) = (x - t_1)(x - t_2)\cdots(x - t_{k-1}) = x^{k-1} + a_{k-2}x^{k-2} + \ldots + a_1 x + a_0$$

can be written as a difference of two polynomials that correspond to tuples for arbitrary different $t_1, \ldots, t_{k-1}$. Each a_i can be written as a differnce of $d_{1,i} \in D_1$ and $d_{2,i} \in D_2$. Then $w(x)$ is the difference of the polynomials

$$z_j(x) = (2 - j)x^{k-1} + d_{j,k-2}x^{k-2} + \ldots + d_{j,1}x + d_{j,0} \quad (j = 1, 2).$$

If $n < p$, then $\pi_{\{1,2,\ldots,n\}}(I)$ is an appropriate representation of $\mathcal{K}_k^n$.

Instead of Chebyshev's theorem (Bertrand's postulate) on the density of primes (between n and $2n$ there is a prime) the statement follows from a theorem of Luo and Yao [9], stating that for the nth prime p_n

$$p_{n+1} - p_n \ll p_n^{6/11+\varepsilon} \tag{39}$$

holds for any $\varepsilon > 0$. $\square$

This improvement shows that complete k-uniform key systems are well representable, even if $k = f(n)$ slightly tends to infinity.

4.2 New Well Representable Key Systems

We can use the idea of Theorem 1.3 to prove well representability for key systems, that differ in not too many elements from $\mathcal{K}_k^n$.

First, let us consider the case when the key system $\mathcal{K}$ contains one element $\{(t_1, \ldots, t_{k-1})\}$ of size $k-1$ and all k element subsets but its supersets. One can construct sets $D_1^{(c)}$ and $D_2^{(c)}$, such that $D_1^{(c)} - D_2^{(c)} = GF(p) \backslash \{c\}$, for each $c \in GF(p)$. Let $P(x) = (x-t_1)(x-t_2) \cdots (x-t_{k-1}) = x^{k-1} + c_{k-2}x^{k-2} + \ldots + c_0$. Then the polynomials will be:

$$x^{k-1} + d_1^{(k-2)}x^{k-2} + d_{1,k-3}x^{k-3} + \ldots + d_{1,0},$$
$$x^{k-1} + d_{1,k-2}x^{k-2} + d_1^{(k-3)}x^{k-3} + \ldots + d_{1,0}, \ldots,$$
$$x^{k-1} + d_{1,k-2}x^{k-2} + d_{1,k-3}x^{k-3} + \ldots + d_1^{(0)},$$
$$d_2^{(k-2)}x^{k-2} + d_{2,k-3}x^{k-3} + \ldots + d_{2,0},$$
$$d_{2,k-2}x^{k-2} + d_2^{(k-3)}x^{k-3} + \ldots + d_{2,0}, \ldots,$$
$$d_{2,k-2}x^{k-2} + d_{2,k-3}x^{k-3} + \ldots + d_2^{(0)},$$

where $d_i^{(t)} \in D_i^{(c_t)}, (i = 1, 2; \; 0 \le t \le k-2), d_{i,j} \in D_i (i = 1, 2)$ (see Theorem 4.1). So we get each $(k-1)$-degree polynomial, except for $P(x)$. The size of this relation is about $2(k-1)p^{(k-1)/2}$.

Based on the idea above one can construct a good representation for a key system having $o(n^{1/4})$ elements of cardinality $k-1$ and containing all k element sets but their supersets. Let $\nabla(A) = \{B | B \supseteq A, |B| = |A| + 1\}$ and $\nabla(\mathcal{A}) = \bigcup_{A \in \mathcal{A}} \nabla(A)$.

Theorem 4.2. Let $\mathcal{K}_n = \mathcal{A}_n \cup \mathcal{B}_n$ be a Sperner system, such that $\mathcal{A}_n \subseteq \mathcal{K}_{k-1}^n$, $\nabla(\mathcal{A}_n) \cup \mathcal{B}_n = \mathcal{K}_k^n$, and $|\mathcal{A}_n| = o(n^{1/4})$. Then

$$s(\mathcal{K}_n) \le 2(k-1)n^{(k-1)/2} + o(n^{(k-1)/2}). \tag{40}$$

Proof. It follows from the conditions, that $\mathcal{K} \subseteq \mathcal{K}_{k-1}^n \cup \mathcal{K}_k^n$ and $\mathcal{K}^{-1} \subseteq \mathcal{K}_{k-2}^n \cup \mathcal{K}_{k-1}^n$. Let $\mathcal{A}_n = \{(t_1^{(1)}, \ldots, t_{k-1}^{(1)}), \ldots, (t_1^{(m)}, \ldots, t_{k-1}^{(m)})\}$ and $p > n$. For $1 \le r \le m$ let us consider the polynomial

$$w_r(x) = (x - t_1^{(r)})(x - t_2^{(r)}) \cdots (x - t_{k-1}^{(r)}) = x^{k-1} + c_{k-2}^{(r)}x^{k-2} + \ldots + c_1^{(r)}x + c_0^{(r)}.$$

For each $0 \le h \le k-2$ let $J_1^h = [a_1^h, b_1^h], J_2^h = [a_2^h, b_2^h], \ldots, J_{m_h}^h = [a_{m_h}^h, b_{m_h}^h]$ $(m_h \le m)$ be the (possibly 1-length) intervals of (the cyclically ordered) $GF(p)$, such that $\{a_1^h, \ldots, a_{m_h}^h\} = \{c_h^{(1)}, \ldots, c_h^{(m)}\}$ and $\{b_1^h, \ldots, b_{m_h}^h\} = \{c_h^{(1)} - 1, \ldots, c_h^{(m)} - 1\}$. So $a_1^h \le b_1^h = a_2^h - 1, \ldots, a_{m_h}^h \le b_{m_h}^h = a_1^h - 1$. Let $l_i^h = b_i^h - a_i^h + 1$ be the length of J_i^h.

Let $R_t^h = \{1 \le i \le m_h \mid p^{1/2^t} < l_i^h \le p^{1/2^{t-1}}\}$ and $r_t = \lfloor p^{1/2^t} \rfloor$ ($t = 1, 2, \ldots, \lceil \log p \rceil$). Let $D_t^h = \bigcup_{i \in R_t^h} \{a_i^h + r_t, a_i^h + 2r_t, \ldots, \lfloor (l_i^h - 1)/r_t \rfloor r_t, b_i^h\}$ and $E_t = \{0, 1, 2, \ldots, r_t - 1\}$. Furthermore let $D = \{0, r_1, 2r_1, \ldots, r_1^2, \}$ and $E = E_1$.

We construct a representation I_n of $\mathcal{K}_n$. We give p-tuples of three types. The coordinates of the tuples triples of form $(P_i(j), h, t)$. For each h and t we take all polynomials P_i over $\mathrm{GF}(p)$ of the form

$$P_i(x) = x^{k-1} + d_{k-2}x^{k-2} + \ldots + d_1 x + d_0,$$

where $d_h \in D_t^h$ and $d_i \in D$ $(0 \le i \le k-2, i \ne h)$.

Tuples of the second type have coordinates of a triple $(Q_i(j), h, t)$, too. For each h and t we take all polynomials Q_i over $\mathrm{GF}(p)$ of the form

$$Q_i(x) = e_{k-2}x^{k-2} + \ldots + e_1 x + e_0,$$

where $e_h \in E_t^h$ and $e_i \in E$ $(0 \le i \le k-2, i \ne h)$.

Tuples of the third type have coordinates of the form $(P(j), i)$, two polynomials for each i, where i corresponds to a $(k-1)$-tuple of coefficients, $(c_{k-2}, c_{k-3}, \ldots c_0)$. All coordinates are a coefficient of some $w_r(x), 1 \le r \le m$, while the polynomial $z(x) = x^{k-1} + c_{k-2}x^{k-2} + \ldots + c_0$ is not among these polynomials. Choose the two polynomials so, that their difference is $z(x)$. (E.g, $z(x)$ and the 0 polynomial.)

We get I_n by deleting the last $p - n$ coordinates of the tuples of the constructed relation. There are no two tuples of I_n having the same value in k coordinates (polynomials of degree $k-1$ can not have k common roots). It is also easy to check that there are no two tuples having the same coordinates in $A, |A| = k-1$ if and only if $A \notin \mathcal{A}_n$.

There are $p^{(k-1)/2} + o(p^{(k-1)/2})$ tuples of type $(P_i(j), h, 1)$ and $(Q_i(j), h, 1)$. The number of $(P_i(j), h, 2)$ and $(Q_i(j), h, 2)$ type tuples are at most $mp^{1/4}p^{(k-2)/2} = o(p^{(k-1)/2})$. Similarly, there is no more than $mp^{1/8}p^{(k-2)/2} = o(p^{(4k-5)/8})$ tuples of type $(P_i(j), h, t)$ and $(Q_i(j), h, t)$ $(t \ge 3)$. Finally, the number of the tuples of type $(P(j), i)$ is at most $m^{k-1} = o(p^{(k-1)/4})$.

The statement follows from (39).

5   Further Research

It remains an open problem to determine the exact value of $T_\ell(m)$. Even in the case of $\ell = 2$ for general m.

The above improvements on the labelled directed tree lemma (Theorem 2.1 and Theorem 2.2) opens a new dimension of the minimum representation problem. Is the log factor needed? If yes/no, what is the exact constant?

Example 5.1. Let $\mathcal{C}_n = \{\{1,2\}, \{2,3\}, \ldots, \{n-1, n\}, \{n, 1\}\}$ be the cycle. We know from [11]

$$|\mathcal{C}_n^{-1}| \le T_2(s(\mathcal{C}_n))$$

So by (1) and we have for $n \ge 5$

$$\frac{2|\mathcal{C}_n^{-1}|}{\log |\mathcal{C}_n^{-1}|} \le s(\mathcal{C}_n) \le |\mathcal{C}_n^{-1}| + 1. \tag{41}$$

Note, that in this case the upper bound cannot be the truth. One can construct an instance in which the rowpairs (see Lemma 1.1) having a non-trivial density, proving say $0.99|\mathcal{C}_n^{-1}|$ for n large enough. □

Improvements similar to (41) can be obtained for the problems considered in [11], maximizing $s(\mathcal{K})$ over $\mathcal{K}$'s from a special class, such as e.g, for all graph.

Theorem 5.1. *Let $\mathfrak{G}_n$ denote the set of all graph on n vertices.*

$$\frac{1}{2^{10^{15}}} \cdot \frac{3^{n/3}}{n^{\log 3}} \leq \max_{\mathcal{G} \in \mathfrak{G}_n} s(\mathcal{G}) \leq 3^{n/3} + 1. \tag{42}$$

It still remains open to show a key system, that is as badly representable as the Demetrovics-Gyepesi probabilistic estimate.

References

1. S. Abiteboul, R. Hull, V. Vianu, Foundations of Databases, Addison-Wesley, Reading, 1995.
2. W.W. Armstrong, *Dependency structures of database relationship,* Information processing 74 (North Holland, Amsterdam 1974) 580–583.
3. I. Csiszár, J. Körner, Information theory. Coding theorems for discrete memoryless systems. Probability and Mathematical Statistics. Academic Press, Inc. (Harcourt Brace Jovanovich, Publishers), New York-London, 1981.
4. J. Demetrovics, *On the equivalence of candidate keys with Sperner systems,* Acta Cybernetica **4** (1979) 247–252.
5. J. Demetrovics, Z. Füredi, G. O. H. Katona, *Minimum matrix representation of closure operations,* Discrete Appl. Math. **11** (1985), no. 2, 115–128.
6. J. Demetrovics, Gy. Gyepesi, *On the functional dependency and some generalizations of it,* Acta Cybernetica **5** (1980/81), no. 3, 295–305.
7. J. Demetrovics, Gy. Gyepesi *A note on minimal matrix representation of closure operations,* Combinatorica **3** (1983), no. 2, 177–179.
8. Z. Füredi, *Perfect error-correcting databases,* Discrete Appl. Math. 28 (1990) 171–176.
9. Lou Shi Tuo, Yao Qi, *A Chebychev's type of prime number theorem in a short interval. II.,* Hardy-Ramanujan J. 15 (1992), 1–33 (1993).
10. Van Lindt, Wilson, A Course in Combinatorics, Cambridge Univ. Press, 1992, Chapter 22.
11. K. Tichler, *Extremal theorems for databases,* Annals of Mathematics and Artificial Intelligence **40** (2004) 165–182.

On Multivalued Dependencies
in Fixed and Undetermined Universes

Sebastian Link*

Information Science Research Centre, Dept. of Information Systems,
Massey University, Palmerston North, New Zealand
s.link@massey.ac.nz

Abstract. The implication of multivalued dependencies (MVDs) in
relational databases has originally been defined in the context of some
fixed finite universe. While axiomatisability and implication problem
have been intensely studied with respect to this notion, almost no research has been devoted towards the alternative notion of implication in
which the underlying universe of attributes is left undetermined.

Based on a set of common inference rules we reveal all axiomatisations in undetermined universes, and all axiomatisations in fixed universes that indicate the role of the complementation rule as a means of
database normalisation. This characterises the expressiveness of several
incomplete sets of inference rules. We also establish relationships between
axiomatisations in fixed and undetermined universes, and study the time
complexity of the implication problem in undetermined universes.

1 Introduction

Relational databases still form the core of most database management systems,
even after more than three decades following their introduction in [12]. The
relational model organises data into a collection of relations. These structures
permit the storage of inconsistent data, inconsistent in a semantic sense. Since
this is not acceptable additional assertions, called dependencies, are formulated
that every database is compelled to obey. There are many different classes of
dependencies which can be utilised for improving the representation of the target
database [17, 33, 36].

Multivalued dependencies (MVDs) [14, 16, 41] are an important class of dependencies. A relation exhibits an MVD precisely when it is decomposable without
loss of information [16]. This property is fundamental to relational database design, in particular 4NF [16], and a lot of research has been devoted to studying
the behaviour of these dependencies. Recently, extensions of multivalued dependencies have been found very useful for various design problems in advanced data
models such as the nested relational data model [18], the Entity-Relationship
model [34], data models that support nested lists [23, 24] and XML [37, 38].

---

* Sebastian Link was supported by Marsden Funding, Royal Society of New Zealand.

J. Dix and S.J. Hegner (Eds.): FoIKS 2006, LNCS 3861, pp. 258–277, 2006.
© Springer-Verlag Berlin Heidelberg 2006

The classical notion of an MVD [16] is dependent on the underlying universe R. This dependence is reflected syntactically by the R-complementation rule which is part of the axiomatisation of MVDs, see for instance [6]. The complementation rule is special in the sense that it is the only inference rule which is dependent on R. Further research on this fact has lead to an alternative notion of semantic implication in which the underlying universe is left undetermined [11]. In the same paper Biskup shows that this notion can be captured syntactically by a sound and complete set of inference rules, denoted by $\mathfrak{S}_0$. If $R\mathfrak{S}_0$ results from adding the R-complementation rule to $\mathfrak{S}_0$, then $R\mathfrak{S}_0$ is R-sound and R-complete for the R-implication of MVDs. In fact, every inference of an MVD by $R\mathfrak{S}_0$ can be turned into an inference of the same MVD in which the R-complementation rule is applied at most once, and if it is applied, then in the last step of the inference ($R\mathfrak{S}_0$ is said to be R-complementary). This indicates that the R-complementation rule simply reflects a part of the decomposition process, and does not necessarily infer semantically meaningful consequences. Interestingly, research has not been continued in this direction but focused on the original notion of R-implication. Since research on MVDs seems to experience a recent revival in the context of other data models [18, 34, 23, 24, 37, 38] it seems desirable to further extend the knowledge on relational MVDs. An advancement of such knowledge may simplify the quest of finding suitable and comprehensible extensions of MVDs to currently popular data models.

In this paper we will further study the alternative notion of implication of an MVD as suggested in [11]. First, we will identify all minimal complete subsets of a set $\mathfrak{S}_U$ of eight common sound inference rules for MVDs. Minimality refers to the fact that none of the rules can be omitted without losing completeness. Essentially, it turns out that apart from the set $\mathfrak{S}_0$ there are two other subsets of $\mathfrak{S}_U$ that are also minimal. It is further shown that a sound set of inference rules is complete if and only if for all relation schemata R that set extended by the R-complementation rule is both R-complete and R-complementary. Subsequently, the time-complexity of the corresponding implication problem is studied. Herein, the classical notion of a dependency basis of an attribute set with respect to a set of MVDs can be rephrased in the context of undetermined universes. It turns out that the "traditional" dependency basis of X with respect to Σ deviates from the dependency basis of X with respect to Σ in undetermined universes by at most one set. This set, however, can be described by the notion of the Σ-scope of an attribute set X, which is the union of all those attribute sets Y such that $X \twoheadrightarrow Y$ is implied by Σ. The Σ-scope itself can be computed in time linear in the total number of attributes that occur in Σ. The computation is similar to computing the closure of an attribute set with respect to a set of functional dependencies [5]. The problems studied in this paper are not just of theoretical interest. In practice one does not necessarily want to generate all consequences of a given set of MVDs but only some of them. Such a task can be accomplished by using incomplete sets of inference rules. However, it is then essential to explore the power of such incomplete sets.

The paper is structured as follows. Section 2 summarises notions from the relational model of data. Section 3 identifies all minimal axiomatisations in $\mathfrak{S}_U$ for the implication of MVDs in undetermined universes. A rather general result is proven in Section 5 which shows an equivalence between axiomatisations of MVDs in fixed universes and those in undetermined ones. Finally, the implication problem of MVDs in undetermined universes is studied in Section 6. The paper concludes with some open problems in Section 7.

2   Multivalued Dependencies in Relational Databases

Let $\mathfrak{A} = \{A_1, A_2, \ldots\}$ be a (countably) infinite set of attributes. A *relation schema* is a finite set $R = \{A_1, \ldots, A_n\}$ of distinct symbols, called *attributes*, which represent column names of a relation. Each attribute A_i of a relation schema is associated an infinite domain $dom(A_i)$ which represents the set of possible values that can occur in the column named A_i. If X and Y are sets of attributes, then we may write XY for $X \cup Y$. If $X = \{A_1, \ldots, A_m\}$, then we may write $A_1 \cdots A_m$ for X. In particular, we may write simply A to represent the singleton $\{A\}$. A *tuple* over $R = \{A_1, \ldots, A_n\}$ (R-tuple or simply tuple, if R is understood) is a function $t : R \to \bigcup_{i=1}^{n} dom(A_i)$ with $t(A_i) \in dom(A_i)$ for $i = 1, \ldots, n$. For $X \subseteq R$ let $t[X]$ denote the restriction of the tuple t over R on X, and $dom(X) = \prod_{A \in X} dom(A)$ the Cartesian product of the domains of attributes in X. A *relation* r over R is a finite set of tuples over R. The relation schema R is also called the domain $Dom(r)$ of the relation r over R. Let $r[X] = \{t[X] \mid t \in r\}$ denote the *projection* of the relation r over R on $X \subseteq R$. For $X, Y \subseteq R$, $r_1 \subseteq dom(X)$ and $r_2 \subseteq dom(Y)$ let $r_1 \bowtie r_2 = \{t \in dom(XY) \mid \exists t_1 \in r_1, t_2 \in r_2 \text{ with } t[X] = t_1[X] \text{ and } t[Y] = t_2[Y]\}$ denote the *natural join* of r_1 and r_2. Note that the 0-ary relation $\{()\}$ is the projection $r[\emptyset]$ of r on $\emptyset$ as well as left and right identity of the natural join operator.

2.1   MVDs in Fixed Universes

Functional dependencies (FDs) between sets of attributes have always played a central role in the study of relational databases [12, 13, 5, 7, 8], and seem to be central for the study of database design in other data models as well [1, 21, 27, 24, 32, 39, 40]. The notion of a functional dependency is well-understood and the semantic interaction between these dependencies has been syntactically captured by Armstrong's well-known axioms [2, 3]. A *functional dependency* (FD) [13] on the relation schema R is an expression $X \to Y$ where $X, Y \subseteq R$. A relation r over R *satisfies* the FD $X \to Y$, denoted by $\models_r X \to Y$, if and only if every pair of tuples in r that agrees on each of the attributes in X also agrees on the attributes in Y. That is, $\models_r X \to Y$ if and only if $t_1[Y] = t_2[Y]$ whenever $t_1[X] = t_2[X]$ holds for any $t_1, t_2 \in r$.

FDs are incapable of modelling many important properties that database users have in mind. Multivalued dependencies (MVDs) provide a more general notion and offer a response to the shortcomings of FDs. A *multivalued dependency* (MVD) [14, 16, 41] on R is an expression $X \twoheadrightarrow Y$ where $X, Y \subseteq R$. A relation r over R *satisfies* the MVD $X \twoheadrightarrow Y$, denoted by $\models_r X \twoheadrightarrow Y$, if and only if for all $t_1, t_2 \in r$ with $t_1[X] = t_2[X]$ there is some $t \in r$ with $t[XY] = t_1[XY]$ and $t[X(R - Y)] = t_2[X(R - Y)]$. Informally, the relation r satisfies $X \twoheadrightarrow Y$ when the value on X determines the set of values on Y independently from the set of values on $R - Y$. This actually suggests that the relation schema R is overloaded in the sense that it carries two independent facts XY and $X(R-Y)$. More precisely, it is shown in [16] that MVDs "provide a necessary and sufficient condition for a relation to be decomposable into two of its projections without loss of information (in the sense that the original relation is guaranteed to be the join of the two projections)". This means that $\models_r X \twoheadrightarrow Y$ if and only if $r = r[XY] \bowtie r[X(R - Y)]$. This characteristic of MVDs is fundamental to relational database design and 4NF [16]. A lot of research has therefore been devoted to studying the behaviour of these dependencies.

For the design of a relational database schema dependencies are normally specified as semantic constraints on the relations which are intended to be instances of the schema. During the design process one usually needs to determine further dependencies which are logically implied by the given ones. In order to emphasise the dependence of implication from the underlying relation schema R we refer to R-implication.

Definition 2.1. *Let R be a relation schema, and let $\Sigma = \{X_1 \twoheadrightarrow Y_1, \ldots, X_k \twoheadrightarrow Y_k\}$ and $X \twoheadrightarrow Y$ be MVDs on R, i.e., $X \cup Y \cup \bigcup_{i=1}^{k}(X_i \cup Y_i) \subseteq R$. Then Σ R-implies $X \twoheadrightarrow Y$ if and only if each relation r over R that satisfies all MVDs in Σ also satisfies $X \twoheadrightarrow Y$.* $\square$

In order to determine all logical consequences of a set of MVDs one can use the following set of inference rules for the R-implication of multivalued dependencies [6]. Note that we use the natural complementation rule [10] instead of the complementation rule that was originally proposed [6].

$$\frac{}{X \twoheadrightarrow Y}\, Y \subseteq X \qquad \frac{X \twoheadrightarrow Y}{XU \twoheadrightarrow YV}\, V \subseteq U \qquad \frac{X \twoheadrightarrow Y, Y \twoheadrightarrow Z}{X \twoheadrightarrow Z - Y}$$
$$\text{(reflexivity, } \mathcal{R}) \qquad \text{(augmentation, } \mathcal{A}) \qquad \text{(pseudo-transitivity, } \mathcal{T})$$

$$\frac{X \twoheadrightarrow Y}{X \twoheadrightarrow R - Y}$$
$$\text{(R-complementation, } \mathcal{C}_R)$$

$$\frac{X \twoheadrightarrow Y, X \twoheadrightarrow Z}{X \twoheadrightarrow YZ} \qquad \frac{X \twoheadrightarrow Y, X \twoheadrightarrow Z}{X \twoheadrightarrow Z - Y} \qquad \frac{X \twoheadrightarrow Y, X \twoheadrightarrow Z}{X \twoheadrightarrow Y \cap Z}$$
$$\text{(union, } \mathcal{U}) \qquad \text{(difference, } \mathcal{D}) \qquad \text{(intersection, } \mathcal{I})$$

Beeri et al. [6] prove that this set of inference rules is both R-sound and R-complete for the R-implication of MVDs, for each relation schema R. Let

$\Sigma \cup \{\sigma\}$ be a set of MVDs on the relation schema R. Let $\Sigma \vdash_{\mathfrak{S}} \sigma$ denote the inference of σ from a set Σ of dependencies with respect to the set $\mathfrak{S}$ of inference rules. Let $\Sigma_{\mathfrak{S}}^+ = \{\sigma \mid \Sigma \vdash_{\mathfrak{S}} \sigma\}$ denote the *syntactic hull* of Σ under inference using only rules from $\mathfrak{S}$. The set $R\mathfrak{S}$ is called *R-sound* for the R-implication of MVDs if and only if for every set Σ of MVDs on the relation schema R we have $\Sigma_{R\mathfrak{S}}^+ \subseteq \Sigma_R^* = \{\sigma \mid \Sigma \ R\text{-implies } \sigma\}$. The set $R\mathfrak{S}$ is called *R-complete* for the R-implication of MVDs if and only if for every set Σ of MVDs on R we have $\Sigma_R^* \subseteq \Sigma_{R\mathfrak{S}}^+$. Furthermore, the set $R\mathfrak{S}$ is called complete (sound) for the R-implication of MVDs if and only if it is R-complete (R-sound) for the R-implication of MVDs for all relation schemata R.

An interesting question is now whether all the rules of a certain set of inference rules are really necessary to capture the R-implication of MVDs for every relation schema R. More precisely, an inference rule $\mathfrak{R}$ is said to be *R-independent* from the set $R\mathfrak{S}$ if and only if there is some set $\Sigma \cup \{\sigma\}$ of MVDs on the relation schema R such that $\sigma \notin \Sigma_{R\mathfrak{S}}^+$, but $\sigma \in \Sigma_{R\mathfrak{S}\cup\{\mathfrak{R}\}}^+$. Moreover, an inference rule $\mathfrak{R}$ is said to be *independent* from $R\mathfrak{S}$ if and only if there is some relation schema R such that $\mathfrak{R}$ is R-independent from $R\mathfrak{S}$. A complete set $R\mathfrak{S}$ is called *minimal* for the R-implication of MVDs if and only if every inference rule $\mathfrak{R} \in R\mathfrak{S}$ is independent from $R\mathfrak{S} - \{\mathfrak{R}\}$. This means that no proper subset of $R\mathfrak{S}$ is still complete. It was shown by Mendelzon [29] that $R\mathfrak{M} = \langle \mathcal{R}, \mathcal{C}_R, \mathcal{T} \rangle$ forms such a minimal set for the R-implication of MVDs. The R-complementation rule $\mathcal{C}_R$ plays a special role as it is the only rule which depends on the underlying relation schema R. In the same paper, Mendelzon further motivates the study of the independence of inference rules and comments in more detail on the special role of the R-complementation rule.

2.2   MVDs in Undetermined Universes

Consider the classical example [16] in which the MVD *Employee* $\twoheadrightarrow$ *Child* is specified, i.e., the set of children is completely determined by an employee, independently from the rest of the information in any schema. If the relation schema R consists of the attributes *Employee*, *Child* and *Salary*, then we may infer the MVD *Employee* $\twoheadrightarrow$ *Salary* by means of the complementation rule. However, if the underlying relation schema R consists of the four attributes *Employee*, *Child*, *Salary* and *Year*, then the MVD *Employee* $\twoheadrightarrow$ *Salary* is no longer R-implied. Note the fundamental difference of the MVDs

$$Employee \ \twoheadrightarrow \ Child \qquad \text{and} \qquad Employee \ \twoheadrightarrow \ Salary.$$

The first MVD has been specified to establish the relationship of employees and their children as a fact due to a set-valued correspondence. The second MVD does not necessarily correspond to any semantic information, but simply results from the context in which *Employee* and *Child* are considered. If the context changes, the MVD disappears.

It may therefore be argued that consequences which are dependent on the underlying relation schema are in fact no consequences. This implies, however, that the notion of R-implication is not suitable.

Biskup introduced the following notion of implication [11]. An MVD is a syntactic expression $X \twoheadrightarrow Y$ with $X, Y \subseteq \mathfrak{A}$. The MVD $X \twoheadrightarrow Y$ is satisfied by some relation r if and only if $X \cup Y \subseteq Dom(r)$ and $r = r[XY] \bowtie r[X \cup (Dom(r) - Y)]$.

Definition 2.2. *The set* $\Sigma = \{X_1 \twoheadrightarrow Y_1, \ldots, X_k \twoheadrightarrow Y_k\}$ *of MVDs implies the single MVD* $X \twoheadrightarrow Y$ *if and only if for each relation* r *with* $X \cup Y \cup \bigcup_{i=1}^{k}(X_i \cup Y_i) \subseteq Dom(r)$ *the MVD* $X \twoheadrightarrow Y$ *is satisfied by* r *whenever* r *already satisfies all MVDs in* Σ. $\qquad\square$

In this definition, the underlying relation schema is left undetermined. The only requirement is that the MVDs must apply to the relations. If $X \cup Y \cup \bigcup_{i=1}^{k}(X_i \cup Y_i) \subseteq R$, then it follows immediately that $\Sigma = \{X_1 \twoheadrightarrow Y_1, \ldots, X_k \twoheadrightarrow Y_k\}$ R-implies $X \twoheadrightarrow Y$ whenever Σ implies $X \twoheadrightarrow Y$. The converse, however, is false [11].

A set $\mathfrak{S}$ of inference rules is called *sound* for the implication of MVDs if and only if for every finite set Σ of MVDs we have $\Sigma_{\mathfrak{S}}^+ \subseteq \Sigma^* = \{\sigma \mid \Sigma \text{ implies } \sigma\}$. The set $\mathfrak{S}$ is called *complete* for the implication of MVDs if and only if for every finite set Σ of MVDs we have $\Sigma^* \subseteq \Sigma_{\mathfrak{S}}^+$. An inference rule $\mathfrak{R}$ is said to be *independent* from the set $\mathfrak{S}$ if and only if there is some finite set $\Sigma \cup \{\sigma\}$ of MVDs such that $\sigma \notin \Sigma_{\mathfrak{S}}^+$, but $\sigma \in \Sigma_{\mathfrak{S} \cup \{\mathfrak{R}\}}^+$. A complete set $\mathfrak{S}$ of inference rules is called *minimal* if and only if every inference rule $\mathfrak{R}$ in $\mathfrak{S}$ is independent from $\mathfrak{S} - \{\mathfrak{R}\}$. This means that no proper subset of $\mathfrak{S}$ is still complete for the implication of MVDs.

While the singletons $\mathcal{R}, \mathcal{A}, \mathcal{T}, \mathcal{U}, \mathcal{D}, \mathcal{I}$ are all sound, the R-complementation rule and R-axiom are R-sound, but not sound [11]. In fact, the main result of [11] shows that the following set $\mathfrak{B}_0$ (denoted by $\mathfrak{S}_0$ in [11]) of inference rules

$$\frac{}{\emptyset \twoheadrightarrow \emptyset} \qquad \frac{X \twoheadrightarrow Y}{XU \twoheadrightarrow YV}\, V \subseteq U \qquad \frac{X \twoheadrightarrow Y, Y \twoheadrightarrow Z}{X \twoheadrightarrow Z - Y}$$

$$\text{(empty-set-axiom,} \mathcal{R}_\emptyset) \qquad\qquad (\mathcal{A}) \qquad\qquad\qquad (\mathcal{T})$$

$$\frac{X \twoheadrightarrow Y, Y \twoheadrightarrow Z}{X \twoheadrightarrow YZ} \qquad\qquad \frac{X \twoheadrightarrow Y, W \twoheadrightarrow Z}{X \twoheadrightarrow Y \cap Z}\, Y \cap W = \emptyset$$

$$\text{(additive transitivity, } \mathcal{T}^*) \qquad\qquad \text{(subset, } \mathcal{S})$$

is sound and complete for the implication of MVDs. The major proof argument shows that every inference of an MVD $X \twoheadrightarrow Y$ using the set $R\mathfrak{B}_0 = \{\mathcal{R}_\emptyset, \mathcal{A}, \mathcal{T}, \mathcal{T}^*, \mathcal{S}, \mathcal{C}_R\}$ can be turned into an inference of $X \twoheadrightarrow Y$ which applies the R-complementation rule $\mathcal{C}_R$ at most once, and if it is applied, then it is applied in the last step (a set of inference rules with this property is said to be *R-complementary*). This shows that

$$X \twoheadrightarrow Y \in \Sigma_{R\mathfrak{B}_0}^+ \quad \text{iff} \quad X \twoheadrightarrow Y \in \Sigma_{\mathfrak{B}_0}^+ \text{ or } X \twoheadrightarrow (R - Y) \in \Sigma_{\mathfrak{B}_0}^+ \qquad (2.1)$$

where $\Sigma = \{X_1 \twoheadrightarrow Y_1, \ldots, X_k \twoheadrightarrow Y_k\}$ and $X \cup Y \cup \bigcup_{i=1}^{k}(X_i \cup Y_i) \subseteq R$.

3 Minimal Axiomatisations in Undetermined Universes

In this section, all minimal complete subsets of

$$\mathfrak{S}_U = \{\mathcal{R}, \mathcal{S}, \mathcal{A}, \mathcal{T}, \mathcal{T}^*, \mathcal{U}, \mathcal{D}, \mathcal{I}\}$$

will be revealed. We will start with some independence proofs. These proofs
have been checked using simple GNU Pascal programs (which offer good set
arithmetic) which compute the closure $\Sigma_{\mathfrak{S}}^+$ of a set Σ of MVDs under some set
$\mathfrak{S}$ of inference rules (neglecting trivial MVDs).

Lemma 3.1. *Let* $\mathfrak{S} = \{\mathcal{S}, \mathcal{A}, \mathcal{T}, \mathcal{T}^*, \mathcal{U}, \mathcal{D}, \mathcal{I}\}$. *The reflexivity axiom* $\mathcal{R}$ *is independent from* $\mathfrak{S}$.

Proof. $\mathcal{R}$ is the only axiom. If $\Sigma = \emptyset$ and $\sigma = \emptyset \twoheadrightarrow \emptyset$, then $\sigma \notin \Sigma_{\mathfrak{S}}^+$, but
$\sigma \in \Sigma_{\mathfrak{S} \cup \{\mathcal{R}\}}^+$. □

The subset rule $\mathcal{S}$ is independent from $\{\mathcal{R}, \mathcal{A}, \mathcal{T}, \mathcal{T}^*, \mathcal{U}, \mathcal{D}, \mathcal{I}\}$. This strengthens
the result of Biskup who has shown [11–Theorem 2] that the subset rule is
independent from $\{\mathcal{R}_\emptyset, \mathcal{A}, \mathcal{T}, \mathcal{T}^*\}$.

Lemma 3.2. *Let* $\mathfrak{S} = \{\mathcal{R}, \mathcal{A}, \mathcal{T}, \mathcal{T}^*, \mathcal{U}, \mathcal{D}, \mathcal{I}\}$. *The subset rule* $\mathcal{S}$ *is independent from* $\mathfrak{S}$.

Proof (Sketch). Let $\Sigma = \{A \twoheadrightarrow BC, D \twoheadrightarrow CD\}$, and $\sigma = A \twoheadrightarrow C$. Since $\sigma \notin \Sigma_{\mathfrak{S}}^+$,
but $\sigma \in \Sigma_{\mathfrak{S} \cup \{\mathcal{S}\}}^+$ we have found witnesses Σ and σ for the independence of $\mathcal{S}$
from $\mathfrak{S}$. □

Lemma 3.1 and 3.2 show that any complete subset of $\mathfrak{S}_U$ must include both
reflexivity axiom $\mathcal{R}$ and subset rule $\mathcal{S}$. The following result has been proven in
[22, Lemma 2].

Lemma 3.3. *The augmentation rule* $\mathcal{A}$ *is derivable from* $\{\mathcal{R}, \mathcal{T}, \mathcal{U}\}$. □

Lemma 3.4. *The additive transitivity rule* $\mathcal{T}^*$ *is derivable from the pseudo-transitivity rule* $\mathcal{T}$ *and the union rule* $\mathcal{U}$.

Proof.

$$\mathcal{T}: \cfrac{X \twoheadrightarrow Y \quad Y \twoheadrightarrow Z}{\mathcal{U}: \cfrac{X \twoheadrightarrow Z - Y \quad X \twoheadrightarrow Y}{X \twoheadrightarrow Y \cup Z}}$$

This concludes the proof. □

Recall that $\mathfrak{B}_0 = \langle \mathcal{R}_\emptyset, \mathcal{A}, \mathcal{T}, \mathcal{T}^*, \mathcal{S} \rangle$ is complete. Applying Lemma 3.3 and 3.4 to
$\mathfrak{B}_0$ result in the first new complete set.

Theorem 3.1. *The set* $\mathfrak{L}_1 = \langle \mathcal{R}, \mathcal{S}, \mathcal{T}, \mathcal{U} \rangle$, *consisting of reflexivity axiom, subset rule, pseudo-transitivity rule and union rule, is sound and complete for the implication of MVDs.*

Proof. It suffices to show that the empty-set-axiom $\mathcal{R}_\emptyset$, augmentation rule $\mathcal{A}$ and additive transitivity rule $\mathcal{T}^*$ are derivable from $\mathfrak{L}_1$. $\mathcal{R}_\emptyset$ is a very weak form of the reflexivity axiom $\mathcal{R}$. Lemma 3.3 and 3.4 show that $\mathcal{A}$ and $\mathcal{T}^*$ are derivable from $\mathfrak{L}_1$ as well. This concludes the proof. $\qquad\square$

Lemma 3.5. *The pseudo-transitivity rule $\mathcal{T}$ can be derived from the additive pseudo-transitivity rule $\mathcal{T}^*$ and the difference rule $\mathcal{D}$.*

Proof. Recall that $(Y \cup Z) - Y = Z - Y$.

$$\mathcal{D}: \cfrac{X \twoheadrightarrow Y \quad \mathcal{T}^*: \cfrac{X \twoheadrightarrow Y \quad Y \twoheadrightarrow Z}{X \twoheadrightarrow Y \cup Z}}{X \twoheadrightarrow \underbrace{(Y \cup Z) - Y}_{=Z-Y}}$$

This concludes the proof. $\qquad\square$

Lemma 3.6. *The union rule $\mathcal{U}$ is derivable from the reflexivity axiom $\mathcal{R}$, the additive pseudo-transitivity rule $\mathcal{T}^*$ and the difference rule $\mathcal{D}$.*

Proof. Note that $YZ = XYZ - (X - YZ)$.

$$\mathcal{D}: \cfrac{\mathcal{T}^*: \cfrac{\mathcal{T}^*: \cfrac{\mathcal{R}: \overline{X \twoheadrightarrow X} \quad X \twoheadrightarrow Y}{X \twoheadrightarrow XY} \quad \mathcal{T}^*: \cfrac{\mathcal{R}: \overline{XY \twoheadrightarrow X} \quad X \twoheadrightarrow Z}{XY \twoheadrightarrow XZ}}{X \twoheadrightarrow XYZ} \quad \mathcal{R}: \overline{X \twoheadrightarrow (X - (YZ))}}{X \twoheadrightarrow \underbrace{XYZ - (X - (YZ))}_{=YZ}}$$

This concludes the proof. $\qquad\square$

The next theorem shows that one can achieve completeness without using the pseudo-transitivity rule nor the union rule.

Theorem 3.2. *The set $\mathfrak{L}_2 = \langle \mathcal{R}, \mathcal{S}, \mathcal{T}^*, \mathcal{D} \rangle$, consisting of reflexivity axiom, subset rule, additive pseudo-transitivity rule and difference rule, is sound and complete for the implication of MVDs.*

Proof. The completeness of $\mathfrak{L}_2$ follows from Theorem 3.1, Lemma 3.5 and Lemma 3.6. $\qquad\square$

We will now show that $\mathfrak{L}_1$, $\mathfrak{L}_2$ and $\mathfrak{B} = \langle \mathcal{R}, \mathcal{S}, \mathcal{A}, \mathcal{T}, \mathcal{T}^* \rangle$ are the only minimal complete subsets of $\mathfrak{S}_U$.

Lemma 3.7. *Let $\mathfrak{S} = \{\mathcal{R}, \mathcal{S}, \mathcal{T}, \mathcal{T}^*, \mathcal{I}\}$. The union rule $\mathcal{U}$ is independent from $\mathfrak{S}$.*

Proof (Sketch). Let $\Sigma = \{AB \twoheadrightarrow BC, AB \twoheadrightarrow CD\}$ and $\sigma = AB \twoheadrightarrow BCD$. Since $\sigma \notin \Sigma_{\mathfrak{S}}^+$, but $\sigma \in \Sigma_{\mathfrak{S} \cup \{\mathcal{U}\}}^+$ we have found witnesses Σ and σ for the independence of $\mathcal{U}$ from $\mathfrak{S}$. $\qquad\square$

For the next lemma note that $\mathfrak{L}_2 = \langle \mathcal{R}, \mathcal{S}, \mathcal{T}^*, \mathcal{D} \rangle$ is complete, i.e., the pseudo-transitivity rule $\mathcal{T}$ is derivable from any of its supersets, in particular.

Lemma 3.8.

1. *The pseudo-transitivity rule $\mathcal{T}$ is independent from $\mathfrak{S}_1 = \{\mathcal{R}, \mathcal{S}, \mathcal{A}, \mathcal{T}^*, \mathcal{U}, \mathcal{I}\}$.*
2. *The pseudo-transitivity rule $\mathcal{T}$ is independent from $\mathfrak{S}_2 = \{\mathcal{R}, \mathcal{S}, \mathcal{A}, \mathcal{U}, \mathcal{D}, \mathcal{I}\}$.*

Proof (Sketch). For the first statement let $\Sigma = \{A \twoheadrightarrow B, B \twoheadrightarrow C\}$ and $\sigma = A \twoheadrightarrow C$. Since $\sigma \notin \Sigma^+_{\mathfrak{S}_1}$, but $\sigma \in \Sigma^+_{\mathfrak{S}_1 \cup \{\mathcal{T}\}}$ we have found witnesses Σ and σ for the independence of $\mathcal{T}$ from $\mathfrak{S}_1$.

For the second statement let $\Sigma = \{A \twoheadrightarrow B, B \twoheadrightarrow C\}$ and $\sigma = A \twoheadrightarrow C$. Since $\sigma \notin \Sigma^+_{\mathfrak{S}_2}$, but $\sigma \in \Sigma^+_{\mathfrak{S}_2 \cup \{\mathcal{T}\}}$ we have found witnesses Σ and σ for the independence of $\mathcal{T}$ from $\mathfrak{S}_2$.   □

Lemma 3.9. *Let $\mathfrak{S} = \{\mathcal{R}, \mathcal{S}, \mathcal{A}, \mathcal{T}, \mathcal{D}, \mathcal{I}\}$. The additive transitivity rule $\mathcal{T}^*$ is independent from $\mathfrak{S}$.*

Proof (Sketch). Let $\Sigma = \{A \twoheadrightarrow B, B \twoheadrightarrow C\}$ and $\sigma = A \twoheadrightarrow BC$. Since $\sigma \notin \Sigma^+_{\mathfrak{S}}$, but $\sigma \in \Sigma^+_{\mathfrak{S} \cup \{\mathcal{T}^*\}}$ we have found witnesses Σ and σ for the independence of $\mathcal{T}^*$ from $\mathfrak{S}$.   □

Theorem 3.3. *The only minimal, sound and complete subsets of $\mathfrak{S}_U$ for the implication of MVDs are $\mathfrak{B}$, $\mathfrak{L}_1$ and $\mathfrak{L}_2$.*

Proof (Sketch). Lemma 3.1 and 3.2 show that $\mathcal{R}$ and $\mathcal{S}$ must be part of any complete subset of $\mathfrak{S}_U$. For a complete subset $\mathfrak{S} \subseteq \mathfrak{S}_U$ every inference rule in $\mathfrak{S}_U - \mathfrak{S}$ must be derivable from $\mathfrak{S}$. We consider every subset $\mathfrak{S}$ of $\mathfrak{S}_U$ that includes at least $\mathcal{R}$ and $\mathcal{S}$. If $\mathfrak{S}$ is not a superset of $\mathfrak{B}$, $\mathfrak{L}_1$ or $\mathfrak{L}_2$, then at least one of Lemma 3.7 or 3.8 or 3.9 shows that there is some inference rule in $\mathfrak{S}_U - \mathfrak{S}$ that is independent from $\mathfrak{S}$, i.e., $\mathfrak{S}$ cannot be complete.

It follows from Theorems 3.1 and 3.2 as well as [11] that $\mathfrak{B}$, $\mathfrak{L}_1$ and $\mathfrak{L}_2$ are complete. The minimality of all three sets follows from the fact that no proper subset of $\mathfrak{B}$, or $\mathfrak{L}_1$ or $\mathfrak{L}_2$, respectively, is still complete.   □

4   Weakening Reflexivity

Although the set $\mathfrak{B} = \langle \mathcal{R}, \mathcal{S}, \mathcal{A}, \mathcal{T}, \mathcal{T}^* \rangle$ is minimal, the reflexivity axiom $\mathcal{R}$ can be replaced by the much weaker empty-set-axiom $\mathcal{R}_\emptyset$ resulting in the complete set $\mathfrak{B}_0$. This is because $\mathcal{R}$ is derivable from $\mathcal{R}_\emptyset$ and $\mathcal{A}$.

Instead of using the reflexivity axiom $\mathcal{R}$ within $\mathfrak{L}_1$ we may also use the empty-set-axiom $\mathcal{R}_\emptyset$ together with the membership-axiom $\mathcal{M}$ which is $\dfrac{}{X \twoheadrightarrow A} A \in X$.

Theorem 4.1. *The set $\mathfrak{L}_3 = \langle \mathcal{R}_\emptyset, \mathcal{M}, \mathcal{S}, \mathcal{T}, \mathcal{U} \rangle$, consisting of empty-set-axiom, membership-axiom, subset rule, pseudo-transitivity rule and union rule, is sound and complete for the implication of MVDs.*

Proof. We show that the reflexivity axiom $\mathcal{R}$ is derivable from $\{\mathcal{R}_\emptyset, \mathcal{M}, \mathcal{T}, \mathcal{U}\}$. If $X = \emptyset$, then the only instance of the reflexivity axiom is the empty-set-axiom $\mathcal{R}_\emptyset$. We may therefore assume that $X \neq \emptyset$. We proceed by induction on the number n of attributes in Y. If $n = 0$, then we obtain the following inference:

$$\mathcal{T} : \frac{\mathcal{M} : \overline{X \twoheadrightarrow A}^{A \in X} \quad \mathcal{M} : \overline{A \twoheadrightarrow A}^{A \in \{A\}}}{X \twoheadrightarrow \emptyset}.$$

Suppose $Y = \{A_1, \ldots, A_n, A_{n+1}\}$. Note that $\{A_1, \ldots, A_n\} \subseteq X$ and $A_{n+1} \in X$ as $Y \subseteq X$. We then have the following inference

$$\mathcal{U} : \frac{\mathcal{R}_{\text{hypothesis}} : \overline{X \twoheadrightarrow \{A_1, \ldots, A_n\}}^{\{A_1, \ldots, A_n\} \subseteq X} \quad \mathcal{M} : \overline{X \twoheadrightarrow A_{n+1}}^{A_{n+1} \in X}}{X \twoheadrightarrow Y}.$$

This completes the proof. □

We may also replace $\mathcal{R}$ in $\mathfrak{L}_2$ by $\mathcal{R}_\emptyset$ and $\mathcal{M}$.

Theorem 4.2. *The set* $\mathfrak{L}_4 = \langle \mathcal{R}_\emptyset, \mathcal{M}, \mathcal{T}^*, \mathcal{D}, \mathcal{U}, \mathcal{S} \rangle$, *which consists of empty-set-axiom, membership axiom, additive pseudo-transitivity rule, difference rule, union rule, and subset rule, is sound and complete for the implication of MVDs.*

Proof. Theorem 3.1 and Lemma 3.5 show that $\langle \mathcal{R}, \mathcal{S}, \mathcal{T}^*, \mathcal{D}, \mathcal{U} \rangle$ is complete. The proof of Theorem 4.1 shows then that $\mathcal{R}$ can be replaced by $\mathcal{R}_\emptyset$ and $\mathcal{M}$ still maintaining completeness. However, while the union rule $\mathcal{U}$ is derivable from $\{\mathcal{R}, \mathcal{T}^*, \mathcal{D}\}$, it is independent from $\mathfrak{S} = \{\mathcal{R}_\emptyset, \mathcal{M}, \mathcal{T}^*, \mathcal{D}, \mathcal{S}\}$. Namely, let $\Sigma = \emptyset$ and $\sigma = AB \twoheadrightarrow AB$. Since $\sigma \notin \Sigma_\mathfrak{S}^+$, but $\sigma \in \Sigma_{\mathfrak{S} \cup \{\mathcal{U}\}}^+$ we have found witnesses Σ and σ for the independence of $\mathcal{U}$ from $\mathfrak{S}$. □

5   Axiomatisations for Fixed Universes

Let $\mathfrak{S}$ be a set of inference rules that is sound for the implication of MVDs. Let $R\mathfrak{S}$ denote the set $\mathfrak{S} \cup \{\mathcal{C}_R\}$ which is therefore sound for R-implication. The set $R\mathfrak{L}_1$ is complete for the R-implication of MVDs which follows immediately from the completeness of $R\mathfrak{M}$ [29]. In fact, something much stronger holds. We call a set $R\mathfrak{S}$ *complementary* iff it is R-complementary for the R-implication of MVDs for all relation schemata R.

Theorem 5.1. *Let* $\mathfrak{S}$ *be a sound set of inference rules for the implication of MVDs. The set* $\mathfrak{S}$ *is complete for the implication of MVDs if and only if the set* $R\mathfrak{S} = \mathfrak{S} \cup \{\mathcal{C}_R\}$ *is complete and complementary for the R-implication of MVDs.*

Proof. We show first that if $\mathfrak{S}$ is complete for the implication of MVDs, then for each relation schema R the set $R\mathfrak{S} = \mathfrak{S} \cup \{\mathcal{C}_R\}$ is both R-complete and R-complementary for the R-implication of MVDs.

Let R be arbitrary. We know that $R\mathfrak{B}_0$ is R-complete, i.e., $\Sigma_R^* \subseteq \Sigma_{R\mathfrak{B}_0}^+$. Moreover, $\mathfrak{S}$ and $\mathfrak{B}_0$ are both sound and complete, i.e., $\Sigma_{\mathfrak{B}_0}^+ = \Sigma^* = \Sigma_\mathfrak{S}^+$. Let

$X, Y \subseteq R$ and $X \twoheadrightarrow Y \in \Sigma_R^*$. Since $R\mathfrak{B}_0$ is R-complete it follows that $X \twoheadrightarrow Y \in \Sigma_{R\mathfrak{B}_0}^+$. Equation (2.1) shows that $X \twoheadrightarrow Y \in \Sigma_{\mathfrak{B}_0}^+$ or $X \twoheadrightarrow (R - Y) \in \Sigma_{\mathfrak{B}_0}^+$ holds. Since $\Sigma_{\mathfrak{B}_0}^+ = \Sigma_{\mathfrak{G}}^+$ this is equivalent to $X \twoheadrightarrow Y \in \Sigma_{\mathfrak{G}}^+$ or $X \twoheadrightarrow (R - Y) \in \Sigma_{\mathfrak{G}}^+$. However, $Y = R - (R - Y)$ and therefore $X \twoheadrightarrow Y \in \Sigma_{R\mathfrak{G}}^+$. This shows that $\Sigma_R^* \subseteq \Sigma_{R\mathfrak{G}}^+$, i.e., $R\mathfrak{G}$ is R-complete. Moreover, $\Sigma_{R\mathfrak{G}}^+ = \Sigma_{R\mathfrak{B}_0}^+$ and equation (2.1) implies that

$$X \twoheadrightarrow Y \in \Sigma_{R\mathfrak{G}}^+ \quad \text{if and only if} \quad X \twoheadrightarrow Y \in \Sigma_{\mathfrak{G}}^+ \text{ or } X \twoheadrightarrow (R - Y) \in \Sigma_{\mathfrak{G}}^+$$

whenever $X, Y \subseteq R$ and Σ is a set of MVDs on R. That is, every inference of an MVD $X \twoheadrightarrow Y$ using $R\mathfrak{G}$ can be turned into an inference of $X \twoheadrightarrow Y$ in which the R-complementation rule is applied at most once, and if it is applied, then as the last rule of the inference. Since R was arbitrary the set $R\mathfrak{G}$ is R-complete and R-complementary. That is, $R\mathfrak{G}$ is complete and complementary for the R-implication of MVDs.

It remains to show that $\mathfrak{G}$ is complete for the implication of MVDs whenever for all relation schemata R the set $R\mathfrak{G} = \mathfrak{G} \cup \{\mathcal{C}_R\}$ is both R-complete and R-complementary for the R-implication of MVDs. We need to show that $\Sigma^* \subseteq \Sigma_{\mathfrak{G}}^+$ holds for every finite set Σ of MVDs. Let $\Sigma = \{X_1 \twoheadrightarrow Y_1, \ldots, X_k \twoheadrightarrow Y_k\}$ and $X \twoheadrightarrow Y \in \Sigma^* = \Sigma_{\mathfrak{B}_0}^+$. Let $T := X \cup Y \cup \bigcup_{i=1}^{k}(X_i \cup Y_i)$ and R be some relation schema such that T is properly contained in R, i.e., $T \subset R$. Fact 1 in [11] shows that $X \twoheadrightarrow Y \in \Sigma_R^*$. The R-completeness of $R\mathfrak{G}$ implies further that $X \twoheadrightarrow Y \in \Sigma_{R\mathfrak{G}}^+$. Since $R\mathfrak{G}$ is also R-complementary we must have $X \twoheadrightarrow Y \in \Sigma_{\mathfrak{G}}^+$ or $X \twoheadrightarrow (R - Y) \in \Sigma_{\mathfrak{G}}^+$. Assume that $X \twoheadrightarrow (R - Y) \in \Sigma_{\mathfrak{G}}^+$. The soundness of $\mathfrak{G}$ implies that $X \twoheadrightarrow (R - Y) \in \Sigma^* = \Sigma_{\mathfrak{B}_0}^+$. Furthermore, the derivability of the union rule from $\mathfrak{B}_0$ implies that $X \twoheadrightarrow R \in \Sigma_{\mathfrak{B}_0}^+$. However, we obtain the contradiction $R \subseteq T \subset R$ by Lemma 5 of [11]. Consequently, $X \twoheadrightarrow Y \in \Sigma_{\mathfrak{G}}^+$ must hold, and this shows the completeness of $\mathfrak{G}$.   □

Corollary 5.1. *The sets $R\mathfrak{L}_1, R\mathfrak{L}_2, R\mathfrak{L}_3, R\mathfrak{L}_4$ are sound, complete and complementary for the R-implication of MVDs.*   □

If $\mathfrak{G}$ is a minimal, sound and complete set of inference rules for the implication of MVDs, then it is not necessarily true that the set $R\mathfrak{G}$ is minimal for the R-implication of MVDs. In fact, the next theorem indicates that complete and complementary sets $R\mathfrak{G}$ may not be minimal.

Theorem 5.2. *The complete sets $R\mathfrak{B}$, $R\mathfrak{L}_1$, $R\mathfrak{L}_2$, $R\mathfrak{L}_3$, and $R\mathfrak{L}_4$ are not minimal for the R-implication of MVDs.*

Proof (Sketch). $R\mathfrak{B}$ and $R\mathfrak{L}_1$ are not minimal as $\mathcal{S}$ is derivable from $\{\mathcal{C}_R, \mathcal{A}, \mathcal{T}\}$ according to [11, Theorem 1]. $R\mathfrak{L}_2$ is not minimal since $\mathcal{A}$ is derivable from $\{\mathcal{R}, \mathcal{T}^*, \mathcal{D}\}$. $R\mathfrak{L}_3$ is not minimal since $\mathcal{S}$ is derivable from $\{\mathcal{C}_R, \mathcal{R}_\emptyset, \mathcal{M}, \mathcal{T}, \mathcal{U}\}$. $R\mathfrak{L}_4$ is not minimal since $\mathcal{D}$ is derivable from $\{\mathcal{C}_R, \mathcal{U}\}$.   □

In particular, in order to make the minimal set $R\mathfrak{M}$ also complementary one may add the subset rule $\mathcal{S}$ and the union rule $\mathcal{U}$ in order to obtain $R\mathfrak{L}_1$.

Corollary 5.2. *Let $\mathfrak{S}$ be a sound set of inference rules for the implication of MVDs. The set $\mathfrak{S}$ is minimal and complete for the implication of MVDs if and only if the set $R\mathfrak{S}$ is complete and complementary for the R-implication of MVDs, and there is no inference rule $\mathfrak{R} \in \mathfrak{S}$ such that $R(\mathfrak{S} - \{\mathfrak{R}\})$ is still both complete and complementary for the R-implication of MVDs.* □

The last corollary helps us finding all subsets of $R\mathfrak{S}_U$ that are complete and complementary for the R-implication of MVDs. The next lemma is an extension of [29, Lemma 1].

Lemma 5.1. *The R-complementation rule $\mathcal{C}_R$ is independent from $\mathfrak{S}_U$.*

Proof (Sketch). Let $R = A$, $\Sigma = \emptyset$ and $\sigma = \emptyset \twoheadrightarrow A$. Since $\sigma \notin \Sigma^+_{\mathfrak{S}_U}$, but $\sigma \in \Sigma^+_{\mathfrak{S}_U \cup \{\mathcal{C}_R\}}$ we have found witnesses R, Σ and σ for the independence of $\mathcal{C}_R$ from $\mathfrak{S}_U$. □

The next corollary is a consequence of Theorem 3.3, Corollary 5.2 and Lemma 5.1.

Corollary 5.3. *There are no proper subsets of $R\mathfrak{B}, R\mathfrak{L}_1$ and $R\mathfrak{L}_2$ which are both complete and complementary for the R-implication of MVDs.* □

6 The Implication Problem

The R-implication problem for MVDs is to decide whether an arbitrary set Σ of MVDs R-implies a single MVD σ in the sense of Definition 2.1. This problem has been well-studied [4, 15, 19, 20, 25, 26, 30, 31, 35]. The fundamental notion is that of a dependency basis for an attribute set $X \subseteq R$ with respect to a set Σ of MVDs [4]. Given some set $\Sigma = \{X_1 \twoheadrightarrow Y_1, \ldots, X_k \twoheadrightarrow Y_k\}$ of MVDs with $\bigcup_{i=1}^{k} (X_i \cup Y_i) \subseteq R$ the set $Dep_R(X) = \{Y \mid X \twoheadrightarrow Y \in \Sigma^+_{R\mathfrak{B}_0}\}$ consists of all those attribute sets $Y \subseteq R$ such that $X \twoheadrightarrow Y$ is derivable from Σ using some R-sound and R-complete set of inference rules for the R-implication of MVDs, in this case $R\mathfrak{B}_0$. The structure $(Dep_R(X), \subseteq, \cup, \cap, -, \emptyset, R)$ is a finite Boolean powerset algebra due to the derivability of union, intersection and difference rule from $R\mathfrak{B}_0$. Recall that an element $a \in P$ of a poset $(P, \sqsubseteq, 0)$ with least element 0 is called an *atom* of $(P, \sqsubseteq, 0)$ [9] if and only if every element $b \in P$ with $b \sqsubseteq a$ satisfies $b = 0$ or $b = a$. $(P, \sqsubseteq, 0)$ is called *atomic* if and only if for every element $b \in P$ with $b \neq 0$ there is an atom $a \in P$ with $a \sqsubseteq b$. In particular, every finite Boolean algebra is atomic. The set $DepB_R(X)$ of all atoms of $(Dep_R(X), \subseteq, \emptyset)$ is called the dependency basis of X with respect to Σ.

We will now study the implication problem for MVDs. The problem is to decide whether an arbitrary finite set Σ of MVDs implies a single MVD σ in the sense of Definition 2.2. Therefore, the following definition introduces the notion of a dependency basis for undetermined universes.

Definition 6.1. *Let* $\Sigma = \{X_1 \twoheadrightarrow Y_1, \ldots, X_k \twoheadrightarrow Y_k\}$ *be a set of MVDs, and* $X \subseteq \mathfrak{A}$ *some attribute set. Let* $Dep_U(X) = \{Y \mid X \twoheadrightarrow Y \in \Sigma^+_{\mathfrak{B}_0}\}$ *be the set of all attribute sets* Y *such that* $X \twoheadrightarrow Y$ *is derivable from* Σ *using* $\mathfrak{B}_0$. *The dependency basis* $DepB_U(X)$ *of* X *with respect to* Σ *is the set of all atoms of* $(Dep_U(X), \subseteq, \emptyset)$. $\square$

Note that $Dep_U(X)$ is invariant under different choices of sound and complete sets of inference rules for the implication of MVDs. More precisely, if $\mathfrak{S}_1$ and $\mathfrak{S}_2$ are two different sound and complete sets of inference rules for the implication of MVDs, then $\{Y \mid X \twoheadrightarrow Y \in \Sigma^+_{\mathfrak{S}_1}\} = \{Y \mid X \twoheadrightarrow Y \in \Sigma^+_{\mathfrak{S}_2}\}$. We will now introduce the notion of the Σ-scope for an attribute set X.

Definition 6.2. *Let* Σ *be a finite set of MVDs and* $X \subseteq \mathfrak{A}$ *some set of attributes. The set* $X^S = \bigcup Dep_U(X)$ *is called the scope of* X *with respect to* Σ *(or short* Σ-*scope of* X*).* $\square$

If $\Sigma = \{X_1 \twoheadrightarrow Y_1, \ldots, X_k \twoheadrightarrow Y_k\}$, then $X^S \subseteq R_{\min} := X \cup \bigcup\limits_{i=1}^{k} (X_i \cup Y_i)$ according to [11, Lemma 5]. It follows immediately that $X^S \in Dep_U(X)$ since the union rule is derivable from any complete set of inference rules for the implication of MVDs. Moreover, if $Y \in Dep_U(X)$, then $Y \subseteq X^S$, i.e., X^S is the maximal element of $Dep_U(X)$ with respect to $\subseteq$.

In fact, union, intersection and difference rule are all derivable from any complete set of inference rules. Therefore, $(Dep_U(X), \subseteq, \cup, \cap, -, \emptyset, X^S)$ is again a finite Boolean algebra, with top-element X^S. The existence and uniqueness of $DepB_U(X)$ follow from the fact that every finite Boolean algebra is atomic.

The notion of a dependency basis gains its importance from the fact that $X \twoheadrightarrow Y$ is logically R-implied by Σ if and only if Y is the union of some sets of $DepB_R(X)$ [4]. In order to solve the R-implication problem, it is therefore sufficient to find an algorithm for computing the dependency basis. The following theorem shows that the same is true for $DepB_U(X)$, i.e., for implication in undetermined universes.

Theorem 6.1. *Let* Σ *be a finite set of MVDs. Then* $X \twoheadrightarrow Y \in \Sigma^+_{\mathfrak{B}_0}$ *if and only if* $Y = \bigcup \mathcal{Y}$ *for some* $\mathcal{Y} \subseteq DepB_U(X)$.

Proof. If $X \twoheadrightarrow Y \in \Sigma^+_{\mathfrak{B}_0}$, then $Y \in Dep_U(X)$. That means $Y = \bigcup \mathcal{Y}$ for some $\mathcal{Y} \subseteq DepB_U(X)$ since $DepB_U(X)$ consists of all atoms of $(Dep_U(X), \subseteq, \emptyset)$.

Vice versa, if $Y = \bigcup \mathcal{Y}$ for some $\mathcal{Y} \subseteq DepB_U(X)$, then $Y \in Dep_U(X)$ according to the derivability of the union rule from $\mathfrak{B}_0$. It follows that $X \twoheadrightarrow Y \in \Sigma^+_{\mathfrak{B}_0}$ holds. $\square$

Considerable effort has been devoted to finding fast algorithms that compute $DepB_R(X)$ given X and given Σ, see for instance [4, 15, 19, 20, 25, 26, 30, 31, 35]. Currently, the best upper bound for solving $\Sigma \models X \twoheadrightarrow Y$ is $\mathcal{O}((1 + \min\{s, \log p\}) \cdot n)$ from [19] where s denotes the number of dependencies in Σ, p the number of sets in $DepB_R(X)$ that have non-empty intersection with Y and n denotes the total number of occurrences of attributes in Σ.

We will show now that an extension of any algorithm that computes $DepB_R(X)$ for any R with $R_{\min} \subseteq R$ can be used to compute $DepB_U(X)$. The followin theorem shows that $DepB_U(X)$ and $DepB_R(X)$ deviate in at most one element, namely $R - X^S$. Intuitively, that makes perfect sense since X^S is $\subseteq$-maximal among the attribute sets Y with $X \twoheadrightarrow Y \in \Sigma^+_{\mathfrak{B}_0}$ and, given that $X^S \subset R$, the R-complement $R - X^S$ of X^S is an atom of $(Dep_R(X), \subseteq, \emptyset)$.

Theorem 6.2. *Let Σ be a finite set of MVDs, $X \subseteq \mathfrak{A}$ some attribute set, and R some relation schema with $R_{min} \subseteq R$. Then*

$$DepB_U(X) = \begin{cases} DepB_R(X) & , \text{if } X^S = R \\ DepB_R(X) - \{R - X^S\} & , \text{if } X^S \subset R \end{cases} .$$

Proof. Let $X^S = R$, i.e., in particular $X^S = R_{\min}$. We show that $Dep_U(X) = Dep_R(X)$ holds.

For $Y \in Dep_U(X)$ follows $X \twoheadrightarrow Y \in \Sigma^+_{\mathfrak{B}_0}$ and therefore also $X \twoheadrightarrow Y \in \Sigma^+_{R\mathfrak{B}_0}$ since $Y \subseteq X^S \subseteq R$. This, however, means that $Y \in Dep_R(X)$.

If $Y \in Dep_R(X)$, then $X \twoheadrightarrow Y \in \Sigma^+_{R\mathfrak{B}_0}$. Consequently, $X \twoheadrightarrow Y \in \Sigma^+_{\mathfrak{B}_0}$ or $X \twoheadrightarrow (R - Y) \in \Sigma^+_{\mathfrak{B}_0}$ by equation (2.1). Since $X^S = R$ we have in the latter case that $X \twoheadrightarrow (X^S - Y) \in \Sigma^+_{\mathfrak{B}_0}$. According to the derivability of the union rule from $\mathfrak{B}_0$ we have $X \twoheadrightarrow X^S \in \Sigma^+_{\mathfrak{B}_0}$, and due to the derivability of the difference rule from $\mathfrak{B}_0$ we also know that $X \twoheadrightarrow (X^S - (X^S - Y)) \in \Sigma^+_{\mathfrak{B}_0}$ holds. Since, $(X^S - (X^S - Y)) = Y$, it follows that $Y \in Dep_U(X)$. Therefore, $Dep_U(X) = Dep_R(X)$ holds indeed and this shows that $DepB_U(X) = DepB_R(X)$ whenever $X^S = R$.

Let now be $X^S \subset R$, i.e., $R - X^S \neq \emptyset$. We show first that $R - X^S \in DepB_R(X)$. First, $X^S \in Dep_R(X)$ by derivability of the union rule from $R\mathfrak{B}_0$. Then $R - X^S \in Dep_R(X)$ by application of the R-complementation rule from $R\mathfrak{B}_0$. Suppose there is some $Y \in Dep_R(X)$ with $Y \subseteq (R - X^S)$. We know again by equation (2.1) that $Y \in Dep_U(X)$ or $(R - Y) \in Dep_U(X)$ holds. In the first case we have $Y \subseteq X^S$ and therefore $Y \subseteq X^S \cap (R - X^S)$, i.e., $Y = \emptyset$. In the latter case we have $(R - Y) \subseteq X^S$, i.e., $(R - X^S) \subseteq Y$, i.e., $Y = (R - X^S)$. We have shown that every $Y \in Dep_R(X)$ with $Y \subseteq (R - X^S)$ satisfies $Y = \emptyset$ or $Y = R - X^S$. That means $R - X^S$ is an atom of $(Dep_R(X), \subseteq, \emptyset)$, i.e., $R - X^S \in DepB_R(X)$.

We show now that $DepB_U(X) \subseteq DepB_R(X)$. If $Y \in DepB_U(X)$, then $X \twoheadrightarrow Y \in \Sigma^+_{\mathfrak{B}_0}$ in particular. Consequently, $X \twoheadrightarrow Y \in \Sigma^+_{R\mathfrak{B}_0}$. Suppose there is some Z with $\emptyset \neq Z \subset Y$ and $X \twoheadrightarrow Z \in \Sigma^+_{R\mathfrak{B}_0}$. Then either $X \twoheadrightarrow Z \in \Sigma^+_{\mathfrak{B}_0}$ which contradicts the assumption that Y is an atom of $(Dep_U(X), \subseteq, \emptyset)$, or $X \twoheadrightarrow (R - Z) \in \Sigma^+_{\mathfrak{B}_0}$. Consequently, $R - Z \subseteq X^S$ and since $R - Y \subseteq R - Z$ we have $R - Y \subseteq X^S$, too. However, from $Y \subseteq X^S$ follows then that $Y \cup (R - Y) \subseteq X^S$, i.e., $R = X^S$, a contradiction to our assumption. Therefore, $Y \in DepB_R(X)$. So far, we have shown that $DepB_U(X) \cup \{R - X^S\} \subseteq DepB_R(X)$.

As $R - X^S \notin DepB_U(X)$, it remains to show that $DepB_R(X) \subseteq DepB_U(X) \cup \{R - X^S\}$ holds as well. Let Y be some atom of $(Dep_R(X), \subseteq, \emptyset)$. In particular, $Y \in Dep_U(X)$ or $(R - X^S) \subseteq Y$. In the first case, Y must also be an atom of $Dep_U(X)$ since $Dep_U(X) \subseteq Dep_R(X)$. In the second case assume that $Y \neq R - X^S$. Then we have $R - X^S \neq \emptyset$ and $R - X^S \subset Y$. However, $X \twoheadrightarrow (R - X^S) \in \Sigma^+_{R\mathfrak{B}_0}$ since $X \twoheadrightarrow X^S \in \Sigma^+_{\mathfrak{B}_0}$ and the R-complementation rule is in $R\mathfrak{B}_0$. This contradicts our assumption that Y is an atom of $Dep_R(X)$. Hence, $Y = R - X^S$. We have therefore shown that every atom of $Dep_R(X)$ is either an atom of $Dep_U(X)$ or equals $R - X^S$.   □

The following example illustrates Theorem 6.2.

Example 6.1. Suppose $\Sigma = \{A \twoheadrightarrow BC\}$ and the underlying relation schema $R = \{A, B, C, D, E\}$. The Boolean algebra of right-hand sides Y of those MVDs $A \twoheadrightarrow Y$ which are R-implied by Σ is illustrated in Figure 1. The dependency basis $DepB_R(A) = \{A, BC, DE\}$ consists of the atoms of the algebra. Since $A^S = ABC$ we obtain $DepB_U(A) = \{A, BC\}$ and the Boolean algebra of right-hand sides Y of those MVDs $A \twoheadrightarrow Y$ which are implied by Σ is printed boldly in Figure 1.   □

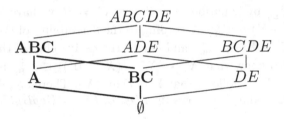

Fig. 1. Dependency Basis in fixed and undetermined Universe

Theorem 6.2 provides us with a strategy for computing the dependency basis $DepB_U(X)$ of X with respect to $\Sigma = \{X_1 \twoheadrightarrow Y_1, \ldots, X_k \twoheadrightarrow Y_k\}$. In fact, one may use any algorithm for computing $DepB_{R_{\min}}(X)$ with respect to Σ. Having computed the Σ-scope X^S of X, one removes $R_{\min} - X^S$ from $DepB_{R_{\min}}(X)$ if and only if $X^S \subset R_{\min}$. Therefore, it remains to compute the Σ-scope X^S.

Recall from [5] the notion of a closure $\overline{X} = \bigcup\{Y \mid X \rightarrow Y \in \Sigma^+\}$ of an attribute set X with respect to a set Σ of functional dependencies. In order to compute $\overline{X}$ all FDs $Y \rightarrow Z \in \Sigma$ are inspected, and whenever $Y \subseteq \overline{X}$, then $\overline{X}$ is replaced by $\overline{X} \cup Z$. This process is repeated until no further attributes have been added to $\overline{X}$ after inspecting all FD in Σ. The correctness of this algorithm is due to the soundness of the following inference rule: if $Y \rightarrow Z \in \Sigma^+$ and $Y \subseteq X$, then $X \rightarrow X \cup Z \in \Sigma^+$.

The definition of $\overline{X}$ is very similar to that of the Σ-scope X^S. In fact, if $Y \twoheadrightarrow Z \in \Sigma^+$ and $Y \subseteq X$, then we also have $X \twoheadrightarrow X \cup Y \in \Sigma^+$ according to the soundness of the reflexivity rule, additive transitivity rule and union rule. The idea is therefore to compute the Σ-scope X^S of X by employing essentially the linear-time algorithm from [5] for computing the closure of X with respect to a set of FDs.

Algorithm 1 (Dependency Basis)

Input: $\Sigma = \{X_1 \twoheadrightarrow Y_1, \ldots, X_k \twoheadrightarrow Y_k\}$, and a set X of attributes

Output: $DepB_U(X)$ with respect to Σ

Method:

VAR $R_{\min}, X_{\text{new}}^S, X_{\text{old}}^S, X_{\text{alg}}^S$: Set of attributes; *MVDList*: List of MVDs;

(1) $R_{\min} := X \cup \bigcup\limits_{i=1}^{k}(X_i \cup Y_i)$;

(2) Use the Algorithm in [19] to compute $DepB_{R_{\min}}(X)$;

(3) $X_{\text{new}}^S := X$;

(4) *MVDList*:= List of MVDs in Σ;

(5) REPEAT

(6) $X_{\text{old}}^S := X_{\text{new}}^S$;

(7)   Remove all attributes in X_{new}^S from the LHS of all MVDs in *MVDList*;

(8)   FOR all MVDs $\emptyset \twoheadrightarrow Y$ in *MVDList* LET $X_{\text{new}}^S := X_{\text{new}}^S \cup Y$;

(9) UNTIL $X_{\text{new}}^S = X_{\text{old}}^S$;

(10) $X_{\text{alg}}^S := X_{\text{new}}^S$;

(11) IF $X_{\text{alg}}^S = R_{\min}$ THEN RETURN($DepB_{R_{\min}}(X)$)

(12)   ELSE RETURN($DepB_{R_{\min}}(X) - \{R_{\min} - X^S\}$);   □

The next theorem verifies essentially the correctness of Algorithm 1 in computing the Σ-closure X^S of X (i.e. the part between line (4) and line (10)).

Theorem 6.3. *Algorithm 1 computes $DepB_U(X)$ with respect to Σ in time $\mathcal{O}((1 + \min\{s, \log \bar{p}\}) \cdot n)$ where s denotes the number of dependencies in Σ, $\bar{p}$ the number of sets in $DepB_U(X)$ and n denotes the total number of occurrences of attributes in Σ.*

Proof. If we show that the part between line (4) and line (10) of Algorithm 1 computes indeed the scope of X with respect to Σ, i.e. $X_{\text{alg}}^S = X^S$, then the correctness of Algorithm 1 follows from the correctness of the algorithm in [19] and Theorem 6.2.

In order to show that $X_{\text{alg}}^S = X^S$ we prove first that $X \twoheadrightarrow X_{\text{alg}}^S \in \Sigma^+$, i.e., $X_{\text{alg}}^S \subseteq X^S$ holds. We proceed by induction on the number j of runs through the REPEAT loop between line (5) and (9). If $j = 0$, then $X_{\text{new}}^S = X$ and $X \twoheadrightarrow X \in \Sigma^+$ by reflexivity. For $j > 0$ we assume that $X \twoheadrightarrow X_{\text{new}}^S \in \Sigma^+$ for X_{new}^S after the jth run through the REPEAT loop. Suppose there is some MVD $Z \twoheadrightarrow Y \in \Sigma$ such that $Z - X_{\text{new}}^S = \emptyset$, i.e., $Z \subseteq X_{\text{new}}^S$ (otherwise there is nothing to show). By derivability of the augmentation rule we infer $X_{\text{new}}^S \twoheadrightarrow Y \in \Sigma^+$. An application of the additive transitivity rule shows that $X \twoheadrightarrow X_{\text{new}}^S \cup Y \in \Sigma^+$. Therefore, $X \twoheadrightarrow X_{\text{new}}^S \in \Sigma^+$ holds also after the $j + 1$st run through the REPEAT loop.

It remains to show that $X^S \subseteq X_{\text{alg}}^S$. Consider the chain $\Sigma = \Sigma_0 \subset \Sigma_1 \subset \cdots \subset \Sigma_k = \Sigma^+$ where Σ_{i+1} is generated from Σ_i by an application of a single inference rule from, say, the complete set $\mathfrak{L}_1$. We show by induction on i that if $U \twoheadrightarrow V \in \Sigma_i$ and $U \subseteq X_{\text{alg}}^S$, then also $V \subseteq X_{\text{alg}}^S$.   (6.2)

For $i = k$ it follows that if $U \twoheadrightarrow V \in \Sigma^+$ and $U \subseteq X_{\text{alg}}^S$, then also $V \subseteq X_{\text{alg}}^S$. The proof concludes for $U = X$ and $V = X^S$ since $X \twoheadrightarrow X^S \in \Sigma^+$ by the derivability of the union rule. It remains to show (6.2). For $i = 0$ we have $U \twoheadrightarrow V \in \Sigma$ and $U \subseteq X_{\text{alg}}^S$ means that $U \subseteq X_{\text{new}}^S$ at some point in time. This implies that $\emptyset \twoheadrightarrow V$ occurs in the MVDList after line (7). Consequently, after line (8) we have $V \subseteq X_{\text{new}}^S \subseteq X_{\text{alg}}^S$ as well. If $i > 0$, then $\Sigma_{i+1} - \Sigma_i$ contains exactly one $U \twoheadrightarrow V$. There is nothing to show for the MVDs in Σ_i (hypothesis). Thus, it suffices to consider $U \twoheadrightarrow V$, and we distinguish between 4 different cases.

1. If $U \twoheadrightarrow V$ results from an application of the reflexivity axiom, then $V \subseteq U \subseteq X_{\text{alg}}^S$.
2. If $U \twoheadrightarrow V$ results from an application of the subset rule, then $U \twoheadrightarrow Z, Y \twoheadrightarrow W \in \Sigma_i$ with $Z \cap Y = \emptyset$ and $V = Z \cap W$. From $U \subseteq X_{\text{alg}}^S$ follows $Z \subseteq X_{\text{alg}}^S$ since $U \twoheadrightarrow Z \in \Sigma_i$. However, that means $V = Z \cap W \subseteq Z \subseteq X_{\text{alg}}^S$.
3. If $U \twoheadrightarrow V$ results from an application of the pseudo-transitivity rule, then $U \twoheadrightarrow W, W \twoheadrightarrow Z \in \Sigma_i$ and $V = Z - W$. From $U \subseteq X_{\text{alg}}^S$ follows $W \subseteq X_{\text{alg}}^S$ since $U \twoheadrightarrow W \in \Sigma_i$. Consequently, $W \subseteq X_{\text{alg}}^S$ implies $Z \subseteq X_{\text{alg}}^S$ since $W \twoheadrightarrow Z \in \Sigma_i$. That means, $V = Z - W \subseteq Z \subseteq X_{\text{alg}}^S$.
4. If $U \twoheadrightarrow V$ results from an application of the union rule, then $U \twoheadrightarrow W, U \twoheadrightarrow Z \in \Sigma_i$ and $V = W \cup Z$. From $U \subseteq X_{\text{alg}}^S$ follows $W, Z \subseteq X_{\text{alg}}^S$. Consequently, $V = W \cup Z \subseteq X_{\text{alg}}^S$.

The time complexity of Algorithm 1 is essentially the time complexity of the algorithm in [19]. The code between line (4) and (10) to compute the Σ-scope of X can be implemented in time $\mathcal{O}(n)$, see [5]. Therefore, the time-complexity is $\mathcal{O}((1 + \min\{s, \log \bar{p}\}) \cdot n)$.

$\square$

Example 6.2. Consider a classical example [4]. Let $\Sigma = \{AB \twoheadrightarrow DEFG, CGJ \twoheadrightarrow ADHI\}$, and suppose we want to compute $DepB_U(\{A, C, G, J\})$ with respect to Σ. We obtain $R_{\min} = \{A, B, C, D, E, F, G, H, I, J\}$ and

$$DepB_{R_{\min}}(\{A, C, G, J\}) = \{\{A\}, \{C\}, \{G\}, \{J\}, \{D\}, \{H, I\}, \{B, E, F\}\} \, [4].$$

The Σ-scope of $\{A, C, G, J\}$ is $\{A, C, G, J\}^S = \{A, C, D, G, I, H, J\}$. Therefore, $\{A, C, G, J\}^S \subset R_{\min}$ and $R_{\min} - \{A, C, G, J\}^S = \{B, E, F\}$. That is,

$$DepB_U(\{A, C, G, J\}) = \{\{A\}, \{C\}, \{G\}, \{J\}, \{D\}, \{H, I\}\}$$

and the MVD $ACGJ \twoheadrightarrow BDEF$ which is $R_{\min}$-implied by Σ is not implied by Σ.

$\square$

Theorem 6.4. *Let Σ be a finite set of MVDs. The MVD $X \twoheadrightarrow Y$ is implied by Σ if and only if $Y \subseteq X^S$ and $X \twoheadrightarrow Y$ is $R_{\min}$-implied by Σ.*

Proof. If $X \twoheadrightarrow Y$ is implied by Σ, then Y is the union of some subset of $DepB_U(X) \subseteq DepB_{R_{\min}}(X)$. Consequently, $Y \subseteq X^S$ and $X \twoheadrightarrow Y$ is $R_{\min}$-implied by Σ.

If $X \twoheadrightarrow Y$ is $R_{\min}$-implied by Σ, then Y is the union of some subset of $DepB_{R_{\min}}(X)$. If $R_{\min} = X^S$, then $DepB_{R_{\min}}(X) = DepB_U(X)$ by Theorem 6.2 and $X \twoheadrightarrow Y$ is also implied by Σ according to Theorem 6.1. Otherwise $X^S \subset R_{\min}$. Since $Y \subseteq X^S$ holds as well, we have $Y \cap (R_{\min} - X^S) = \emptyset$, i.e., Y is in fact the union of some subset of $DepB_{R_{\min}}(X) - \{R_{\min} - X^S\} = DepB_U(X)$. That means, $X \twoheadrightarrow Y$ is implied by Σ according to Theorem 6.1. $\qquad\square$

Corollary 6.1. *The implication problem $\Sigma \models X \twoheadrightarrow Y$ can be decided in time $\mathcal{O}((1 + \min\{s, \log p\}) \cdot n)$ where s denotes the number of dependencies in Σ, p the number of sets in $DepB_U(X)$ that have non-empty intersection with Y and n denotes the total number of occurrences of attributes in Σ.*

Proof. This follows from the time bound in [19], Theorem 6.3 and Theorem 6.1. It can be decided in time $\mathcal{O}(n)$ whether $Y \subseteq X^S$ holds. If $Y \not\subseteq X^S$, then $X \twoheadrightarrow Y$ is not implied by Σ. Otherwise, we can decide in time $\mathcal{O}((1 + \min\{s, \log p\}) \cdot n)$ whether Y is the union of some subset of $DepB_U(X)$. Note that there is at most one more element in $DepB_{R_{\min}}(X)$ than in $DepB_U(X)$, namely $R_{\min} - X^S$. However, Y and $R_{\min} - X^S$ are disjoint since $Y \subseteq X^S$. $\qquad\square$

If there is a linear-time algorithm for computing the dependency basis in a fixed universe, then Algorithm 1 also provides a linear-time algorithm for computing the dependency basis in undetermined universes. Vice versa, if there is a linear-time algorithm for computing the dependency basis in undetermined universes, then Theorem 6.2 shows that the dependency basis can also be computed in linear time in a fixed universe.

7   Open Problems

The paper concludes by listing some open problems.

1. Are there any complete sets (not a subset of $\mathfrak{S}_U$) in which the subset rule $\mathcal{S}$ does not occur?
2. Lien studies MVDs in the presence of null values [28], but only in fixed universes. Clarify the role of the $\mathcal{R}$-complementation rule in this context and investigate MVDs in the presence of null values in undetermined universes!
3. Consider MVDs in undetermined universes together with FDs. What are minimal complete sets for the implication of FDs and MVDs in undetermined universes?
4. In [29], Mendelzon identifies all minimal subsets of $\{\mathcal{R}, \mathcal{A}, \mathcal{T}, \mathcal{U}, \mathcal{D}, \mathcal{I}, \mathcal{C}_R\}$. Do the same for $\{\mathcal{R}\text{-axiom}, \mathcal{R}, \mathcal{R}_\emptyset, \mathcal{A}, \mathcal{T}, \mathcal{U}, \mathcal{D}, \mathcal{I}, \mathcal{T}^*, \mathcal{S}, \mathcal{C}_R\}$!
5. Are there any minimal sets of inference rules that are also complementary?

References

1. M. Arenas and L. Libkin. A normal form for XML documents. *Tansactions on Database Systems (ToDS)*, 29(1):195–232, 2004.
2. W. W. Armstrong. Dependency structures of database relationships. *Information Processing*, 74:580–583, 1974.
3. W. W. Armstrong, Y. Nakamura, and P. Rudnicki. Armstrong's axioms. *Journal of formalized Mathematics*, 14, 2002.
4. C. Beeri. On the membership problem for functional and multivalued dependencies in relational databases. *ACM Tans. Database Syst.*, 5(3):241–259, 1980.
5. C. Beeri and P. A. Bernstein. Computational problems related to the design of normal form relational schemata. *ACM Tans. Database Syst.*, 4(1):30–59, 1979.
6. C. Beeri, R. Fagin, and J. H. Howard. A complete axiomatization for functional and multivalued dependencies in database relations. In *Proceedings of the SIGMOD International Conference on Management of Data*, pages 47–61. ACM, 1977.
7. P. Bernstein. Synthesizing third normal form relations from functional dependencies. *ACM Tans. Database Syst.*, 1(4):277–298, 1976.
8. P. A. Bernstein and N. Goodman. What does Boyce-Codd normal form do? In *Proceedings of the 6th International Conference on Very Large Data Bases*, pages 245–259. IEEE Computer Society, 1980.
9. G. Birkhoff. *Lattice Theory*. American Mathematical Society, 1940.
10. J. Biskup. On the complementation rule for multivalued dependencies in database relations. *Acta Informatica*, 10(3):297–305, 1978.
11. J. Biskup. Inferences of multivalued dependencies in fixed and undetermined universes. *Theor. Comput. Sci.*, 10(1):93–106, 1980.
12. E. F. Codd. A relational model of data for large shared data banks. *Commun. ACM*, 13(6):377–387, 1970.
13. E. F. Codd. Further normalization of the database relational model. In *Courant Computer Science Symposia 6: Data Base Systems*, pages 33–64. Prentice-Hall, 1972.
14. C. Delobel. Normalisation and hierarchical dependencies in the relational data model. *ACM Tans. Database Syst.*, 3(3):201–222, 1978.
15. J. Diederich and J. Milton. New methods and fast algorithms for database normalization. *ACM Tans. Database Syst.*, 13(3):339–365, 1988.
16. R. Fagin. Multivalued dependencies and a new normal form for relational databases. *ACM Tans. Database Syst.*, 2(3):262–278, 1977.
17. R. Fagin and M. Y. Vardi. The theory of data dependencies: a survey. In *Mathematics of Information Processing: Proceedings of Symposia in Applied Mathematics*, pages 19–71. American Mathematical Society, 1986.
18. P. C. Fischer, L. V. Saxton, S. J. Thomas, and D. Van Gucht. Interactions between dependencies and nested relational structures. *J. Comput. Syst. Sci.*, 31(3):343–354, 1985.
19. Z. Galil. An almost linear-time algorithm for computing a dependency basis in a relational database. *J. ACM*, 29(1):96–102, 1982.
20. K. Hagihara, M. Ito, K. Taniguchi, and T. Kasami. Decision problems for multivalued dependencies in relational databases. *SIAM J. Comput.*, 8(2):247–264, 1979.
21. C. Hara and S. Davidson. Reasoning about nested functional dependencies. In *Proceedings of the 18th SIGACT-SIGMOD-SIGART Symposium on Principles of Database Systems*, pages 91–100. ACM, 1999.

22. S. Hartmann and S. Link. On a problem of Fagin concerning multivalued dependencies in relational databases. accepted for Theoretical Computer Science (TCS).

23. S. Hartmann and S. Link. Multi-valued dependencies in the presence of lists. In *Proceedings of the 23rd SIGACT-SIGMOD-SIGART Symposium on Principles of Database Systems*, pages 330–341. ACM, 2004.

24. S. Hartmann, S. Link, and K.-D. Schewe. Axiomatisations of functional dependencies in the presence of records, lists, sets and multisets. accepted for Theoretical Computer Science (TCS).

25. M. Ito, M. Iwasaki, K. Taniguchi, and K. Kasami. Membership problems for data dependencies in relational expressions. *Theor. Comput. Sci.*, 34:315–335, 1984.

26. V. Lakshmanan and C. VeniMadhavan. An algebraic theory of functional and multivalued dependencies in relational databases. *Theor. Comput. Sci.*, 54:103–128, 1987.

27. M. Levene and G. Loizou. Axiomatisation of functional dependencies in incomplete relations. *Theoretical Computer Science (TCS)*, 206(1-2):283–300, 1998.

28. Y. E. Lien. On the equivalence of data models. *Journal of the ACM*, 29(2):333–363, 1982.

29. A. Mendelzon. On axiomatising multivalued dependencies in relational databases. *J. ACM*, 26(1):37–44, 1979.

30. D. Parker and C. Delobel. Algorithmic applications for a new result on multivalued dependencies. In *Proceedings of the 5th International Conference on Very Large Data Bases*, pages 67–74. IEEE Computer Society, 1979.

31. Y. Sagiv. An algorithm for inferring multivalued dependencies with an application to propositional logic. *J. ACM*, 27(2):250–262, 1980.

32. Z. Tari, J. Stokes, and S. Spaccapietra. Object normal forms and dependency constraints for object-oriented schemata. *ACM Tans. Database Syst.*, 22:513–569, 1997.

33. B. Thalheim. *Dependencies in Relational Databases*. Teubner-Verlag, 1991.

34. B. Thalheim. Conceptual treatment of multivalued dependencies. In *Conceptual Modeling - ER 2003, 22nd International Conference on Conceptual Modeling, Chicago, IL, USA, October 13-16, 2003, Proceedings*, number 2813 in Lecture Notes in Computer Science, pages 363–375. Springer, 2003.

35. M. Vardi. Inferring multivalued dependencies from functional and join dependencies. *Acta Inf.*, 19(2):305–324, 1983.

36. M. Y. Vardi. Fundamentals of dependency theory. In E. Börger, editor, *Trends in Theoretical Computer Science*, pages 171–224. Computer Science Press, 1987.

37. M. Vincent and J. Liu. Multivalued dependencies in XML. In *Proceedings of the 20th British National Conference on Databases*, number 2712 in Lecture Notes in Computer Science, pages 4–18. Springer, 2003.

38. M. Vincent, J. Liu, and C. Liu. A redundancy free 4NF for XML. In *Proceedings of the 1st International XML Database Symposium*, number 2824 in Lecture Notes in Computer Science, pages 254–266. Springer, 2003.

39. G. Weddell. Reasoning about functional dependencies generalized for semantic data models. *ACM Tans. Database Syst.*, 17(1):32–64, 1992.

40. J. Wijsen. Temporal FDs on complex objects. *ACM Tans. Database Syst.*, 24(1):127–176, 1999.

41. C. Zaniolo. *Analysis and Design of Relational Schemata for Database Systems*. PhD thesis, UCLA, Tech. Rep. UCLA-ENG-7769, 1976.

Preference-Based Query Tuning Through Refinement/Enlargement in a Formal Context

Nicolas Spyratos[1] and Carlo Meghini[2]

[1] Université Paris-Sud, Laboratoire de Recherche en Informatique,
Orsay Cedex, France
spyratos@lri.fr
[2] Consiglio Nazionale delle Ricerche,
Istituto della Scienza e delle Tecnologie della Informazione,
Pisa, Italy
meghini@isti.cnr.it

Abstract. The user of an information system rarely knows exactly what he is looking for, but once shown a piece of information he can quickly tell whether it is what he needs. Query tuning is the process of searching for the query that best approximates the information need of the user. Typically, navigation and querying are two completely separate processes, and the user usually has to switch often from one to the other–a painstaking process producing a frustrating experience. In this paper, we propose an approach to query tuning that integrates navigation and querying into a single process, thus leading to a more flexible and more user friendly method of query tuning. The proposed approach is based on formal concept analysis, and models the directory of an information source as a formal context in which the underlying concept lattice serves for navigation and the attributes of the formal context serve for query formulation. In order to support the user in coping with a possibly overwhelming number of alternative query tunings, preferences are introduced.

1 Introduction

The work reported in this paper originates in the following basic observation: the user of an information system rarely knows exactly what he is looking for, but once shown a piece of information he can quickly tell whether it is what he needs.

Query tuning is the process of searching for the query that best approximates the information need of the user. We note that, finding the query that best expresses a given information need is important not only for retrieving the information satisfying the current need, but also in order to name and store the query for use at some later time (without having to re-invent it over and over).

In advanced information systems, query tuning proceeds in two steps: (a) the user navigates the information space until he finds a subspace of interest and (b) in that subspace, the user issues a query. If the answer to the query is satisfactory then the session terminates, otherwise a new navigation step begins.

J. Dix and S.J. Hegner (Eds.): FoIKS 2006, LNCS 3861, pp. 278–293, 2006.

However, navigation and querying are two completely separate processes, and the user usually has to switch often from one to the other -a painstaking process usually producing a frustrating experience.

In this paper, we propose an approach to query tuning that interleaves, or integrates navigation and querying into a single process, thus leading to a more flexible and more user friendly method of query tuning.

The proposed approach is based on Formal Concept Analysis (FCA, from now on), and models the directory of an information source (IS, for short) as a formal context in which the objects represent documents and the attributes represent indexing terms [16]. As a result, the concepts of the underlying concept lattice represent meaningful classes of documents, in the sense that all documents in a class share the same set of indexing terms. Therefore, we propose to use the concept lattice as the basic navigation tool.

We assume the user queries to be Boolean combinations of indexing terms, and more specifically conjunctions of indexing terms. Indeed, the objective of this paper is not to propose a new, more powerful query language but, rather, use an existing (simple) language in order to illustrate our approach to query tuning.

Our approach is user-controlled, and proceeds as a sequence of refine/enlarge commands until a user-approved query is obtained. More precisely, query tuning in our approach proceeds roughly as follows:

- *Query mode.* The user formulates a query to the IS and receives an answer; if the answer is satisfactory then the session terminates and the issued query is considered tuned, otherwise the user can issue a Refine or an Enlarge command.
- *Refine.* A Refine command returns all maximal concepts (from the concept lattice) that refine the user query, in the sense that their extent is strictly included in the answer to the user query; then the user can decide to either return to query mode and query one or more of those concepts, or terminate.
- *Enlarge.* An Enlarge command returns all minimal concepts (from the concept lattice) that subsume the user query, in the sense that their extent strictly includes the answer to the user query; then the user can decide to either return to query mode and query one or more of those concepts, or terminate.

In what follows, we firstly relate our work to existing results; then we introduce the IS model (Section 3) and recall the basic notions needed from formal concept analysis (Section 4). Our query tuning approach is presented in Section 5 and illustrated through a running example. The preference-based ordering of query tunings is given in Section 6. Finally, we offer some concluding remarks in Section 7.

2   Related Work

The use of FCA in information system is not new. The structuring of information that FCA supports has inspired work on browsing [13, 3], clustering [4], and ranking [6, 15]. A basic drawback of these approaches is that they require the computation of the whole concept lattice, whose size may be exponential in that of the context, as it will be argued below. An integrated approach to browsing and querying that uses only part of the lattice, and thus can be computed efficiently, is presented in [5].

Preferences, on the other hand, are enjoying a vast popularity, due to their ability of capturing user requirements. In general, preferences can be captured either quantitatively, or qualitatively as formulas inducing orderings [9, 14]. Our approach subscribes to the latter view.

Our approach extends efficient, FCA-based query tuning by considering qualitatively expressed preferences.

3   Information Sources

Given the foundational nature of our work, we deliberately adopt a simple model of an information source, close in spirit to that of a *digital library,* (or DL for short). Essentially, a DL serves a network of providers willing to share their documents with other providers and/or consumers (hereafter, collectively called "users"). Each document resides at the local repository of its provider, so all providers' repositories, collectively, can be seen as a distributed repository of documents spread over the network. The DL system acts as a mediator, supporting transparent access to all sharable documents by the library users. Existing DL systems are consistent with this view [7, 8].

Two of the basic services supported by the library are *document registration* and *querying.*

3.1   Document Registration

When a provider wishes to make a document sharable over the network of users he must register it at the library. To do so he must provide two items to the library:

- the document identifier
- the document description

We assume that the document identifier is a global identifier, such as a URI, or just the URL where the document can be accessed, (however, for convenience of notation, we use integers as document identifiers in our examples). As for the document description, we consider only content description and we assume that such a description is given by selecting a set of terms from a controlled vocabulary. For example, the document description {QuickSort, Java} would indicate that the document in question is about the quick sort algorithm and Java.

Therefore, to register a document, its provider submits to the library an identifier i and a set of terms D. We assume that registration of the document by the library is done by storing a pair (i, t) in the library repository, for each term t in D. In our previous example, if i is the document identifier, the library will store two pairs: (i,QuickSort) and (i, Java). The set of all such pairs (i, t) is what we call the *library directory,* or simply *directory* (the well known Open Directory [2] is an example of such a directory). Clearly the directory is a binary relation between document identifiers and terms, *i.e.* a formal context in the sense defined in the next section.

3.2   Querying

Library users access the library in search of documents of interest, either to use them directly (*e.g.*, as learning objects) or to reuse them as components in new documents that they intend to compose. Search for documents of interest is done by issuing queries to the library management system, and the library management system uses its directory to return the identifiers (*i.e.*, the URIs) of all documents satisfying the query.

The query language that we use is a simple language in which a query is just a Boolean combination of terms:

$$q ::= t \mid q_1 \wedge q_2 \mid q_1 \vee q_2 \mid q_1 \wedge \neg q_2 \mid (q)$$

where t is any term.

The answer to a query q is defined recursively as follows:

if q *is a term then* $ans(q) = \{i \mid (i, q)$ *is in the directory*$\}$
  else begin
    if q *is* $q_1 \wedge q_2$ *then* $ans(q) = ans(q_1) \cap ans(q_2)$
    if q *is* $q_1 \vee q_2$ *then* $ans(q) = ans(q_1) \cup ans(q_2)$
    if q *is* $q_1 \wedge \neg q_2$ *then* $ans(q) = ans(q_1) \setminus ans(q_2)$
  end

In other words, to answer a query, the underlying digital library management system simply replaces each term appearing in the query by its extension from the directory, and performs the set theoretic operations corresponding to the Boolean connectives.

The reader familiar with logic will have recognized documents as individuals, terms as unary predicate symbols and the library directory as an interpretation of the resulting logic language; in other words, the presence of a pair (i, t) in the library directory means that the individual i is in the interpretation of term t. Query answering can then be seen as based on the notion of satisfaction: an individual i is returned in response to a query q just in case i is in the extension of q (in the current interpretation).

4   Formal Concept Analysis

Formal concept analysis (hereafter FCA for short) is a mathematical tool for the analysis of data based on lattice theory [11, 10, 12, 1].

Let $\mathcal{O}$ be a set of objects and $\mathcal{A}$ a set of attributes. A *formal context,* or simply *context* over $\mathcal{O}$ and $\mathcal{A}$ is a triple $(\mathcal{O}, \mathcal{A}, C)$, where $C \subseteq \mathcal{O} \times \mathcal{A}$, is a binary

	A	B	C	D	E	F
1	x		x	x	x	
2		x	x			
3	x		x	x		x
4		x	x	x	x	x
5	x	x			x	x

Fig. 1. A Formal Context

relation between $\mathcal{O}$ and $\mathcal{A}$, called the *incidence* of the formal context. Figure 1 shows a formal context in tabular form, in which objects correspond to rows and attributes correspond to columns. The pair (o, a) is in C (that is, object o has attribute a) if and only if there is a x in the position defined by the row of object o and the column of attribute a.

Let i and e be respectively the functions *intension* and *extension* as they are normally used in information systems, that is:

$$\text{for all } o \in \mathcal{O}, \quad i(o) = \{a \in \mathcal{A} \mid (o, a) \in C\}$$
$$\text{for all } a \in \mathcal{A}, \quad e(a) = \{o \in \mathcal{O} \mid (o, a) \in C\}.$$

In Figure 1, the intension of an object o consists of all attributes marked with a x in the row of o; the extension of an attribute a consists of all objects marked with a x in the column of a. Now define:

$$\text{for all } O \subseteq \mathcal{O}, \quad \varphi(O) = \bigcap\{i(o) \mid o \in O\}$$
$$\text{for all } A \subseteq \mathcal{A}, \quad \psi(A) = \bigcap\{e(a) \mid a \in A\}.$$

A pair $(O, A), O \subseteq \mathcal{O}$ and $A \subseteq \mathcal{A}$, is a *formal concept* of the context $(\mathcal{O}, \mathcal{A}, C)$ if and only if $O = \psi(A)$ and $A = \varphi(O)$. O is called the *extent* and A the *intent* of concept (O, A). In the formal context shown in Figure 1, $(\{1, 3, 4\}, \{C, D\})$ is a concept, while $(\{1, 3\}, \{A, D\})$ is not. The computation of φ for a given set of objects O requires the intersection of $|O|$ sets, thus it requires $O(|\mathcal{O}| \cdot |\mathcal{A}|^2)$ time. Analogously, the computation of ψ for a given set of attributes A requires $O(|\mathcal{A}| \cdot |\mathcal{O}|^2)$ time. These are both upper bounds that can be reduced by adopting simple optimization techniques, such as ordering. In addition, they are worst case measures.

The concepts of a given context are naturally ordered by the *subconcept-superconcept* relation defined by:

$$(O_1, A_1) \leq (O_2, A_2) \text{ iff } O_1 \subseteq O_2 (\text{iff } A_2 \subseteq A_1)$$

This relation induces a lattice on the set of all concepts of a context. For example, the concept lattice induced by the context of Figure 1 is presented in Figure 2.

There is an easy way to "read" the extent and intent of every concept from the lattice. To this end, we define two functions γ and μ, mapping respectively objects and attributes into concepts, as follows:

$$\gamma(o) = (\psi(\varphi(\{o\})), \varphi(\{o\})) \quad \text{for all } o \in \mathcal{O}$$
$$\mu(a) = (\psi(\{a\}), \varphi(\psi(\{a\}))) \quad \text{for all } a \in \mathcal{A}.$$

It is easy to see that $\gamma(o)$ and $\mu(a)$ are indeed concepts. In addition, $(o, a) \in C$ is equivalent to $\gamma(o) \leq \mu(a)$. The functions γ and μ are represented in Figure 2 by labeling the node corresponding to the concept $\gamma(o)$ with o as a subscript, and the node corresponding to concept $\mu(a)$ with a as a superscript. Finally, it can be proved that for any concept $c = (O_c, A_c)$, we have:

$$O_c = \{o \in \mathcal{O} \mid \gamma(o) \leq c\}$$
$$A_c = \{a \in \mathcal{A} \mid c \leq \mu(a)\}.$$

Thus, the extent of a concept c is given by the objects that label the concepts lower than c in the concept lattice. For example, the extent of the concept labelled by A, that is $\mu(A)$, is $\{1, 3, 5\}$. Analogously, the intent of c is given by the attributes that label concepts higher than c in the lattice, just $\{A\}$ for $\mu(A)$. It follows that

$$\mu(A) = (\{1, 3, 5\}, \{A\}).$$

By the same token, it can be verified that

$$\gamma(3) = (\{3\}, \{A, C, D, F\}).$$

Notice that this reading is consistent with the reading of class hierarchies in object-oriented languages. In such hierarchies, objects are inherited "upward", *i.e.* by the classes that are more general than the classes where they belong, while attributes are inherited "downward", *i.e.* by the classes that are more special than the classes that define them.

It is easy to see the following:

1. $O_1 \subseteq O_2$ implies $\varphi(O_1) \supseteq \varphi(O_2)$; $A_1 \subseteq A_2$ implies $\psi(A_1) \supseteq \psi(A_2)$; $O \subseteq \psi \circ \varphi(O)$; and $A \subseteq \varphi \circ \psi(A)$; hence, φ and ψ form a Galois connection between the powerset of $\mathcal{O}$, $\mathcal{P}(\mathcal{O})$, and that of $\mathcal{A}$, $\mathcal{P}(\mathcal{A})$.

2. $O \subseteq \psi \circ \varphi(O)$; $O_1 \subseteq O_2$ implies $\psi \circ \varphi(O_1) \subseteq \psi \circ \varphi(O_2)$; and $\psi \circ \varphi(\psi \circ \varphi(O)) = O$. The above 3 properties of $\psi \circ \varphi$ tell us that $\psi \circ \varphi$ is a closure operator on $\mathcal{P}(\mathcal{O})$, and similarly $\varphi \circ \psi$ is a closure operator on $\mathcal{P}(\mathcal{A})$.

Since closure systems may be exponentially large in the size of their domain, a concept lattice may have an exponential number of concepts in the size of the underlying context. For this reason, any approach that requires the computation of the whole concept lattice is bound to be intractable, in general. On the other hand, our approach only requires the computation of small portions of the lattice, namely that consisting of single concepts and their lower or upper neighbors. This guarantees the tractability of the involved operations, as it will be argued in due course.

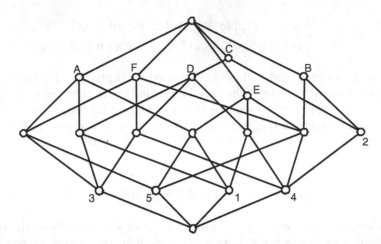

Fig. 2. A concept lattice

4.1   Digital Libraries and FCA

We are now in a position to make precise the relationship between DLs and FCA. The directory $\mathcal{D}$ of a DL can be seen as a formal context, defined as follows:

- The objects of the formal context are the identifiers of the documents registered in $\mathcal{D}$; we call the set of all such identifiers the *domain* of $\mathcal{D}$, and denote it by $dom(\mathcal{D})$.
- The attributes of the formal context are the terms which appear in at least one document description; we call the set of all such identifiers the *range* of $\mathcal{D}$, and denote it by $ran(\mathcal{D})$.
- The incidence of the formal context is just $\mathcal{D}$, the DL directory.

So, corresponding to the DL directory $\mathcal{D}$ we have the formal context $(dom(\mathcal{D}), ran(\mathcal{D}), \mathcal{D})$. Each concept of this context may be understood as a *class* of documents. The extent of the concept gives the instances of the class, while its intent gives the attributes (terms) defining the class. By definition of concept, the extent is the largest set of documents having the attributes in the intent and, dually, the intent is the largest set of attributes shared by the documents in the extent. This is compliant with the notion of class as it stems from object-oriented information modelling, thus it is a merit of FCA to offer a mathematically well-founded and intuitively appealing criterion to model information.

In addition, unlike the traditional approach in which classes are manually derived in the design of the information system, here classes are automatically derived solely on the base of the available information. In this way, they do not run the risk of becoming a straightjacket: they are a perfect fit for the underlying information at any point of the system lifetime, since their definition evolves with the evolving of the information. This is very valuable in the context of DLs, for which the identification of classes is more difficult than in a traditional information system, due to the complexity and the heterogeneity of the documents, and to the multiplicity of their usage.

Our work is a first exploitation of the notion of formal concept in DLs, hinging on a basic link between concepts and conjunctive queries. Indeed, the intent of each concept in the context $(dom(\mathcal{D}),\ ran(\mathcal{D}),\ \mathcal{D})$ is actually the most specific conjunctive query whose answer is the extent of that concept. In our example, on Figure 1 it can be verified that the queries: $(C \wedge F)$, $(D \wedge F)$ and $(C \wedge D \wedge F)$ are the ones (and the only ones) having the set of documents $\{3,4\}$ as an answer. The most specific of these queries is $(C \wedge D \wedge F)$ and in fact $(\{3,4\}, \{C, D, F\})$ is a formal concept.

Conversely, any answer to a conjunctive query is the extent of some concept. For example, the set of documents $\{1, 2, 3\}$ is not the extent of a concept and in fact no conjunctive query can produce it as an answer.

It follows that, by navigating the concept lattice, the user can be guided to the best result he can obtain with a conjunctive query. These facts and observations lie at the heart of our query tuning approach that we present in the next section.

5 Query Tuning

During user interaction with the digital library, query tuning is obtained by using four commands: *Query, Terminate, Refine* and *Enlarge*. In this section we define each of these commands separately and illustrate them in our running example. We recall that we restrict our attention to conjunctive queries only.

5.1 Query

The user issues a query to the system. The system returns the answer, obtained by evaluating $\psi(A)$ where A is the set of terms in the query. In fact:

$$ans(\bigwedge A) = \bigcap \{ans(a) \mid a \in A\} \quad \text{(by definition of } ans \text{ on conjunctions)}$$

$$= \bigcap \{\{o \in \mathcal{O} \mid (o, a) \in C\} \mid a \in A\} \quad \text{(by definition of } ans \text{ on terms)}$$

$$= \bigcap \{e(a) \mid a \in A\} \quad \text{(by definition of } e)$$

$$= \psi(A) \quad \text{(by definition of } \psi).$$

The system also shows to the user the most precise (*i.e.* the most specific) query that can be used to obtain the same result; this query is the conjunction of all terms in $\varphi(\psi(A))$. In other words, upon evaluating a query $\bigwedge A$ the system "places" itself on the concept $(\psi(A), \varphi(\psi(A)))$, which becomes the current concept. This placement can be obtained using a polynomial amount of time, since it requires the computation of ψ on the set of terms making up the query, followed by a computation of φ on the result.

In our example, let us assume the user poses a query consisting of a single term, say D. In response, the system returns the answer $\psi(D) = \{1, 3, 4\}$ and shows the query $C \wedge D$, since $\varphi(\psi(D)) = \{C, D\}$. The current concept is therefore $\mu(D)$ in Figure 2.

5.2 Terminate

The user is satisfied by the answer and issues a *Terminate* command. The system switches to next-query mode; otherwise, the user performs one of the two actions described next.

5.3 Refine

The user judges the answer to be too rich, *e.g.* the cardinality of the answer set is too big or, upon inspection, there are too many irrelevant answers in the answer set; and issues a *Refine* command. The system then computes and returns to the user the following pair of information items, for each concept *max* whose extent O_{max} is a maximal subset of the current answer:

1. The intent A_{max}
2. The objects in the current answer lying outside O_{max}.

Let us explain further these two items that are returned to the user. Any concept like *max* above is a maximal sub-concept of the current concept, or, in other words, a lower neighbor of the current concept in the concept lattice. By definition, the intent of such a concept is a superset of the current concept intent. The system computes each such concept *max* and shows its intent to the user, by simply presenting him the additional terms of the intent with respect to the current concept intent. This computation can be done in polynomial time with respect to the size of the context. Let us see how in our example.

Let us assume that the current answer is $\{1, 3, 4\}$, and that the user considers this answer to be too large, so he executes a *Refine*. To compute the terms to be shown to the user, the system looks at a smaller context, consisting of the objects in the extent of the current concept, and of the attributes outside the intent of the current concept. The context we are looking at is:

	A	B	E	F
1	x		x	
3	x			x
4		x	x	x

In order to achieve maximality, we select the terms which have maximal extension in this context; these are A, E and F. Each of these terms t must be shown to the user, since it leads to a maximal sub-concept of the current concept, given by:

$$(\psi(A_c \cup \{t\}), \varphi(\psi(A_c \cup \{t\})))$$

where A_c is the intent of the current concept. In our example, we recall that the current concept is $\mu(D) = (\{1, 3, 4\}, \{C, D\})$. Then, term A leads to concept $(\{1, 3\}, \{A, C, D\})$, term E leads to concept $(\{1, 4\}, \{C, D, E\})$, term F leads to concept $(\{3, 4\}, \{C, D, F\})$. All these concepts are maximal sub-concepts of the current concept; as a consequence, each has a larger intent, which means that its associated query is a refinement of the query originally posed by the user.

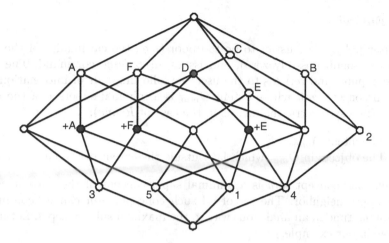

Fig. 3. The maximal sub-concepts of the current concept

Figure 3 shows the 3 maximal sub-concepts (green) of the current concept (light blue) in the concept lattice; next to each one, the additional term is shown.

The second item shown to the user is the set of objects in the extent of the current concept lying outside O_{max}. Intuitively, these are the objects that will be "lost" by selecting the corresponding refinement. For instance, along with the term F, the user is shown the object set $\{1\}$ containing the answers which will no longer be answers if the query is refined by adding the term F to it.

The complete answer that the user gets in response to a *Refine* in our example is reported in Table 1. For convenience, the Table also shows the refined query and the concept corresponding to each solution.

Table 1. Result of a *Refine*

Option	Added Terms	Lost Objects	Refined Query	Concept
1	$\{A\}$	$\{4\}$	$A \wedge C \wedge D$	$(\{1,3\}, \{A,C,D\})$
2	$\{E\}$	$\{3\}$	$C \wedge D \wedge E$	$(\{1,4\}, \{C,D,E\})$
3	$\{F\}$	$\{1\}$	$C \wedge D \wedge F$	$(\{3,4\}, \{C,D,F\})$

Upon deciding whether to accept a proposed refinement, the user can figure out the attributes he gains by inspecting the added terms, or the answers he looses by inspecting the lost objects. If the user does decide to move, then the concept corresponding to the selected solution becomes the current concept, and a new interaction cycle begins (by querying, refining and enlarging).

Notice that if no maximal sub-concept exists (*i.e.* the current concept is the least concept of the lattice), then the system returns empty and, subsequently, the user may issue an *Enlarge* command (see below) or try a new query.

5.4   Enlarge

The user judges the answer to be too poor (*e.g.*, the cardinality of the answer set is too small, possibly zero), and issues an *Enlarge* command. The system then computes and returns to the user the following pair of information items, for each concept *min* whose extent O_{min} is a minimal superset of the current answer (in a symmetric manner as in the case of *Refine*):

1. The intent A_{min}.
2. The objects in O_{min} which lay outside the extent of the current context.

Each such concept *min* is a minimal super-concept of the current concept, or an upper neighbor. The set of all such concepts *min* can be computed in polynomial time in an analogous way to the maximal sub-concepts. Let us again see how in our example.

Let us assume that the user refines the initial query by selecting option 1 in Table 1, thus making $(\{1, 3\}, \{A, C, D\})$ the current concept, and that he then asks to enlarge this set. To compute the objects leading to a minimal super-concept of the current concept, we look at a smaller context, consisting of just the attributes in the intent of the current context and of the objects outside the extent of the current context. That is:

	A	C	D
2	x		
4		x	x
5	x		

From this context we select the objects with maximal intention, that is 4 and 5. Each of these objects o leads to a minimal super-concept of the current concept, given by:

$$(\psi(\varphi(O_c \cup \{o\})), \varphi(O_c \cup \{o\}))$$

where O_c is the extent of the current concept. In our example, object 4 leads back to concept $(\{1, 3, 4\}, \{C, D\})$ while object 5 leads to concept $(\{1, 3, 5\}, \{A\})$, which are all minimal super-concepts of the current concept. Figure 4 shows the 2 minimal super-concepts (red) of the current concept (light blue) in the concept lattice; next to each one, the additional term is shown. Notice that the query associated to each such concepts is a relaxation of the query associated to the current concept.

The complete answer that the user gets in response to an *Enlarge* in our example is reported in Table 2. For each option, the Table shows the terms that are lost in the query enlargement, the added objects, the enlarged query and the corresponding concept. Upon deciding whether to accept a proposed enlargement, the user can figure out the attributes he looses or the answers he gains. If the user does decide to move, then the concept corresponding to the selected option becomes the current concept, and a new interaction cycle begins (by querying, refining and enlarging).

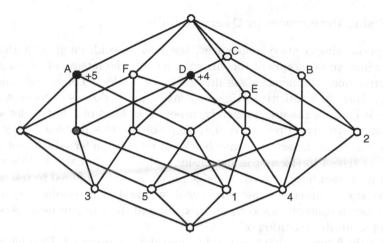

Fig. 4. The minimal super-concepts of the current concept

Table 2. Result of an *Enlarge*

Option	Lost Terms	Added Objects	Enlarged Query	Concept
1	A	4	$C \wedge D$	$(\{1,3,4\}, \{C,D\})$
2	C, D	5	A	$(\{1,3,5\}, \{A\})$

Notice that if no minimal super-concept exists (*i.e.* the current concept is the greatest concept of the lattice), then the system returns empty and, subsequently, the user may issue a *Refine* command or try a new query.

6   Introducing Preferences

Preferences are a way of capturing user specific requirements in the usage of a DL. As the amount of information in a DL and the number of its users may grow very rapidly, capturing user preferences may be a very valuable tool to make the DL usable.

We will model the preferences of a given user u, as a partial order P_u (simply P when there is no ambiguity) over the terms employed for content description. If terms a and b are in the P relation, *i.e.* $(a, b) \in P$, we say that a is preferred over b (by user u).

P may be derived in several ways, for instance by having the user declare preferences, or by mining them from user actions, or both. In addition, while reflexivity and transitivity seem reasonable properties for a preference relation, there is strong evidence that antisymmetry is not so. However, we have assumed it because it is a well-known mathematical fact that a partial order can be uniquely derived from a pre-order by considering term equivalence. We will do this below for term set, thus for the time being suffice it to say that there is no loss in generality by considering P to be a partial order.

6.1   Using Preferences in Query Refining

Upon performing a query refinement, the user is confronted with the set of
the maximal sub-concepts of the current concept, the intents of which give the
alternative query refinements, as illustrated in Table 1. Let us call this latter
set A_{ref}. Now, in a realistic setting A_{ref} may contain dozens of concept intents,
making it hard for the user to merely inspect the result of a *Refine*, let alone to
select an alternative. Preferences may help solving this problem. They can be
used to partition the set A_{ref} into blocks of equally preferred concept intents,
which are shown to the user in decreasing order of preference. In this way, the
output is divided into more consumable portions, and is offered to the user in a
sensible way. In order to achieve this goal, we need to determine: (a) a partial
order between concepts based on preferences; and (b) a way to use this order for
defining a suitable partition of A_{ref}.

As for the former goal, it is natural to consider a concept (A, B) preferred over
a concept (A', B') based on a criterion involving the intents B and B', which are
sets of terms and can therefore directly reflect preferences. Since no two concepts
can have the same intents, any order defined on intents can be understood also
as an order defined on concepts. Moreover, since the set of concept intents is a
subset of $\mathcal{P}(\mathcal{A})$, any partial order defined on the latter carries over the former
by inheritance. It follows that goal (a) above amounts to define a partial order
between term sets.

Given two term sets $B, B' \subseteq \mathcal{A}$, B is said to be *preferred over* B', written as
$B \preceq B'$, if for all terms $b \in B$ there exists a term $b' \in B'$ such that $(b, b') \in P$.
It is not difficult to see that $\preceq$ is a pre-order on $\mathcal{P}(\mathcal{A})$: both its reflexivity and
transitivity are implied by those of P. Antisymmetry does not hold for $\preceq$: it
suffices to consider the term sets $\{a, b\}$ and $\{a, b, c\}$ such that (c, b) is in P.

Let us define two term sets B and B' *preferentially equivalent*, written $B \equiv B'$,
if $B \preceq B'$ and $B' \preceq B$. Now $\equiv$ is an equivalence relation on $\mathcal{P}(\mathcal{A})$, whose induced
equivalence classes we denote as $[B]$, *i.e.* $[B] = \{X \subseteq \mathcal{A} \mid X \equiv B\}$. $\preceq$ can be
extended to the so defined equivalence classes as follows:

$$[B] \preceq [B'] \text{ iff } B \preceq B'.$$

As it can be verified, $\preceq$ is a partial order on equivalence classes. Thus by replacing
the notion of term set with that of class of equivalent term sets we have a partial
order on term sets, and have so accomplished goal (a) above. Let us resume
the example on refinement developed in Section 5.3, whose result is shown in
Table 1. In this example, we have

$$A_{ref} = \{\{A, C, D\}, \{C, D, E\}, \{C, D, F\}\}.$$

Assuming that the Hasse diagram of P includes only the pairs (A, E) and (A, F),
the only comparable classes amongst those relevant to A_{ref} are:

$$[\{A, C, D\}] \preceq [\{C, D, E\}] \text{ and}$$
$$[\{A, C, D\}] \preceq [\{C, D, F\}].$$

In order to accomplish goal (b), below we describe how the set A_{ref} of concept intents resulting from a *Refine* is partitioned, leaving aside for simplicity the other information returned to the user. The partition in question is given by the sets $A_1, A_2, \ldots$, defined as ($k \geq 1$):

$$A_k = \min_{\preceq} S_k$$

where S_k consists of the elements of A_{ref} which have not yet been inserted in any block, and is given by ($k \geq 1$):

$$S_1 = A_{ref}$$
$$S_{k+1} = S_k \setminus A_k$$

Thus, A_1 consists of the minimal elements of A_{ref}, A_2 of the minimal elements of A_{ref} after the removal of the minimal elements, and so on.

In our example:

$$S_1 = \{\{A, C, D\}, \{C, D, E\}, \{C, D, F\}\}$$
$$A_1 = \{\{A, C, D\}\}$$
$$S_2 = \{\{C, D, E\}, \{C, D, F\}\}$$
$$A_2 = \{\{C, D, E\}, \{C, D, F\}\}$$
$$S_i = A_i = \emptyset \quad \text{for } i \geq 3.$$

Accordingly, the user is first shown the first row of Table 1 and then the second and third ones.

Formally, it can be shown that, for any set of term sets A_{ref} and partial ordering on $\mathcal{A}$, the following hold:

1. there exists $m \geq 1$ such that for all $j \geq m$, $S_j = A_j = \emptyset$.
2. $\{A_j \mid 1 \leq j < m\}$ is a partition of A_{ref}.
3. for all pairs of term sets $B, B' \in A_{ref}$ such that $[B] \preceq [B']$ and $[B] \neq [B']$, there exist i, j such that $B \in A_i$, $B' \in A_j$ and $i < j$.

The first statement can be made more precise by proving that m actually equals the length of the longest path in the graph $(G, \preceq|_G)$, where $G = \{[B] \mid B \in A_{ref}\}$.

We conclude by observing that the ordering criterion defined above preserves the tractability of the original method. Indeed, the computation of each element of the sequence $A_1, \ldots, A_m$ requires a polynomial amount of time in the size of the Hasse diagram of P, A_{ref} and $\mathcal{A}$. Since also the length m of the sequence, as just argued, is polynomially bound by the size of A_{ref}, we have the polynomial time complexity of the method.

6.2 Using Preferences in Query Enlargement

The result of an *Enlarge* is (see Table 2) the set of the minimal super-concepts of the current concept, the intents of which represent enlargements of the underlying query. Evidently, also in this case the size of this set, hence the amount of information to be displayed to the user may be overwhelming, and ordering can result in a significant benefit. To this end, the same method described in the previous section can be applied.

7   Concluding Remarks

We have seen an approach to query tuning that combines navigation and querying into a single process thus providing a more flexible and more user friendly interaction between the users and the information system.

In the traditional approach, the interaction proceeds by repeating the following two steps (in some order): (1) Query and Terminate, or (2) Navigate. In our approach, the interaction proceeds by repeating the following three steps (in some order) before terminating: (1) Query, (2) Refine, or (3) Enlarge. Here, Refine and Enlarge represent navigation steps that might be interleaved with the Query step and might be repeated several times before termination, *e.g.*, Query-Enlarge-Query-Refine-Query-Enlarge- ... -Terminate.

In order to support the user in selecting, or even just in examining, the possibly many alternative query tunings, we have used preferences. The adopted preferential relation is very liberal, thus resulting into a pre-ordering rather than into a partial ordering as customary for preferences. This problem had been circumvented by considering the ordering induced on equivalence classes.

All the involved problems have been shown to be computationally tractable.

Acknowledgements. The authors gratefully acknowledge the DELOS Network of Excellence on Digital Libraries for partially supporting this work, under Task 2.10 "Modeling of User Preferences in Digital Libraries" of JPA2.

References

1. Formal concept analysis homepage. http://www.upriss.org.uk/fca/fca.html.
2. The open directory project. http://dmoz.org/about.html.
3. C. Carpineto and G. Romano. Information retrieval through hybrid navigation of lattice representations. *International Journal of Human-Computer Studies*, 45(5):553–578, 1996.
4. C. Carpineto and G. Romano. A lattice conceptual clustering system and its application to browsing retrieval. *Machine Learning*, 24(2):95–122, 1996.
5. C. Carpineto and G. Romano. Effective reformulation of boolean queries with concept lattices. In *Proceedings of International Conference on Flexible Query Answering Systems*, number 1495 in Lecture Notes in Artificial Intelligence, pages 83–94, Roskilde, Denmark, May 1998. Springer Verlag.
6. C. Carpineto and G. Romano. Order-theoretical ranking. *Journal of American Society for Information Science*, 51(7):587–601, 2000.
7. Hsinchun Chen, editor. *Journal of the American Society for Information Science. Special Issue: Digital Libraries: Part 1*, volume 51. John Wiley & Sons, Inc., 2000.
8. Hsinchun Chen, editor. *Journal of the American Society for Information Science. Special Issue: Digital Libraries: Part 2*, volume 51. John Wiley & Sons, Inc., 2000.

9. Jan Chomicki. Preference formulas in relational queries. *University at Buffalo, Buffalo, New York ACM Transactions on Database Systems*, 28(4), December 2003.

10. B.A. Davey and H.A. Priestley. *Introduction to lattices and order*, chapter 3. Cambridge, second edition, 2002.

11. B. Ganter and R. Wille. Applied lattice theory: Formal concept analysis. http://www.math.tu.dresden.de/~ganter/psfiles/concept.ps.

12. Bernhard Ganter and Rudolf Wille. *Formal Concept Analysis: Mathematical Foundations*. Springer Verlag, 1st edition, 1999.

13. R. Godin, J. Gecsei, and C. Pichet. Design of a browsing interface for information retrieval. In *Proceedings of SIGIR89, the Twelfth Annual International ACM Conference on Research and Development in Information Retrieval*, pages 32–39, Cambridge, MA, 1989.

14. Andrka H., Ryan M., and Schobbens P-Y. Operators and laws for combining preference relations. *Journal of Logic and Computation*, 12(1):13–53, February 2002.

15. Uta Priss. Lattice-based information retrieval. *Knowledge Organization*, 27(3):132–142, 2000.

16. P. Rigaux and N. Spyratos. Metadata inference for document retrieval in a distributed repository. In *Proceedings of ASIAN'04, 9th Asian Computing Science Conference*, number 3321 in LNCS, Chiang-Mai, Thailand, 8-10 December 2004. Invited Paper.

Processing Ranked Queries with the Minimum Space

Yufei Tao[1] and Marios Hadjieleftheriou[2]

[1] Department of Computer Science, City University of Hong Kong,
Tat Chee Avenue, Hong Kong
taoyf@cs.cityu.edu.hk
[2] Department of Computer Science, Boston University, Boston, USA
marioh@cs.bu.edu

Abstract. Practical applications often need to rank multi-variate records by assigning various priorities to different attributes. Consider a relation that stores students' grades on two courses: *database* and *algorithm*. Student performance is evaluated by an "overall score" calculated as $w_1 \cdot g_{db} + w_2 \cdot g_{alg}$, where w_1, w_2 are two input "weights", and g_{db} (g_{alg}) is the student grade on *database* (*algorithm*). A "top-k ranked query" retrieves the k students with the best scores according to specific w_1 and w_2.

We focus on top-k queries whose k is bounded by a constant c, and present solutions that guarantee low worst-case query cost by using provably the minimum space. The core of our methods is a novel concept, "minimum covering subset", which contains only the necessary data for ensuring correct answers for all queries. Any 2D ranked search, for example, can be processed in $O(\log_B(m/B) + c/B)$ I/Os using $O(m/B)$ space, where m is the size of the minimum covering subset, and B the disk page capacity. Similar results are also derived for higher dimensionalities and approximate ranked retrieval.

1 Introduction

Practical applications often need to rank multi-variate records by assigning various priorities to different attributes. Consider a relation that stores students' grades on two courses: *database* and *algorithm*. Student performance is evaluated by an "overall score" calculated as $w_1 \cdot g_{db} + w_2 \cdot g_{alg}$, where w_1, w_2 are two input "weights", and g_{db} (g_{alg}) is the student's grade on *database* (*algorithm*). A common operation is to retrieve the best k students according to specific weights. For example, a top-k search with $w_1 = 1$ and $w_2 = 0$ returns students with the highest *database* grades, while a query with $w_1 = w_2 = 0.5$ selects students by the sum of their grades on the two courses. In this paper, we consider supporting ranked queries with low worst-case overhead for all user-defined weights.

1.1 Problem Statements

Consider a d-dimensional space, where each axis has a domain $[0, \infty)$. A *weight vector* $w = \{w[1], w[2], ..., w[d]\}$ specifies a positive weight $w[i]$ on each dimension $1 \leq i \leq d$. Given such a vector w, the *score* of a point p in the data space equals $\sum_{i=1}^{d}(w[i] \cdot p[i])$, where $p[i]$ is the coordinate of p on the i-th axis ($1 \leq i \leq d$).

J. Dix and S.J. Hegner (Eds.): FoIKS 2006, LNCS 3861, pp. 294–312, 2006.

Problem 1. Let $\mathcal{D}$ be a set of d-dimensional points. Given a weight vector w, a *top-k ranked query* returns the k points from $\mathcal{D}$ with the highest scores. Let c be a constant (by far) smaller than the cardinality of $\mathcal{D}$. The goal is to minimize the worst-case cost of processing any top-k query with $k \leq c$.

The motivation behind introducing the constant c is that a typical query in practice aims at finding only the "best-few" objects [11], e.g., the 10 best students from the whole university with a huge student population.

Many applications accept approximate answers with low (bounded) error, especially if computing such results requires less (time and space) overhead than the precise ones. Hence, we also consider a novel variation of ranked retrieval, called "top-(k,K) search". For example, a top-$(1,10)$ query reports a single point whose score is at most the 10-th largest in the dataset $\mathcal{D}$. Hence, the query result is *not* unique, since it can be any of the objects whose scores are the 1st, 2nd, ..., 10th highest in $\mathcal{D}$. However, the quality of the result is guaranteed — in the worst case, the retrieved point is the "10-th best" in $\mathcal{D}$. Similarly, a legal outcome of a top-$(3,10)$ query may consist of any 3 points in the top-10 set, and therefore, the number of permissible results equals $\binom{10}{3}$. We are ready to define the second problem tackled in this paper.

Problem 2. A *top-(k,K) ranked query* specifies a weight vector w, and two integers k, K with $k \leq K$. The query result includes any k objects in the top-K set for w. Let c and C be two constants (by far) smaller than the dataset cardinality, and $c \leq C$. The goal is to minimize the worst-case cost of any top-(k,C) query with $k \leq c$.

1.2   Previous Results

While Problem 1 has received considerable attention [10][4][5][7][8][9][11] in the database literature, the previous approaches mostly rely on heuristics which have poor worst-case performance. In particular, they require accessing the entire database to answer a single query [10][4][5][7], or consume space several times the dataset size [8][9].

The only exception is due to Tsaparas et al [11]. They propose an index that occupies $O(c^2 \cdot s/B)$ space and answers a 2D ranked query in $O(\log_B(s/B) + \log_B(c) + c/B)$ I/Os, where c is as defined in Problem 1, s is the size of the "c-skyline" of the dataset, and B is the disk page capacity. Specifically, a *c-skyline* consists of the objects that are not dominated by c other objects (an object p dominates another p' if the coordinates of p are larger on all dimensions). In the dataset of Figure 1, a 1-skyline contains p_2, p_3, p_7, p_4, p_5. Point p_1, for example, is not in the skyline because it is dominated by p_3.

The solutions of [11] are not applicable to higher dimensionalities. To the best of our knowledge, no previous results on Problem 2 exist.

1.3   Our Results

Any method that correctly solves Problem 1 must store a *minimum covering subset*, which is the result union of all the possible (an infinite number of) ranked queries with $k \leq c$. For example, in Figure 1, as clarified later the minimum covering subset for $c = 1$ contains p_2, p_3, p_4, p_5 — the top-1 object for any weight vector must be captured in this subset. Note that, the subset is smaller than the 1-skyline which, as mentioned earlier, also includes p_7 and p_6.

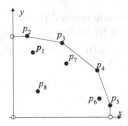

Fig. 1. An example

We propose polynomial-time algorithms for extracting minimum covering subsets in arbitrary dimensionalities. Notice that *the discovery of such a subset immediately improves the worst-case behavior of any previous approaches*. In particular, instead of applying a technique on the original dataset, we can deploy it on the minimum subset directly. Hence, any ranked query can be processed without considering other data not in the subset.

As a second step, we pre-process them into appropriate structures for performing ranked retrieval effectively. Specifically, any 2D ranked query can be answered in $O(\log_B (m/B) + c/B)$ I/Os using $O(m/B)$ space, where m is the size of the minimum covering subset. These bounds significantly improve those of [11]. For higher dimensionalities, a query can be solved in $O(m/B)$ I/Os by storing $O(m/B)$ information. Note that our methods in both scenarios require asymptotically the smallest amount $O(m/B)$ of space.

For Problem 2, there also exists a corresponding "minimum subset" containing the necessary data for ensuring correct results for all queries. If this subset has size m, we develop an index that occupies $O(m'/B)$ space, and processes an approximate 2D ranked query in $O(\log_B(m'/B)+c/B)$ I/Os, where m' is bounded by $(\ln m+1) \cdot m$. In higher-dimensional space, a query can be answered in $O(m'/B)$ I/Os with $O(m'/B)$ space.

The rest of the paper is organized as follows. Section 2 elaborates the definition of minimum covering subsets, and Section 3 discusses their computation in arbitrary dimensionality. Section 4 explains an "incremental" approach for deriving minimum covering subsets. Section 5 presents an index structure that optimizes exact ranked search, while Section 6 discusses approximate retrieval. Section 7 concludes the paper with directions for future work.

2 Minimum Covering Subsets

Let $\mathcal{D}_{\subseteq}$ be a subset of $\mathcal{D}$. We say that $\mathcal{D}_{\subseteq}$ *covers* an exact top-k query if $\mathcal{D}_{\subseteq}$ contains all the k points in its result. Given a constant c, $\mathcal{D}_{\subseteq}$ is a *c-covering subset* if it covers *all* possible top-k queries whose k is at most c. For instance, a "1-covering subset" includes the results of all top-1 queries, *regardless of their weight vectors*, while a "3-covering subset" covers all top-1, top-2, and top-3 queries. Among all the c-covering subsets, the one with the *smallest* size is called the *minimum c-covering subset*, represented as $min\mathcal{D}_{\subseteq}$.

As an example, consider a 1D dataset $\mathcal{D}$ with 10 records $\{1, 2, ..., 10\}$. Accordingly, a weight vector contains a single number w. A top-k query reports the k tuples $p \in \mathcal{D}$ that maximize $w \cdot p$. Clearly, the result simply consists of the k largest numbers in $\mathcal{D}$. Indeed, when the dimensionality equals 1, all top-k queries with the same k produce identical results, independent of w. Therefore, for any integer c, the minimum c-covering subset $min\mathcal{D}_\subseteq$ includes the c largest tuples. For instance, for $c = 1$, $min\mathcal{D}_\subseteq = \{10\}$, and for $c = 2$, $min\mathcal{D}_\subseteq = \{10, 9\}$, etc. Note that $\{10, 9\}$ is also a 1-covering subset, but not a minimal one.

The importance of $min\mathcal{D}_\subseteq$ lies in the fact that it can replace the original dataset $\mathcal{D}$ to correctly answer any top-k query, as long as the parameter k is not larger than c. This replacement significantly reduces the space consumption because, even if the cardinality of $\mathcal{D}$ is huge, $min\mathcal{D}_\subseteq$ may contain only a small number of points. Furthermore, notice that $min\mathcal{D}_\subseteq$ must be stored by any technique that aims at providing correct results to all top-k queries. *Hence, the size of $min\mathcal{D}_\subseteq$ corresponds to the lower bound for the space consumption of ranked retrieval.*

As stated in Problem 2, a top-(k,K) query returns k points in the top-K set for w. As mentioned in Section 1.1, the result is not unique — since any k objects in the top-K set forms a "legal" result, the number of possible results equals $\binom{K}{k}$. Given a subset $\mathcal{D}_\subseteq$ of the dataset $\mathcal{D}$, we say that $\mathcal{D}_\subseteq$ *covers* a top-(k,K) query, if there exist k points in $\mathcal{D}_\subseteq$ that constitute *one* of the $\binom{K}{k}$ legal results. We define the *(c,C)-covering subset* to be a subset of $\mathcal{D}$ that covers all top-(k,K) queries with $k \leq c$ and $K = C$. The (c,C)-covering subset with the smallest size is the *minimum (c,C)-covering subset $min\mathcal{D}_\subseteq^*$*, where the asterisk differentiates it from the notation $min\mathcal{D}_\subseteq$ of a minimum c-covering subset.

We illustrate these concepts using again a 1D dataset $\mathcal{D}$ storing data 1, 2, ..., 10. A possible minimum $(1,5)$-covering subset of $\mathcal{D}$ can involve a single tuple $\{6\}$. Indeed, for any top-$(1,5)$ query, the tuple 6 is always a legal result since it is in the top-5 set. In fact, a minimum $(1,5)$-covering subset can involve any *single* record chosen from $\{6, 7, 8, 9, 10\}$. Similarly, a minimum $(3,5)$-covering subset can be $\{6, 7, 8\}$, or in general, any subset of $\{6, 7, 8, 9, 10\}$ with 3 elements.

The minimum (c,C)-covering subset $min\mathcal{D}_\subseteq^*$ can substitute the original database $\mathcal{D}$ to support top-(k,C) ranked search whose parameter k is at most c. Next, we discuss the computation of minimum c-covering subsets, while (c,C)-covering subsets are the topic of Section 6.

3   Finding Minimum C-Covering Subsets

In this section, we analyze extracting the minimum c-covering subset in arbitrary dimensionality. Section 3.1 first presents some fundamental results, based on which Section 3.2 elaborates the concrete algorithms.

3.1   Basic Results

Lemma 1. *Let $\mathcal{D}$ be a set of multi-dimensional points, and $\mathcal{D}_\subseteq$ be a subset of $\mathcal{D}$. If $\mathcal{D}_\subseteq$ covers all top-c queries on $\mathcal{D}$, then it also covers all top-k queries, for any $1 \leq k \leq c$.*

The lemma has an important corollary:

Corollary 1. *The minimum c-covering subset $minD_\subseteq$ of $\mathcal{D}$ is the union of the results of all (exact) top-c queries.*

Hence, the computation of $minD_\subseteq$ would be simple if we *were* able to execute an infinite number of top-c queries with all possible weight vectors w. In the sequel, we present polynomial-time approaches based on several inherent connections between $minD_\subseteq$ and the "positive convex hull" $\mathcal{PCH}$ of dataset $\mathcal{D}$. To formally define $\mathcal{PCH}$, let us formulate a set $\mathcal{D}'$, which contains all the objects from $\mathcal{D}$, together with $d+1$ "dummy" points. The first one is the origin of the data space, and there is also a dummy point on each dimension, whose coordinate on this dimension equals the maximum coordinate of the points in $\mathcal{D}$ on this axis, and 0 on the others. All the $d+1$ dummy records must appear in the convex hull of $\mathcal{D}'$. The *positive convex hull* $\mathcal{PCH}$ of the original dataset $\mathcal{D}$ includes the non-dummy points in the convex hull of $\mathcal{D}'$.

Figure 1 illustrates an example where $\mathcal{D}$ contains 8 points $p_1, p_2, ..., p_8$. The augmented dataset $\mathcal{D}'$ involves 3 dummy points represented as white dots. The convex hull of $\mathcal{D}'$ contains all the dummy points, together with p_2, p_3, p_4, and p_5. Hence, the positive convex hull $\mathcal{PCH}$ of $\mathcal{D}$ consists of p_2, p_3, p_4, and p_5. Clearly, the time needed to compute the $\mathcal{PCH}$ of any dataset $\mathcal{D}$ is bounded by the cost of obtaining a complete convex hull of $\mathcal{D}$.

It is known [4] that the result of any top-1 query on $\mathcal{D}$ can be found in the $\mathcal{PCH}$ of $\mathcal{D}$, namely:

Lemma 2. *The minimum 1-covering subset of $\mathcal{D}$ is the positive convex hull of $\mathcal{D}$.*

It is natural to wonder whether this lemma can be trivially extended to capture minimum c-covering subsets for arbitrary c. Specifically, should the minimum 2-covering subset of $\mathcal{D}$ be the union of $\mathcal{PCH}(D)$ and $\mathcal{PCH}(D - \mathcal{PCH}(D))$? That is, can we obtain the minimum 2-covering subset by combining the minimum 1-covering subset, and the positive convex hull of the remaining data of $\mathcal{D}$ after excluding the points in $\mathcal{PCH}(D)$ (this corresponds to the Onion technique in [4])? Unfortunately, the answer is negative. For example, in Figure 1, $\mathcal{PCH}(D)$ equals $\{p_2, p_3, p_4, p_5\}$. After removing $\mathcal{PCH}(D)$, the positive convex hull of the remaining objects consists of p_1, p_7, p_6. However, as clarified in the next section, the minimum 2-covering subset of $\mathcal{D}$ has the same content as the minimum 1-covering subset.

3.2   Algorithm for Arbitrary Dimensionality

We prove a reduction that transforms the problem of discovering minimum c-covering subsets $minD_\subseteq$ to finding positive convex hulls. As a result, $minD_\subseteq$ can be obtained using any existing algorithm [3] for computing convex hulls. Given a value c and dataset $\mathcal{D}$, we represent the minimum c-covering subset as $minD_\subseteq(c, \mathcal{D})$. According to Lemma 2, $minD_\subseteq(1, \mathcal{D})$ is equivalent to $\mathcal{PCH}(\mathcal{D})$.

Theorem 1. *For any $c \geq 2$, a minimum c-covering subset can be computed in a recursive manner:*

$$minD_\subseteq(c, \mathcal{D}) = \mathcal{PCH}(D) \bigcup \left(\bigcup_{\forall p \in \mathcal{PCH}(\mathcal{D})} minD_\subseteq(c-1, \mathcal{D} - \{p\}) \right) \quad (1)$$

Algorithm *find-minD* $(c, \mathcal{D})$
1. if $c = 1$
2.   return $\mathcal{PCH}(\mathcal{D})$ //using any convex-hull method
3. else
4. $S = \mathcal{PCH}(\mathcal{D})$
5.   for each point $p \in \mathcal{PCH}(\mathcal{D})$
6. $S = S \cup$ *find-minD* $(c - 1, \mathcal{D} - \{p\})$
7.   return S

Fig. 2. Finding $minD_\subseteq(c, \mathcal{D})$ in any dimensionality

Based on Theorem 1, Figure 2 describes the algorithm *find-minD* for retrieving $minD_\subseteq(c, \mathcal{D})$ in any dimensionality. We illustrate the idea of *find-minD* by using it to find the minimum-2-covering subset on the dataset $\mathcal{D}$ in Figure 1. First, *find-minD* invokes the selected convex-hull algorithm to extract $\mathcal{PCH}(\mathcal{D}) = \{p_2, p_3, p_4, p_5\}$, i.e., the content of $minD_\subseteq(1, \mathcal{D})$. All the points in this set must belong to $minD_\subseteq(2, \mathcal{D})$. To find the other objects in $minD_\subseteq(2, \mathcal{D})$, *find-minD* removes a point, say p_2, from $\mathcal{D}$, and computes the $\mathcal{PCH}$ of the remaining data. The result of this computation is $\mathcal{PCH}(\mathcal{D} - \{p_2\}) = \{p_3, p_4, p_5\}$. Next, $minD_\subseteq(2, \mathcal{D})$ is updated to the union of its current content and $\mathcal{PCH}(\mathcal{D} - \{p_2\})$, which incurs no change to $minD_\subseteq(2, \mathcal{D})$. *Find-minD* performs the above operations with respect to every other point p_3, p_4, and p_5 of $\mathcal{PCH}(\mathcal{D})$ in turn. Namely, for each $i = 3, 4, 5$, it obtains $\mathcal{PCH}(\mathcal{D} - \{p_i\})$, and updates $minD_\subseteq(2, \mathcal{D})$ by union-ing it with $\mathcal{PCH}(\mathcal{D} - \{p_i\})$. It can be easily verified that no modification to $minD_\subseteq(2, \mathcal{D})$ happens, leaving the final result $minD_\subseteq(2, \mathcal{D}) = \{p_2, p_3, p_4, p_5\}$, that is, same as $\mathcal{PCH}(\mathcal{D})$.

Note that *find-minD* needs to compute the positive convex hull $\mathcal{PCH}$ of several different datasets. For example, in our earlier example of computing $minD_\subseteq(2, \mathcal{D})$ in Figure 1, we calculated the $\mathcal{PCH}$ of $\mathcal{D}$, and $\mathcal{D} - \{p_i\}$ for each integer i in the range $2 \le i \le 5$. Hence, we have:

Theorem 2. *Let α be the highest cost of each $\mathcal{PCH}$ computation in* find-minD *(line 4), and β be the maximum number of $\mathcal{PCH}$ points retrieved in each execution of line 4. Then, the cost of* find-minD *is $O(\alpha \cdot \beta^{c-1})$.*

The factor α in Theorem 2 corresponds to the efficiency of the algorithm used by *find-minD* to compute $\mathcal{PCH}$.

Thus, a direct corollary of the theorem is:

Corollary 2. *The execution time of* find-minD *is worse than that of the deployed algorithm for computing convex hulls by at most a polynomial factor β^{c-1}.*

Notice that β is $O(|\mathcal{D}|)$ (the database cardinality) in the worst case. This happens in the very rare case where almost all the points in $\mathcal{D}$ belong to the positive convex hull. For practical datasets, β is fairly small. For instance, for uniform data, β is at the order of $(\ln |\mathcal{D}|)^{d-1}/(d-1)!$ [2], where d is the dimensionality of the dataset. For anti-correlated data (where most points lie around the major diagonal of the data space), β is expected to be a constant. In this case, *find-minD* is asymptotically as fast as computing the convex hull.

4   Incremental Computation of $minD_\subseteq$

Before computing the minimum c-covering subset $minD_\subseteq$, the algorithm in Figure 2 requires that the value of c should be known in advance. In the sequel, we present an alternative method that obtains the $minD_\subseteq$ in a faster "incremental" manner. Specifically, it first finds the minimum 1-covering subset, which is used to discover the 2-covering subset, then the 3-covering, and so on. Our discussion focuses on 2D space in this section, while the extension to higher dimensionalities is presented Section 5.2.

4.1   Slope Space and Its Decomposition

Given a 2D weight vector $w = \{w[1], w[2]\}$, we refer to $\frac{w[2]}{w[1]}$ as the *weight slope* λ. Clearly, λ ranges in the *slope space* $[0, \infty)$. The following lemma is proved in [11]:

Lemma 3. *For an arbitrary integer k, top-k queries with the same weight slope λ return identical results. In particular, if we project each data point onto a ray shot from the origin with slope λ, the query result consists of the k objects whose projected points are the farthest from the origin.*

For example, the top-k objects produced by a query with weight vector $\{10, 20\}$ are the same as those reported by a query with weight vector $\{1, 2\}$ — both queries have a weight slope 2. Figure 3 shows a dataset that contains 3 data points p_1, p_2, and p_3. Ray l has a slope 2, and the projection of point p_i onto l is p_i' for each integer $i \in [1, 3]$. Since p_1' (p_2') is the farthest (2nd farthest) from the origin among the 3 projections, object p_1 (p_2) has the highest (2nd highest) score for the weight slope 2.

In the sequel we characterize a top-k query by the value of k and its weight slope λ. By Corollary 1, $minD_\subseteq(c, D)$ corresponds to the union of the results of top-c queries for all $\lambda \in [0, \infty)$. Imagine that we slowly increase λ from 0, and meanwhile continuously monitor the corresponding top-c set. Since the size of $minD_\subseteq(c, D)$ is finite, there can be only a finite number of top-c changes as λ travels from 0 to ∞. Therefore, we can decompose the slope space $[0, \infty)$ into a set of *disjoint* intervals such that the results are identical for those queries whose weight slopes are in the same interval.

We call the set HD of intervals thus obtained the *top-c homogeneous decomposition* of the slope space. The size of HD equals the total number of times the top-c objects incur a "change" as λ grows from 0 to ∞. A change here means that a data point is

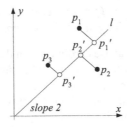

Fig. 3. Deciding score relationship from projections on l

removed from the current top-c set, and another point is added. Note that, no change is generated if only the scores of the existing top-c points switch their relative order. For instance, assume that the top-2 result for the current λ contains p_1, p_2, and the score of p_1 is higher. As λ increases to some value, the score of p_2 becomes larger, but p_1, p_2 still have the 2 largest scores among all the data points. This is *not* counted as a result change.

4.2   Computing Top-1 Homogeneous Decomposition

Let $I = [I_\vdash, I_\dashv)$ be an interval in $\mathcal{HD}$, where $I_\vdash$ ($I_\dashv$) is the *starting* (*ending*) slope of I. For any weight slope in $[I_\vdash, I_\dashv)$, the top-c set is the same, and is represented as $I.S$. To compute the top-1 homogeneous decomposition, we first obtain the minimum 1-covering subset $minD_\subseteq(1, \mathcal{D})$ (i.e., the positive convex hull $\mathcal{PCH}$ of $\mathcal{D}$). Assume, without loss of generality, that the $\mathcal{PCH}$ contains m points p_1, p_2, ..., p_m sorted in descending order of their x-coordinates, where m is the size of $\mathcal{PCH}$. By the definition of the $\mathcal{PCH}$, if we start from p_1 and walk on the hull boundary (passing vertices p_2, ..., p_m in this order), we will make a left turn every time a vertex is encountered. This implies that the slopes of hull edges monotonically decrease in the order that we walk through them. Figure 4 shows an example where the $\mathcal{PCH}$ contains 4 points p_1, ..., p_4. The slope of edge p_1p_2 is larger than that of p_2p_3 (note that both slopes are negative), which in turn is greater than the slope of p_3p_4.

Let us shoot a ray l_i from the origin of the data space, vertically to the hull boundary p_ip_{i+1} for each $i \in [1, m-1]$. The slopes of the $m-1$ rays l_1, l_2, ..., l_{m-1} must increase monotonically. For example, in Figure 4 where $m = 3$, we obtain 3 rays l_1, l_2, and l_3. Since the edge p_1p_2 has a larger slope than p_2p_3, the slope of l_1 is smaller than that of l_2. This can be verified easily with the fact that the product of the slopes of two mutually orthogonal lines equals -1. Similarly, l_2 has a smaller slope than l_3.

Lemma 4. *Assume that the positive convex hull of $\mathcal{D}$ contains m points p_1, p_2, ..., p_m, sorted in descending order of their x-coordinates. Let i be any integer in the range $[1, m-1]$, and x be the slope of the ray shot from the origin perpendicular to the segment connecting p_i and p_{i+1}. Then, p_i has a larger score than p_{i+1} for any weight slope in $[0, x)$, while p_{i+1} has a higher score for any weight slope in (x, ∞).*

For example, let λ_1 be the slope of l_1 in Figure 4. The score of p_1 is larger than that of p_2 (for any weight slope) in the range $[0, \lambda_1)$, while p_2 has a greater score than p_1 in (λ_1, ∞). Similarly, if the slope of l_2 is λ_2, the score of p_2 is larger than that of p_3 for any weight slope smaller than λ_2, while p_3 has a higher score for slopes larger than λ_2.

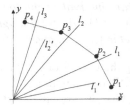

Fig. 4. Top-1 homogeneous decomposition

Based on Lemma 4, the top-1 homogeneous decomposition $\mathcal{HD}$ can be decided as follows. Given $\mathcal{PCH} = \{p_1, ..., p_m\}$ and $m - 1$ rays $l_1, ..., l_{m-1}$ as described earlier, the first interval I_1 in $\mathcal{HD}$ is $[I_{1\vdash}, I_{1\dashv})$, where $I_{1\vdash} = 0$, and $I_{1\dashv}$ equals the slope of l_1. The second interval I_2 starts at $I_{1\dashv}$, and terminates at the slope of ray l_2, and so on. The last interval I_m starts at the slope of I_{m-1}, and ends at ∞. The top-1 object for all weight slopes in interval I_i ($1 \le i \le m$) is point p_i, i.e., $I_i.S = \{p_i\}$.

Lemma 5. *Assume that the positive convex hull of $\mathcal{D}$ contains m points p_1, p_2, ..., p_m, sorted in descending order of their x-coordinates. θ_1, θ_2 are two arbitrary weight slopes, and $\theta_1 < \theta_2$. Let p_i (p_j) be the top-1 object for slope θ_1 (θ_2). Then, the union of top-1 objects for all weight slopes in $[\theta_1, \theta_2]$ equals $\{p_i, p_{i+1}, ..., p_j\}$.*

For example, let θ_1 (θ_2) be the slope of the ray l_1' (l_2') in Figure 4. Point p_1 (p_3) is the object that has the largest score at θ_1 (θ_2). Then, for any weight slope in the range $[\theta_1, \theta_2]$, the top-1 object must be found in the set $\{p_1, p_2, p_3\}$.

4.3 Computing Top-k Homogeneous Decomposition

The subsequent analysis shows that the top-$(k+1)$ homogeneous decomposition $\mathcal{HD}(k + 1, \mathcal{D})$ can be derived efficiently from the top-k counterpart $\mathcal{HD}(k, \mathcal{D})$.

Tail Sets. Consider an arbitrary interval $I = [I_\vdash, I_\dashv) \in \mathcal{HD}(k, \mathcal{D})$. $I.S$ contains k points (in $\mathcal{D}$) with the highest scores for any weight slope in $[I_\vdash, I_\dashv)$. To derive $\mathcal{HD}(k + 1, \mathcal{D})$, we need to decide the union of the top-$(k + 1)$ objects produced by all weight slopes in $[I_\vdash, I_\dashv)$. Let $(\mathcal{D} - I.S)$ be the set of points in $\mathcal{D}$ after excluding those in $I.S$. We define $p_\vdash'$ ($p_\dashv'$) to be the object in $(\mathcal{D} - I.S)$ that has the largest score for the weight slope $I_\vdash$ ($I_\dashv$). By Lemma 2, both $p_\vdash'$ and $p_\dashv'$ appear on the positive convex hull of $(\mathcal{D} - I.S)$. Then, the *tail set $I.TS$ of I contains the vertices of this hull between (but including)* $p_\vdash'$ *and* $p_\dashv'$.

Figure 5a shows an example where the black dots represent the data points in $\mathcal{D}$. Assume an interval $I \in \mathcal{HD}(k, \mathcal{D})$ whose starting value $I_\vdash$ (ending value $I_\dashv$) equals the slope of ray $l_\vdash$ ($l_\dashv$). Objects $p_1, p_2, ..., p_k$ are in the top-k set $I.S$ for all weight slopes in $[I_\vdash, I_\dashv)$ ($p_3, ..., p_{k-1}$ are omitted from the figure). Point $p_\vdash'$ ($p_\dashv'$) is the object that has the highest score at weight slope $I_\vdash$ ($I_\dashv$) among the data in $(\mathcal{D} - I.S)$. Objects $p_\vdash'$, p_2', p_3', $p_\dashv'$ are the vertices between $p_\vdash'$ and $p_\dashv'$ on the positive convex hull of $(\mathcal{D} - I.S)$; hence, they constitute the tail set $I.TS$ of I. According to the following lemma, the result of any top-$(k+1)$ query with weight slope in $[I_\vdash, I_\dashv)$ must be included in the union of $I.S$ and $\{p_\vdash', p_2', p_3', p_\dashv'\}$.

Lemma 6. *Let I be an arbitrary interval in the top-k homogeneous decomposition $\mathcal{HD}(k, \mathcal{D})$. The union of the top-$(k+1)$ objects for weight slopes in $[I_\vdash, I_\dashv)$ equals $I.S \cup I.TS$, where $I.S$ is the top-k set in $[I_\vdash, I_\dashv)$, and $I.TS$ the tail set of I.*

An important step in extracting the tail set of I is to identify points $p_\vdash'$ and $p_\dashv'$, which are the top-1 objects in $(\mathcal{D} - I.S)$ at slopes $I_\vdash$ and $I_\dashv$, respectively. Both points can be efficiently obtained as follows. First, we sort all the intervals in the top-k homogeneous decomposition $\mathcal{HD}(k, \mathcal{D})$ in ascending order of their starting slopes. To decide $p_\vdash'$ ($p_\dashv'$),

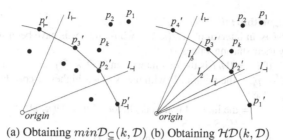

(a) Obtaining $minD_\subseteq(k, D)$ (b) Obtaining $HD(k, D)$

Fig. 5. The tail set and its application

we check whether I is the first (last) interval in the sorted list. If the answer is positive, $p'_\vdash$ ($p'_\dashv$) corresponds to the point in D with the $(k+1)$-st largest x- (y-) coordinate. Otherwise, we take the interval I' that ranks just before (after) I in the sorted list. By the definition of $HD(k, D)$ (see Section 4.1), $I'.S$ involves one object that is not in $I.S$, i.e., the top-k result change at the boundary between I and I'. Then, $p'_\vdash$ ($p'_\dashv$) is set to this object. Figure 6 shows the formal procedures of computing the tail set.

After computing the tail sets of all intervals in $HD(k, D)$, we can obtain the minimum $(k+1)$-covering subset of the dataset D immediately:

Theorem 3. *Given the top-k homogeneous decomposition $HD(k, D)$, the minimum $(k+1)$-covering subset can be decided as:*

$$minD_\subseteq(k + 1, D) = \left(\bigcup_{I \in HD(k, D)} I.S \cup I.TS \right) \qquad (2)$$

Computing $HD(k + 1, D)$. Next we clarify the derivation of $HD(k + 1, D)$ from the tail sets of the intervals in $HD(k, D)$. Before presenting the detailed algorithm, we first discuss the general idea using a concrete example in Figure 5b, which is based on Figure 5a. As mentioned earlier, $I = [I_\vdash, I_\dashv)$ is an interval in $HD(k, D)$, where $I_\vdash$ ($I_\dashv$) equals the slope of ray $l_\vdash$ ($l_\dashv$). The tail set $I.TS$ consists of points $p'_1, p'_2, ..., p'_4$, sorted in descending order of their x-coordinates (note that p'_1 and p'_4 are equivalent to $p'_\vdash$ and $p'_\dashv$ in Figure 5a, respectively). For each segment $p'_i p'_{i+1}$ ($1 \leq i \leq 3$), we shoot a ray l_i from the origin perpendicularly to it. The slope of λ_i of l_i must be larger than $I_\vdash$ and but smaller than $I_\dashv$ (note that l lies between $l_\vdash$ and $l_\dashv$). The 3 numbers λ_1, λ_2, and λ_3 divide $[I_\vdash, I_\dashv)$ into 4 pieces with different top-$(k+1)$ results. To facilitate illustration, let us denote $\lambda_0 = I_\vdash$ and $\lambda_4 = I_\dashv$. Then, for $i \in [0, 3]$, the top-$(k+1)$ set contains p'_i and the objects in $I.S$ at any weight slope in $[\lambda_i, \lambda_{i+1})$. Recall that $I.S = \{p_1, ..., p_k\}$ is the top-k result at any weight slope in I.

Formally, given an interval $I \in HD(k, D)$, the algorithm *break-interval* in Figure 7 divides I into a set $HD(I)$ of disjoint pieces. Each interval $I' \in HD(I)$ is associated with a set $I'.S$ that is the top-$(k+1)$ result for any weight slope in I'. In order to obtain $HD(k+1, D)$, we execute *break-interval* for every interval $I \in HD(k, D)$, after which $HD(k + 1, D)$ corresponds to the union of all $HD(I)$ produced.

There is one minor detail worth mentioning. As will be shown in an example, the $HD(k + 1, D)$ thus decided may contain multiple intervals I whose associated top-$(k+1)$ sets $I.S$ are the same. These intervals should be combined into a single one

Algorithm *tail-set* $(I, \mathcal{HD}(k, \mathcal{D}))$
/* I is an interval in $\mathcal{HD}(k, \mathcal{D})$, whose intervals have been sorted
by their starting values */
1. if $I_{\vdash} = 0$ //I is the first interval in $\mathcal{HD}(k, \mathcal{D})$
2. $p'_{\vdash}$ = the point in $\mathcal{D}$ with the $(k{+}1)$-st highest x-coordinate
3. else
4. I' = the interval that ranks before I in $\mathcal{HD}(k, \mathcal{D})$
5. $p'_{\vdash} = I'.S - I.S$
6. if $I_{\dashv} = \infty$ //I is the last interval in $\mathcal{HD}(k, \mathcal{D})$
7. $p'_{\dashv}$ = the point in $\mathcal{D}$ with the $(k{+}1)$-st highest y-coordinate
8. else
9. I' = the interval that ranks after I in $\mathcal{HD}(k, \mathcal{D})$
10. $p'_{\dashv} = I'.S - I.S$
11. $I.TS$ = the set of vertices between (including) $p'_{\vdash}$ and $p'_{\dashv}$ on
    the convex hull of $\mathcal{D} - I.S$

Fig. 6. Algorithm for computing the tail set

Algorithm *break-interval* $(I, I.S, I.TS)$
/* I is an interval in $\mathcal{HD}(k, \mathcal{D})$; $I.S$ is the top-k set at any
weight slope in I; $I.TS$ is the tail set of I */
1. $\theta_{last} = I_{\vdash}$; $\mathcal{HD}(\mathcal{I}) = \emptyset$; m = number of points in $I.TS$
2. sort the points in $I.TS$ in descending order of their
    x-coordinates; let the sorted order be $\{p'_1, p'_2, ..., p'_m\}$
3. for $i = 1$ to $m - 1$
4.   shoot a ray from the origin perpendicularly to the
       segment connecting p'_i and p'_{i+1}
5. λ = the slope of the ray
6.   create an interval $I_i = [I_{i\vdash}, I_{i\dashv})$ with $I_{i\vdash} = \theta_{last}$,
       $I_{i\dashv} = \lambda$, and $I_i.S = p'_i \cup I.S$
7. $\mathcal{HD}(\mathcal{I}) = \mathcal{HD}(\mathcal{I}) \cup \{I_i\}$, and $\theta_{last} = \lambda$
8. add to $\mathcal{HD}(\mathcal{I})$ $I_m = [\theta_{last}, I_{\dashv})$ with $I_m.S = p'_m \cup I.S$
9. return $\mathcal{HD}(\mathcal{I})$

Fig. 7. Algorithm for breaking an interval in $\mathcal{HD}(k, \mathcal{D})$ into ones in $\mathcal{HD}(k + 1, \mathcal{D})$

which spans all their respective slope ranges. This *duplicate-removal* process, as well as the overall algorithm for computing $\mathcal{HD}(k + 1, \mathcal{D})$ is shown in Figure 8.

An Example. Consider Figure 9, where p_1 and p_2 are the vertices on the positive convex hull of a dataset $\mathcal{D}$. Hence, they constitute the minimum 1-covering subset $minD_{\subseteq}(1, \mathcal{D})$. Let λ_1 be the slope of ray l_1, which passes the origin and is vertical to segment p_1p_2. The top-1 homogeneous decomposition $\mathcal{HD}(1, \mathcal{D})$ contains two intervals $I_1 = [0, \lambda_1)$ and $I_2 = [\lambda_1, \infty)$. The top-1 object (for any weight slope) in I_1 is $I_1.S = \{p_1\}$, and that in I_2 is $I_2.S = \{p_2\}$.

Next we compute $minD_{\subseteq}(2, \mathcal{D})$ and $\mathcal{HD}(2, \mathcal{D})$, by considering each interval of $\mathcal{HD}(1, \mathcal{D})$ in turn. For the first interval $I_1 = [0, \lambda_1)$, its tail set $I_1.TS$ consists of points p_4 and p_1. Accordingly, the algorithm *break-interval* in Figure 7 divides I_1 into two

Algorithm *ho-decomp* ($\mathcal{HD}(k,\mathcal{D})$)
1. $minD_\subseteq(k+1,\mathcal{D}) = \emptyset$; $\mathcal{HD}(k+1,\mathcal{D}) = \emptyset$
2. assume that $\mathcal{HD}(k,\mathcal{D})$ contains m intervals $I_1, ..., I_m$
   sorted in ascending order of their starting values
3. for $i = 1$ to m
4. $I.TS = tail\text{-}set\ (I_i, \mathcal{HD}(k,\mathcal{D}))$
5. $minD_\subseteq(k+1,\mathcal{D})\cup = I.S \cup I.TS$
6. $\mathcal{HD}(k+1,\mathcal{D})\cup = break\text{-}interval\ (I, I.S, I.TS)$
7. $\mathcal{HD}(k+1,\mathcal{D}) = remove\text{-}duplicate\ (\mathcal{HD}(k+1,\mathcal{D}))$
8. return $\mathcal{HD}(k+1,\mathcal{D})$

Algorithm *remove-duplicate* ($\mathcal{HD}(k+1,\mathcal{D})$)
//assume that $\mathcal{HD}(k+1,\mathcal{D})$ contains m' intervals $I'_1, ..., I'_{m'}$
1. $\delta\mathcal{HD}_- = \delta\mathcal{HD}_+ = \emptyset$
2. sort all intervals in $\mathcal{HD}(k+1,\mathcal{D})$ in ascending order
   of their starting slopes;
3. $i = 1$
4. while $i \le m - 1$
5. j = the largest integer s.t. $I'_i.S = I'_{i+1}.S = ... = I'_j.S$
6.   add $I'_i, I'_{i+1}, ..., I'_j$ into $\delta\mathcal{HD}_-$
7.   create a new interval $I = \cup_{x=i}^{j} I'_x$ with $I.S = I'_i.S$
8. $i = j + 1$
9. $\mathcal{HD}(k+1,\mathcal{D}) = \mathcal{HD}(k+1,\mathcal{D}) - (\delta\mathcal{HD}_-) \cup (\delta\mathcal{HD}_+)$

Fig. 8. Algorithm for computing $\mathcal{HD}(k+1,\mathcal{D})$

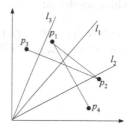

Fig. 9. Illustration of the algorithm in Figure 8

intervals $I'_1 = [0, \lambda_2)$ and $I'_2 = [\lambda_2, \lambda_1)$, where λ_2 is the slope of ray l_2 perpendicular to segment p_1p_4. The top-2 result in I'_1 (I'_2) is $I'_1.S = \{p_2, p_4\}$ ($I'_2.S = \{p_2, p_1\}$). Let $\mathcal{HD}(I_1)$ be the set $\{I'_1, I'_2\}$.

Similarly, we examine the second interval $I_2 = [\lambda_1, \infty)$ of $\mathcal{HD}(1, \mathcal{D})$. Its tail set $I_2.TS$ includes objects p_2, p_3. *Break-interval* divides I_2 at the slope λ_3 of ray l_3 orthogonal to segment p_2p_3. Specifically, I_2 is broken into $\mathcal{HD}(I_2) = \{I'_3, I'_4\}$, where $I'_3 = [\lambda_1, \lambda_3)$ and $I'_4 = [\lambda_3, \infty)$. The top-2 result in I'_3 is $I'_3.S = \{p_2, p_1\}$, and that in I'_4 is $I'_4.S = \{p_3, p_1\}$.

Theorem 3 shows that the minimum 2-covering subset involves all the data in Figure 9, which equals the union of $I_1.S$, $I_1.TS$, $I_2.TS$ and $I_2.TS$. Furthermore, by merging $\mathcal{HD}(I_1)$ and $\mathcal{HD}(I_2)$, we obtain a top-2 homogeneous decomposition $\mathcal{HD}(2, \mathcal{D})$ with 4 intervals $I'_1, ..., I'_4$. The top-2 results in I'_2 and I'_3, however, are both

$\{p_2,\ p_1\}$. Therefore, algorithm *remove-duplicate* in Figure 8 combines these two intervals into one $I_5' = I_2' \cup I_3' = [\lambda_2, \lambda_3)$, with $I_5'.S = \{p_2, p_1\}$. Therefore, the final $\mathcal{HD}(2, \mathcal{D})$ involves only 3 elements: I_1', I_5', and I_3'.

4.4   Analysis

The subsequent discussion aims at bounding (i) the size of the top-c homogeneous decomposition $\mathcal{HD}(c, \mathcal{D})$, and (ii) the time of computing $\mathcal{HD}(c, \mathcal{D})$. We have:

Theorem 4. *The number of intervals in the top-c homogeneous decomposition $\mathcal{HD}(c, \mathcal{D})$ is asymptotically the same as the number of points in the corresponding minimum c-covering subset $minD_\subseteq(c, \mathcal{D})$.*

Now we analyze the cost of computing top-c homogeneous decompositions.

Theorem 5. *Let α be the time of computing the convex hull of $\mathcal{D}$, and α_i the highest cost of computing the tail set of an interval in $\mathcal{HD}(i, \mathcal{D})$ ($2 \leq i \leq c$). $\mathcal{HD}(c, \mathcal{D})$ and $minD_\subseteq(c, \mathcal{D})$ can be computed in $O(\alpha + \sum_{i=2}^{c}(\alpha_i \cdot |minD_\subseteq(i - 1, \mathcal{D})|))$ time, where $|minD_\subseteq(i - 1, \mathcal{D})|$ is the number of objects in $minD_\subseteq(i - 1, \mathcal{D})$.*

Factor α_i ($1 \leq i \leq c$) depends on the concrete convex-hull algorithm for calculating the tail set. The theorem indicates that $minD_\subseteq(c, \mathcal{D})$ can be computed in shorter time in the 2D space than the algorithm presented in Section 3 (which applies to any dimensionality). To better illustrate this, we utilize the fact that $|minD_\subseteq(i, \mathcal{D})|$ is bounded by $|minD_\subseteq(c, \mathcal{D})|$ for any $i < c$, which results in:

Corollary 3. *The 2D minimum c-covering subset can be computed in $O(\alpha + \alpha_{max} \cdot |minD_\subseteq(c - 1, \mathcal{D})| \cdot (c - 1))$ time, where α_{max} equals the maximum of $\alpha_2, \alpha_3, ..., \alpha_c$ defined in Theorem 5.*

5   Ranked Indexes

Once the minimum c-covering subset $minD_\subseteq(c, \mathcal{D})$ has been discovered, we can correctly answer any top-k query, with arbitrary weight slope and $k \leq c$, by performing $O(m/B)$ I/Os, where m is the number of points in $minD_\subseteq(c, \mathcal{D})$, and B is the size of a disk page. Although being relatively straightforward, this method constitutes the first solution for ranked retrieval in any dimensionality that does not require examining the entire dataset in the worst case.

In the next section, we show how to pre-process $minD_\subseteq(c, \mathcal{D})$ to further reduce query cost in the 2D space. In Section 5.2, we present pessimistic results that explain why a similar approach is not feasible for dimensionalities $d \geq 3$.

5.1   A 2D Solution

As an obvious approach, we could extract the top-c homogeneous decomposition $\mathcal{HD}(c, \mathcal{D})$. Assume that it contains m intervals $I_1, I_2, ..., I_m$, sorted in ascending order of their starting values $I_{1\vdash}, ..., I_{m\vdash}$, which are indexed by a B-tree. The B-tree entry for

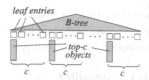

Fig. 10. Indexing the top-c homog. decomp.

each $I_{i\vdash}$ ($1 \leq i \leq m$) is associated with a pointer to a sequence of $O(c/B)$ pages, storing the c objects in $I_i.S$. Given a top-k query with weight slope λ, we first locate the leaf entry I_{i-} in the B-tree that is the largest among all the leaf entries smaller than λ. (i.e., λ falls in the corresponding interval I_i). This requires $O(\log_B m)$ I/Os. Following the pointer stored with I_{i-}, we retrieve $I_i.S$ (using $O(c/B)$ I/Os), and report the k records in $I_i.S$ with the highest scores at λ. The total query cost is $O(\log_B m + c/B)$ I/Os.

The space consumption of this approach is $O(c \cdot m/B)$ pages. We present an alternative solution that achieves the same query time but with only $O(m/B)$ space. Recall that for two consecutive intervals I_i and I_{i+1}, $I_i.S$ differs from $I_{i+1}.S$ by exactly one point. Motivated by this, we store $I.S$ in a compact way as illustrated in Figure 10. For the first interval I_1, c/B pages are used to record the c objects of $I_1.S$ in the same way as the previous method. For the next $c - 1$ intervals I_i ($2 \leq i \leq c$), we do not materialize the full $I_i.S$. Instead, we keep only the difference between $I_i.S$ and $I_{i-1}.S$ in the (B-tree) leaf entry I_i using $O(1)$ space. The above process is repeated for the next c intervals $I_{c+1}, I_{c+2}, ..., I_{2c}$. Specifically, we write to the disk the complete $I_{c+1}.S$. For I_i with $c + 2 \leq i \leq 2c$, only the $O(1)$ result changes (between $I_i.S$ and $I_{i-1}.S$) are kept. Then, a similar process is performed for still the next c intervals, and so on. Since $O(c/B)$ space is allocated for every c intervals, the total space consumption is $O(\frac{m}{c} \cdot c/B) = O(m/B)$ pages.

Given a top-k query with weight slope λ, we first identify the largest leaf entry $I_{i\vdash}$ smaller than λ in the same way as in the previous solution. First, we scan, at the leaf level, the preceding leaf entries $I_{(i-1)\vdash}, I_{(i-2)\vdash}, ...$ (in this order), until finding the first entry $I_{i'\vdash}$ ($i' \leq i$) whose $I_{i'\vdash}.S$ is completely stored. Then, we create a copy S of $I_{i'\vdash}.S$ in the memory, and re-visit the leaf entries (that were just scanned) in a *reverse* order: $I_{(i'+1)\vdash}, I_{(i'+2)\vdash}, ..., I_{i\vdash}$. At each $I_{j\vdash}$ for $i' + 1 \leq j \leq i$, we update S to the top-c result $I_j.S$ (of any weight slope) in interval I_j by applying the $O(1)$ result changes recorded in $I_{j\vdash}$. Hence, when we arrive at $I_{i\vdash}$, the content of S becomes $I_i.S$, from which the k records with the highest scores at the query slope λ are returned. Since we need to trace (from $I_{i\vdash}$ to $I_{i'\vdash}$) at most $O(c)$ leaf entries (by accessing $O(c/B)$ pages), the query cost is bounded by $O(\log_B m + c/B)$ I/Os.

By Theorem 4, the size m of $\mathcal{HD}(c, \mathcal{D})$ is asymptotically the same as the number of points in $min\mathcal{D}_\subseteq(c, \mathcal{D})$. Therefore, we have:

Theorem 6. *Given a constant c, we can pre-process a 2D dataset $\mathcal{D}$ into a structure that consumes $O(m/B)$ space, and answers any top-k query with $k \leq c$ in $O(\log_B (m/B) + c/B)$ I/Os, where m is the size of the minimum c-covering subset.*

Evidently, the 2D ranked index presented earlier consumes the smallest amount of space, since any solution that ensures correct answers for all ranked queries must store at least $minD_{\subseteq}(c, D)$. In practice the disk page size B is usually fairly large (a typical value is 4k bytes), and is asymptotically comparable to c (i.e., $c = O(B)$). In this case, the query performance of our solution becomes $\log_B(m/B)$.

5.2   Discussion on Higher Dimensionalities

The technique in the previous section can be extended to dimensionality $d > 2$ for achieving logarithmic query cost. Such an extension, however, turns out to be purely of theoretical interests because it requires expensive space consumption for large c (and hence violates our goal of using the minimum space to support ranked retrieval). In the sequel, we discuss this in the 3D space, since the analysis for even higher dimensionalities is similar.

Given a 3D weight vector $w = \{w[1], w[2], w[3]\}$, we define its *slope vector* λ as a 2D vector $\{\frac{w[1]}{w[3]}, \frac{w[2]}{w[3]}\}$. Hence, λ can be regarded as a point in a 2D *slope space*, where both axes have domain $[0, \infty)$. Similar to Lemma 3, two weight vectors with the same slope vector have identical top-k sets (for any k). Therefore, in the sequel, we represent a top-k query equivalently using k and its slope vector λ.

Lemma 7. *Given two 3D points p_1 and p_2, let pl be a function of 2D slope vector $\lambda = \{\lambda[1], \lambda[2]\}$:*

$$pl(\lambda) = \sum_{i=1}^{2} (\lambda[i] \cdot (p_1[i] - p_2[i])) + (p_1[3] - p_2[3]) \qquad (3)$$

Then, p_1 has a higher score than p_2 for all slope vectors λ in a half-plane $pl(\lambda) > 0$, while the score of p_2 is higher in $pl(\lambda) < 0$. The scores of the two points are identical for λ on the line $pl(\lambda) = 0$.

Given the c-covering minimum subset $minD_{\subseteq}$ of a 3D dataset D, the slope space can be divided into a set of disjoint regions, such that the top-c sets for all slope vectors in a region are the same. Following the terminology in Section 4, we call such a division a *top-c homogeneous decomposition*, and denote it as $HD(c, D)$. In the sequel, we first analyze the top-1 homogeneous decomposition $HD(1, D)$, and then generalize the discussion to top-c.

Let $minD_{\subseteq}$ contain m points $p_1, p_2, ..., p_m$. To compute $HD(1, D)$, for each p_i $(1 \le i \le m)$, we obtain $m - 1$ half-planes in the form of $pl(\lambda) > 0$ by defining pl as in equation 3 using p_i and every other point in $minD_{\subseteq}$. Then, the intersection of these $m - 1$ half-planes includes all the slope vectors for which p_i has the highest score among all points in the original dataset D. We refer to this intersection area as the *valid region* of p_i. Observe that the m valid regions of all objects in $minD_{\subseteq}$ are disjoint and cover the entire slope space. Since each region is bounded by at most m edges, the total complexity of all regions equals $O(m^2)$.

Therefore, a 3D top-1 search is reduced to a 2D point-location problem. Specifically, we first obtain the 2D slope vector λ of the query, and identify the valid region that contains λ. Then, the point associated with the region is the query result. Using an efficient point-location structure [1], we can answer a top-1 query in $O(\log_B(m/B))$ I/Os using $O(m^2/B)$ space.

The problem with this technique is that the space consumption is no longer minimum (i.e., size m of $minD_\subseteq$). The situation is even worse for top-c queries with $c > 1$. In this case, we can generalize the above derivation and prove that any top-c query in the 3D space can be answered in $O(c \cdot log_B(m/B))$ I/Os using $O(m^{c+1}/B)$ space, where m is the size of the minimum c-covering subset $minD_\subseteq$. As c increases, the space consumption may become prohibitive, rendering this solution impractical. In this case, a better method is to answer a query by simply scanning $minD_\subseteq$ without using any structure, as mentioned at the beginning of Section 5.

6 Approximate Ranked Retrieval

In this section, we consider approximate ranked queries formulated in Problem 2. Specifically, given a weight vector w, a top-(k,K) query returns k objects among the K data points with the highest scores for w. Section 6.1 first elaborates the characteristics of (c,C)-covering subsets, based on which Section 6.2 discusses their computation in 2D space. Section 6.3 presents query processing algorithms for all dimensionalities.

6.1 Properties of (c,C)-Covering Subsets

As discussed in Section 4.1, in the 2D space, the relative order of objects' scores is determined by the slope of the corresponding weight vector. Hence, we will characterize a top-(k,K) query using three values: k, K, and the weight slope. Recall that a (c,C)-covering subset covers any top-(k,C) query with $k \le c$. Specifically, for any value k and weight slope λ, the subset always contains c objects in the top-C set at λ. Among all these subsets, the smallest one is called the *minimum (c,C)-covering subset*, represented as $minD_\subseteq^*(c, C, D)$.

It is easy to see that the minimum C-covering subset $minD_\subseteq (C, D)$ (see the previous sections) is a (c,C)-covering subset. In fact, for each weight slope λ, $minD_\subseteq(C, D)$ includes the corresponding top-C set. Hence, a top-(k,C) query with slope λ can be answered by returning any k points in the top-C set. The following lemma reveals the relationship between $minD_\subseteq(C, D)$ and $minD_\subseteq^*(c, C, D)$.

Lemma 8. $minD_\subseteq^*(c, C, D) \subseteq minD_\subseteq(C, D)$.

The lemma motivates a strategy for computing $minD_\subseteq^*(c, C, D)$. Specifically, we first retrieve $minD_\subseteq(C, D)$, and then eliminate the points of $minD_\subseteq(C, D)$ that are not needed for top-(c,C) processing. The remaining data constitute $minD_\subseteq^*(c, C, D)$. In order to identify the objects necessary for top-(c,C) queries, we resort to an interesting connection between (c,C)-covering subsets and the top-C homogeneous decomposition $HD(C, D)$, illustrated below.

Assume that $HD(C, D)$ contains m intervals $I_1, I_2, ..., I_m$. Each $I_i = [I_{i\vdash}, I_{i\dashv})$, for $1 \le i \le m$, is associated with a set $I_i.S$ consisting of the C objects having the highest scores at any weight slope in $[I_{i\vdash}, I_{i\dashv})$. Therefore, any c points in $I_i.S$ form a legal result for a top-(c,C) query whose weight slope falls in $[I_{i\vdash}, I_{i\dashv})$.

For each object $p \in minD_\subseteq(C, D)$, we construct a *legal set* $p.LS$, containing the intervals of $HD(C, D)$ whose top-C sets involve p. For example, consider Figure 9 where, as discussed at the end of Section 4.3, $minD_{\subseteq}(2, \mathcal{D})$ involves all the points shown in the figure. $\mathcal{HD}(2, \mathcal{D})$ includes three intervals $I_1 = [0, \lambda_2)$, $I_2 = [\lambda_2, \lambda_3)$, and $I_3 = [\lambda_3, \infty)$, where λ_2 (λ_3) is the slope of ray l_2 (l_3). The top-2 set $I_1.S$ of I_1 equals $\{p_2, p_4\}$, while $I_2.S = \{p_2, p_1\}$, and $I_3.S = \{p_3, p_1\}$. Hence, the legal set $p_1.LS$ of p_1 contains intervals I_2 and I_3 since their top-2 sets include p_1. Similarly, $p_2.LS = \{I_1, I_2\}$, $p_3.LS = \{I_3\}$, and $p_4.LS = \{I_1\}$.

Lemma 9. *Let $\mathcal{D}_{\subseteq}$ be a subset of $minD_{\subseteq}(C, \mathcal{D})$, and also a (c,C)-covering subset. Then, each interval in $\mathcal{HD}(C, \mathcal{D})$ is included in the legal sets of at least c points in $\mathcal{D}_{\subseteq}$.*

In Figure 9, for instance, $\{p_1\}$ is *not* a (1,2)-covering subset since interval I_3, which is in $\mathcal{HD}(2, \mathcal{D})$, does not belong to $p_1.LS$. Set $\{p_1, p_2\}$, on the other hand, is a (1,2)-covering subset because each interval in $\mathcal{HD}(2, \mathcal{D})$ is in the legal set of either p_1 or p_2 — I_1 is in $p_2.LS$, while I_2 and I_3 belong to $p_1.LS$. Similarly, $\{p_2, p_3\}$ is also a (1,2)-covering subset.

6.2 Computing (c,C)-Covering Subsets

Lemmas 8 and 9 indicate that finding the minimum (c,C)-covering subset is equivalent to extracting the smallest number of points from $minD_{\subseteq}(C, \mathcal{D})$, such that each interval in $\mathcal{HD}(C, \mathcal{D})$ is included in the legal sets of at least c retrieved points. This is an instance of the *minimum set cover* problem which, unfortunately, is NP-hard [6].

We provide a greedy solution which computes a (c,C)-covering subset, whose cardinality is larger than the minimum size by only a small factor. The pseudo-code of the algorithm, called *find-cCsub*, is presented in Figure 11, which assumes that the legal sets of all points in $minD_{\subseteq}(C, \mathcal{D})$ have been obtained. *Find-cCsub* maintains a set $\mathcal{D}_{\subseteq}$, which is empty at the beginning of the algorithm, and becomes the produced (c,C)-covering subset at termination. For each interval I in $\mathcal{HD}(c, \mathcal{D})$, we keep a counter $I.cnt$, which equals the number of points in the current $\mathcal{D}_{\subseteq}$ whose legal sets include I. The counter is set to 0 initially.

Find-cCsub executes in iterations. In each iteration, it identifies the point p in $minD_{\subseteq}(C, \mathcal{D})$ with the largest legal set $p.LS$, and incorporates p into $\mathcal{D}_{\subseteq}$. For each interval I in $P.LS$, the counter $I.cnt$ is increased by 1, reflecting the fact that a point whose legal set includes I has been newly added to $\mathcal{D}_{\subseteq}$. Once $I.cnt$ reaches c, I is removed from the legal set of every point in $minD_{\subseteq}(C, \mathcal{D})$ — it does not need to be considered in the remaining execution. Finally, the object p inserted to $\mathcal{D}_{\subseteq}$ in this iteration is eliminated from $minD_{\subseteq}(C, \mathcal{D})$. The algorithm terminates if the legal sets of all remaining data in $minD_{\subseteq}(C, \mathcal{D})$ are empty. Otherwise, it performs another iteration.

Theorem 7. *The size of the subset returned by the algorithm in Figure 11 is at most $(\ln \gamma + 1)$ times larger than that of the minimum (c,C)-covering subset, where γ is the maximum cardinality of the legal sets of the points in $minD_{\subseteq}(C, \mathcal{D})$.*

Note that $\gamma = |\mathcal{HD}(C, \mathcal{D})|$ in the worst case, when the legal set of an object in $minD_{\subseteq}(C, \mathcal{D})$ contains all the intervals in $\mathcal{HD}(C, \mathcal{D})$. Since $|\mathcal{HD}(C, \mathcal{D})| = O(|minD_{\subseteq}(C, \mathcal{D})|)$ (Theorem 4), the size of the subset produced by *find-cCsub* is larger than that of the minimum (c,C)-covering subset by a factor of at most

Algorithm *find-cCsub* $(c, minD_\subseteq(C, D), HD(C, D))$
/* the legal set of every point in $minD_\subseteq(C, D)$ has been computed */
1. $D_\subseteq = \emptyset$; $I.cnt = 0$ for each interval $I \in HD(C, D)$
2. while (a point in $minD_\subseteq(C, D)$ has a non-empty legal set)
3. p = point in $minD_\subseteq(C, D)$ with the largest legal set
4. $D_\subseteq = D_\subseteq \cup \{p\}$
5.   for each interval I in the legal set $p.LS$
6. $I.cnt = I.cnt + 1$
7.   if $I.cnt = c$ then remove I from the legal set of
           every point in $minD_\subseteq(C, D)$
8. $minD_\subseteq(C, D) = minD_\subseteq(C, D) - \{p\}$
9. return $D_\subseteq$

Fig. 11. Finding a (c,C)-covering subset

$\ln |minD_\subseteq(C, D)| + 1$. In practice, we expect γ to be much smaller than $|HD(C, D)|$, and as a result, the subset obtained by *find-cCsub* has a cardinality close to the theoretically minimum value.

Theorem 8. *The algorithm in Figure 11 finishes $O(m^2)$ time, where m is the number of points in $minD_\subseteq(C, D)$.*

6.3 Query Processing

Let $D_\subseteq$ be a (c,C)-covering subset computed by the method in Figure 11. Given a top-(k,C) query ($k \leq c$) with weight slope λ, we simply return the k points in $D_\subseteq$ having the highest scores at λ. Due to the properties of (c,C)-covering subsets, these k objects are guaranteed to be a legal result. The query performance can be optimized using exactly the same techniques as in Section 4, by replacing the original dataset D with $D_\subseteq$. All the bounds on the execution cost and space consumption in Section 4 are still valid.

The above analysis applies to dimensionality 2. For higher-dimensional space, we can compute the minimum c-covering subset $minD_\subseteq(c, D)$ using the algorithm in Figure 2. Note that $minD_\subseteq(c, D)$ is a (c,C)-covering subset, and hence, can be deployed to substitute the original dataset D to support top-(k,C) queries ($k \leq c$). This leads to an approach that solves any such query in $O(|minD_\subseteq(c, D)| / B)$ I/Os. It is important to note that we computed (c,C)-covering subset in the 2D space based on $minD_\subseteq(C, D)$ (note the capitalized C), as is required by Lemma 8.

7 Conclusions

This paper introduced the concept of "minimum covering subset", which is the smallest subset of the database that must be stored by any ranked-retrieval algorithms to ensure correct results for all queries. For 2D space, we developed a technique that consumes $O(m/B)$ space and solves any top-k query in $O(\log_B(m/B) + c/B)$ I/Os, where c is the upper bound of k, m the size of the minimum c-covering subset, and B the disk page capacity. For higher dimensionality, our approach requires $O(m/B)$ space and query time.

As a second step, we provided the first solutions for approximate ranked retrieval that do not require inspecting the entire database in the worst case. In the 2D scenario, our method occupies $O(m'/B)$ space and solves a top-(k,C) query with $k \leq c$ in $O(\log_B(m'/B)+c/B)$ where m' is larger than the size of the minimum (c,C)-covering subset by a small bounded factor. For higher dimensionality, we showed that a top-(k,C) $(k \leq c)$ query can be answered in $O(m/B)$ I/Os and space, where m is the size of the minimum c-covering subset.

This work lays down a foundation for continued investigation of ranked queries. A promising direction for future work is to study faster algorithms for computing the minimum covering subsets, utilizing the properties presented earlier. Another interesting problem is the dynamic maintenance of minimum subsets at the presence of data updates. Specialized approaches may be developed to explore the tradeoff between update efficiency and the space overhead.

Acknowledgement

This work was fully supported by RGC Grant CityU 1163/04E from the government of HKSAR.

References

1. L. Arge, A. Danner, and S.-M. Teh. I/o-efficient point location using persistent b-trees. In *ALENEX*, pages 82–92, 2003.
2. J. L. Bentley, H. T. Kung, M. Schkolnick, and C. D. Thompson. On the average number of maxima in a set of vectors and applications. *J. ACM*, 25(4):536–543, 1978.
3. M. Berg, M. Kreveld, M. Overmars, and O. Schwarzkopf. *Computational Geometry: Algorithms and Applications*. Springer, 2000.
4. Y.-C. Chang, L. D. Bergman, V. Castelli, C.-S. Li, M.-L. Lo, and J. R. Smith. The onion technique: Indexing for linear optimization queries. In *SIGMOD*, pages 391–402, 2000.
5. S. Chaudhuri and L. Gravano. Evaluating top-k selection queries. In *VLDB*, pages 397–410, 1999.
6. T. H. Cormen, C. Stein, R. L. Rivest, and C. E. Leiserson. *Introduction to Algorithms*. McGraw-Hill Higher Education, 2001.
7. D. Donjerkovic and R. Ramakrishnan. Probabilistic optimization of top n queries. In *VLDB*, pages 411–422, 1999.
8. V. Hristidis, N. Koudas, and Y. Papakonstantinou. Prefer: a system for the efficient execution of multi-parametric ranked queries. In *SIGMOD*, pages 259–270, 2001.
9. V. Hristidis and Y. Papakonstantinou. Algorithms and applications for answering ranked queries using ranked views. *The VLDB Journal*, 13(1):49–70, 2004.
10. A. Marian, N. Bruno, and L. Gravano. Evaluating top-k queries over web-accessible databases. *ACM Trans. Database Syst.*, 29(2):319–362, 2004.
11. P. Tsaparas, T. Palpanas, Y. Kotidis, N. Koudas, and D. Srivastava. Ranked join indices. In *ICDE*, 2003.

Hybrid Minimal Spanning Tree and Mixture of Gaussians Based Clustering Algorithm

Agnes Vathy-Fogarassy[1], Attila Kiss[2], and Janos Abonyi[3]

[1] University of Veszprem, Department of Mathematics and Computing,
P.O. Box 158, Veszprém, H-8201 Hungary
vathya@almos.vein.hu

[2] Eötvös Lóránd University, Department of Information Systems,
P.O. Box 120, H-1518 Budapest, Hungary
kiss@ullman.inf.elte.hu

[3] University of Veszprem, Department of Process Engineering,
P.O. Box 158, Veszprém, H-8201 Hungary
abonyij@fmt.vein.hu

Abstract. Clustering is an important tool to explore the hidden structure of large databases. There are several algorithms based on different approaches (hierarchical, partitional, density-based, model-based, etc.). Most of these algorithms have some discrepancies, e.g. they are not able to detect clusters with convex shapes, the number of the clusters should be *a priori* known, they suffer from numerical problems, like sensitiveness to the initialization, etc. In this paper we introduce a new clustering algorithm based on the sinergistic combination of the hierarchial and graph theoretic *minimal spanning tree based clustering* and the partitional *Gaussian mixture model-based clustering* algorithms. The aim of this hybridization is to increase the robustness and consistency of the clustering results and to decrease the number of the heuristically defined parameters of these algorithms to decrease the influence of the user on the clustering results. As the examples used for the illustration of the operation of the new algorithm will show, the proposed algorithm can detect clusters from data with arbitrary shape and does not suffer from the numerical problems of the Gaussian mixture based clustering algorithms.

1 Introduction

Nowadays the amount of data doubles almost every year. Hence, there is an urgent need for a new generation of computational techniques and tools to assist humans in extracting useful information (knowledge) from the rapidly growing volume of data. Data mining is one of the most useful methods for exploring large datasets. Clustering is one of the most commonly used methods for discovering the hidden structure of the considered dataset. The aim of cluster analysis is to partition a set of N objects in c clusters such that objects within clusters should be similar to each other and objects in different clusters should be dissimilar from

J. Dix and S.J. Hegner (Eds.): FoIKS 2006, LNCS 3861, pp. 313–330, 2006.

each other. Clustering can be used to quantize the available data, to extract a set of cluster prototypes for the compact representation of the dataset, to select the relevant features, to segment the dataset into homogenous subsets, and to initialize regression and classification models. Therefore, clustering is a widely applied technique, e.g. clustering methods have been applied in DNA analysis [4, 24], in medical diagnosis [6], on astronomical data [3, 23], in web applications [7, 16], etc.

Clustering, as an unsupervised learning is mainly carried out on the basis of the data structure itself, so the influence of the user should be minimal on the results of the clustering. However, due to the huge variety of problems and data, it is a difficult challenge to find a general and powerful method that is quite robust and that does not require the fine-tuning of the user. The large variety of clustering problems resulted in several algorithms that are based on various approaches (hierarchical, partitional, density-based, graph-based, model-based, etc.). Most of these algorithms have some discrepancies. For example the basic partitional methods are not able to detect convex clusters; when using hierarchical methods the number of the clusters should be a priori known, and they are not efficient enough for large datasets; while linkage-based methods often suffer from the chaining effect. Partitional clustering algorithms obtain a single partition of the data instead of a complex clustering structure, such as the dendrogram produced by a hierarchical technique. Partitional methods have advantages in applications involving large datasets for which the construction of a dendrogram is computationally prohibitive. Furthermore, the clusters are represented by easily interpretable cluster prototypes, that is a significant benefit when clustering is used to extract interesting information from data.

A problem accompanying the use of a partitional algorithm is that the number of the desired clusters should be given in advance. A seminal paper [9] provides guidance on this key design decision. The partitional techniques usually produce clusters by optimizing a criterion function defined either locally (on a subset of the patterns) or globally (defined over all of the patterns). Combinatorial search of the set of possible labelings for an optimum value of a criterion is clearly computationally prohibitive. In practice, the algorithms are typically run multiple times with different starting states, and the best configuration obtained from these runs is used as the output clustering.

A common limitation of partitional clustering algorithms based on a fixed distance norm, like k-means or fuzzy c-means clustering, is that such a norm induces a fixed topological structure and forces the objective function to prefer clusters of spherical shape even if it is not present. Generally, different cluster shapes (orientations, volumes) are required for the different clusters (partitions), but there is no guideline as to how to choose them a priori. The norm-inducing matrix of the cluster prototypes can be adapted by using estimates of the data covariance, and can be used to estimate the statistical dependence of the data in each cluster. The Gaussian mixture based fuzzy maximum likelihood estimation algorithm (Gath–Geva algorithm (GG) [12]) is based on such an adaptive distance measure, it can adapt the distance norm to the underlying distribution of the data which is reflected in the different sizes of the clusters, hence it is able to detect clusters with different orientation and volume (see the Appendix of the paper for the description of the algorithm). The mixture resolving approach to cluster analysis has been addressed in a number of ways. The underlying assumption is that the patterns to be clustered are drawn from one of several distributions, and the goal is to identify the parameters of each and (perhaps) their number. Most of the work in this area has assumed that the individual components of the mixture density are Gaussians, and in this case the parameters of the individual Gaussians are to be estimated by the procedure. Traditional approaches to this problem involve obtaining (iteratively) a maximum likelihood estimate of the parameter vectors of the component densities [17]. More recently, the Expectation Maximization (EM) algorithm (a general purpose maximum likelihood algorithm [8] for missing-data problems) has been applied to the problem of parameter estimation. A recent book [19] provides an accessible description of the technique. In the EM framework, the parameters of the component densities are unknown, as are the mixing parameters, and these are estimated from the patterns. The EM procedure begins with an initial estimate of the parameter vector and iteratively rescores the patterns against the mixture density produced by the parameter vector. The rescored patterns are then used to update the parameter estimates. In a clustering context, the scores of the patterns (which essentially measure their likelihood of being drawn from particular components of the mixture) can be viewed as hints at the class of the pattern. Those patterns, placed (by their scores) in a particular component, would therefore be viewed as belonging to the same cluster. A hard clustering algorithm allocates each pattern to a single cluster during its operation and in its output. A fuzzy clustering method assigns degrees of membership in several clusters to each input pattern. The fuzzy maximum likelihood estimates clustering algorithm employs a distance norm based on the fuzzy maximum likelihood estimates, proposed by Bezdek and Dunn. Unfortunately the GG algorithm is very sensitive to initialization, hence often it cannot be directly applied to the data.

Contrary, clustering algorithms based on linkage approach are able to detect clusters of various shapes and sizes, they can work with multivariate data, and they do not require initialization. There are several methods for detecting clusters by linking data patterns to each other. The hierarchical clustering approaches are related to graph-theoretic clustering. Single-link clusters are subgraphs of the minimum spanning tree of the data [15] which are also the connected components [14]. Complete-link clusters are maximal complete subgraphs, and are related to the node colorability of graphs [2]. The maximal complete subgraph was considered the strictest definition of a cluster in [1, 22]. One of the best-known graph-based divisive clustering algorithm is based on the construction of the minimal spanning tree (MST) of the objects [3, 11, 13, 20, 24]. The elimination of any edge from the MST we get subtrees which correspond to clusters.

Figure 1 depicts the minimal spanning tree obtained from 75 two-dimensional points distributed into three clusters. The objects belonging to different clusters are marked with different dot notations. Clustering methods using a minimal

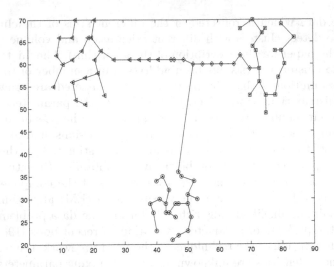

Fig. 1. Example of a minimal spanning tree

spanning tree take advantages of the MST. For example building the minimal spanning tree of a dataset does not need any *a priori* information about the underlying data. Moreover, as the MST ignores many possible connections between the data patterns, the cost of clustering can be decreased. However the use of minimal spanning trees in clustering algorithms also raises some interesting questions. How can we determine the edges at which the best cluster separations might be made? For finding the best clusters, when should we stop our algorithm?

In this paper we propose a synergistic hybridization of the MST and GG clustering algorithms that can automatically handle these questions. In the following, in Section 2, this new algorithm will be described. Section 3. contains application examples based on synthetic and real life datasets to illustrate the usefulness of the proposed method. Section 4. concludes the paper.

2   Clustering Based on Minimal Spanning Tree

2.1   Generation and Partition of the Minimal Spanning Tree

Using a minimal spanning tree for clustering was initially proposed by Zahn [25]. A minimal spanning tree is a weighted connected graph, where the sum of the weights is minimal. A graph G is a pair (V, E), where V is a finite set of the elements, called vertices, and E is a collection of unordered pairs of V. An element of E, called edge, is $e_{i,j} = (v_i, v_j)$, where $v_i, v_j \in V$. In a weighted graph a weight function w is defined, which function determines a weight $w_{i,j}$ for each edge $e_{i,j}$. The complete graph K_N on a set of N vertices is the graph that has all the $\binom{N}{2}$ possible edges. Creating the minimal spanning tree means, that we are searching the $G' = (V, E')$, the connected subgraph of G, where $E' \subset E$ and the cost is computed in the following way:

$$\sum_{e \in E'} w(e) \tag{1}$$

where $w(e)$ denotes the weight of the edge $e \in E$. In a graph G, where the number of the vertices is N, MST has exactly $N - 1$ edges.

A minimal spanning tree can be efficiently computed in $O(N^2)$ time using either Prim's [21] or Kruskal's [18] algorithm. Prim's algorithm starts with an arbitrary vertex as the root of a partial tree. In each step of the algorithm the partial tree grows by iteratively adding an unconnected vertex to it using the lowest cost edge, until no unconnected vertex remains. Kruskal's algorithm begins with the connection of the two nearest objects. In each step the nearest objects placed in different trees are connected. So the Kruskal's algorithm iteratively merges two trees (or a tree with a single object) in the current forest into a new tree. The algorithm continues until a single tree remains only, connecting all points.

A minimal spanning tree can be used in clustering in the following way: let $V = \{\mathbf{x}_1, \mathbf{x}_2, ..., \mathbf{x}_N\}$ be a set of the data with N distinct objects which we want to distribute in different clusters. $\mathbf{x}_i$ denotes the i-th object, which consists n measured variables, grouped into an n-dimensional column vector $\mathbf{x}_i = [x_{1,i}, x_{2,i}, ..., x_{n,i}]^T$, $\mathbf{x}_i \in R^n$. Let $d_{i,j} = d(\mathbf{x}_i, \mathbf{x}_j)$ be the distance defined between any $\mathbf{x}_i$ and $\mathbf{x}_j$. This distance can be computed in different ways (e.g. Euclidean distance, Manhattan distance, Mahalanobis distance, mutual neighbour distance, etc.). The drawback of the direct use of the Euclidean metrics is the tendency of the largest-scaled features to dominate the others. To avoid the greater influence of some features of the objects an initialization step is recommended. As a solution for this problem we suggest the normalization of the continuous features to a common range or variance. So in the graph $G = (V, E)$ V represents the objects and the function w defined over the edges $e \in E$ representing the distances between the data points ($w_{i,j} = d_{i,j}$).

Removing edges from the MST leads to a collection of connected subgraphs of G, which can be considered as clusters. Using MST for clustering we are interested in finding the edges, which elimination lead to the best clustering result. Such edges is called *inconsistent edges*.

Clustering by minimal spanning tree can be viewed as a hierarchical clustering algorithm which follows the divisive approach. Using this method firstly we construct a linked structure of the objects, and then the clusters are recursively divided into subclusters. The major steps of the clustering methods using a minimal spanning tree are the following:

- **Step 1.** Construct the minimal spanning tree so that the edges weights are the distances between the data points.
- **Step 2.** Remove the inconsistent edges to get a set of connected components (clusters).
- **Step 3.** Repeat Step 2 until a terminating criterion is not satisfied.

Elimination of k edges from a minimal spanning tree results in $k+1$ disconnected subtrees. In the simplest recursive theories $k = 1$. Denote δ the length of the deleted edge, and let V_1, V_2 be the sets of the points in the resulting two clusters. In the set of clusters we can state that there are no pairs of points $(\mathbf{x}_1, \mathbf{x}_2)$, $\mathbf{x}_1 \in V_1$, $\mathbf{x}_2 \in V_2$ such that $d(\mathbf{x}_1, \mathbf{x}_2) < \delta$. There are several ways to define the distance between two disconnected groups of individual objects (minimum distance, maximum distance, average distance, distance of centroids, etc.). Defining the separation between V_1 and V_2, we have the result that the separation is at least δ.

In the case of clustering using MST firstly a minimal spanning tree of all the nodes is given. Thus initially only one cluster is available. In the second phase we must choose the edge or the edges which will be removed. The major advantage of clustering based on a minimal spanning tree appears in this step. Because the MST has only $N - 1$ edges, we can choose the inconsistent edge (or edges) by revising only $N - 1$ values, instead of checking the $\binom{N}{2}$ possible connections between all of the objects. The identification of the inconsistent edges causes problems in the MST clustering algorithms. There exist numerous ways to divide clusters successively, but there is not a suitable choice for all cases. In special cases the second step is not executed recursively. In these cases a global parameter is used, which determines the edges to be removed from the MST. When the second step of the algorithm is repeated, we must determine a terminating criterion, when the running of the algorithm is finished, and the current trees can be seen as clusters. Determination of the terminating criterion is also a difficult challenge. The methods which use recursive cutting define some possible terminating criteria.

In the next subsection we will overview some well-known cutting conditions and terminating criteria, then we introduce our suggestions for using the minimal spanning tree for clustering with new cutting criteria.

2.2   Conditions for Cutting the Minimal Spanning Tree

Criterion-1: The simplest way to delete edges from the minimal spanning tree is based on the distance between the vertices. By deleting the longest edge in each iteration step we get a nested sequence of subgraphs. As other hierarchical methods, this approach also requires a terminating condition. Several ways are known to stop the algorithms, for example the user can define the number of clusters, or we can give a threshold value on the length also.

Similarly to Zahn [25] we suggest a global threshold value δ, which considers the distribution of the data in the feature space. In [25] this threshold (δ) is based on the average weight (distances) of the MST :

$$\delta = \lambda \frac{1}{N-1} \sum_{e \in E'} w(e) \qquad (2)$$

where λ is a user defined parameter.

Of course, λ can be defined in several manner. E.g. it can be derived from a physical model. Let us consider the data points in the features space as planets in space. When each data point has the same meaning for the clustering, we can suppose, that the planets representing the data points has equal mass. As planets in space, the objects in the feature space affect each other. Using this theory we can calculate a threshold index. This index, called *attraction threshold* (T_{at}) is calculated in the following way:

$$T_{at} = \frac{1}{2}\sqrt{\frac{N(N-1)}{\sum_{i=1}^{N}\sum_{j=i+1}^{N}\frac{1}{d(\mathbf{x}_i, \mathbf{x}_j)^2}}} \tag{3}$$

where N is the number of the objects, $\mathbf{x}_i$, $\mathbf{x}_j$ are two objects, and $d(\mathbf{x}_1, \mathbf{x}_2)$ defines the distance between the objects calculated with using Euclidean metric. The attraction threshold means a global parameter, which takes into consideration the placement of all objects in the normalized features space. In our approach we suppose if the inequality

$$d(\mathbf{x}_i, \mathbf{x}_j) > T_{at} \tag{4}$$

holds, then the objects $\mathbf{x}_i$ and $\mathbf{x}_j$ belong to different clusters. The attraction threshold can be seen as a cutting criterion in the minimal spanning tree. In this case we must delete all of the edges whose weight is greater than the attraction threshold. So our approach is a linkage based clustering method with the use of MST. In the basic form our approach is not a hierarchical clustering method. The *Criterion-1at* can be considered as an improvement of *Criterion-1*, by which the influence of the user is eliminated. For the calculation of the T_{at} it is needed to compute the distance between all pairs of objects. The computation cost is not increased, because the construction of minimal spanning tree also needs these values. So the previously computed values stored in a matrix can be used for calculating T_{at}.

Long edges of the MST need not be associated with cluster separation. For a unimodal density, even with compact support, some of the longest edges of the MST will connect the outlying points in the low density region to the 'main body' of data. Hence, this first criteria can also be used to detect outliers.

Criterion-2: Zahn [25] proposed also an idea to detect the hidden separations in the data. Zahn's suggestion is based on the distance of the separated subtrees. He suggested, that an edge is inconsistent if its length is at least f times as long as the average of the length of nearby edges. The input parameter f must be adjusted by the user. To determine which edges are 'nearby' is another question. It can be determined by the user, or we can say, that point $\mathbf{x}_i$ is nearby point of $\mathbf{x}_j$ if point $\mathbf{x}_i$ is connected to the point $\mathbf{x}_j$ by a path in a minimal spanning tree containing k or fewer edges. This method has the advantage of determining clusters which have different distances separating one another. Another use of the MST based clustering based on this criterion is to find dense clusters embedded in a sparse set of points. All that has to be done is to remove all edges longer than some predetermined length in order to extract clusters which are closer than the specified length to each other. If the length is chosen accordingly, the dense clusters are extracted from a sparse set of points easily. The drawback of this method is that the influence of the user is significant at the selection of the f and k parameters.

Several clustering methods based on linkage approach suffer from some discrepancies. In these cases the clusters are provided by merging or splitting of the objects or clusters using a distance defined between them. Occurrence of a data chain between two clusters can cause that these methods can not separate these clusters. This also happens with the basic MST clustering algorithm. To solve the chaining problem we suggest a new complementary condition for cutting the minimal spanning tree.

Criterion-3: In another approaches the separation is specified with the goodness of the obtained partitions. Cluster validity refers to the problem whether a given partition fits to the data all. The clustering algorithm always tries to find the best fit for a fixed number of clusters and the parameterized cluster shapes. However this does not mean that even the best fit is meaningful at all. Either the number of clusters might be wrong or the cluster shapes might not correspond to the groups in the data, if the data can be grouped in a meaningful way at all. Two main approaches to determining the appropriate number of clusters in data can be distinguished:

- Starting with a sufficiently large number of clusters, and successively reducing this number by merging clusters that are similar (compatible) with respect to some predefined criteria. This approach is called *compatible cluster merging*.
- Clustering data for different values of c, and using *validity measures* to assess the goodness of the obtained partitions. This can be done in two ways:
  - The first approach defines a validity function which evaluates a complete partition. An upper bound for the number of clusters must be estimated (c_{max}), and the algorithms have to be run with each $c \in \{2, 3, \ldots, c_{max}\}$. For each partition, the validity function provides a value such that the results of the analysis can be compared indirectly.
  - The second approach consists of the definition of a validity function that evaluates individual clusters of a cluster partition. Again, c_{max} has to be estimated and the cluster analysis has to be carried out for c_{max}. The resulting clusters are compared to each other on the basis of the validity function. Similar clusters are collected in one cluster, very bad clusters are eliminated, so the number of clusters is reduced. The procedure can be repeated until there are clusters that not satisfy the predefined criterion.

Different scalar validity measures have been proposed in the literature, but none of them is perfect on its own. For example partition index [5] is the ratio of the sum of compactness and separation of the clusters. Separation index [5] uses a minimum distance separation for partition validity. Dunn's index [10] is originally proposed to be used at the identification of compact and well separated clusters. The problems of the Dunn index are: i) its considerable time complexity, ii) its sensitivity to the presence of noise in datasets. Three indices, are proposed in the literature that are more robust to the presence of noise. They are known as Dunn-like indices since they are based on Dunn index. One of the three indices uses for the definition the concepts MST. Let c_i be a cluster and the complete graph G_i whose vertices correspond to the vectors of c_i. The weight $w(e)$ of an edge e of this graph equals the distance between its two end points, $\mathbf{x}$, $\mathbf{y}$. Let e_i^{MST} be the edge with the maximum weight. Then the diameter of the cluster c_i is defined as the weight of e_i^{MST}. With the use of this notation the Dunn-like index based on the concept of the MST is given by equation:

$$D_{n_c} = \min_{i=1,\dots,n_c} \left\{ \min_{j=i+1,\dots,n_c} \left(\frac{d(c_i, c_j)}{\max_{k=1,\dots,n_c} diam\,(c_k)} \right) \right\} \tag{5}$$

where $d(c_i, c_j)$ is the dissimilarity function between two clusters c_i and c_j defined as $\min_{\mathbf{x}\in c_i, \mathbf{y}\in c_j} d(\mathbf{x}, \mathbf{y})$, and $diam(c)$ is the diameter of a cluster, which may be considered as a measure of clusters dispersion. The diameter of a cluster c_i can be defined as follows: $\max_{\mathbf{x},\mathbf{y}\in c_i} d(\mathbf{x}, \mathbf{y})$. n_c denotes the number of clusters. The number of clusters at which D_{n_c} takes its maximum value indicates the number of clusters in the underlying data.

Varma and Simon [24] used the Fukuyama-Sugeno clustering measure for deleting different edges from the MST. Denote S the sample index set, and let S_1, S_2 be partitions of S. N_k denotes the number of the objects for each S_k. The Fukuyama-Sugeno clustering measure is defined in the following way:

$$FS(S) = \sum_{k=1}^{2} \sum_{j=1}^{N_k} [\|\mathbf{x}_j^k - \mathbf{v}_k\|^2 - \|\mathbf{v}_k - \mathbf{v}\|^2] \tag{6}$$

where $\mathbf{v}$ denotes the global mean of all objects, $\mathbf{v}_k$ denotes the mean of the objects in S_k. The symbol $\mathbf{x}_j^k$ refers to the j-th object in the cluster S_k. If the value of $FS(S)$ is small, it indicates tight clusters with large separations between them. Varma and Simon found, that the Fukuyama-Sugeno measure gives the best performance in a dataset with a large number of noisy features.

In this paper the second approach of the automatic determination of the number of clusters is followed. A validity function that evaluates the individual clusters of a cluster partition has been used. Since the clusters of the MST will be approximated by multivariate Gaussians, the *fuzzy hyper volume* validity index is applied. Let $\mathbf{F}_j$ the fuzzy covariance matrix of the j-th cluster defined as

$$\mathbf{F}_j = \frac{\sum_{i=1}^{N} (\mu_{ij})^m (\mathbf{x}_i - \mathbf{v}_j)(\mathbf{x}_i - \mathbf{v}_j)^T}{\sum_{i=1}^{N} (\mu_{ij})^m}, \tag{7}$$

where μ_{ij} denotes the degree of membership of the $\mathbf{x}_i$ in cluster c_j. The symbol m is the fuzzifier parameter of the fuzzy clustering algorithm and indicates the fuzzyness of clustering result. The *fuzzy hyper volume* of i-th cluster is given by the equation

$$V_i = \sqrt{det\,(\mathbf{F}_i)} \tag{8}$$

The total fuzzy hyper volume is given by the equation

$$HV = \sum_{i=1}^{c} V_i \tag{9}$$

The $det(\mathbf{F}_i)$ represents the volume of the clusters (ellipsoids). The resulting clusters can be compared to each other on the basis of their volume. Very bad clusters with large volumes can be further partitioned, so the number of clusters can be recursively increased. The procedure can be repeated until there are bad clusters.

2.3 Hybrid MST-GG Clustering Algorithm

The previous subsections presented the main building blocks of the proposed hybrid clustering algorithm. In the following the whole algorithm will be described.

The proposed algorithm (called Hybrid MST-GG algorithm) is based on the construction of the minimal spanning tree of the objects and uses the fuzzy hyper volume validity measure. Additionally, by the use of the Gath-Geva algorithm the Gaussian mixture representation of the refined outcome is permitted also.

In the initialization step the creation of the minimal spanning tree of the considered objects is required. This structure will be partitioned based on the following approaches: (i) classical use of the MST in the clustering algorithm, (ii) the application of the fuzzy hyper volume validity measure to eliminate edges from the MST. The proposed Hybrid MST-GG algorithm iterative builds the possible clusters. In each iteration step a binary split is performed. The use of the cutting criteria results in a hierarchical tree, in which the nodes denote partitions of the objects. To refine the partitions evolved in the previous step, firstly the calculation of the volumes of the obtained clusters is needed. In each iteration step the cluster (a leave of the binary tree) having the largest hyper volume is selected for the cutting. For the elimination of edges from the MST firstly the cutting conditions Criterion-1 and Criterion-2 are applied, which were previously introduced. The use of the classical MST-based clustering method detects well separated clusters, but does not solve the typical problem of the graph-based clustering algorithms (chaining affect). To dissolve this discrepancy we apply the fuzzy hyper volume measure. If the cutting of the partition having the largest hyper volume can not executed based on Criterion-1 or Criterion-2, then the cut is performed based on the measure of the total fuzzy hyper volume. If this partition has N objects, we must check up $N - 1$ possibilities. Each of the $N - 1$ possibilities results in a binary splitting, hereby the objects placed in the cluster with the largest hyper volume are distributed into two subclusters. We must choose the binary splitting which results in the least total fuzzy hyper volume. The whole process is carried out until a termination criterion is satisfied (e.g. the predefined number of clusters, or the minimal number of the objects of the partitions is reached).

The application of this hybrid cutting criterion can be seen as a divisive hierarchical method. Figure 2 demonstrates a possible result after applying the two different cutting methods on the MST. The partitions marked by the continuous dark line are resulted by the applying of the classical MST-based clustering method, and the partitions having gray dotted notations are arising from the application of the fuzzy hyper volume criterion.

When the compact parametric representation of the result of the clustering is needed, a Gaussian mixture model based clustering can be performed, where the number of the Gaussians is equal to the termination nodes, and the iterative algorithm is initialized based on the partition obtained at the previous step. This approach is really fruitful, since it is well known that the Gath-Geva algorithm is sensitive to the initialization of the partitions. The previously obtained clusters give an appropriate starting-point for the GG algorithm. Hereby the iterative application of the Gath-Geva algorithm results in a good and compact representation of the clusters.

Algorithm 2.1 (Hybrid MST-GG Clustering Algorithm)

Step 0 *Create the minimal spanning tree of the objects.*

Repeat *Iteration*

   Step 1 *Node selection. Select the node (i.e., sub-cluster) with the largest hyper volume V_i from the so-far formed hierarchical tree. Perform a cutting on this node based on the following criteria.*

   Step 2 *Binary Splitting. If the cluster having the largest hyper volume can not be cut by Criterion-1 or Criterion-2, then use Criterion-3 for the splitting. All of the edges of the selected sub-MST is cut. With each cut a binary split of the objects is formed. If the current node includes N_i objects then $N_i - 1$ such splits are formed. The two sub-clusters, formed by the binary split, plus the clusters formed so far (excluding the current node) compose the potential partition. The hyper volume (HV) of all formed $N_i - 1$ potential partitions are computed. The one that exhibits the lowest HV is selected as the best partition of the objects in the current node.*

Until *Termination criterion. Following a depth-first tree-growing process, steps 1, 2 and 3a-b are iteratively performed. The final outcome is a hierarchical clustering tree, where the termination nodes are the final clusters. Special parameters control the generalization level of the hierarchical clustering tree (e.g., minimum number of objects in each sub-cluster).*

Step 3 *When the compact parametric representation of the result of the clustering is needed, a Gaussian mixture model based clustering is performed, where the number of the Gaussians is equal to the termination nodes, and the iterative algorithm is initialized based on the partition obtained at the previous step.*

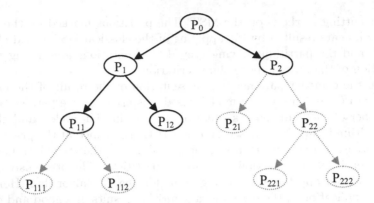

Fig. 2. A possible binary tree given by the proposed MST-GG algorithm

3   Application Examples

In this section we present the results obtained on the clustering of some tailored data and well-known data sets.

3.1   Handling the Chaining Effect

The first example is intended to illustrate that the proposed cluster volume based splitting extension of the basic MST based clustering algorithm is able to handle (avoid) the chaining phenomena of the classical single linkage scheme. As Figure 3 illustrates, for this toy example the classical MST based algorithm detects only two clusters.

With the use of the volume-based partitioning criterion, the first cluster has been split (Figure 1).

This short example illustrates the main benefit of the incorporation of the cluster validity based criterion into the classical MST based clustering algorithm.

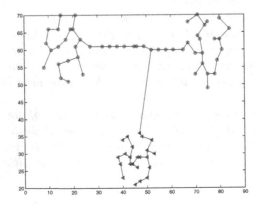

Fig. 3. Clusters obtained by the classical MST based algorithm

In the following it will be shown how the resulting nonparametric clusters can be approximated by mixture of Gaussians, and how this approach is beneficial for the initialization of these iterative partitional algorithms.

3.2   Handling the Convex Shapes of Clusters - Effect of the Initialization

Let us consider a more complex clustering problem with convex shape of clusters. As Figure 4 shows, the proposed MST based clustering algorithm is able to detect properly cluster of this data. The partitioning of the clusters has not been stopped at the detection of the well separated clusters (Criterion-1 and Criterion-2), but the resulting clusters have been further split to get clusters with small volumes, (Criterion-3).

The main benefit of the resulted partitioning is that it can be easily approximated by mixture of multivariate Gaussians (ellipsoids). This approximation is useful since the obtained Gaussians give a compact and parametric descriptions of the clusters, and the result of the clustering is soft (fuzzy).

Figure 5 shows the results of the clustering we obtained after performing the iteration steps of the Gaussian mixtures based EM algorithm given in the Appendix. In this figure the dots represent the data points and the 'o' markers are the cluster centers. The membership values are also shown, since the curves represent the isosurfaces of the membership values that are inversely proportional to the distances. As can be seen, the clusters provide an excellent description of the distribution of the data. The clusters with complex shape are approximated by a set of ellipsoids. It is interesting to note, that this clustering step only slightly modifies the placement of the clusters.

In order to demonstrate the effectiveness of the proposed initialization scheme, Figure 6 illustrates the result of the Gaussian mixture based clustering, where the clustering was initialized by the classical fuzzy c-means algorithm. As can

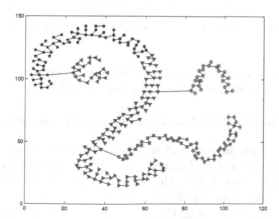

Fig. 4. Top: Minimal spanning tree created by the proposed MST-GG algorithm

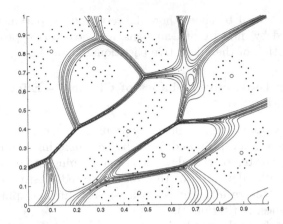

Fig. 5. Result of the proposed MST-GG algorithm after performing the Gaussian mixture based clustering

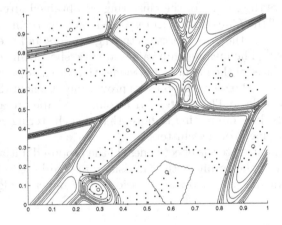

Fig. 6. Result of the GG algorithm initialized from the fuzzy c-means algorithm

be seen, this widely applied approach failed to find the proper clustering of the data set, only a sub-optimal solution has been found.

3.3   Application to Real Life Data: Classification of Iris Flowers

The previous example showed that it is possible to obtain a properly clustered representation by the proposed mapping algorithm. However, the real advantage of the algorithm was not shown. This will be done by the following real clustering problem, the clustering of the well known Iris data, coming from the UCI Repository of Machine Learning Databases (http://www.ics.uci.edu).

The Iris data set contains measurements on three classes of Iris flowers. The data set was made by measurements of sepal length and width and petal length

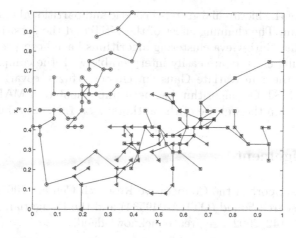

Fig. 7. Two Dimensional Plot of the Minimal Spanning Tree of the Iris Data

and width for a collection of 150 irises. The problem is to distinguish the three different types (Iris setosa, Iris versicolor and Iris virginica). These data have been analyzed many times to illustrate various methods.

The MST clustering based on Criterion-1 and Criterion-2 detects only two clusters. In this case the third cluster is formed only after the application of the cluster volume-based splitting criterion (Criterion-3), see Figure 7. The resulting three clusters correspond to the three classes of the Iris flowers. At the analysis of the distribution of the classes in the clusters, we found only 3 misclassification errors. The mixture of Gaussians density model is able to approximate this cluster arrangement. The resulting fuzzy clustering (**Step 3**) can be converted to a hard clustering by assigning each pattern to the cluster with the largest measure of membership. After this fine-tuning clustering step, we found only 5 misclassifications. This means 96.67% classification correctness, that is a quite good result for this classification problem.

4 Conclusions

A divisive clustering method begins with all patterns in a single cluster and performs splitting until a stopping criterion is met. The best-known graph-theoretic divisive clustering algorithm is based on construction of the minimal spanning tree (MST) of the data [25], and then deleting the MST edges with the largest lengths to generate clusters. In this paper, we presented a new splitting criterion to improve the performance of this clustering algorithm. The algorithm is based on the calculation of the hyper volume of the clusters that are approximated by a multivariate Gaussian functions. The result of this clustering can be effectively used for the initialization of Gaussian mixture model based clustering algorithms. The approach is demonstrated by some tailored data and by the well-known iris benchmark classification problem. The results showed the advantages of the hybridization of the hierarchial graph-theoretic and partitional model based clustering algorithm. The chaining effect of the MST and the sensitivity to the initialization of the Gath-Geva clustering algorithms have been properly handled, and the resulting clusters are easily interpretable with the compact parametric description of the multivariate Gaussian clusters (fuzzy covariance matrices). The proposed MST-GG algorithm has been implemented in MATLAB, and it is downloadable from the website of the authors: www.fmt.veim.hu/softcomp.

Acknowledgement

The financial support of the Cooperative Research Center (2005-III-1) and the Hungarian Research Found (OTKA-049534) and the Hungarian Ministry of education (IKTA 142/2002) are greatly acknowledged.

References

1. J.G. Augustson, J. Minker, An analysis of some graph theoretical clustering techniques. J. ACM, Vol.17, 4, 571-588, 1970.
2. F.B. Backer, L.J. Hubert, A graph-theoretic approach to goodness-of-fit in complete-link hierarchical clustering. J. Am. Stat. Assoc., Vol. 71, 870-878, 1976.
3. J.D. Barrow, S.P. Bhavsar, D.H. Sonoda, Minimal spanning trees, filaments and galaxy clustering. In Monthly Notices of the Royal Astronomical Society, 216:17-35, 1985.
4. A. Ben-Dor and Z. Yakhini, Clustering gene expression patterns. In Proceedings of the 3rd Annual International Conference on Computational Molecular Biology (RECOMB99), 11-14, 1999.
5. J.C. Bezdek, L.P. Clarke, M.L. Silbiger, J.A. Arrington, A.M. Bensaid, L.O. Hall, R.F. Murtagh, Validity-guided (re)clustering with applications to image segmentation. IEEE Transactions on Fuzzy Systems, 4:112-123, 1996.
6. G. Castellano, A.M. Fanelli, C. Mencar, A fuzzy clustering approach for mining diagnostic rules. In Proceedings of 2003 IEEE International Conference on Systems, Man & Cybernetics (IEEE SMC'03), Vol. 1, 2007-2012, 2003.
7. R. Cooley, B. Mobasher, J. Srivastava, Data preparation for mining world wide web browsing. In Journal of Knowledge Information Systems, 1, 1, 5-32, 1999.
8. A.P. Dempster, N.M. Laird, D.B. Rubin, Maximum likelihood from incomplete data via the EM algorithm. Journal of the Royal Statistical Society, Series B (Methodological), 39(1):1-38, 1977.
9. R.C. Dubes, How many clusters are best? – an experiment. Pattern Recogn., 20(6):645-663, 1987.
10. J.C. Dunn, Well separated clusters and optimal fuzzy partitions. Journal Cybernetics, 4:95-104, 1974.
11. M. Forina, C. Oliveros, M. Concepcion, C. Casolino, M. Casale, Minimum spanning tree: ordering edges to identify clustering structure. In Analytica Chimica Acta 515: 43-53, 2004.
12. I. Gath, A.B. Geva, Unsupervised Optimal Fuzzy Clustering. IEEE Trans. on Pattern Analysis and Machine Intelligence, vol. 11, pp. 773-781, 1989.

13. J.M. Gonzáles-Barrios and A.J. Quiroz, A clustering procedure based on the com-
    parsion between the k nearest neighbors graph and the minimal spanning tree. In
    Statistics & Probability Letter, 62:23-34, 2003.
14. C.C. Gotlieb, S. Kumar, Semantic Clustering of Index Terms. J. ACM, 15(4):493-
    513, 1968.
15. J.C. Gower, G.J.S. Ross, Minimal Spanning Trees and Single Linkage Cluster Anal-
    ysis. Applied Statistics, Vol. 18, 54-64, 1969.
16. J. Heer and E. Chi, Identification of Web user traffic composition using multimodal
    clustering and information scent. 1st SIAM ICDM, Workshop on Web Mining,
    Chicago, IL., 51-58, 2001.
17. A.K. Jain and R.C. Dubes, Algorithms for Clustering Data. Prentice-Hall advanced
    reference series, Prentice-Hall, Inc., 1988.
18. J.B. Kruskal, On the shortest spanning subtree of a graph and the traveling sales-
    man problem. In American Mathematical Society, 7:48-50, 1956.
19. T. Mitchell, Machine Learning. McGraw-Hill, Inc., New York, NY, 1997.
20. N. Päivinen, Clustering with a minimum spanning tree of scale-free-like structure.
    In Pattern Recognition Letters, 26(7):921-930, 2005.
21. R. Prim, Shortest connection networks and some generalizations. Bell System Tech-
    nical Journal, 36:1389-1401, 1957.
22. V.V. Raghavan, C.T. Yu, A comparison of the stability characteristics of some
    graph theoretic clustering methods. IEEE Trans. Pattern Anal. Mach. Intell., 3,
    3-402, 1981.
23. J. Sander, M. Ester, H.-P. Kriegel, X. Xu, Density-based clustering in spatial
    databases: the algorithm GDBSCAN and its applications. In Data Mining and
    Knowledge Discovery, 2, 2, 169-194, 1998.
24. S. Varma, R. Simon, Iterative class discovery and feature selection using Minimal
    Spanning Trees. BMC Bioinformatics, 5:126-134, 2004.
25. C.T. Zahn, Graph-theoretical methods for detecting and describing gestalt clusters.
    IEEE Transaction on Computers C20:68-86, 1971.

Appendix: Gath-Geva Clustering Algorithm

The fuzzy maximum likelihood estimates clustering algorithm employs a distance
norm based on the fuzzy maximum likelihood estimates, proposed by Bezdek and
Dunn:

$$D_{ik}^2(\mathbf{x}_k, \mathbf{v}_i) = \frac{\sqrt{\det(\mathbf{F}_i)}}{\alpha_i} \exp\left(\frac{1}{2}\left(\mathbf{x}_k - \mathbf{v}_i^{(l)}\right)^T \mathbf{F}_i^{-1}\left(\mathbf{x}_k - \mathbf{v}_i^{(l)}\right)\right) \qquad (10)$$

Note that, contrary to the Gustaffson-Kessel (GK) algorithm, this distance norm
involves an exponential term and thus decreases faster than the inner-product
norm. $\mathbf{F}_i$ denotes the fuzzy covariance matrix of the i-th cluster, given by the
equation (7). The reason for using this m exponent is to enable to generalize
this expression. The α_i is the prior probability of selecting cluster i, given by:

$$\alpha_i = \frac{1}{N} \sum_{k=1}^{N} \mu_{ik}. \qquad (11)$$

The membership degrees μ_{ik} are interpreted as the posterior probabilities of selecting the i-th cluster given the data point $\mathbf{x}_k$. Gath and Geva reported that the fuzzy maximum likelihood estimates clustering algorithm is able to detect clusters of varying shapes, sizes and densities. This is because the cluster covariance matrix is used in conjunction with an "exponential" distance, and the clusters are not constrained in volume. However, this algorithm is less robust in the sense that it needs a good initialization, since due to the exponential distance norm, it converges to a near local optimum. The minimum of the cost function is sought by the alternating optimization (AO) method (Gath-Geva clustering algorithm).

Algorithm 4.1 (Gath-Geva algorithm)

Given a set of data $\mathbf{X}$ *specify c, choose a weighting exponent* $m > 1$ *and a termination tolerance* $\epsilon > 0$. *Initialize the partition matrix.*

Repeat *for* $l = 1, 2, \ldots$

Step 1 *Calculate the cluster centers:* $\mathbf{v}_i^{(l)} = \dfrac{\sum\limits_{k=1}^{N} (\mu_{ik}^{(l-1)})^m \mathbf{x}_k}{\sum\limits_{k=1}^{N} (\mu_{ik}^{(l-1)})^m}, \; 1 \le i \le c$

Step 2 *Compute the distance measure* D_{ik}^2.

The distance to the prototype is calculated based the fuzzy covariance matrices of the cluster

$$\mathbf{F}_i^{(l)} = \dfrac{\sum\limits_{k=1}^{N} (\mu_{ik}^{(l-1)})^m \left(\mathbf{x}_k - \mathbf{v}_i^{(l)}\right)\left(\mathbf{x}_k - \mathbf{v}_i^{(l)}\right)^T}{\sum\limits_{k=1}^{N} (\mu_{ik}^{(l-1)})^m}, \; 1 \le i \le c \qquad (12)$$

The distance function is chosen as

$$D_{ik}^2(\mathbf{x}_k, \mathbf{v}_i) = \dfrac{(2\pi)^{\left(\frac{n}{2}\right)} \sqrt{det(\mathbf{F}_i)}}{\alpha_i} exp\left(\frac{1}{2}\left(\mathbf{x}_k - \mathbf{v}_i^{(l)}\right)^T \mathbf{F}_i^{-1}\left(\mathbf{x}_k - \mathbf{v}_i^{(l)}\right)\right) \qquad (13)$$

with the a priori probability $\alpha_i = \frac{1}{N}\sum_{k=1}^{N}\mu_{ik}$

Step 3 *Update the partition matrix*

$$\mu_{ik}^{(l)} = \dfrac{1}{\sum_{j=1}^{c}\left(D_{ik}\left(\mathbf{x}_k, \mathbf{v}_i\right)/D_{jk}\left(\mathbf{x}_k, \mathbf{v}_j\right)\right)^{2/(m-1)}}, 1 \le i \le c, 1 \le k \le N \qquad (14)$$

until $\|\mathbf{U}^{(l)} - \mathbf{U}^{(l-1)}\| < \epsilon$.

Author Index

Lecture Notes in Computer Science

For information about Vols. 1–3770

please contact your bookseller or Springer

Vol. 3817: M. Faundez-Zanuy, L. Janer, A. Esposito, A. Satue-Villar, J. Roure, V. Espinosa-Duro (Eds.), Nonlinear Analyses and Algorithms for Speech Processing. XII, 380 pages. 2006. (Sublibrary LNAI).

Vol. 3816: G. Chakraborty (Ed.), Distributed Computing and Internet Technology. XXI, 606 pages. 2005.

Vol. 3815: E.A. Fox, E.J. Neuhold, P. Premsmit, V. Wuwongse (Eds.), Digital Libraries: Implementing Strategies and Sharing Experiences. XVII, 529 pages. 2005.

Vol. 3814: M. Maybury, O. Stock, W. Wahlster (Eds.), Intelligent Technologies for Interactive Entertainment. XV, 342 pages. 2005. (Sublibrary LNAI).

Vol. 3813: R. Molva, G. Tsudik, D. Westhoff (Eds.), Security and Privacy in Ad-hoc and Sensor Networks. VIII, 219 pages. 2005.

Vol. 3810: Y.G. Desmedt, H. Wang, Y. Mu, Y. Li (Eds.), Cryptology and Network Security. XI, 349 pages. 2005.

Vol. 3809: S. Zhang, R. Jarvis (Eds.), AI 2005: Advances in Artificial Intelligence. XXVII, 1344 pages. 2005. (Sublibrary LNAI).

Vol. 3808: C. Bento, A. Cardoso, G. Dias (Eds.), Progress in Artificial Intelligence. XVIII, 704 pages. 2005. (Sublibrary LNAI).

Vol. 3807: M. Dean, Y. Guo, W. Jun, R. Kaschek, S. Krishnaswamy, Z. Pan, Q.Z. Sheng (Eds.), Web Information Systems Engineering – WISE 2005 Workshops. XV, 275 pages. 2005.

Vol. 3806: A.H. H. Ngu, M. Kitsuregawa, E.J. Neuhold, J.-Y. Chung, Q.Z. Sheng (Eds.), Web Information Systems Engineering – WISE 2005. XXI, 771 pages. 2005.

Vol. 3805: G. Subsol (Ed.), Virtual Storytelling. XII, 289 pages. 2005.

Vol. 3804: G. Bebis, R. Boyle, D. Koracin, B. Parvin (Eds.), Advances in Visual Computing. XX, 755 pages. 2005.

Vol. 3803: S. Jajodia, C. Mazumdar (Eds.), Information Systems Security. XI, 342 pages. 2005.

Vol. 3802: Y. Hao, J. Liu, Y.-P. Wang, Y.-m. Cheung, H. Yin, L. Jiao, J. Ma, Y.-C. Jiao (Eds.), Computational Intelligence and Security, Part II. XLII, 1166 pages. 2005. (Sublibrary LNAI).

Vol. 3801: Y. Hao, J. Liu, Y.-P. Wang, Y.-m. Cheung, H. Yin, L. Jiao, J. Ma, Y.-C. Jiao (Eds.), Computational Intelligence and Security, Part I. XLI, 1122 pages. 2005. (Sublibrary LNAI).

Vol. 3799: M. A. Rodríguez, I.F. Cruz, S. Levashkin, M.J. Egenhofer (Eds.), GeoSpatial Semantics. X, 259 pages. 2005.

Vol. 3798: A. Dearle, S. Eisenbach (Eds.), Component Deployment. X, 197 pages. 2005.

Vol. 3797: S. Maitra, C. E. V. Madhavan, R. Venkatesan (Eds.), Progress in Cryptology - INDOCRYPT 2005. XIV, 417 pages. 2005.

Vol. 3796: N.P. Smart (Ed.), Cryptography and Coding. XI, 461 pages. 2005.

Vol. 3795: H. Zhuge, G.C. Fox (Eds.), Grid and Cooperative Computing - GCC 2005. XXI, 1203 pages. 2005.

Vol. 3794: X. Jia, J. Wu, Y. He (Eds.), Mobile Ad-hoc and Sensor Networks. XX, 1136 pages. 2005.

Vol. 3793: T. Conte, N. Navarro, W.-m.W. Hwu, M. Valero, T. Ungerer (Eds.), High Performance Embedded Architectures and Compilers. XIII, 317 pages. 2005.

Vol. 3792: I. Richardson, P. Abrahamsson, R. Messnarz (Eds.), Software Process Improvement. VIII, 215 pages. 2005.

Vol. 3791: A. Adi, S. Stoutenburg, S. Tabet (Eds.), Rules and Rule Markup Languages for the Semantic Web. X, 225 pages. 2005.

Vol. 3790: G. Alonso (Ed.), Middleware 2005. XIII, 443 pages. 2005.

Vol. 3789: A. Gelbukh, Á. de Albornoz, H. Terashima-Marín (Eds.), MICAI 2005: Advances in Artificial Intelligence. XXVI, 1198 pages. 2005. (Sublibrary LNAI).

Vol. 3788: B. Roy (Ed.), Advances in Cryptology - ASIACRYPT 2005. XIV, 703 pages. 2005.

Vol. 3787: D. Kratsch (Ed.), Graph-Theoretic Concepts in Computer Science. XIV, 470 pages. 2005.

Vol. 3785: K.-K. Lau, R. Banach (Eds.), Formal Methods and Software Engineering. XIV, 496 pages. 2005.

Vol. 3784: J. Tao, T. Tan, R.W. Picard (Eds.), Affective Computing and Intelligent Interaction. XIX, 1008 pages. 2005.

Vol. 3783: S. Qing, W. Mao, J. Lopez, G. Wang (Eds.), Information and Communications Security. XIV, 492 pages. 2005.

Vol. 3782: K.-D. Althoff, A. Dengel, R. Bergmann, M. Nick, T.R. Roth-Berghofer (Eds.), Professional Knowledge Management. XXIII, 739 pages. 2005. (Sublibrary LNAI).

Vol. 3781: S.Z. Li, Z. Sun, T. Tan, S. Pankanti, G. Chollet, D. Zhang (Eds.), Advances in Biometric Person Authentication. XI, 250 pages. 2005.

Vol. 3780: K. Yi (Ed.), Programming Languages and Systems. XI, 435 pages. 2005.

Vol. 3779: H. Jin, D. Reed, W. Jiang (Eds.), Network and Parallel Computing. XV, 513 pages. 2005.

Vol. 3778: C. Atkinson, C. Bunse, H.-G. Gross, C. Peper (Eds.), Component-Based Software Development for Embedded Systems. VIII, 345 pages. 2005.

Vol. 3777: O.B. Lupanov, O.M. Kasim-Zade, A.V. Chaskin, K. Steinhöfel (Eds.), Stochastic Algorithms: Foundations and Applications. VIII, 239 pages. 2005.

Vol. 3776: S.K. Pal, S. Bandyopadhyay, S. Biswas (Eds.), Pattern Recognition and Machine Intelligence. XXIV, 808 pages. 2005.

Vol. 3775: J. Schönwälder, J. Serrat (Eds.), Ambient Networks. XIII, 281 pages. 2005.

Vol. 3774: G. Bierman, C. Koch (Eds.), Database Programming Languages. X, 295 pages. 2005.

Vol. 3773: A. Sanfeliu, M.L. Cortés (Eds.), Progress in Pattern Recognition, Image Analysis and Applications. XX, 1094 pages. 2005.

Vol. 3772: M.P. Consens, G. Navarro (Eds.), String Processing and Information Retrieval. XIV, 406 pages. 2005.

Vol. 3771: J.M.T. Romijn, G.P. Smith, J. van de Pol (Eds.), Integrated Formal Methods. XI, 407 pages. 2005.